Ballistics

Beat P. Kneubuehl

# Ballistics

## Theory and Practice

Translated by Steve Rawcliffe

With 256 figures and 51 tables

Springer

Beat P. Kneubuehl
bpk consultancy GmbH
Thun, Switzerland

*Translated by* Steve Rawcliffe, BA, PgDip. Int. Trans. F-01170 Gex, France

ISBN 978-3-662-69236-3 ISBN 978-3-662-69237-0 (eBook)
https://doi.org/10.1007/978-3-662-69237-0

Translation from the German language edition: “Ballistik. Theorie und Praxis” by Beat P. Kneubuehl, © Springer Verlag GmbH Heidelberg Berlin 2022. Published by Springer Heidelberg. All Rights Reserved.



This Springer imprint is published by the registered company Springer-Verlag GmbH, DE, part of Springer Nature.
The registered company address is: Heidelberger Platz 3, 14197 Berlin, Germany

# Preface

This is the English translation of the first edition of *Ballistik, Theorie und Praxis*, summarizing over 40 years' work on the range and at my desk. The aim of this book is to provide a solid foundation for those who wish to become acquainted with ballistics, and to share with those who are already active in the field – professionally or privately – the experience I have acquired in dealing with these topics and through my involvement in countless experiments.

The German original of my book has already enjoyed considerable success. Correspondence received from English-speaking and other countries, and the comments of a number of acquaintances, indicated that an English version would be of value, and I therefore decided to have it translated.

Steve Rawcliffe had previously translated my book *Wound Ballistics, Basics and Applications* (*Wundballistik: Grundlagen und Anwendungen*) and he once again took on the translation of a complex text covering several disciplines. I am extremely grateful to him for undertaking this challenging task. My thanks also to Springer-Verlag for agreeing to publish the English translation. It was once again a pleasure to work with this publisher.

Thun, Switzerland, March 2024 Beat P. Kneubuehl

# Foreword

Ballistics is not a particularly well known branch of physics. One can probably count on the fingers of one hand the universities and colleges at which it is possible to study the subject, making it virtually impossible to enter this field directly. After an apprenticeship in the construction industry, a degree in maths and physics at the University of Bern and several years as a teacher at a sixth-form college, I had seen enough of teaching and was looking for a new career. At the time, the only physics-related job for a mathematician within reasonable travelling distance from my home was at the Swiss defence ministry's testing centre for ballistics, weapons and munitions. So it was that I entered the field of ballistics quite by chance. Back then, I knew nothing about the subject other than what I had learned in physics classes – how to calculate a trajectory in a vacuum.

After half a dozen courses at the Carl-Cranz-Gesellschaft in Weil am Rhein, Germany, and at ENSTA Paris (the *École Nationale des Techniques avancées*, or National School of Advanced Techniques), plus many hours and days spent on shooting ranges, I realized that I had ended up in one of the most varied and exciting areas of physics – at least as far as I was concerned.

Exterior ballistics alone requires not only an understanding of conventional mechanics but also an in-depth awareness of the structure of the atmosphere and knowledge of meteorology, aerodynamics and the theory of gyroscopes. Interior ballistics is based on principles of mechanical, thermodynamic and fluid dynamics, and requires some basic knowledge of explosives. Terminal ballistics, where phenomena last only a few milliseconds or microseconds, opens up a virtually unresearched field of study, as materials react to an extremely high load applied for an extremely short period of time differently to the manner with which we are familiar from a quasi-static approach.

Acquiring a thorough understanding of all these areas of ballistics has enabled me to apply a physics/ballistics approach to other areas of science and knowledge, particularly the testing and assessment of ballistic protection, criminology, forensic medicine and surgery. In the case of surgery, the greatest benefit has stemmed from empirical simulation of short-duration penetration phenomena, which has enabled significantly enhanced assessment and treatment of gunshot and war wounds.

The aim of this book is to provide a solid foundation for those who wish to become acquainted with ballistics, and to share with those who are already active in the field – professionally or privately – the experience I have acquired through more than 40 years of dealing with these topics and my involvement in countless experiments.

Writing a book on such a wide-ranging topic as ballistics would not be possible without the assistance of various organizations and specialists. Above all, I would like to thank armasuisse Science and Technology S+T (Thun, Switzerland), and its director Dr Thomas Rothacher for their unconditional support and for allowing me to use in the present work all the reports and other documents I produced in my 33 years in the department for ballistics and shooting tests. They also authorized me to include illustrations resulting from my work with them.

I should also like to thank Nitrochemie (Wimmis, Switzerland) and in particular Dr Beat Vogelsanger und Pascal Beer for their support regarding explosives and other matters related to interior ballistics. Dr Rudi Heiser (Weil am Rhein, Germany) provided skilled assistance regarding the description of the gas dynamics model. Dr Patrick Folly carried out a careful and thorough content review of everything related to interior ballistics. I am grateful to both of them for their support.

Volker Fleck (Freiburg im Breisgau, Germany) kindly made available his documentation regarding exterior ballistics models, which I used for the sections on models with 4 and 6 degrees of freedom. I am grateful to Peter Näf for carefully reviewing the chapter on exterior ballistics.

As I had virtually no experience regarding the hydrodynamic penetration process in terminal ballistics, Wilhelm Odermatt kindly wrote the section on that topic, which is his specific area of expertise.

I am grateful to Dr Fabiano Riva of the Institute of Forensic Medicine, University of Bern, for his active collaboration on the section regarding the forensic applications of ballistics and to the Institute itself for the illustrations from the Forensic Physics and Ballistics Centre.

I received documentation from Henri Habegger of the Swiss Army Museum Society, Dr Tobias Schmid of Rheinmetall WM GmbH and Philippe Drapela und Martin Zumstein of armasuisse S+T.

I was once again able to count on Ueli Geiger (Rüti, Zürich) for a thorough and expert review of the entire German manuscript. My grateful thanks to him. Special thanks go to my wife for her patience during the creation of this work, and for checking the German of the original version.

Warm thanks go to my publisher, Springer, with whom it was a pleasure to work and who are responsible for the excellent appearance of the book.

Thun, Switzerland, September 2018 Beat P. Kneubuehl

# Outline of contents

# Table of content

# Table of symbols

## General

This book uses SI units and units derived from them. Dimensionless quantities are indicated by [-]. Where no dimension is possible, the corresponding space is left blank. Where possible, internationally-recognized symbols have been used (e.g. A for area, F for force, a for acceleration, etc.).

## Symbols used throughout this book

| | | |
|---|---|---|
| $A$ | Area | [$m^2$] |
| $A_0$ | Reference cross-sectional area | [$m^2$] |
| $C_D$ | Drag coefficient | [-] |
| $C_L$ | Lift force coefficient | [-] |
| $C_M$ | overturning moment coefficient | [-] |
| $C_{PR}$ | Pressure coefficient of a body | [-] |
| $C_p$ | Specific heat capacity at constant pressure | [J/(kg·K)] |
| $C_v$ | Specific heat capacity at constant volume | [J/(kg·K)] |
| CG | Centre of gravity | |
| CM | Centre of mass | |
| $E$ | Energy | [J] |
| $E_{pr}$ | Pressure energy | [J] |
| $E_{kin}$ | Kinetic energy | [J] |
| $E_{pot}$ | Potential energy | [J] |
| $E_{rot}$ | Rotational kinetic energy | [J] |
| $F$ | Force | [N] |
| $F_D$ | Drag force | [N] |
| $F_L$ | Lift force | [N] |
| $F_T$ | Thrust | [N] |
| $F_W$ | Weight | [N] |
| $G$ | Gravitational constant | [$m^3/(kg \cdot s^2)$] |
| $I$ | Momentum | [kg·m/s] |
| $J$ | Moment of inertia | [$kg \cdot m^2$] |
| $J_a$ | Axial moment of inertia | [$kg \cdot m^2$] |
| $J_r$ | Radial moment of inertia | [$kg \cdot m^2$] |
| $L$ | Angular momentum | [N·m·s] |
| $M$ | Torque | [N·m] |
| Ma | Mach number | [-] |
| $M_M$ | Torque associated with $F_A$ | [N·m] |
| $R$ | Special gas constant | [J/(kg·K)] |
| $R_E$ | Earth's radius | [m] |

| | | |
|---|---|---|
| $R_L$ | Special gas constant of air | [J/(kg·K)] |
| Re | Reynolds number | [-] |
| T | Temperature | [K] |
| $T_C$ | Temperature in degrees Celsius | [°C] |
| Q | Thermal energy | [J] |
| U | Internal energy | [J] |
| V | Volume | [m³] |
| W | Work | [J] |
| We | Weber number | [-] |
| a | Acceleration | [m/s²] |
| $a_c$ | Coriolis acceleration | [m/s²] |
| c | Speed of sound | [m/s] |
| d | Calibre (sometimes "diameter of projectile") | [mm], [m] |
| g | Acceleration due to gravity | [m/s²] |
| $\ell$ | Length | [m] |
| m | Mass | [kg] |
| $m_E$ | Earth mass | [kg] |
| p | Pressure | [Pa], [bar] |
| q | Sectional density | [kg/m²] |
| r | Radius, radial distance | [m] |
| s | Distance | [m] |
| t | Time | [s] |
| v | Velocity | [m/s] |
| $v_0$ | Muzzle velocity | [m/s] |
| (x, y, z) | Position coordinates | [m] |
| $\gamma$ | Heat capacity ratio | [-] |
| $\rho$ | Density | [kg/m³] |
| $\tau$ | Temperature gradient | [K/m] |
| $\omega$ | Angular velocity | [1/s] |
| $\theta$ | Trajectory angle | [mil] |
| $\theta_0$ | Angle of departure (elevation) | [mil] |

***Indices***

| | |
|---|---|
| E | Final (final distance, final time, etc.) |
| c | Charge |
| i | Inside, interior, inner |
| p | Projectile |
| b | Barrel |
| x, y, z | Component in x, y and z direction respectively |
| 0 | Initial value, starting value |

## Symbols – Chapter 2

| | | |
|---|---|---|
| $E_{mech}$ | Mechanical energy | [J] |
| $\underset{\sim}{H}$ | Enthalpy | [J] |
| $\hat{J}$ | Tensor of inertia | [kg·m²] |
| S | Entropy | [J/K] |
| $a_{cf}$ | Centrifugal acceleration | [m/s²] |
| $a_{cp}$ | Centripetal acceleration | [m/s²] |
| b | Wingspan | [m] |
| $d_0$ | Reference distance (lever arm) for torque | [m] |
| $p_i$ | Pressure in container (internal pressure) | [bar] |
| $p_m$ | Nozzle pressure | [bar] |
| $p_0$ | Pressure of undisturbed flow | [bar] |
| $p_\lambda$ | Laval pressure | [bar] |
| $v_r$ | Radial velocity | [m/s] |
| $v_t$ | Tangential velocity | [m/s] |
| $\Gamma$ | Circulation in turbulent flow | [m²/s] |
| $\alpha$ | Angular acceleration (mechanics) | [1/s²] |
| $\alpha$ | Mach angle (fluid dynamics) | [-] |
| $\delta$ | Corner angle (of a flow) | [-] |
| $\eta$ | Dynamic viscosity | [Pa·s] |
| $\kappa$ | Compressibility | [1/Pa] |
| $\nu$ | Kinematic viscosity | [m²/s] |
| $\nu_N$ | Nutation frequency of a gyroscope | [1/s] |
| $\sigma$ | Oblique shock angle | [-] |
| $\sigma_s$ | Shear stress | [N/m²] |
| $\omega_p$ | Angular velocity of precession | [1/s] |

## Symbols – Chapter 3

| | | |
|---|---|---|
| $A_c$ | Propellant grain surface area | [m²] |
| $A_e$ | Nozzle end cross-section | [m²] |
| $A_b$ | Barrel cross-section | [m²] |
| $A_m$ | Muzzle cross-section | [m²] |
| $A_0$ | Initial surface area (propellant) | [m²] |
| B | Burn rate coefficient | [1/m] |
| $B_A$ | Vivacity | [1/s] |
| $C_S$ | Thrust coefficient | [N] |
| D | Spring constant | [N/m] |
| $E_c$ | Energy of the propellant accelerated with the projectile | [J] |
| $F_L$ | Total rifling groove force | [N] |

| | | |
|---|---|---|
| $F_b$ | Friction in the barrel | [N] |
| $F_{se}$ | Specific energy of a propellant ("propellant force") | [J/kg] |
| $F_{tf}$ | Tension force (spring weapon) | [N] |
| K | Klemmung, usually represented as Kn | [-] |
| $Q_{ex}$ | Combustion heat | [J/g], [kJ/kg] |
| $T_{ex}$ | Explosion temperature (manometric bomb) | [K] |
| $V_{ch}$ | Chamber volume | [m³] |
| $a_b$ | Acceleration during propulsion phase (rocket) | [m/s²] |
| $a_f$ | Deceleration during midcourse phase (rocket) | [m/s²] |
| e | Thickness of burnt propellant | [mm] |
| $\ell_r$ | Barrel length | [m] |
| $m_a$ | Mass ejected | [kg] |
| $m_b$ | Barrel mass | [kg] |
| $m_{ca}$ | Carrier mass (spring weapons) | [kg] |
| $m_f$ | Mass of the fragments (fragment acceleration) | [kg] |
| $m^+$ | Displaced mass (projectile plus gases) | [kg] |
| $p_e$ | Pressure at the nozzle exit | [bar] |
| $p_\mu$ | Gas pressure as the projectile passes through the muzzle | [bar] |
| $p_{br}$ | Pressure at the action face (breech face) | [bar] |
| $p_p$ | Pressure at the base of the projectile | [bar] |
| $p_0$ | Starting pressure for the equations of motion | [bar] |
| r | Burning rate | [m/s] |
| s | Draw length (spring weapons) | [m] |
| $t_{LR}$ | Time taken for the projectile to pass through the barrel | [s], [ms] |
| $t_s$ | Time from primer firing to muzzle exit | [ms], [s] |
| u | Fluid velocity in the barrel | [m/s] |
| $v_a$ | Outflowing gas velocity | [m/s] |
| $v_{exp}$ | Detonation rate of an explosive | [m/s] |
| z | Relative residual propellant volume | [m³] |
| $\Gamma$ | Angle of twist | [-] |
| $\Lambda$ | Twist length | [m] |
| $\Psi$ | Lost energy | [J] |
| $\beta$ | Linear burning rate | [m/s] |
| $\varepsilon$ | Sébert's factor (propellant) | [-] |
| $\eta$ | Efficiency | [-] |
| $\eta^+$ | Mechanical efficiency of a weapon | [-] |
| $\phi(z)$ | Form function | [-] |

## Symbols – Chapter 4

| | | |
|---|---|---|
| $A_m$ | Muzzle cross-section | [m²] |
| $E_R$ | Recoil energy | [J] |
| $E_d$ | Energy of rotation during recoil | [J] |
| $E_t$ | Translational component of $E_R$ | [J] |
| $I_N$ | Additional momentum | [N·s] |
| $I_R$ | Recoil momentum | [N·s] |
| $I_d$ | Rotational component of recoil momentum | [N·s] |
| $I_m$ | Muzzle impulse | [N·s] |
| $I_n$ | Impulse of the gases behind the projectile | [N·s] |
| $I_p$ | Projectile momentum | [N·s] |
| $J_w$ | Moment of inertia of the weapon | [kg·m²] |
| $m_p$ | Mass of projectile | [g], [kg] |
| $m_w$ | Mass of weapon | [kg] |
| $p_i$ | Internal pressure | [bar] |
| $v_a$ | Exhaust velocity | [m/s] |
| $v_m$ | Mean exhaust velocity | [m/s] |
| $v_R$ | Recoil speed | [m/s] |

## Symbols – Chapter 5

| | | |
|---|---|---|
| $A_0$ | Reference area | [m²] |
| BC | Ballistic coefficient | [-] |
| $F_A$ | Aerodynamic forces | [N] |
| $F_D$ | Drag force | [N] |
| $F_{DP}$ | Dynamic pressure resistance | [N] |
| $F_E$ | Pitch damping force | [N] |
| $F_K$ | Magnus force | [N] |
| $F_{KP}$ | Pitch Magnus force | [N] |
| $F_L$ | Lift force | [N] |
| $F_N$ | Normal force on the projectile | [N] |
| $F_T$ | Tangential force on the projectile | [N] |
| $F_X$ | Axial force | [N] |
| $M_H$ | Pitch damping moment | [N·m] |
| $M_I$ | Roll damping moment | [N·m] |
| $M_J$ | Magnus moment | [N·m] |
| $M_{JP}$ | Pitch Magnus moment | [N·m] |
| $M_M$ | Overturning moment | [N·m] |
| P, Q | Quaternions | |
| Q | Piton-Bressant number | [1/m] |
| $R_A$ | Special gas constant of air | [J/(kg·K)] |

| | | |
|---|---|---|
| $R_{A,d}$ | Special gas constant of dry air | [J/(kg·K)] |
| $R_{A,m}$ | Special gas constant of moist air | [J/(kg·K)] |
| $R_{wa}$ | Special gas constant of water vapour | [J/(kg·K)] |
| $R_E$ | Radius of Earth | [m] |
| S | Cross-sectional area of projectile | [$m^2$] |
| V | Volume of projectile | [$m^3$] |
| $C_{D,W}$ | Wave drag coefficient | [-] |
| $C_{D,F}$ | Friction coefficient related to the reference area | [-] |
| $C_{D,B}$ | Base drag coefficient related to the reference area | [-] |
| $C_E$ | Pitch damping force coefficient | [-] |
| $C_F$ | Friction coefficient related to the surface | [-] |
| $C_H$ | Pitch damping moment coefficient | [-] |
| $C_J$ | Magnus moment coefficient | [-] |
| $C_K$ | Magnus force coefficient | [-] |
| $\vec{c}, \vec{p}, \vec{q}$ | Unit vectors, body-fixed coordinate system | [m] |
| $a_f$ | Centrifugal acceleration | [$m/s^2$] |
| $d_z$ | Lateral acceleration in the point-mass model | [$m/s^2$] |
| $e_A$ | Lever arm of aerodynamic force | [m] |
| f | Tractability number (exterior ballistics) | [-] |
| $g_n$ | Acceleration due to gravity (ICAO standard) | [$m/s^2$] |
| $\vec{i}, \vec{j}, \vec{k}$ | Unit vectors, earth-fixed coordinate system | [m] |
| $\ell_p$ | Length of projectile | [mm] |
| $p_{wa}$ | Saturation vapour pressure of water | [mbar] |
| r | Radius of the rotating body | [m] |
| $r_h$ | Relative humidity (relative to 1) | [-] |
| s | Gyroscopic stability number (exterior ballistics) | [-] |
| $\hat{s}$ | Molitz (dynamic) stability factor | [-] |
| s(t) | Arc length (trajectory, differential geometry) | [m] |
| $\vec{t}, \vec{s}, \vec{h}$ | Unit vectors, trajectory tangent system | [m] |
| u | Fluid velocity | [m/s] |
| v | Velocity | [m/s] |
| $v_W$ | Wind velocity | [m/s] |
| $v_{WX}$ | Wind velocity in direction of flight | [m/s] |
| $v_{WZ}$ | Wind velocity crossways to the direction of flight | [m/s] |
| $\Phi$ | Potential function of flow velocity | [$m^2/s$] |
| $\Theta$ | Tensor of inertia of the projectile | [$kg \cdot m^2$] |
| $\vec{\Omega}$ | Vector of angular velocity | [1/s] |
| $\alpha$ | Azimuth with respect to north, measured clockwise | [mil] |
| $\alpha_i$ | (i = 0,1,3) Real coefficients of the differential equation for $\zeta$ | [-] |
| $\beta$ | Geographical latitude of the location | [-] |
| $\beta_i$ | (i = 0,1,3) Imaginary coefficients of the differential equation for $\zeta$ | [-] |

| | | |
|---|---|---|
| $\delta$ | Angle of attack | [-] |
| $\phi$ | Interference potential of flow velocity | [$m^2/s$] |
| $\lambda$ | Angle of sight (between horizontal and target line) | [mil] |
| $\sigma$ | Angle of shot | [mil] |
| $\tau$ | Temperature gradient | [K/m] |
| $\omega_E$ | Angular velocity of Earth | [1/s] |
| $\omega_H$ | Angle of descent relative to the horizontal | [mil] |
| $\omega_{TL}$ | Angle of descent relative to the target line | [mil] |
| $\xi$ | Angle between earth-fixed and wind fixed coordinate systems | [-] |
| $\psi$ | Angular displacement of the projectile axis about the tangent to the trajectory | [-] |
| $\zeta$ | Complex angle of attack | [-] |

## Symbols – Chapter 6

| | | |
|---|---|---|
| A | Projectile-material contact area | [$m^2$] |
| $C_S$ | Shear fracture stress | [$N/m^2$] |
| D | Plate thickness | [m] |
| $D_g$ | Length of geometric projectile path | [m] |
| ED | Energy density | [$J/mm^2$] |
| $E_P$ | Energy expended for perforation | [J] |
| $E_{thr}$ | Maximum energy at which perforation does not occur | [J] |
| $E_S$ | Shear energy | [J] |
| $HBW_T$ | Brinell hardness of the target material | [-] |
| L | Penetrator length | [m] |
| P | Penetration depth | [m] |
| $P_A$ | Acceptance probability | [-] |
| $P_{AT}$ | Attack probability | [-] |
| $P_P$ | Perforation probability | [-] |
| $P_R$ | Rejection probability | [-] |
| $P_S$ | Protection probability | [-] |
| $Y_P$ | Strength of the penetrator | [$N/m^2$] |
| $R_T$ | Strength of the target material | [$N/m^2$] |
| W(x) | Damage potential at point x | [J/cm] |

| | | |
|---|---|---|
| a, b, $b_{0.1}$ | Experimentally determined parameters for relative penetration depth | [-] |
| c, $c_{0.1}$ | Constant parameters | [-] |
| u | Penetration velocity | [m/s] |
| x | Penetration depth | [cm], [m] |
| $s_e$ | Penetration depth | [cm], [m] |
| u | Penetration velocity | [m/s] |

| | | |
|---|---|---|
| $v_{50}$ | Average perforation velocity | [m/s] |
| $\alpha$ | Angle of incidence | [-] |
| $\beta$ | NATO angle | [-] |
| $\delta$ | Angular displacement of the plane of impact with respect to the direction of fire | [-] |
| $\kappa$ | Angle of inclination of the plane of impact with respect to the direction of fire | [-] |
| $\rho_P$ | Density of penetrator | [kg/m$^3$] |
| $\rho_T$ | Density of target material | [kg/m$^3$] |

## Symbols – Chapter 7

| | | |
|---|---|---|
| $E_u$ | Energy after impact | [J] |
| $F_y$ | Contact pressure | [N] |
| Q | Deformation work during impact | [J] |
| $Q_T$ | Loss of kinetic energy | [J] |
| $Q_w$ | Remaining deformation work | [J] |
| $R_x$ | Friction | [N] |
| T | Torque shock | [N·m] |
| b | Shock eccentricity | [m] |
| h | Nose length | [m] |
| k | Restitution coefficient | [-] |
| $s_x$ | Length of ricochet mark | [mm] |
| $t_s$ | Impact time (projectile-target contact time) | [ms], [s] |
| u | Velocity after impact | [m/s] |
| v | Velocity before impact | [m/s] |
| $\alpha$ | Angle of incidence | [-] |
| $\beta$ | Angle of departure | [-] |
| $\gamma$ | Horizontal angle of deflection | [-] |
| $\varepsilon$ | Change in angle of projectile axis resulting from torque shock | [-] |
| $\lambda$ | Angle of deflection | [-], [mil] |
| $\theta$ | Moment of inertia number | [-] |
| $\mu$ | Mass ratio $m_1/m_2$ | [-] |
| $\nu$ | Frequency of rotation (nutation) | [1/s] |
| $\sigma$ | Friction number (not to be confused with coefficient of friction) | [-] |

Indices

| | |
|---|---|
| n | Normal |
| t | Tangential |
| 1 | Body 1 |
| 2 | Body 2 |

# Conversions

## U.S. → metric

| | U.S. | Symbol | Definition | Metric | |
|---|---|---|---|---|---|
| Length | inch | in | | 25.4 | mm |
| | foot | ft | 12 in | 0.3048 | m |
| | yard | yd | 3 ft | 0.9144 | m |
| Area | square inch | $in^2$ | | 645.16 | $mm^2$ |
| Velocity | foot/second | ft/s | | 0.3048 | m/s |
| Mass | grain | gr | 1/7000 lb | 0.0647989 | g |
| | pound | lb | | 0.4535924 | kg |
| Energy | foot pound weight | ft lb wt | | 1.35582 | J |
| Force | pound weight | lb wt | | 4.448221 | N |
| Pressure | pound/square inch | psi | | 0.0689476 | bar |

## Metric → U.S.

| | Metric | Symbol | Definition | U.S. | |
|---|---|---|---|---|---|
| Length | millimetre | mm | 1/1000 m | 0.03937 | in |
| | metre | m | Base unit | 3.28084 | ft |
| | metre | m | | 1.0936133 | yd |
| Area | square millimetre | $mm^2$ | | 0.001550 | $in^2$ |
| Velocity | metre/second | m/s | | 3.28084 | ft/s |
| Mass | gram | g | 1/1000 kg | 15.43236 | gr |
| | kilogram | kg | Base unit | 2.204622 | lbs |
| Energy | joule | J | 1 N·m | 0.737561 | ft lb wt |
| Force | newton | N | 1 kg·m/s² | 0.224809 | lb wt |
| | bar | bar | $10^5$ Pa | 14.503774 | psi |

## Prefix symbols for multiples and submultiples of SI units

| Factor | Name | Symbol | Factor | Name | Symbol |
|---|---|---|---|---|---|
| $10^{12}$ | tera | T | $10^{-1}$ | deci | d |
| $10^9$ | giga | G | $10^{-2}$ | centi | c |
| $10^6$ | mega | M | $10^{-3}$ | milli | m |
| $10^3$ | kilo | k | $10^{-6}$ | micro | µ |
| $10^2$ | hecto | h | $10^{-9}$ | nano | n |
| $10^1$ | deca | da | $10^{-12}$ | pico | p |

# 1 Introduction

People – especially those of Greek culture – were already studying the flights of stones and arrows several hundred years before our era. The Greek word for "to throw" ("βάλλειν", *ballein*) is the root of our word "ballistics".

The arrival of firearms and the acceleration of a projectile in a tube presented new problems. For a long time, these were addressed by means of trial and (often painful) error. Not until the major advances in physics and measuring techniques of the 18th Century did study in this area become more systematic, with ballistics splitting into two sub-areas: the acceleration of the projectile in the barrel and the deceleration of the projectile during its flight through the air.

As firearms became more common, armour became obsolete, because it provided no protection against bullets. At the end of the 19th Century, the arrival of motor vehicles provoked renewed interest in protection, and a new topic emerged within ballistics: the interaction between projectiles and materials, including human tissue.

Today, ballistics is the science of bodies in flight, encompassing all the physical phenomena involved and the movement of the projectile. A distinction is still made between the four sub-topics that have emerged from the history of ballistics, and they are often studied separately. This book aims to study ballistics as a whole, covering the individual disciplines and their principal topics.

*Interior ballistics* will look at explosive materials – especially propellants – and the methods currently used to calculate barrel pressure and the velocity of the projectile in the barrel (the thermodynamic and gas dynamic models). We shall also attempt to gain an understanding of the interior ballistics of open tubes (in such devices as bird-scarers and blank-firing pistols). We shall present alternative ways of accelerating projectiles and look at how to determine the initial velocity of fragments.

The chapter on *intermediate ballistics* examines the processes that take place immediately before and after the projectile leaves the barrel. In addition to describing the phenomena, we shall present an equation for estimating the exit velocity of the gases and methods for calculating the angle of jump.

This book devotes considerable attention to *exterior ballistics*, not least because it is the most tangible discipline. In addition to presenting the various models and discussing trajectories in some detail, we shall present various solutions to the



B. P. Kneubuehl, *Ballistics*,
https://doi.org/10.1007/978-3-662-69237-0_1

boundary value problem, examine projectile aerodynamics and drag-optimized projectile designs, and explain how to calculate drag.

The interaction between projectiles and materials does not lend itself to mathematical modelling, and so the chapter on *terminal ballistics* primarily summarizes and comments on facts and data that have been acquired empirically. One of the main topics of this chapter will be the problems associated with evaluating ballistic protection. *Ricochets* are part of terminal ballistics, but this topic has its own chapter because of its significance, especially in criminology. The creation of gunshot wounds and the physics involved are covered in another work.[1]

The final chapter lists a number of *applications* of ballistics to forensics that are rarely mentioned and discusses shooting range safety.

To save the reader needing consulting a separate text book on *physics*, the present work starts with an introduction to the topic.

The rules of ballistics apply regardless of calibre. The explanations in this book apply equally to a 155 mm artillery shell and a 2 mm shot pellet. In practice, however, one often sees weapons divided into small, medium and large calibre, frequently with divergent terminology. We shall underline the general applicability of statements made regarding ballistics by using consistent terminology throughout this book, regardless of calibre.

Finally, ballistics is not a matter of equations. It is a matter of understanding what are generally extremely rapid movements. If you are not entirely familiar with the language of equations, just skip them. In most cases, they simply express in short form what is explained in the text.

[1] B. P. KNEUBUEHL (ed.), *Wound Ballistics. Basics and Applications.*, Springer Nature Verlag, Heidelberg, 2nd edition, 2022.

# 2 Physical basics

## 2.1 Coordinates, systems of units and notation

To describe physical phenomena simply, one needs a suitable system of coordinates. For ballistic phenomena, we generally use the *ballistic coordinate system*, which is defined as follows: the x- and y-axes between them define a vertical plane, with the y-axis pointing in the opposite direction to the Earth's gravitational pull. The z-axis completes the system, creating a right-handed three-dimensional system (see Fig. 2.1). The movements of a body are described by reference to a body-fixed frame. The origin of that system is located at the centre of mass of the body concerned and its principal axis is aligned with the direction of movement of the centre of mass at any given point in time (making it a tangential system). This system is also right-handed.

This book uses the units of the SI system (*Système Internationale d'Unités*), which is the official system in many countries. Length is measured in metres, mass in kilograms and time in seconds, often modified by prefixes such as milli, kilo or mega. Other, derived units will be explained the first time the corresponding parameter is mentioned.

Imperial/US units are still widely used for small-calibre weapons (handguns and shoulder weapons). The factors for converting between these units and SI units are at the end of the table of symbols (p. XXXI).

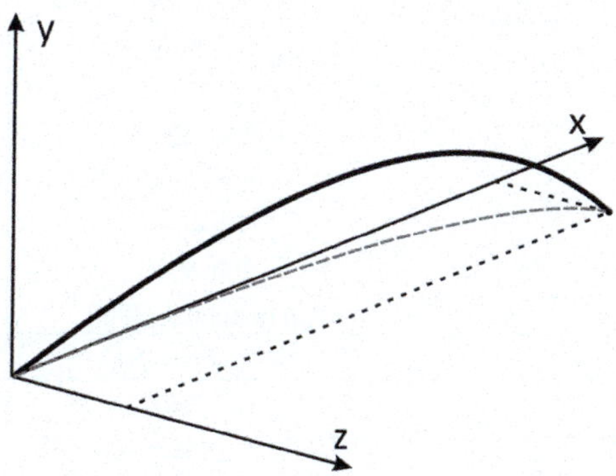


**Fig. 2.1**. Ballistic coordinate system
x-axis in the direction of shot, y-axis upwards, z-axis to the right


B. P. Kneubuehl, *Ballistics*,
https://doi.org/10.1007/978-3-662-69237-0_2

For the differential quotients used in definitions and equations, we shall use the notation generally followed in physics: derivatives over time are written with a dot over the symbol, while derivatives over distance are indicated using a prime.

(2.1:1) $$\frac{dx}{dt} \Leftrightarrow \dot{x}, \quad \frac{dv_x}{dt} \Leftrightarrow \dot{v}_x, \quad \frac{dv_x}{dx} \Leftrightarrow v'_x .$$

## 2.2 Mechanics

### 2.2.1 Kinematics and linear motion

Kinematics is the study of the motion of a body in space, disregarding the specific characteristics of the body concerned. The primary role of kinematics is to describe the path, or trajectory of the body. For these purposes we ignore the size of the body, treating it as a point. The most important kinematic parameter is *velocity*. Velocity is a vector value, of which the components are distance covered per unit time along the three axes:

(2.2:1) $$v_x = \frac{dx}{dt}, \quad v_y = \frac{dy}{dt}, \quad v_z = \frac{dz}{dt} . \qquad [m/s]$$

The magnitude of velocity (also known as speed) is given by:

(2.2:2) $$v = \sqrt{v_x^2 + v_y^2 + v_z^2} . \qquad [m/s]$$

Table 2.1 gives the orders of magnitude of typical velocities encountered in ballistics.

Change in velocity per unit time is termed *acceleration* if velocity increases and *deceleration* (or negative acceleration) if velocity decreases.

**Table 2.1.** Typical velocities

| System | Projectile | Velocity [m/s] |
|---|---|---|
| Tennis racquet | Tennis ball | Up to approx. 70 |
| Golf club | Golf ball | 80 |
| Small arms | Handgun bullet | 250–400 |
| | Rifle bullet | 600–1000 |
| Mortar | Mortar bomb | 70–400 |
| Artillery gun/howitzer | Shell | 300–900 |
| Shell | Fragment | Up to approx. 2000 |
| **Object** | **Medium** | **Velocity [m/s]** |
| Sound waves | Air (15 °C) | 340 |
| | Water (20 °C) | 1483 |
| | Steel or glass | Approx. 5200 |

(2.2:3) $$a_x = \frac{dv_x}{dt}, \quad a_y = \frac{dv_y}{dt}, \quad a_z = \frac{dv_z}{dt}.$$ [m/s²]

We can express the magnitude of acceleration in a manner similar to that already discussed for speed:

(2.2:4) $$a = \sqrt{a_x^2 + a_y^2 + a_z^2}.$$ [m/s²]

Table 2.2 lists some typical acceleration values encountered in ballistics.

Because velocity is a vector, a change in direction is also an acceleration, even if speed does not change.

If acceleration is constant, we can readily calculate speed and distance using the following formula:

(2.2:5) $$v = v_0 + a \cdot t,$$ [m/s]

(2.2:6) $$x = x_0 + v_0 \cdot t + \tfrac{1}{2} \cdot a \cdot t^2,$$ [m]

where $v_0$ is the speed at time $t = 0$ and $x_0$ is the distance that the object has already moved at that moment in time.

Movement along a curved path can always be expressed as movement along a segment of a circle. This rotational movement is described in kinematic terms by the *angular velocity*, with angular displacement $\varphi$ measured in radians (one complete revolution = $2\pi$ radians, which equates to 360°).

(2.2:7) $$\omega = \frac{d\phi}{dt}.$$ [1/s]

We obtain the *frequency of rotation* $\nu$ (the number of revolutions per second) from the angular velocity, thus:

**Table 2.2.** Typical acceleration values in ballistics

| Object | Acceleration [m/s²] |
|---|---|
| Acceleration due to gravity (standard value) | –9.80665 |
| Table tennis ball in air | –450 |
| Golf ball in air | –10 |
| Rifle bullet in air | –400 |
| Artillery shell in air | –30 |
| Fragment in air | –60,000 |
| Golf ball when struck | 200,000 |
| Rifle bullet in barrel on firing (average) | 500,000 |
| Rifle bullet in barrel on firing (maximum) | 1,500,000 |
| Artillery shell in barrel on firing (average) | 50,000 |
| Saturn V rocket during launch | 2–40 |

**Table 2.3.** Typical values for angular velocity ω, speed of rotation ν and circumferential velocity v.

| Type of projectile | ω [rad/s] | ν [1/s] | v [m/s] |
|---|---|---|---|
| Tennis ball with topspin (maximum) | 440 | 70 | 15 |
| Golf ball (depending on stroke, average) | 380 | 60 | 8 |
| Pistol bullet (9 mm Luger) | 8800 | 1400 | 40 |
| Revolver bullet (.44 Rem. Mag.) | 5440 | 866 | 30 |
| Rifle bullet (7.62 mm NATO) | 17,100 | 2721 | 67 |
| Artillery shell (155 mm, maximum charge) | 1265 | 200 | 98 |

(2.2:8) $$\nu = \frac{\omega}{2 \cdot \pi} \, .$$ [1/s]

Rotation at a constant angular velocity is known as uniform circular motion. The tangential velocity is given by:

(2.2:9) $$v = r \cdot \omega \, .$$ [m/s]

where r is the radius of rotation, i.e. the distance between the point under consideration and the centre of rotation.

This type of motion involves acceleration, because while the speed of the object does not change, its direction is continuously changing. The acceleration (*centripetal acceleration*) is perpendicular to the direction of travel at any given moment, and is directed away from the centre of the circle. Centripetal acceleration is given by the following equation:

(2.2:10) $$a = v \cdot \omega = \frac{v^2}{r} = r \cdot \omega^2 \, .$$ [m/s²]

Conversion between the various representations is performed using Eqn 2.2:9. Changes in angular velocity are expressed as *angular acceleration* α:

(2.2:11) $$\alpha = \frac{d\omega}{dt} \, .$$ [1/s²]

One case of rotation in ballistics is that of a spin-stabilized bullet, which rotates about its longitudinal axis. The angular velocities and speeds of rotation involved are quite high, and although the radius of a bullet is small, the circumferential velocities attained are considerable (see Table 2.3).

### 2.2.2 Mass, momentum and force

Mass is one of the basic characteristics of a body. It expresses itself in two ways:

- a body resists any change in state of movement (*inertia*);
- a body alters the state of movement of other bodies (*gravitation*).

Physical phenomena are often independent of the dimensions of a body. In such cases, it is easier to describe the phenomenon if we treat the body as a point, with the entire mass of the body concentrated in that point. This point – the *centre of mass* – is chosen such that even when forces act upon the body there is no difference between the path of the actual body (of non-zero dimensions) and that of the point mass.

In SI units, mass is expressed in kilograms or grams. Grains and ounces are still used in some English-speaking countries (see conversion factors on p. XXXI).

The velocity of a body, v, multiplied by its mass, m, is its *momentum*, I. Momentum is a vector, of which the direction is identical to that of the body's velocity.

(2.2:12) $$I = m \cdot v \, . \qquad [\mathrm{kg \cdot m/s}]$$

The physical parameter responsible for a change in the state of movement of a body, or deformation of that body, is known as *force*. Force is defined by *NEWTON's laws.*

1. The velocity of a body (or, more accurately, its momentum) remains constant unless a force acts upon it.
2. Force equals change in momentum over time, so:

(2.2:13) $$F = \frac{dI}{dt} = \frac{d}{dt}(m \cdot v) = \frac{dm}{dt} \cdot v + m \cdot \frac{dv}{dt} \, . \qquad [\mathrm{N}]$$

The unit of measurement for force is the Newton, and 1 Newton [N] = 1 $\mathrm{kg \cdot m/s^2}$. Though not an SI unit, the kilopond [kp] is also in widespread use. In some English-speaking countries, force is also measured in pounds-force. Conversion formulae for units of force are to be found on p. XXXI.

If the mass of a body remains constant (dm/dt = 0), we obtain the well-known formula for force from (2.2:13):

(2.2:14) $$F = m \cdot \frac{dv}{dt} = m \cdot a \, . \qquad [\mathrm{N}]$$

3. For every force, there is an opposing force of the same magnitude (action/reaction).

Please see Table 2.4 (next page) for a list of typical forces encountered in ballistics.

As acceleration is a vector, it follows from Eqn 2.2:14 that force also behaves as a vector. In order to describe a force, we must therefore know three things: its *magnitude*, its *direction* and its *point of application*. The best-known force is weight, W. Weight is the force that occurs when a body is subjected to the acceleration due to Earth's gravity, g. From Eqn 2.2:14:

(2.2:15) $$W = m \cdot g \, . \qquad [\mathrm{N}]$$

In physics, it is often useful to express force in relation to the area on which it is acting. When a force acts perpendicular to a surface, it is termed either a *tensile* or a *compressive* force. If a force acts parallel to a surface, it is known as a *shear* force. Force per unit area is known as *stress*.

**Table 2.4.** Typical forces encountered in ballistics

| Type of force | Force [N] |
|---|---|
| Air drag acting on a table tennis ball | 1.25 |
| Air drag acting on a golf ball | 0.46 |
| Air drag acting on a rifle bullet | 4 |
| Air drag acting on an artillery shell | 1380 |
| Mean drag acting on a bullet fired into water | 5000 |
| Maximum force acting on a bullet in a rifle barrel | 12,000 |
| Average force acting on an artillery shell in the barrel | 2,300,000 |
| Maximum force acting on an artillery shell in the barrel | 6,680,000 |

Stress (generally represented by the symbol $\sigma$) is often more important than the corresponding force. As a result, it has its own unit of measurement, the Pascal (Pa). 1 Pa ⇔ 1 N/m². Compressive stress may also be expressed in bar. This unit is widely used in ballistics.

1 bar ⇔ $10^5$ Pa.

In some English-speaking countries, pressure is also measured in pounds per square inch (lbs/in²). Conversion formulae for units of pressure are to be found on p. XXXI.

## 2.2.3 Work and energy

Work and energy, and their relationships to velocity, force, momentum and power, play a crucial role in ballistics. We shall therefore examine these terms in detail, the more so as they are not always used correctly in everyday speech.

*Work* W is defined as the force F acting in a given direction, multiplied by the distance covered under the influence of that force, s:

(2.2:16) $$W = F \cdot s \,. \qquad [J]$$

Any irregular movement, and any change in the structure of a material (deformation, destruction, etc.) requires work. The amount of work required depends not only on the force, but also on the distance covered. Even if the force is large, the work done will be small if the force acts over a short distance (which often equates to a short exposure time).

Given that work is defined as force times distance, the unit is the Newton-metre [N·m]. However, as work is such an important term it has its own unit, the Joule [J].

1 J ⇔ 1 N·m

Certain English-speaking countries still use the foot-pound-weight [ft lb-wt]. See p. XXXI for conversion factors.

In order to move a body against the acceleration due to the Earth's gravity, a force is required that corresponds to the weight of the body. From Eqn 2.2:16, we can calculate the work done W using the following formula:

$$(2.2{:}17) \qquad W = m \cdot g \cdot y\,, \qquad [J]$$

where y is the distance that the body is moved against the force of gravity.

If, however, the body undergoes acceleration as a result of the force acting upon it, the work done results in movement. Substituting Equations 2.2:14, 2.2:6 and 2.2:5 in Eqn 2.2:16 (the latter two equations with $x_0 = 0$ and $v_0 = 0$) we obtain

$$(2.2{:}18) \qquad W = \tfrac{1}{2} \cdot m \cdot v^2\,. \qquad [J]$$

If we ignore non-mechanical phenomena, the work done on a body therefore results either in an increase in distance within the gravitational field, or in movement. In the first instance, this can be expressed in terms of the weight of the body and the increase in distance (Eqn 2.2:17) and in the second case in terms of the mass of the body and the velocity it attains (Eqn 2.2:18).

Both a body that has been raised from its initial position and a body in motion are capable of performing work. The ability to perform work is called *energy*. In the case of a raised body we speak of *potential energy*, while in the case of a moving body the term is *kinetic energy* (see Table 2.5 for examples). In physics, therefore, work and energy are equivalent. Neither work nor energy can be created or destroyed. All that can happen is that the one is converted into the other. From Eqn 2.2:18, we can therefore derive an analogous formula for the kinetic energy of a body in motion, with the same unit [J]

$$(2.2{:}19) \qquad E_{kin} = \tfrac{1}{2} \cdot m \cdot v^2\,. \qquad [J]$$

Using Eqn 2.2:17, we can express potential energy

$$(2.2{:}20) \qquad E_{pot} = m \cdot g \cdot y\,. \qquad [J]$$

In ballistics, we can view energy from two points of view. Kinetic energy is needed to move a projectile from the weapon to the target, and the projectile needs energy to achieve an effect on the target. Projectiles are therefore divided into those that obtain all their energy from the weapon, in kinetic form, and those that receive from the weapon only the energy that they need to travel from the weapon to the target,

**Table 2.5.** Typical levels of kinetic energy encountered in ballistics

| Object | Initial energy [kJ] |
|---|---|
| Golf ball | 0.13 |
| Baseball | 0.18 |
| Arrow, crossbow bolt | 0.04–0.15 |
| Handgun bullet | 0.3–1.2 |
| Rifle bullet | 1.6–16.5 |
| 20 mm cannon shell | 75 |
| 81 mm mortar bomb | Up to 110 |
| 155 mm artillery shell | 13,500 |

carrying with them the energy that they require to achieve an effect on the target in some other form (e.g. in chemical form as an explosive) and only releasing that energy at the target.

### 2.2.4 Rotational movements

If a force acts on a system capable of rotation, at a point away from its axis of rotation, the system will begin to rotate. The angular velocity will depend on the force and on the distance between the point of application and the axis of rotation (the lever arm). This phenomenon can be represented thus:

$$M = F \cdot r \,, \qquad \text{(2.2:21)} \quad [\text{N·m}]$$

where M is the *torque* and r is the distance between the point of application and the centre of rotation (i.e. the length of the lever arm).

As a vector, torque always acts perpendicular to both the direction of the force and the lever arm, and hence is always parallel to the axis of rotation.

From Equations 2.2:14, 2.2:9 and 2.2:11 we obtain

$$M = m \cdot \frac{dv}{dt} \cdot r = m \cdot \frac{r \cdot d\omega}{dt} \cdot r = m \cdot r^2 \cdot \alpha \,. \qquad \text{(2.2:22)} \quad [\text{N·m}]$$

From this equation, we can see that there is a linear relationship between the torque and the angular acceleration it produces, as is the case for force and acceleration. The factor that links the two is the *moment of inertia* J, a characteristic of the body related to the axis of rotation.

$$M = J \cdot \alpha \,. \qquad \text{(2.2:23)} \quad [\text{N·m}]$$

The moment of inertia J therefore has the same relationship to rotation as does mass to linear motion.

The moment of inertia of a body can be calculated by analytical or numerical integration over the volume:

$$J = \int r^2 \cdot dm \,. \qquad \text{(2.2:24)} \quad [\text{kg·m}^2]$$

The moment of inertia can also be measured, using a moment of inertia balance. The formulae for basic, regular bodies are in some cases quite simple.

Using the definition of force in Eqn 2.2:13, we can derive the following relationship from Eqn 2.2:23:

$$M = F \cdot r = \frac{dI}{dt} \cdot r = J \cdot \frac{d\omega}{dt} = J \cdot \alpha \,. \qquad [\text{N·m}]$$

Integrating the two differentials, we obtain an equation with, on one side, the product of the torque and its distance from the axis of rotation. This product is the *angular momentum* or *spin*.

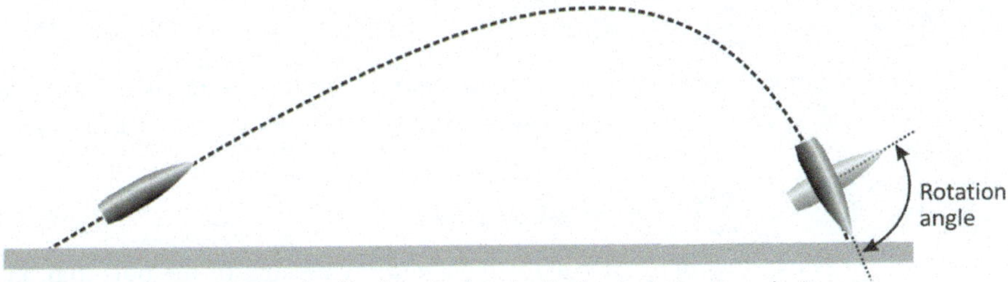


**Fig. 2.2.** Rotation of a projectile about its longitudinal axis during its trajectory.

(2.2:25) $L = I \cdot r = m \cdot v \cdot r = J \cdot \omega \, .$ $[\mathrm{kg{\cdot}m^2/s}]$

When calculating the kinetic energy of a rotating mass (its energy of rotation, or spin energy), we must remember that each mass particle has a velocity that depends on its distance from the axis of rotation. We must therefore integrate the differential energy over the entire volume.

$$E_{rot} = \tfrac{1}{2} \cdot \int_V v^2 \cdot dm = \tfrac{1}{2} \cdot \int_V \omega^2 \cdot r^2 \cdot dm = \tfrac{1}{2} \cdot \omega^2 \cdot \int_V r^2 \cdot dm \, , \qquad [\mathrm{J}]$$

(2.2:26) $E_{rot} = \tfrac{1}{2} \cdot J \cdot \omega^2 \, .$ $[\mathrm{J}]$

It is often forgotten that yaw and partial rotation are also forms of rotation. For instance, in addition to rotating rapidly about its longitudinal axis, a moving bullet makes other rotational movements, such as those about its horizontal transverse axis. See Fig. 2.2.

### 2.2.5 Laws of conservation

All physical phenomena are controlled by a number of *laws of conservation*. This is particularly true of mechanics. The analysis of any physical process starts with the application of these fundamental laws, which generally make it possible to understand a phenomenon quite easily. However, it is important first to delimit the physical system in such a way that one can ignore any interaction with the outside world. Such a system is described as "closed". Laws of conservation apply only within closed systems.

*1. The law of conservation of mass*. The total mass within a closed system remains constant. Mass is neither created nor destroyed (if we disregard atomic physics)

(2.2:27) $m_{tot} = \sum m_i = \text{const} \, .$ $[\mathrm{kg}]$

In a system consisting of several masses, the *centre of mass* of the system also remains constant, regardless of the movements of the individual masses (such as in the case of a shot sheaf).

*2. The law of conservation of momentum*. The total momentum within a closed system (i.e. one on which no external forces are acting) remains constant:

(2.2:28) $$I_{tot} = \sum I_i = \text{const} . \qquad [N \cdot s]$$

*3. The law of conservation of angular momentum.* The total angular momentum within a closed system (i.e. one on which no external moment is acting) remains constant:

(2.2:29) $$L_{tot} = \sum L_i = \text{const} . \qquad [kg \cdot m^2/s]$$

*4. The law of conservation of energy for mechanics.* The sum of the potential, kinetic and rotation energy in a closed, frictionless system (with no energy entering or leaving the system) remains constant:

(2.2:30) $$E_{mech} = \sum E_{pot} + \sum E_{kin} + \sum E_{rot} = \text{const} . \qquad [J]$$

However, the law of conservation of energy goes beyond mechanics. It also applies to other, non-mechanical forms of energy. We can therefore write:

*5. The general law of conservation of energy.* The total energy (i.e. the sum of all forms of energy) in a closed system remains constant.

The conversion of energy from one form to another within a system is not necessarily reversible.

### 2.2.6 Equations of motion

#### 2.2.6.1 General

In physics, the motion of a body is fully described if its position, velocity and orientation in space are known at every point in time. The motion of a body is determined exclusively by the forces acting upon it and the accelerations those forces produce.

Linking the acceleration of a body to its position and velocity yields the equations of motion. These form a system of differential equations that describe the time-dependent velocity and position functions of a body. In the case of a point mass (e.g. the centre of mass of a body), the system consists of six equations – three for the spatial coordinates and three for the velocity components. A rigid body of non-zero volume requires six further equations, which describe the body's orientation in space and the rate at which its orientation changes. In many cases, however, the system of equations of movement can be considerably reduced. This is the case, for instance, when no force is acting in a particular direction.

#### 2.2.6.2 D'ALEMBERT's principle

We can derive equations of movement from the law of conservation of energy or of momentum by differentiation. D'ALEMBERT's principle allows us to take a more direct approach. This principle involves replacing the acceleration of a body by a fictitious force $F_{fict}$ which, in accordance with NEWTON's Third Law of Motion, has to equal the sum of the forces applied:

$$(2.2{:}31)\qquad m \cdot a = F_{fikt} = \sum F_i \ . \qquad [N]$$

The equations of motion utilize another basic principle of mechanics: the *superposition principle*. This principle states that different movements of a body occurring simultaneously do not influence each other. Any movement can therefore be divided into separate components, one for each of the axes of the coordinate system. We can study each component independently of the others and hence can describe each component using simple equations.

#### 2.2.6.3 Trajectory in a vacuum.

One typical example of how equations of movement are developed and solved – and one of particular relevance to ballistics – is that of the trajectory of a body in a vacuum. No forces act along the x-axis, and the only force acting along the y-axis is the weight of the body. In accordance with D'ALEMBERT's principle, we can write the following two equations:

$$(2.2{:}32a)\qquad m \cdot \dot{v}_x = 0 \qquad \text{(x-axis)}, \qquad [N]$$

$$(2.2{:}32b)\qquad m \cdot \dot{v}_y = -m \cdot g \qquad \text{(y-axis)}. \qquad [N]$$

These two differential equations form the system of *equations of movement* describing the trajectory of a body in a vacuum. We can readily integrate the two equations, using the initial values below, derived from the initial velocity $v_0$ and angle of departure $\theta_0$ (the *elevation*):

$$v_{0x} = v_0 \cdot \cos\theta_0 \ , \qquad v_{0y} = v_0 \cdot \sin\theta_0 \qquad [m/s]$$

Integration yields the following two equations:

$$(2.2{:}33a)\qquad x(t) = v_0 \cdot \cos\theta_0 \cdot t \ , \qquad [m]$$

$$(2.2{:}33b)\qquad y(t) = v_0 \cdot \sin\theta_0 \cdot t - \tfrac{1}{2} \cdot g \cdot t^2 \ . \qquad [m]$$

If we eliminate t from both equations, we obtain the equation for the (parabolic) trajectory of a body in a vacuum:

$$(2.2{:}34)\qquad y(x) = \tan\theta_0 \cdot x - \frac{g \cdot x^2}{2 \cdot v_0^2 \cdot \cos^2\theta_0} \ . \qquad [m]$$

Taking y = 0, (the base of the trajectory), the maximum range $x_E$ for a given angle of departure $\theta_0$ is:

$$(2.2{:}35)\qquad x_E = \frac{v_0^2}{g} \cdot \sin(2 \cdot \theta_0) \ . \qquad [m]$$

From this, it follows that the maximum range for any value of $v_0$ is achieved when $\theta_0 = 45°$.

Similarly, we can obtain the corresponding time of flight from Eqn 2.2:33b:

$$(2.2:36) \qquad t_E = \frac{2 \cdot v_0}{g} \cdot \sin\theta_0 \ . \qquad [s]$$

We can readily determine the highest point along the trajectory (the apex) if we bear in mind that the vertical component of velocity must be zero at that point. This gives us the time of flight to the apex. Substituting that into Eqn 2.2:33b we obtain:

$$(2.2:37) \qquad y_G = \frac{v_0^2}{2 \cdot g} \cdot \sin^2\theta_0 \ . \qquad [m]$$

Combining Equations 2.2:36 and 2.2:37 gives us HAUPT's equation for the apex height. This equation is very useful in practice, and to a close degree of approximation is also valid for the trajectory of an object subject to drag.

$$(2.2:38) \qquad y_G = \tfrac{1}{8} \cdot g \cdot t_E^2 \ . \qquad [m]$$

### 2.2.6.4 *Velocity as a function of distance.*

It is often clearer to show velocity as a function of distance instead of time. For these purposes, "distance" generally means the distance the projectile has travelled along its trajectory, although the term may refer to the distance travelled along the x-axis in some cases.

For example: we can say that a light, high-speed hunting bullet with a muzzle velocity of 1050 m/s undergoes a deceleration of 1700 m/s$^2$ shortly after leaving the muzzle, but it is difficult to imagine what this means in practice. If, however, we say that the bullet loses 1.6 m/s per metre travelled, we can understand what is happening much more easily.

We can write the change in distance along the trajectory thus:

$$(2.2:39) \qquad ds = \sqrt{dx^2 + dy^2 + dz^2} \ , \qquad [m]$$

from which we can calculate velocity as follows:

$$v = \frac{ds}{dt} = \sqrt{\left(\frac{dx}{dt}\right)^2 + \left(\frac{dy}{dt}\right)^2 + \left(\frac{dz}{dt}\right)^2} \ , \qquad [m/s]$$

$$v = \sqrt{v_x^2 + v_y^2 + v_z^2} \ . \qquad [m/s]$$

This yields the following formula for converting between velocity as a function of time and velocity as a function of distance (using the typographical convention specified in 2.1):

$$(2.2:40) \qquad \dot{v} = \frac{dv}{dt} = \frac{dv}{ds} \cdot \frac{ds}{dt} = v' \cdot v \ . \qquad [m/s^2]$$

### 2.2.6.5 Movement of a body in a central force field

In the equations of motion derived in 2.2.6.3 (Eqns 2.2:32a and b), it was tacitly assumed that acceleration due to gravity always acted perpendicularly to the x-z plane. However, at longer ranges, where Earth's curvature has a noticeable effect, this assumption may no longer be valid; the projectile will be moving in a central force field.

In a central force field, all acceleration vectors point towards a central point.

A trajectory in Earth's gravitational field (such as the orbit of a satellite) is subject to NEWTON's Law of Gravity, and hence to KEPLER's laws. In a vacuum, such a trajectory will therefore be elliptical.

Our starting point is the equation of motion in a central force field and the conservation of angular momentum with reference to the mass. Using polar coordinates $(r,\psi)$ originating in one of the foci of the ellipse, we obtain

$$m \cdot \ddot{r} - m \cdot r \cdot \dot{\psi}^2 = -\frac{G \cdot m \cdot m_{ER}}{r^2} , \qquad [N]$$

$$r^2 \cdot \dot{\psi} = \text{const} := j , \qquad [m^2/s]$$

and therefore:

$$(2.2{:}41) \qquad \ddot{r} - \frac{j^2}{r^3} = -\frac{G \cdot m_E}{r^2} . \qquad [m/s^2]$$

Where m = mass of projectile, r = distance from the centre of Earth, $\dot{\psi}$ = angular velocity, G = gravitational constant and $m_E$ = Earth mass.

We obtain a differential equation for the trajectory from the time-dependent equation of motion (2.2:41) in a manner analogous to 2.2.6.4, using the following formula:

$$\frac{d}{dt} = \frac{d\psi}{dt} \cdot \frac{d}{d\psi} = \frac{j}{r^2} \cdot \frac{d}{d\psi} . \qquad [1/s]$$

Applying this to (2.2:41)

$$\frac{j}{r^2} \cdot \frac{d}{d\psi}\left(\frac{j}{r^2} \cdot \frac{dr}{d\psi}\right) - \frac{j^2}{r^3} = -\frac{G \cdot m_E}{r^2} , \qquad [m/s^2]$$

$$(2.2{:}42) \qquad \frac{d}{d\psi}\left(\frac{1}{r^2} \cdot \frac{dr}{d\psi}\right) - \frac{1}{r} = -\frac{G \cdot m_E}{j^2} . \qquad [1/m]$$

If we arrange this equation thus:

$$u = \frac{1}{r} , \quad \frac{du}{d\psi} = -\frac{1}{r^2} \cdot \frac{dr}{d\psi} ,$$

then from (2.2:42) we obtain a second-order nonhomogeneous differential equation with constant coefficients for the function u(θ)

$$\text{(2.2:43)} \qquad \frac{d^2u}{d\psi^2} + u = \frac{G \cdot m_E}{j^2} \, . \qquad \text{[1/m]}$$

The corresponding homogeneous equation has the solution $u(\psi) = B \cdot \cos(\psi - \psi_0)$, where B and $\psi_0$ are constants. The right-hand side of Eqn 2.2:43 is also a particular solution to the nonhomogeneous differential equation, giving us the following solution to Eqn 2.2:43:

$$u = \frac{1}{r} = B \cdot \cos(\psi - \psi_a) + \frac{G \cdot m_E}{j^2} \, , \qquad \text{[1/m]}$$

or, with $k := \frac{j^2}{G \cdot m_E}$ and $e = B \cdot k$, we obtain

$$\text{(2.2:44)} \qquad r(\psi) = \frac{k}{e \cdot \cos(\psi - \psi_a) + 1} \quad \text{mit} \quad r(0) = r_0 \, . \qquad \text{[m]}$$

This is a representation of a section through a cone in polar coordinates with its 5.7 origin at one of the foci and with the zero direction passing through the starting point of the trajectory, A. If eccentricity $e < 1$, we have an ellipse in which $\psi_h$ gives the direction of the major axis (see Fig. 2.3).

If $\psi = \psi_h$ and $\psi = \psi_h + \pi$, the major axis is given by

$$\text{(2.2:45)} \qquad h = \frac{k}{1 - e^2} \, . \qquad \text{[m]}$$

With the origin at $\psi = 0$, the initial velocity of the projectile = $v_0$ and the angle of departure (the elevation) = $\theta_0$, we can write

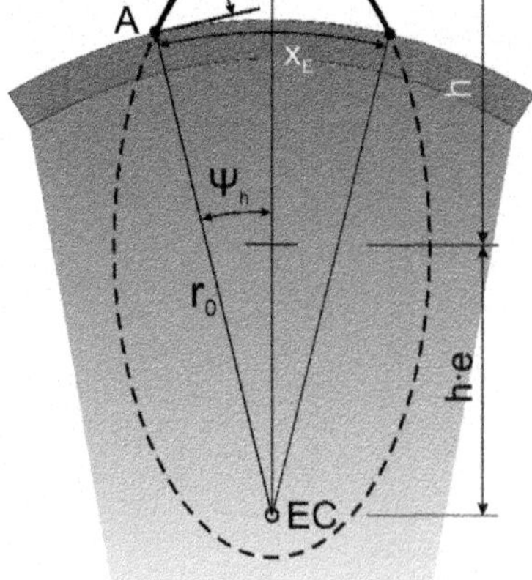


**Fig. 2.3.** Trajectory in a central force field. One focus lies at the centre of the earth, EC. A = the origin, $r_0$ = the distance of the origin from the centre of the earth, h = half the length of the major axis of the ellipse, e = eccentricity and $x_e$ = the distance along the ground between origin and point of impact. $\psi_h$ points in the direction of the major axis.

(2.2:46) $$\begin{aligned} \dot{r}(\psi=0) &= v_0 \cdot \sin\theta_0 \\ r \cdot \dot{\psi}(\psi=0) &= v_0 \cdot \cos\theta_0 \end{aligned} .$$ [m/s]

Under these additional conditions, we can calculate eccentricity e and the direction of the major axis. If the point of departure is $r_0$, $\psi = 0$, the equation for the trajectory Eqn (2.2:44) takes the following form:

(2.2:47a) $$e \cdot \cos(\psi_h) = \frac{k}{r_0} - 1 .$$ [-]

Deriving for time and substituting into Eqn 2.2:46, we obtain

$$e \cdot \sin(\psi_h) \cdot \dot{\psi} = -\frac{k \cdot \dot{r}}{r_0^2} ,$$

(2.2:47b) $$e \cdot \sin(\psi_h) = -\frac{k}{r_0} \cdot \tan\theta_0 .$$ [-]

Equations 2.2:47a and b form a system of equations for the two variables we are trying to calculate, e and $\psi_h$. After a number of operations and re-substituting the constant k, which we obtain by substituting Eqn 2.2:46 thus:

$$k = \frac{r_0^2 \cdot (r_0 \cdot \dot{\psi})^2}{G \cdot m_E} = \frac{r_0^2 \cdot v_0^2 \cdot \cos^2\theta}{G \cdot m_E} ,$$ [m]

we obtain the following expressions for these parameters:

(2.2:48) $$e = \sqrt{1 - \frac{r_0 \cdot v_0^2 \cdot \cos^2\theta_0}{G \cdot m_E} \cdot \left(2 - \frac{r_0 \cdot v_0^2}{G \cdot m_E}\right)} ,$$ [-]

(2.2:49) $$\psi_h = \arcsin \frac{r_0 \cdot v_0^2 \cdot \sin(2 \cdot \theta_0)}{2 \cdot G \cdot m_E \cdot e} .$$ [-]

Because the direction of the major axis $\psi_h$ is also the direction of the apex of the trajectory, then symmetry means that along the entire trajectory

(2.2:50) $$x_E = 2 \cdot r_0 \cdot \psi_h .$$ [m]

This assumes that the point of impact lies at the same altitude (i.e. at the same distance from Earth's centre) as the point of departure.

The height of the apex is the difference between the distance between the upper vertex of the ellipse and Earth's centre and the distance between the point of departure and Earth's centre.

$$h_G = h \cdot (1 + e) - r_0 .$$ [m]

Substituting Eqn 2.2:45 and k, we obtain:

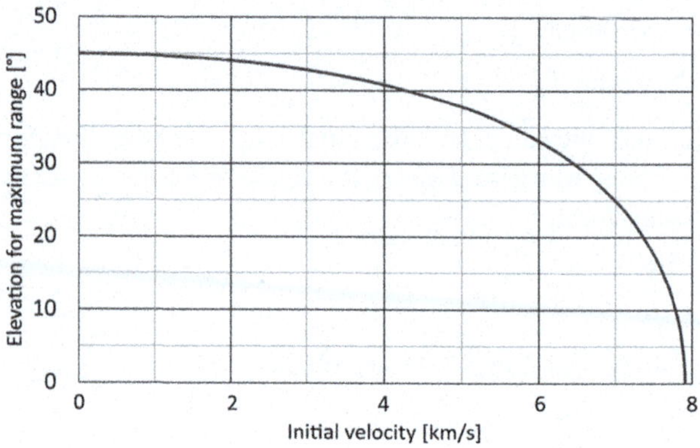


**Fig. 2.4.** Relationship between initial velocity and angle of departure required to achieve maximum range. At the first cosmic velocity, this angle is 0°.

(2.2:51) $$h_G = \frac{r_0^2 \cdot v_0^2 \cdot \cos^2 \theta_0}{G \cdot m_E \cdot (1-e)} - r_0 \ . \qquad [m]$$

Table 2.6 shows range and apex in the central force field of a trajectory, as a function of initial velocity and elevation. As in 2.2.6.3, a vacuum is assumed. For large apices (approx. 30 km and higher), the values are likely to give quite acceptable approximations, as the atmosphere will only have an influence briefly, during the ascending and descending branches of the trajectory.

In a uniform gravitational field, maximum range will always be achieved at an elevation of 45°. However, Table 2.6 shows that in the central force field the angle required for maximum range decreases as initial velocity increases. Fig. 2.4 illustrates this relationship, and shows that at the *first cosmic velocity* the elevation for maximum range is 0°, i.e. tangential to the Earth's circumference.

The first cosmic velocity is the velocity required to keep a projectile in orbit around Earth. At that velocity, weight and centrifugal force are balanced. The corresponding initial velocity is 7910 m/s.

**Table 2.6.** Range $x_E$ (2.2:50) and apex $h_G$ (2.2:51), of a trajectory in the central force field, both in [km], as a function of initial velocity and elevation (calculated with $r_0$ = earth's radius and $G \cdot m_E = 3.986 \cdot 10^{14}$ m³/s²)

| Elev. [°] | | Initial velocity [m/s] 1000 | 2000 | 3000 | 4000 | 5000 | 6000 | 7000 |
|---|---|---|---|---|---|---|---|---|
| 15 | $x_E$ | 51.7 | 216.6 | 529.0 | 1067.4 | 2012.6 | 3837.3 | 8010.9 |
| | $h_G$ | 3.5 | 14.5 | 35.6 | 72.4 | 138.0 | 270.0 | 608.0 |
| 30 | $x_E$ | 89.3 | 370.4 | 888.2 | 1735.1 | 3086.4 | 5263.6 | 8766.0 |
| | $h_G$ | 12.9 | 53.9 | 130.9 | 261.1 | 481.5 | 876.8 | 1661.2 |
| 45 | $x_E$ | 102.6 | 420.6 | 985.5 | 1854.9 | 3117.6 | 4891.3 | 7287.0 |
| | $h_G$ | 25.8 | 106.9 | 256.4 | 501.1 | 893.1 | 1536.7 | 2653.1 |
| 60 | $x_E$ | 88.5 | 358.4 | 822.2 | 1500.4 | 2419.9 | 3608.5 | 5084.9 |
| | $h_G$ | 38.6 | 159.1 | 377.2 | 724.4 | 1258.6 | 2088.8 | 3434.5 |
| 75 | $x_E$ | 51.0 | 204.5 | 462.5 | 827.7 | 1303.3 | 1892.4 | 2596.4 |
| | $h_G$ | 47.9 | 196.7 | 462.9 | 879.1 | 1504.5 | 2448.8 | 3930.7 |

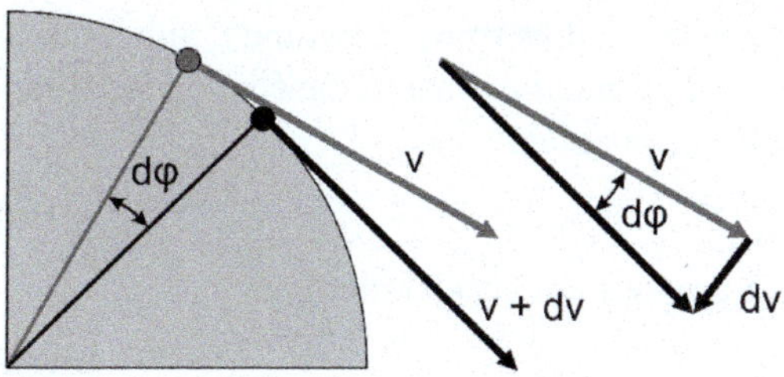


**Fig. 2.5.** Calculating centripetal acceleration in a rotating system.

### 2.2.7 Acceleration in a rotating frame of reference

#### *2.2.7.1 Centrifugal acceleration*

If a body travels along a segment of its orbit at a constant speed v, it is nonetheless undergoing acceleration, as the direction of its speed is constantly changing. This change in velocity is perpendicular to the velocity vector. From Fig. 2.5, we see that

(2.2:52) $$a_{cp} = -\frac{dv}{dt} = -v \cdot \frac{d\phi}{dt} = -v \cdot \omega = -r \cdot \omega^2 \,. \qquad [m/s^2]$$

The value is negative, because the acceleration is directed towards the centre.

If a body is located on a steadily rotating system, and if it is not subject to centripetal acceleration $a_{cp}$ (i.e. it is not fixed to the rotating system) then, because of its inertia, it experiences an acceleration (*centrifugal acceleration*) $a_{cf}$ in the opposite direction to $a_{cp}$, which is only apparent by reference to the rotating system. This is hence referred to as *apparent acceleration.*

(2.2:53) $$a_{cf} = r \cdot \omega^2 \,. \qquad [m/s^2]$$

It always acts perpendicularly to the axis of rotation.

#### *2.2.7.2 Coriolis acceleration*

If a body on a rotating system moves from the centre towards the periphery at a constant speed $v_r$, its tangential velocity will increase in accordance with Eqn 2.2:9 (see Fig. 2.6). The distance the body covers in a tangential direction is given by:

$$s = v_t \cdot t = \omega \cdot r \cdot t \,. \qquad [m]$$

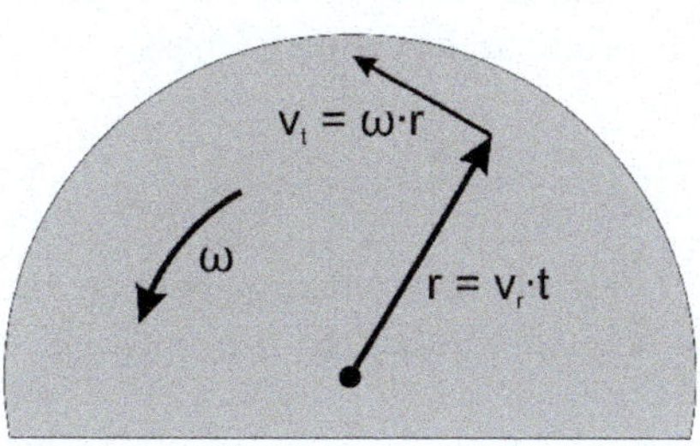


**Fig. 2.6.** Calculating Coriolis acceleration in a rotating system.

At the same time, the body will have reached radius $r = v_r \cdot t$ in time t. Substituting this into the equation for distance above, and equating that to the conventional formula for distance at constant acceleration, we obtain:

$$s = \omega \cdot v_r \cdot t^2 = \tfrac{1}{2} \cdot a_c \cdot t^2 , \qquad [m]$$

which in turn gives us the following expression for acceleration:

$$(2.2{:}54) \qquad a_c = 2 \cdot \omega \cdot v_r . \qquad [m/s^2]$$

This acceleration is known as the *Coriolis acceleration* after its discoverer Gaspard-Gustave DE CORIOLIS. Here again, a body not fixed to the system under acceleration will experience acceleration in the opposite direction because of inertia.

### 2.2.8 Movements of rigid objects

#### *2.2.8.1 Analysis of movements*

In a rigid object, the distance between any pair of points remains constant, regardless of what happens to the object. Unlike a mass point, such a body can not only move but can also rotate about a point. In most cases, the axis of rotation passing through that point will change over time. According to CHASLES' theorem (named after its 19th Century French inventor, mathematician Michel CHASLES), the displacement of any body can be broken down into a translation of its centre of mass and rotation about that point. The translation can be calculated from the position of the centre of mass, the total mass of the body, its momentum and the corresponding forces acting upon it.

As far as translation is concerned, the body can therefore be treated as a point mass.

Rotation of a point mass about a fixed axis was introduced in 2.2.4. We saw that the moment of inertia was an important characteristic of a rotating mass, as it constitutes the proportionality factor between angular momentum and angular velocity, in much the same way as mass links momentum and velocity. The distances of the mass elements of a rigid body from the axis of rotation is an additional factor that makes describing the angular momentum of a rigid body considerably more complex.

However, breaking the movement of a rigid body down into translation and rotation as described makes it possible to consider rotation in isolation. This is generally achieved using a body fixed frame with its origin at the centre of mass.

#### *2.2.8.2 Tensor of inertia*

The velocity of a mass point of a rigid body $\vec{r}_i = (r_{i1}, r_{i2}, r_{i3})$ rotating about an axis that passes through its centre of mass S, at an angular velocity $\vec{\omega} = (\omega_1, \omega_2, \omega_3)$, is given by:

(2.2:55) $\vec{v}_i = \vec{\omega} \times \vec{r}_i \, .$ [m/s]

The angular momentum is the vector product of radius and momentum:

(2.2:56) $\vec{L} = \sum_i \vec{r}_i \times (m_i \cdot \vec{v}_i) = \sum_i m_i \cdot (\vec{r}_i \times \vec{v}_i) \, .$ [kg·m²/s]

Substituting Eqn 2.2:55

$$\vec{L} = \sum_i m_i \cdot \{ \vec{r}_i \times (\vec{\omega} \times \vec{r}_i) \} \, , \qquad [\text{kg·m}^2/\text{s}]$$

and using GRASSMANN's theory of extension, we obtain

(2.2:57) $\vec{L} = \sum_i m_i \cdot \left\{ \vec{\omega} \cdot r_i^2 - \vec{r}_i \cdot (\vec{r}_i \cdot \vec{\omega}) \right\} \, .$ [kg·m²/s]

By breaking the vectors down into their components and using those to calculate the components of the angular momentum, we obtain

(2.2:58)
$$\left| \begin{array}{lcl} L_1 & = & J_1 \cdot \omega_1 + J_{12} \cdot \omega_2 + J_{13} \cdot \omega_3 \\ L_2 & = & J_{21} \cdot \omega_1 + J_2 \cdot \omega_2 + J_{23} \cdot \omega_3 \\ L_3 & = & J_{31} \cdot \omega_1 + J_{32} \cdot \omega_2 + J_3 \cdot \omega_3 \end{array} \right| \, , \qquad [\text{kg·m}^2/\text{s}]$$

or, in vector form:

(2.2:59)
$$\vec{L} = \begin{pmatrix} J_1 & J_{12} & J_{13} \\ J_{21} & J_2 & J_{23} \\ J_{31} & J_{32} & J_3 \end{pmatrix} \cdot \vec{\omega} = \hat{J} \cdot \vec{\omega} \, , \qquad [\text{kg·m}^2/\text{s}]$$

where

$$J_1 = \sum_i m_i \cdot \left( x_{2i}^2 + x_{3i}^2 \right) \, , \qquad [\text{kg·m}^2]$$

$$J_2 = \sum_i m_i \cdot \left( x_{1i}^2 + x_{3i}^2 \right) \, , \qquad [\text{kg·m}^2]$$

$$J_3 = \sum_i m_i \cdot \left( x_{1i}^2 + x_{3i}^2 \right) \, , \qquad [\text{kg·m}^2]$$

$$J_{12} = J_{21} = -\sum_i m_i \cdot x_{1i} \cdot x_{2i} \, , \qquad [\text{kg·m}^2]$$

$$J_{13} = J_{31} = -\sum_i m_i \cdot x_{1i} \cdot x_{3i} \, , \qquad [\text{kg·m}^2]$$

$$J_{23} = J_{32} = -\sum_i m_i \cdot x_{2i} \cdot x_{3i} \, . \qquad [\text{kg·m}^2]$$

For a continuously distributed mass, the summation sign is replaced by a volume integral and the mass by density.

The expression $\hat{J}$ in Eqn 2.2:59 is called the *tensor of inertia*. It can be written as a symmetrical 3×3 matrix. The elements along the main diagonals are the *moments of inertia*. The remaining elements are *deviation moments*. The tensor of inertia is not a constant associated with the body; it also depends on the origin of the body fixed frame and its orientation.

One characteristic of a symmetrical matrix is that for a given origin, it is always possible to orient the reference system in such a way that the matrix takes on diagonal form, and hence that the deviation moments disappear. Such a system is called a *principal inertia axis system*, and the associated moments of inertia are called *principal axes of inertia*. A system of this type is always assumed in ballistics.

#### 2.2.8.3 The free symmetric gyroscope

A rigid body rotating about one of its axes at a specific angular velocity $\vec{\omega}$ is called a *gyroscope*. If two of its three principal moments of inertia are identical (e.g. $J_1 = J_2$), it is deemed to be symmetrical. The principal axis of inertia for $J_3$ is then termed the *figure axis*. If the body is symmetrical about the axis of rotation, the figure axis coincides with the axis of symmetry.

If no external force is acting on the gyroscope (which means that no torque is acting that would tend to rotate it about the point of rotation), it is termed a *free gyroscope*. In such a case, angular momentum and kinetic energy remain constant. However, we can see from Eqn 2.2:59 that even if the deviation moments are eliminated, the angular velocity is generally not parallel to the angular momentum, which means that the orientation of the axis of rotation will change over time.

The resulting movement of a free gyroscope is called *nutation*. This movement depends on the conditions obtaining when it starts. Its frequency $\nu_N$ can be obtained from the equations of motion for a free gyroscope thus:

$$(2.2{:}60) \qquad \nu_N = \frac{\left|J_3 - J_1\right|}{J_1} \cdot \frac{\omega}{2 \cdot \pi} \;. \qquad [1/s]$$

#### 2.2.8.4 The heavy symmetric gyroscope

If a constant torque is applied to a gyroscope, it is termed a *heavy gyroscope*, because weight is often the cause of the torque.

For an elementary understanding of the behaviour of a heavy symmetric gyroscope, one can imagine a classic toy top of mass m, which is set spinning on a level surface and rotates about its figure axis at an angular velocity $\vec{\omega}$. (See Fig. 2.7). The origin of the reference frame is located at the point of support O. Because of symmetry, the centre of gravity CG lies on the figure axis at a distance $\vec{b}$ from O. (See Fig. 2.8). The torque acting on the gyroscope is given by

$$(2.2{:}61) \qquad \vec{M} = \frac{d\vec{L}}{dt} = \vec{b} \times (m \cdot \vec{g}) \;, \qquad [N{\cdot}m]$$

and is therefore perpendicular to both $\vec{b}$ and $\vec{g}$. This is of course also true of the change in angular momentum, which is parallel to the torque, although its compo-

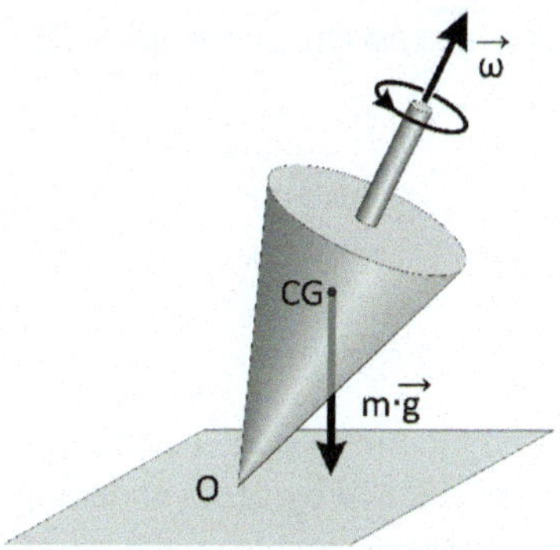


**Fig. 2.7.** Classic representation of a heavy gyroscope (CG = centre of gravity).

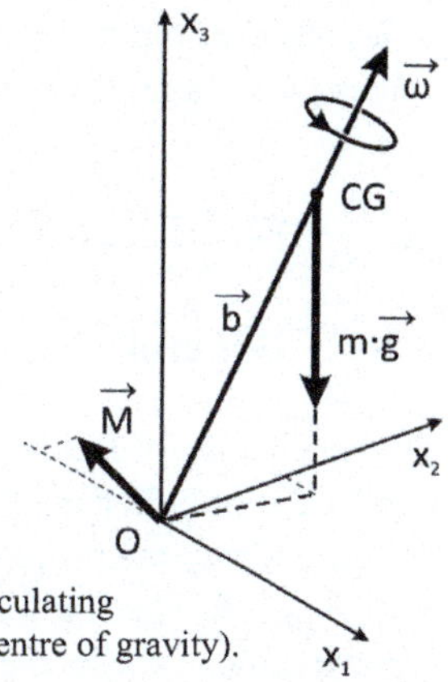


**Fig. 2.8.** Calculating torque (CG = centre of gravity).

nent $L_3$ remains unchanged. The angular momentum vector therefore describes a cone about the axis $e_3$. The resulting movement of the gyroscope is known as *precession.*

The angular velocity of the precession $\omega_p$ is derived from a number of geometrical considerations (see Fig. 2.9). Let $\delta$ be the angle of inclination of the vector of angular momentum $\vec{L}$. If $L_r$ is the radial component of $\vec{L}$ and L its magnitude:

$$L_r = L \cdot \sin\delta \,. \qquad [\text{kg·m}^2/\text{s}]$$

In time dt, $L_r$ moves through an angle $d\alpha$:

$$d\alpha = \frac{dL_r}{L_r} = \frac{dL}{L_r} = \frac{M \cdot dt}{L \cdot \sin\delta}\,, \quad \text{and hence} \qquad [\text{-}]$$

(2.2:62) $$\omega_p = \frac{d\alpha}{dt} = \frac{M}{L \cdot \sin\delta}\,. \qquad [1/\text{s}]$$

We can write this relationship in the following form:

$$\omega_p \cdot L \cdot \sin\delta = M\,, \qquad [\text{N·m}]$$

and because the left-hand side is the magnitude of the vector product of $\vec{\omega}$ and $\vec{L}$, we can also write

(2.2:63) $$\vec{\omega}_p \times \vec{L} = \vec{M}\,. \qquad [\text{N·m}]$$

From this, it is clear that the angular velocity of the precession is independent of the inclination of the axis of the gyroscope.

A nutation is generally superimposed on the precession. As we know, this occurs whenever the vector of angular velocity does not coincide with that of angular momentum. This nutation may be the result of the starting conditions, or may be caused by the brief application

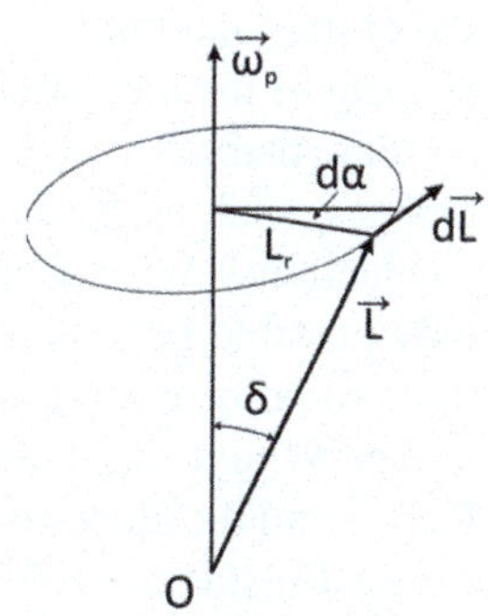


**Fig. 2.9.** Calculating the angular velocity of precession.

of a torque to the gyroscope that deflects the vector of angular momentum. This makes movement of the figure axis highly complex.

## 2.3 Fluid dynamics

### 2.3.1 General

The mechanics of fluids and gases – known as fluid dynamics – play a central role in all areas of ballistics. In general, the projectile is accelerated by gases under high pressure. Once the projectile leaves the barrel, those gases escape from the muzzle at high speed. The flight of the projectile through the air involves the broad field of aerodynamics, while the penetration of a plastically deformable medium by a projectile can often be seen in terms of hydrodynamics.

When we are studying the movement of a body in a medium it makes no difference whether we consider the medium to be stationary and the body in motion, or the body stationary and the medium in motion (in the opposition direction and at the same speed). In terms of physics, the two approaches are equivalent. One can therefore use whichever is clearer in a given context.

Motion in a liquid or gaseous medium always generates heat, whether through friction or as a result of subjecting a gas to pressure. It is therefore necessary to include thermodynamic processes in any consideration of fluid dynamics.

### 2.3.2 Basic concepts in thermodynamics

#### *2.3.2.1 Temperature*

Temperature is one of the basic parameters required to describe processes that involve heat. Temperature describes the thermal state of a body and serves as a measure of the total kinetic energy of the molecules that make up that body. It is independent of the mass and the material composition of the body.

If the thermal state of a material changes, so do a number of its physical characteristics, such as its dimensions (length, volume, etc.), its colour or its electrical conductivity. We can use this behaviour to measure temperature. However, it is only possible to conduct relative measurements. As a result, we are free to select the zero point and the units of a scale at will.

The SI unit for temperature is the Kelvin [K]. The only difference between the Kelvin and the older degree Celsius [°C] is the zero point used for the two systems. On the Celsius scale, 0 °C corresponds to the freezing point of water, whereas 0 K corresponds to "absolute zero," the lowest temperature possible. Degrees Fahrenheit are in widespread use in some English-speaking countries. All scales are calibrated using internationally agreed reference points, corresponding to the boiling

**Table 2.7.** Temperature reference points (at normal pressure)

| Medium | Reference point | [K] | [°C] | [°F] |
|---|---|---|---|---|
| Water | Freezing point | 273.15 | 0.00 | 32.00 |
| | Boiling point | 373.15 | 100.00 | 212.00 |
| Oxygen | Boiling point | 90.18 | -182.97 | -297.35 |
| Gold | Melting point | 1336.15 | 1063.00 | 1945.40 |

and freezing points of various materials at normal pressure (1013.25 mb). See Table 2.7.

Conversion formulae for the various units appear after the table of symbols (p. XXXI).

### 2.3.2.2 *Temperature and heat*

If the temperature of a given quantity of a material rises during a physical process, then heat has been introduced. However, an increase in temperature means that the total kinetic energy of the molecules in the quantity of material under consideration has also increased. Heat is therefore a form of energy and is hence measured in Joule. The quantity of heat added is proportional to the mass and to the increase in temperature. The proportionality factor C is known as the *specific heat capacity* of the material.

(2.3:1) $$\Delta Q = C \cdot m \cdot \Delta T \ .$$ [J]

### 2.3.2.3 *States of matter*

A material can be in a solid, liquid or gaseous state. The more precise terms for these three states are crystalline, amorphous and gaseous. In the solid state, the molecules are arranged in a crystal lattice, and are held in place not only by intermolecular forces of attraction but also by lattice linkage forces. Solid matter has a specific geometrical form and a fixed volume.

If so much heat is applied to a body that its molecules acquire enough kinetic energy to overcome the lattice forces, the material enters the amorphous (liquid) state. All that is keeping the material together at this point is the intermolecular force. Liquid matter has a fixed volume, but generally has no specific geometrical form.

If the intermolecular forces are sufficiently strong, amorphous materials may be shape stable. Examples of this include glass and wax.

If we add enough heat to raise the kinetic energy of the molecules to the point at which the intermolecular forces are overcome, the molecules become free to move. The material enters the gaseous state, in which its volume is limited not by the quantity of matter present, but by the total space available to it. From the above, we see that it is only possible to change the state of a material to the next higher level by adding energy in the form of heat. Similarly, the corresponding quantity of energy is released as the material returns to a lower level.

#### 2.3.2.4 Equation of state for gases

Because of the kinetic energy of its free molecules, a gas exerts a pressure p on the surfaces that delimit its volume. The impacts of the particles against these surfaces result in a mean force per unit area, which can be explained by the law of conservation of momentum. Any change in temperature (i.e. in kinetic energy) or in volume (i.e. in surface area) will affect the pressure. Pressure, temperature and volume are the thermodynamic state variables of the gas in question. The equation that links these variables is known as an equation of state. The best-known equation of state is *BOYLE's law* which, for monatomic gases, takes the following form:

(2.3:2) $$p \cdot V = m \cdot R \cdot T \,. \qquad [J]$$

Where m is the mass of the gas and R is the special (material-dependent) gas constant. If a gas obeys BOYLE's law (Eqn 2.3:2), it is known as an *ideal gas*. If the density of the gas is sufficiently low (i.e. if the pressure is low), many gases behave almost as ideal gases.

In the case of a real gas under high pressure, or one that has larger molecules, we must take account of the volume b of the molecules themselves and of internal pressure, $w/V^2$. The equation that extends Eqn. 2.3:2 to take account of these factors is known as the *Van der Waals equation.*

(2.3:2a) $$\left(p + \frac{w}{V^2}\right) \cdot (V - b) = m \cdot R \cdot T \,. \qquad [J]$$

In gas dynamics, we generally use density (mass per unit volume) rather than mass. Eqn 2.3:2 can then be written:

(2.3:3) $$\frac{p}{\rho} = R \cdot T \,. \qquad [J/kg]$$

If temperature remains constant, the right-hand side of these equations remains constant, and hence so does the left-hand side.

#### 2.3.2.5 Heat, work and internal energy

As the volume of a material can only change when it is in a gaseous state, it follows that materials can only convert heat energy into mechanical work when they are in that state. Devices that perform this conversion are known as *heat engines.*

Heat engines – which include firearms – always use materials in gaseous form. Those materials are usually created by heating or by combustion.

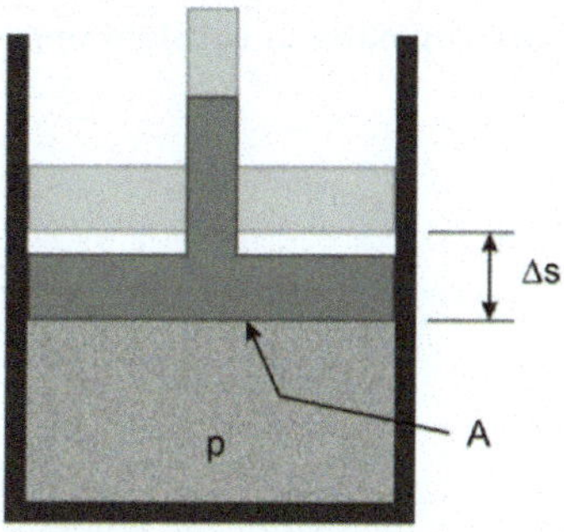


**Fig. 2.10**. Calculating the work done by a gas in expanding by a distance Δs.

If the volume of a gas is increased by heating it, then the gas performs work. According to Eqn 2.2:16, the work done is proportional to the force exerted (pressure times area) and the distance over which it expands (see Fig. 2.10):

(2.3:4) $$\Delta W = p \cdot A \cdot \Delta s = p \cdot \Delta V \, .$$ [J]

As we know, the ability to do work is generally known as potential energy.

The potential energy of a gas of volume V at a pressure p is therefore

(2.3:5) $$E_{pr} = p \cdot V \, .$$ [J]

In this context, the potential energy of the gas is termed *pressure energy*.

When heat energy is added to a system (dQ), two things happen. The first is that the total kinetic energy of the molecules in that system (its *internal energy* U) increases, which means that its temperature also increases. The second is that the system performs mechanical work. The following energy balance therefore applies:

(2.3:6) $$dQ = dU + dW = dU + p \cdot dV \, .$$ [J]

If no heat is added to the system (dQ = 0) then, from Eqn 2.3:6:

$$dU + p \cdot dV = 0 \, ,$$

which means that the sum of internal energy and pressure energy (*enthalpy*, H) remains constant:

(2.3:7) $$H = U + p \cdot V = \text{const} \, .$$ [J]

If no heat (energy) is added, then from the a foregoing and the corresponding law of mechanics (Eqn 2.2:30) we obtain the mechanical thermodynamic law of conservation of energy:

(2.3:8) $$E_{tot} = E_{mech} + H = E_{mech} + U + E_{pr} = \text{const} \, .$$ [J]

In order for the temperature of a gas to rise by a specific amount, the gas must absorb more heat at a constant pressure than at a constant volume. This is a direct consequence of Eqn 2.3:7.

If volume remains constant (dV = 0), all of the additional heat is used to raise the internal energy of the gas (i.e. its temperature), as no mechanical work can be done. At constant pressure, however, heat is "lost" to work, and the temperature increase is smaller.

This means that the specific heat capacity in Eqn 2.3:1 is greater at constant pressure ($C_P$) than at constant volume ($C_V$).

(2.3:9) $$\gamma = \frac{C_p}{C_v} > 1 \,.$$ [-]

The quotient $\gamma$ is known as the *heat capacity ratio*.

### 2.3.3 Characteristics of materials

#### 2.3.3.1 Density and compressibility.

When liquids and gases (fluids) are subjected to a force they change their form and, in many cases, their volume. In order to describe motion, we therefore need physical parameters that describe the main characteristics of the materials involved. In fluid dynamics, *density* $\rho$ is more important that mass:

(2.3:10) $$\rho = \frac{m}{V} \,.$$ [kg/m³]

The *compressibility* of a material is a measure of the relative change in the volume of that material in response to a change in pressure:

(2.3:11) $$\kappa = -\frac{1}{V}\cdot\frac{dV}{dp} \,.$$ [1/Pa]

The minus sign indicates that an increase in pressure brings about a decrease in volume.

Liquids are generally considered incompressible. At low speeds, we can make the same assumption for gases.

#### 2.3.3.2 Viscosity

The absence of lattice force means that a fluid changes its form when subjected to an external force. However, the fluid reacts to that force with a certain inertia. This resistance, caused by the intermolecular forces of attraction, is termed *viscosity*. If a shear stress is applied to the fluid, with a shear force $F_R$, a velocity gradient is created in the area under stress. In Newtonian fluids, this gradient is proportional to the stress. See Fig. 2.11.

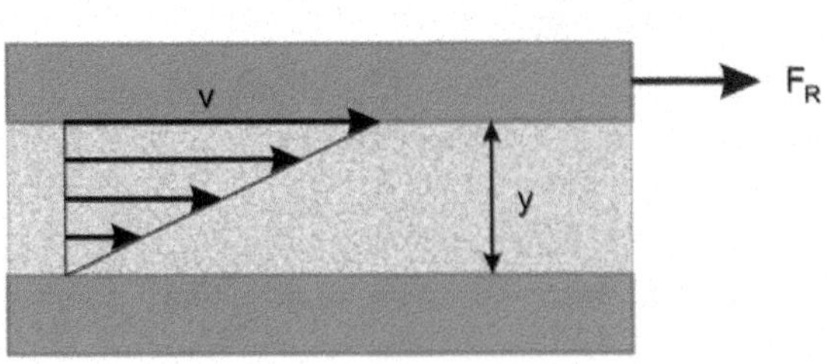


**Fig. 2.11.** The concept of dynamic viscosity.

(2.3:12) $$\sigma_s = \frac{F_R}{A} = \eta \cdot \frac{dv}{dy} . \qquad [N/m^2]$$

The proportionality factor is the *dynamic viscosity*, $\eta$:

(2.3:13) $$\eta = \sigma_s \cdot \frac{dy}{dv} . \qquad [Pa \cdot s]$$

Dynamic viscosity as a function of the density of the fluid is known as *kinematic viscosity*.

(2.3:14) $$\nu = \frac{\eta}{\rho} . \qquad [m^2/s]$$

Viscosity depends very much on temperature, with the viscosity of a gas changing with temperature in exactly the opposite manner to that of a liquid.

As the temperature of a liquid (and hence its internal energy) increases, the effect of the intermolecular forces decreases, and it becomes less viscous. In the case of a gas, internal friction is caused by the exchange of momentum between freely moving molecules. As temperature increases, the speed of those molecules increases, leading to an increase in friction and causing the gas to become more viscous.

If the mass remains constant, a fluid also resists any increase in volume, as such an increase can only occur if the form of the fluid also changes. This effect is known as bulk viscosity, which is to be distinguished from dynamic viscosity.

#### 2.3.3.3 Speed of sound

Changes in pressure and density (disturbances) propagate through a fluid in the form of longitudinal waves.

In the case of longitudinal waves, the disturbances oscillate in the same direction as that in which the wave is moving.

If the disturbances remain small, they propagate at the *speed of sound* c. In liquids, this speed depends on density and compressibility:

(2.3:15) $$c = \sqrt{\frac{1}{\kappa \cdot \rho}} . \qquad [m/s]$$

In gases, it depends on density and pressure:

(2.3:16) $$c = \sqrt{\frac{\gamma \cdot p}{\rho}} = \sqrt{\gamma \cdot R \cdot T} . \qquad [m/s]$$

The symbol $\gamma$ represents the ratio of the specific heat capacities, $C_p/C_V$ (Eqn 2.3:9). For air, its value is 1.4.

The general equation for the propagation speed of small disturbances in a gas is:

$$c^2 = \left(\frac{\partial p}{\partial \rho}\right)_{\Delta S=0} . \qquad (2.3:16a) \qquad [m^2/s^2]$$

The significance of $\Delta S = 0$ is that changes in temperature have no effect on the heat (energy). This is known as constant entropy. In the case of an isentropic gas, this gives us the first part of Eqn 2.3:16.

The use of Eqn 2.3:2 means that the second part of Eqn 2.3:16 only applies to ideal (or nearly ideal) gases. In such cases, the speed of sound depends solely on temperature. As a result, this speed is a suitable reference speed for compressible media. This leads us to the definition of the *Mach number*, Ma:

$$\mathrm{Ma} = \frac{v}{c} . \qquad (2.3:17) \qquad [-]$$

The behaviour of a flow is closely related to the Mach number, which gives us a simple classification criterion. If Ma < 1, the flow is said to be *subsonic*. If Ma > 1, it is described as *supersonic*. If Ma lies in the region of 1, the flow is described as *transonic*.

### 2.3.3.4 Propagation of sound waves. Mach cones.

As mentioned above, a weak disturbance in a fluid (such as that due to a body moving through air) is propagated at the speed of sound.

Strong disturbances, such as shock waves due to detonations, may be propagated at speeds significantly higher than the speed of sound under normal conditions.

If the projectile is moving at less than the speed of sound (Ma < 1), the disturbances will precede it (see Fig. 2.12). If it is travelling faster than the speed of sound (Ma >1), the spacing between the disturbances is reduced along a conical surface;

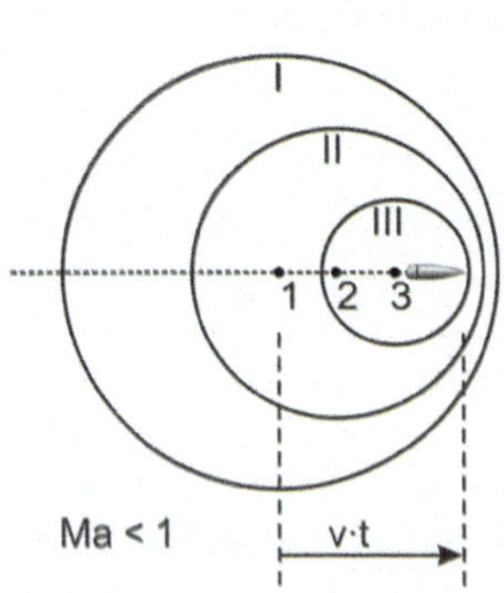


**Fig. 2.12.** Propagation of waves at less than the speed of sound. Sound waves I, II and III, emitted by the projectile at positions 1, 2 and 3, reach the ear consecutively and are virtually inaudible.

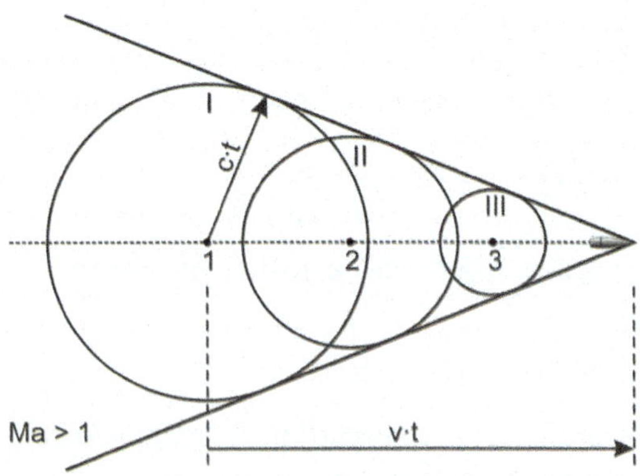


**Fig. 2.13.** Propagation of waves at a speed greater than that of sound. Sound waves emitted at all positions cumulate into a front and result in a loud bang.

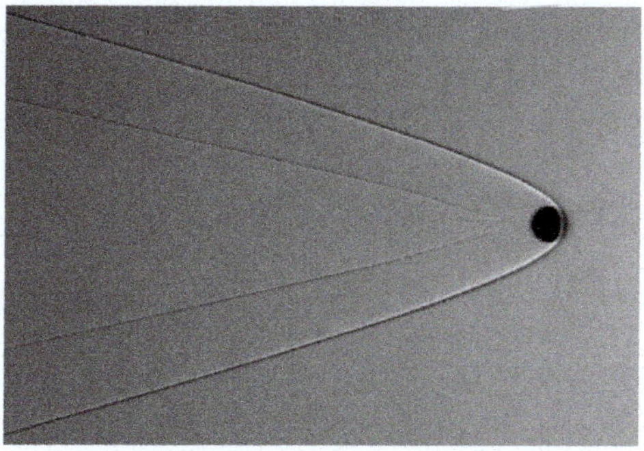

**Fig. 2.14.** A sonic bang rendered visible: A shadowgraph of a 5.5 mm steel ball travelling at approximately 1150 m/s (Ma ≈ 3.4). The densification lines are clearly visible. The bow and tail waves are so close together that the human ear cannot distinguish them. When looking at such an image, one should always remember that it is effectively a cross-section through a cone.

the sound waves are "pushed together" (see Fig. 2.13). This shock wave, also known as a *Mach cone*, produces a sudden increase in pressure, which is heard as a loud bang.

Half the cone angle α of the Mach cone is given by the ratio of the speed of sound c to the speed of the projectile v (see Fig. 2.13). This is also called the *Mach angle*.

(2.3:18) $$\sin\alpha = \frac{c}{v}\,. \qquad [-]$$

It is possible to render the density gradients visible using special photographic techniques. Fig. 2.14 shows flow around a sphere at a velocity of approximately 1150 m/s. It is noticeable that a shock wave occurs not only on the leading side of the projectile but also at the point where the flows merge to the rear. In the case of long objects such as aircraft, one clearly hears two separate bangs.

## 2.3.4 Frictionless flow

### *2.3.4.1 Definition. Laws of conservation.*

The study of flow is much simplified if one can ignore internal friction. In the absence of internal friction, no shear stresses occur, so the fluid does not deform. This assumption also ignores the fact that every fluid adheres to a fixed surface. As a result, external friction is also ignored. Despite this limitation, a surprising number of phenomena can be described and presented in the form of laws.

This is because when a fluid flows over a body, the effect of friction is often limited to a thin layer at the surface of the body, the *boundary layer*. If the boundary layer is very thin compared to the other dimensions of the flow, we can safely ignore the external effects of friction.

A fluid in which no internal friction forces occur is termed frictionless. If, in addition, the fluid is incompressible, it is said to be *ideal*. Flow is also described in terms of its relation to time. If the speed and thermodynamic characteristics of the flow depend only on position, and do not change over time, the flow is described as *stationary*.

The laws of conservation of mass, energy and momentum set out in 2.2.5 also apply to the movements of fluids. Here again, the conservation of mass plays a central role: the change in mass in any fixed volume element corresponds to the difference between the mass entering the system (subscript 1 in the equation below) and the mass leaving it (subscript 2), where $A_1$ and $A_2$ are the cross-sectional areas:

$$(2.3{:}19)\qquad \frac{dm}{dt} = \rho_1 \cdot v_1 \cdot A_1 - \rho_2 \cdot v_2 \cdot A_2 \;. \qquad [kg/s]$$

This relationship is known as a *continuity equation.* At constant mass (dm/dt = 0), Eqn 2.3:19 becomes

$$(2.3{:}19a)\qquad \partial\left(\rho \cdot v \cdot A\right) = \frac{\partial\rho}{\rho} + \frac{\partial v}{v} + \frac{\partial A}{A} = 0 \;. \qquad [m^3/s]$$

In the case of an incompressible fluid (i.e. when ρ is constant), velocity and flow cross-section will be inversely proportional to one another:

$$(2.3{:}19b)\qquad v \cdot A = \text{const} \;. \qquad [m^3/s]$$

The relationship is very clear in the case of a river. If the river-bed is broad, the river flows slowly. If it is narrow, the river flows faster.

The law of conservation of energy applies. If no heat is added, the sum of pressure energy, kinetic energy, potential energy and internal energy remains constant (see Eqn 2.3:8):

$$E_{pr} + E_{kin} + E_{pot} + U = \text{const} \;, \qquad [J]$$

and therefore

$$(2.3{:}20)\qquad p \cdot V + \tfrac{1}{2} \cdot m \cdot v^2 + m \cdot g \cdot y + U = \text{const} \;. \qquad [J]$$

Momentum and angular momentum are also conserved in a flow. As in the case of mechanics, we obtain the equation of motion for flow from the law of conservation of momentum, using D'ALEMBERT's principle. In this context, the equation is known as *EULER's equation.* From Fig. 2.15, we obtain:

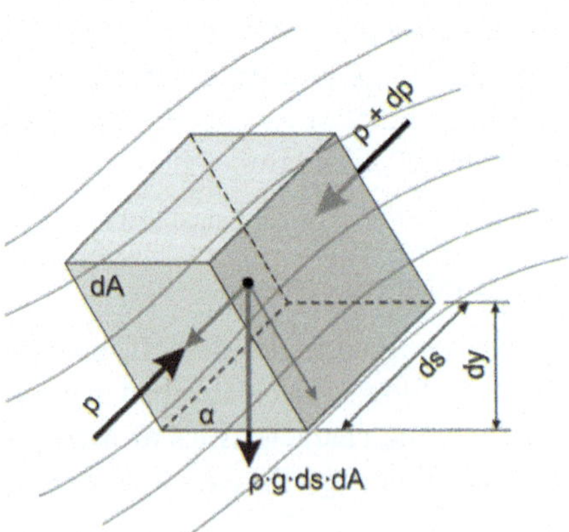


**Fig. 2.15.** Forces acting on a volume element in a flow.

$$p \cdot dA - (p + dp) \cdot dA - \rho \cdot g \cdot ds \cdot \sin\alpha \cdot dA = \rho \cdot ds \cdot dA \cdot \frac{dv}{dt} ,$$

$$- dp \cdot dA - \rho \cdot g \cdot dy \cdot dA = \rho \cdot ds \cdot dA \cdot \frac{dv}{ds} \cdot v ,$$

(2.3:21) $$dp + \rho \cdot v \cdot dv + \rho \cdot g \cdot dy = 0 .$$ [N/m²]

If we divide Eqn 2.3:20 by volume V and calculate the derivative, we also obtain Eqn (2.3:21).

Angular momentum is conserved only in the case of vortices. The rule of conservation of angular momentum implies that if vortices occurs in a flow, identical vortices rotating in the opposite direction must also occur.

#### 2.3.4.2 BERNOUILLI's equation

Integration of the EULER's equation (2.3:21) gives *BERNOUILLI's equation.*

(2.3:22) $$p + \tfrac{1}{2} \cdot \rho \cdot v^2 + \rho \cdot g \cdot y = \text{const} .$$ [N/m²]

At constant density and internal energy U, we can derive (2.3:22) directly from (2.3:20).

If we also assume a constant flow height, we can simplify Eqn 2.3:22 to

(2.3:22a) $$p_1 + \tfrac{1}{2} \cdot \rho_1 \cdot v_1^2 = p_0 + \tfrac{1}{2} \cdot \rho_0 \cdot v_0^2 = \text{const} .$$ [N/m²]

This means that pressure and flow velocity are directly related to one another. If pressure is high, fluid velocity is low. If pressure is low, fluid velocity is high. Acceleration occurs when there is a pressure gradient.

#### 2.3.4.3 Flow resistance

If a flow strikes a flat, stationary plate at right angles, the flow velocity drops to zero at the surface of the plate. If we use index 0 to indicate the values for undisturbed flow and 1 for those at the plate surface, then from Eqn 2.3:22a, for $v_1 = 0$ we obtain

$$p_1 = p_0 + \tfrac{1}{2} \cdot \rho_0 \cdot v_0^2 ,$$

(2.3:23) $$p_1 - p_0 = \tfrac{1}{2} \cdot \rho_0 \cdot v_0^2 .$$ [N/m²]

The left-hand side of Eqn 2.3:23 corresponds to the increase in pressure at the plate surface by comparison with the pressure in undisturbed flow. This is termed the *stagnation pressure* of the plane plate.

In the case of a real body, streaming pressure depends not only on velocity but also on the form and dimensions of the body. The actual values will differ from those calculated using the equation above.

The streaming pressure on a specific body is calculated using the ratio of the actual increase in pressure to the plane stagnation pressure. This (dimensionless) quantity is known as the *pressure coefficient* of the body:

(2.3:24) $$C_{PR} = \frac{p_{eff} - p_0}{\frac{1}{2} \cdot \rho \cdot v^2} .$$ [-]

The force that a flow exerts on a body (the flow resistance) is obtained by multiplying the force by the effective area. However, it is often difficult to determine this area. The effective area is therefore replaced by a fixed reference area $A_0$ which, together with the plane stagnation pressure from Eqn 2.3:23, yields a reference force that is independent of the shape of the body. Dividing the effective resistance $F_D$ by this reference force gives another dimensionless quantity that takes account of all form-specific influences. This is known as the *drag coefficient*.

(2.3:25) $$C_D = -\frac{F_D}{\frac{1}{2} \cdot \rho \cdot v^2 \cdot A_0} ,$$ [-]

(2.3:25a) $$F_D = -C_D \cdot \tfrac{1}{2} \cdot \rho \cdot v^2 \cdot A_0 .$$ [N]

The minus sign indicates that the force acts in the opposite direction to the movement.

A body subject to a flow experiences a lift force ($F_L$) – a force perpendicular to the direction of flow – when the pressure on its two sides is unequal as a result of the flow velocity on the two sides being unequal (BERNOULLI's equation). The lift force is calculated in relation to the same reference force as the resistance/drag. The corresponding ratio is known as the *lift force coefficient*:

(2.3:26) $$C_L = -\frac{F_L}{\frac{1}{2} \cdot \rho \cdot v^2 \cdot A_0} .$$ [-]

As a lift force generally acts at a point away from the centre of mass, it will also induce a rotational moment $M_M$ (the overturning moment), with its own reference distance $d_0$. The corresponding parameter is known as the *overturning moment coefficient*:

(2.3:27) $$C_M = -\frac{M_M}{\frac{1}{2} \cdot \rho \cdot v^2 \cdot A_0 \cdot d_0} .$$ [-]

The reference area $A_0$ and distance chosen $d_0$ differ according to the area of study. In aerodynamics, one often uses either the largest cross-section in the plane of the flow or the wing surface/wingspan. In ballistics, however, one generally uses the calibre cross-sectional area (which is generally smaller than the cross-sectional area of the projectile) and the calibre, although projectile length may also be used. It is therefore important to specify the reference value when stating a coefficient.

Furthermore, coefficients are velocity- and temperature-dependent, and are always expressed as a function of the Mach number.

#### 2.3.4.4 Rotational flow

If a cylinder at rest in a fluid starts to rotate about its longitudinal axis, the fluid begins to rotate with the cylinder, because of the adhesion between fluid and cylinder surface that was mentioned but ignored above, and because of internal friction (which we have likewise ignored until now). Stationary rotational flow occurs. At the surface of the cylinder, the velocity of the flow is the same as the rotational velocity of the cylinder, decreasing steadily with distance from the cylinder. This is known as *rotational flow* (see Fig. 2.16). *Circulation* Γ is used as a measure of the intensity of this rotational flow (its vorticity). It is defined as the line integral around a closed curve of the tangential component of velocity along that curve.

(2.3:28) $$\Gamma = \oint v_t \cdot ds \,.$$ [m²/s]

Along integration path 1 (Fig. 2.16), we obtain the following expression for the circulation of the rotational flow:

$$\Gamma = \oint \omega \cdot r \cdot ds = \int_0^{2\pi} \omega \cdot r^2 \cdot d\alpha = \omega \cdot r^2 \cdot \int_0^{2\pi} d\alpha \,.$$

(2.3:29) $$\Gamma = 2 \cdot \pi \cdot \omega \cdot r^2 = 2 \cdot \pi \cdot r \cdot v \,.$$ [m²/s]

In frictionless parallel flow circulation disappears, as $v_t = v \cdot \cos\alpha$ for a given closed curve:

$$\Gamma = \oint v_t \cdot ds = \int_0^{2\pi} v \cdot \cos\alpha \cdot r \cdot d\alpha = v \cdot r \cdot \int_0^{2\pi} \cos\alpha \cdot d\alpha = 0 \,.$$

Frictionless flow with no circulation is known as *potential flow*. It is irrotational at all points.

Because of the law of conservation of angular momentum, in rotational flow, for a mass particle m at a distance r from the centre of the cylinder:

$$m \cdot r \cdot v(r) = \text{const} \,.$$

In the case of a frictionless rotating flow, this is true for all particles of the same mass at any distance from the centre of the cylinder, which means that from Eqn 2.3:29 we obtain:

(2.3:30) $$r \cdot v(r) = \text{const} = \frac{\Gamma}{2 \cdot \pi} \,,$$ [m²/s]

which allows us to calculate the velocity of the rotational flow as a function of distance from the centre of the cylinder:

(2.3:30a) $$v(r) = \frac{\Gamma}{2 \cdot \pi \cdot r} \,.$$ [m/s]

Given that the velocity in a radial direction is zero, we obtain the following equation for circulation in rotational flow outside the centre (integration path 2 in Fig. 2.16):

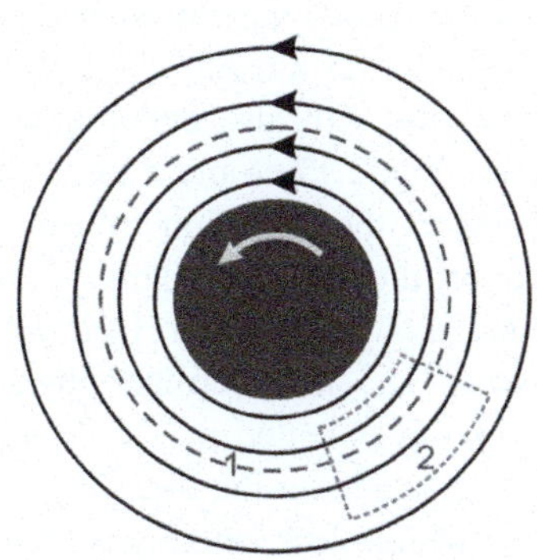


**Fig. 2.16.** Rotational flow caused by a rotating cylinder.

$$\Gamma = \int_0^\beta v_1 \cdot ds + \int_\beta^0 v_2 \cdot ds = \int_0^\beta v_1 \cdot r_1 \cdot d\alpha - \int_0^\beta v_2 \cdot r_2 \cdot d\alpha = 0 ,$$

as $v_1 \cdot r_1 = v_2 \cdot r_2$ because of Eqn 2.3:30. Rotational flow that complies with this equation is irrotational outside the vortex core and is therefore described as an *irrotational* or *free* of vortex.

#### 2.3.4.5 The Magnus effect

If a cylindrical body at rest is subjected to a potential flow perpendicularly to its axis, the resulting decrease in cross-section leads to a change in flow velocity. This results from the continuity equation (Eqn 2.3:19b), which states that at constant density, and assuming conservation of mass, a decrease in cross-section will lead to an increase in velocity. However, this will also lead to a decrease in pressure (see BERNOUILLI's equation, Eqn 2.3:22), which means that both positive and negative pressure regions will occur over the surface of the cylinder. As the flow is symmetrical, these regions will form symmetrically around the body (see Fig. 2.17a), and therefore no net force acts upon it.

If the cylinder now starts to rotate in a uniform manner, then in accordance with 2.3.4.4 rotational flow will occur, superimposed on the potential flow (see Fig. 2.17b). The fluid velocities at the surface of the cylinder will be added together, producing an asymmetric velocity field. As a result, the distribution of pressure will also become asymmetric (see Fig. 2.17 c) and a force perpendicular to the direction of flow will act upon the cylinder. This is named the *Magnus force*, after Heinrich Gustav MAGNUS, the German physicist who first investigated it. The Magnus force always acts in the direction of the side on which flow velocity is greatest.

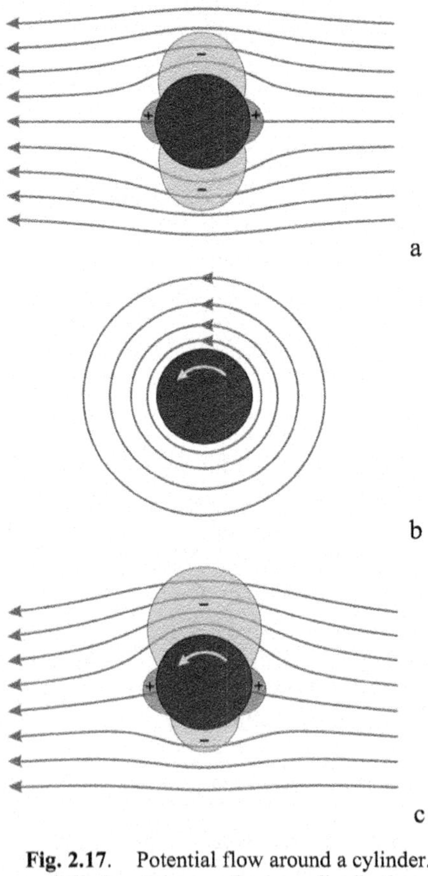


**Fig. 2.17.** Potential flow around a cylinder. **a.** Cylinder stationary. Pressure distribution symmetrical. **b.** Potential vortex. **c.** Flow around a rotating cylinder. Pressure distribution asymmetrical.

**Fig. 2.18.** How an aerofoil works. The asymmetric cross-section causes a difference in velocity between the upper and lower surfaces and hence a difference in pressure.

One practical application of the Magnus force is the Flettner rotor, which is used to power rotor ships. Instead of masts and sails, rotor ships are fitted with large, vertical cylinders. When the wind acts upon them, the Magnus effect creates a forward thrust, as long as the rotors are turning. A 130-metre freighter with four Flettner rotors entered service in 2010.

### *2.3.4.6 The aerofoil principle*

Strictly speaking, aerofoil-equipped projectiles belong to the field of flight mechanics rather than ballistics. However, certain methods for stabilizing projectiles do use the aerofoil principle, so it is worth ballisticians taking a look at the physics behind them.

An aerofoil is a slightly curved, thin, flat body. If such a body is placed in a potential flow, the air particles passing along the curved side have to cover a greater distance to reach the trailing edge than do those that pass along the opposite side. See Fig. 2.18. In order to satisfy the continuity equation, those particles have to increase their velocity temporarily. According to BERNOUILLI's equation, however, higher velocity means lower pressure. As a result, pressure on the curved side is lower, producing a lateral force – lift.

The difference in velocity between the upper and lower surfaces of an aerofoil can be presented by superimposing an irrotational vortex on the parallel flow in such a way that the corresponding velocity conditions occur. As a result, lift $F_A$ can be expressed by means of the corresponding circulation Γ (the Joukowsky transformation). See for instance W. KALIDE, 1980.

$$(2.3:31) \qquad F_A = \rho \cdot v \cdot b \cdot \Gamma , \qquad [N]$$

where b = wingspan. Calculating the circulation requires additional mathematical tools (conformal mapping of complex functions).

## 2.3.5 Flow with friction

### 2.3.5.1 Forces and equations of motion.

In accordance with D'ALEMBERT's principle, equations of motion are derived by equating inertial force and the forces acting on a body. Because density and volume are subject to change, both are expressed as a function of mass.

The following discussion applies to a one-dimensional flow (a streamline). This makes the relationships considerably simpler and clearer, and facilitates comprehension of the subject. For complete versions of the general equations of motion for fluids, please consult the literature on the subject (e.g. J. ZIEREP, 1991).

A particle in a flow can undergo a change in velocity under two circumstances:

- when the velocity field changes over time (*unsteady* flow),
- when, in a (stationary) velocity field, it is displaced to a position at which the flow velocity is higher or lower.

Total acceleration is derived from the combination of the two sources. This acceleration is known as the substantial acceleration.

(2.3:32) $$\frac{dv}{dt} = \frac{\partial v}{\partial t} + v \cdot \frac{\partial v}{\partial x} . \qquad [m/s^2]$$

The two terms on the right can be interpreted as time-dependent or position-dependent inertial forces, expressed as a function of mass.

In calculating the forces, we assume that the streamline under consideration is oriented along the x-axis and has a cross-section of dy × dz (see Fig. 2.19). The compressive force per unit mass on a flow element of length dx is then:

(2.3:33) $$\frac{(p+\partial p)\cdot dy\cdot dz - p\cdot dy\cdot dz}{\rho\cdot dx\cdot dy\cdot dz} = -\frac{1}{\rho}\cdot\frac{\partial p}{\partial x} , \qquad [N/kg]$$

the frictional force is:

(2.3:34) $$\frac{(\sigma+\partial\sigma)\cdot dy\cdot dz - \sigma\cdot dx\cdot dz}{\rho\cdot dx\cdot dy\cdot dz} = \frac{1}{\rho}\cdot\frac{\partial\sigma}{\partial y} . \qquad [N/kg]$$

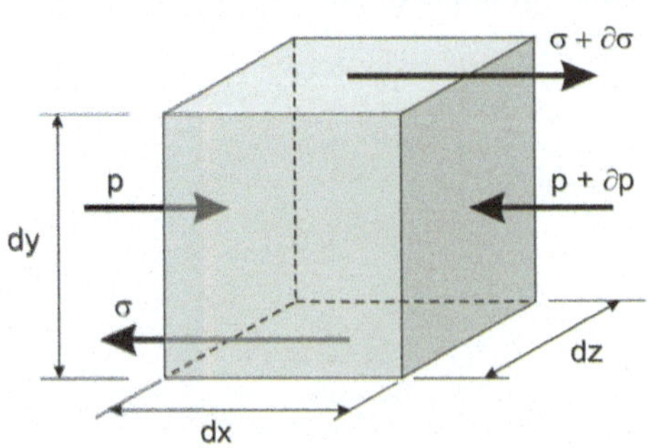


**Fig. 2.19.** Streamline with pressure stresses ($p$) and shear stresses ($\sigma$).

In the case of a Newtonian fluid, we can substitute the shear stress σ from Eqn 2.3:12:

$$(2.3:35)\quad \frac{d\sigma}{dy} = \frac{\partial}{\partial y}\left(\eta \cdot \frac{\partial v}{\partial y}\right) = \eta \cdot \frac{\partial^2 v}{\partial y^2} = \nu \cdot \rho \cdot \frac{\partial^2 v}{\partial y^2}, \qquad [\mathrm{N/m^3}]$$

Where $\eta$ = dynamic viscosity and $\nu$ = kinematic viscosity. Weight is omitted, as it does not exert a force along the x-axis. Nor will we include any external forces, as these would only introduce an additional summand. Setting Eqn 2.3:32 equal to Eqn 2.3:33 and Eqn 2.3:34, and taking into account Eqn 2.3:35, we obtain the following equation of motion:

$$(2.3:36)\quad \frac{\partial v}{\partial t} + v \cdot \frac{\partial v}{\partial x} = -\frac{1}{\rho} \cdot \frac{\partial p}{\partial x} + \nu \cdot \frac{\partial^2 v}{\partial y^2}. \qquad [\mathrm{m/s^2}]$$

The three equations obtained by extending Eqn 2.3:36 to three dimensions are known as the *Navier-Stokes equations* for (incompressible) Newtonian fluids. They constitute the general equations of motion for a Newtonian fluid and only produce unique solutions under very special conditions. To solve real problems, we must use numerical methods (e.g. finite elements).

For stationary ($\partial v/\partial t = 0$) and frictionless ($\nu = 0$) flow, given that

$$v \cdot \frac{\partial v}{\partial x} = \frac{\partial}{\partial x}\left(\tfrac{1}{2} \cdot v^2\right),$$

we obtain the following equation from Eqn 2.3:36:

$$\frac{\partial}{\partial x}\left(\tfrac{1}{2} \cdot \rho \cdot v^2 + p\right) = 0.$$

This corresponds to BERNOULLI's equation (Eqn 2.3:22a) which, as mentioned above, is derived from the equations of motion.

### 2.3.5.2 Taking account of viscosity. Reynolds number.

From the parameters of a flow – length s, velocity v, time t, pressure p and density $\rho$ – it is possible to derive a term for each of the mass-related forces in Eqn 2.3:36. By forming quotients, it is possible to derive a number of dimensionless values from these four terms, which can then be used to describe the forces in the flow.

In order to decide whether it is possible to ignore friction in a flowing fluid, we must know the ratio of inertia forces to friction forces. This (dimensionless) quantity is called the *Reynolds number*. After the Mach number, the Reynolds number is the most important parameter for describing a flow:

$$(2.3:37)\quad Re = \frac{v^2/s}{\nu \cdot v/s^2} = \frac{v \cdot s}{\nu}, \qquad [-]$$

where s is a characteristic length of the object. In the case of a projectile, this is generally the length. $\nu$ = kinematic viscosity.

If Re >> 1, e.g. if speed is high or viscosity is low, it is possible to ignore friction.

Since every flowing fluid adheres to a rigid body, there is a layer at the surface of the body within which the velocity of the flow rises from zero to flow velocity, and within which friction occurs. For large Reynolds numbers this layer (the boundary layer) is very thin, and friction can be ignored. Re values for bullets in air lie between $5 \cdot 10^5$ (for handgun bullets) and $5 \cdot 10^7$ (for artillery shells). As a result, friction plays a secondary role as regards flow behaviour in exterior ballistics. However, friction does make a significant contribution to drag.

If Re ≈ 1 or Re < 1, friction forces must be taken into account in all parts of the flow. This is the case for low speeds, small dimensions or high levels of viscosity.

### 2.3.6 Flow processes and free jets

#### *2.3.6.1 General*

A fluid escaping from a container via a tube can be termed *exit flow*. Exit flow phenomena occur when there is a pressure differential between the container and its surroundings. It is this differential that is responsible for the acceleration of the mass flow. The form of the opening cross-section plays a decisive role. If the cross-section of a tube varies along its length, it is termed a *nozzle*. If the tube is at its narrowest at its exit, it is described as *convergent*.

A *free jet* occurs when a fluid escapes from an opening in a container and penetrates another, stationary fluid at a certain velocity. The density, state and composition of the two fluids can be the same or different. Such a jet always forms a mass flow that is limited in time and space. At high speeds or high densities, it can possess considerable kinetic energy. Under such circumstances, and up to a certain distance, a jet can be seen as a formless projectile, and can have a corresponding effect.

#### *2.3.6.2 Exhaust flow from a convergent nozzle*

The most relevant factors when considering flow from a convergent nozzle are momentum, energy flux and energy-flux density. We can calculate these parameters if we know the exhaust velocity and mass flow rate through the nozzle.

The process by which a gas leaves a nozzle is adiabatic, or nearly so. This means that heat exchange with the surroundings can be ignored. The equation of state for adiabatic processes (including the heat capacity ratio $\gamma$) is:

(2.3:38) $$\frac{p}{p_0} = \left(\frac{\rho}{\rho_0}\right)^{\gamma} . \qquad [-]$$

The exhaust velocity is calculated using BERNOULLI's equation, with density as per Eqn 2.3:38. After some calculation, we obtain:

$$v_m = \sqrt{\frac{2 \cdot \gamma}{\gamma - 1} \cdot \frac{p_i}{\rho_i} \cdot \left[1 - \left(\frac{p_m}{p_i}\right)^{\frac{\gamma-1}{\gamma}}\right]} . \qquad (2.3{:}39) \qquad [\text{m/s}]$$

where $v_m$ = exhaust velocity, $p_m$ = exit pressure and $p_i$ and $\rho i$ = the pressure and density of the fluid in the container respectively.

Using Eqn 2.3:19, and taking Eqn 2.3:38 into account, the mass flow rate for an exhaust cross-sectional area $A_m$ is:

$$\dot{m} = A_m \cdot \sqrt{\frac{2 \cdot \gamma}{\gamma - 1} \cdot p_i \cdot \rho_i \cdot \left[\left(\frac{p_m}{p_i}\right)^{\frac{2}{\gamma}} - \left(\frac{p_m}{p_i}\right)^{\frac{\gamma+1}{\gamma}}\right]} . \qquad (2.3{:}40) \qquad [\text{kg/s}]$$

The mass flow rate therefore depends upon the pressure and density of the fluid in the container, the cross-sectional area of the opening and the ratio of the pressure inside the container to the pressure at the nozzle. From Eqn 2.3:40, it is apparent that there exists a pressure ratio $p_m/p_i$ at which the mass flow rate reaches a maximum.

Eqn 2.3:40 is a continuous positive function, with a value of zero both at $p_m/p_i = 1$ and at $p_m/p_i = 0$. A maximum must therefore occur somewhere between these two values.

That pressure ratio is known as the *critical pressure ratio* and the corresponding nozzle pressure is the *de Laval pressure*, $p_\lambda$. By derivation, and equating the quadratic function Eqn 2.3:40 to zero, we obtain the following expression for the critical pressure ratio:

$$\frac{p_\lambda}{p_i} = \left(\frac{2}{\gamma + 1}\right)^{\frac{\gamma}{\gamma-1}} . \qquad (2.3{:}41) \qquad [-]$$

If the pressure ratio is lower than this figure, the mass flow rate is determined exclusively by the internal pressure.

The corresponding velocity is called the *de Laval velocity*. This is the maximum possible exhaust velocity for a gas leaving a nozzle: By substituting Eqn 2.3:41 into Eqn 2.3:39 and simplifying, we obtain:

$$v_{max} = \sqrt{\frac{2 \cdot \gamma}{\gamma + 1} \cdot \frac{p_i}{\rho_i}} . \qquad (2.3{:}42) \qquad [\text{m/s}]$$

We can calculate the corresponding mass flow rate by substituting Eqn 2.3:41 into Eqn 2.3:40, thus:

$$\dot{m} = A_m \cdot \sqrt{\gamma \cdot p_i \cdot \rho_i \cdot \left(\frac{2}{\gamma + 1}\right)^{\frac{\gamma+1}{\gamma-1}}} . \qquad (2.3{:}43) \qquad [\text{kg/s}]$$

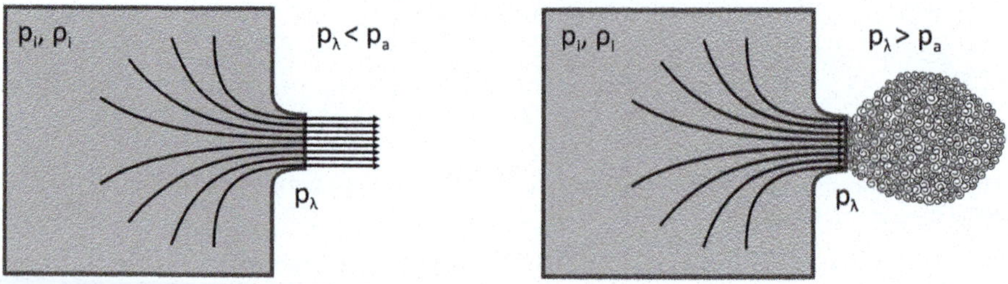


**Fig. 2.20.** Exhaust flow from a nozzle. *Left*: Ambient pressure greater than nozzle pressure (the Laval pressure). A focused jet occurs. *Right*: Ambient pressure significantly lower than the Laval pressure. Expansion from nozzle pressure to ambient pressure will occur explosively, causing completely turbulent flow that is incapable of doing work.

If the ambient pressure is greater than or equal to the de Laval pressure, the escaping gas can expand fully in the nozzle. The jet will be focused and will be capable of transferring kinetic energy. If the ambient pressure is lower than the de Laval pressure, the gas cannot expand fully in the nozzle. Expansion from nozzle pressure to ambient pressure will occur explosively, causing completely turbulent flow that is incapable of doing work (see Fig. 2.20).

### 2.3.6.3 Jet velocity and energy

The exhaust gas velocity and the mass flow rate form the basis for calculating the free jet. These two variables also determine the maximum energy-flux density (i.e. the energy hitting a given area per unit time) that can occur in the jet.

Except at very low velocities, a free jet becomes turbulent very shortly after emerging from the nozzle.

In turbulent flow, irregular fluctuating motion is superposed upon the basic longitudinal motion. As a result, the fluid particles can also move transverse to the direction of flow. In laminar flow, this does not occur.

The velocity of a turbulent flow must therefore be seen as a mean value over time. Because the particles are moving in different directions, there is an apparent increase in internal friction by comparison with the mean movement (see 2.3.5). This *eddy kinematic viscosity* is an important descriptor of turbulent flow.

As a result of the turbulence, the jet partly mixes with the surrounding fluid, dragging particles of that fluid along with it. The mass of fluid picked up by the jet increases with distance, and the jet becomes broader. At the same time, its velocity decreases, as total momentum remains constant.

For the purposes of ballistics, we can limit ourselves to studying round jets, as both the gases that exit the muzzle of a weapon and the propulsion jet of a rocket belong to this category. One advantage of round jets is that the eddy kinematic viscosity is constant throughout the jet. This being so, the equation for the mean velocity distribution in the jet is as follows (only the component acting in the direction of the jet being relevant):

(2.3:44) $$v_x(x,r) = \frac{3}{8\cdot\pi}\cdot\frac{I}{\rho\cdot\varepsilon\cdot x}\cdot\left(\frac{1}{1+\frac{1}{4}\cdot\beta^2}\right)^2 ,$$ [m/s]

(2.3:44a) $$\beta = \sqrt{\frac{3}{16\cdot\pi}}\cdot\sqrt{\frac{I}{\rho}}\cdot\frac{r}{\varepsilon\cdot x} .$$ [-]

Where:
x = distance from opening
I = total momentum (per unit time)
ε = eddy kinematic viscosity
r = distance from axis of jet
ρ = density

Along the axis of the jet (r = 0), β is zero, which simplifies Eqn 2.3:44 considerably. If we know the velocity, we can use the mass flow rate from Eqn 2.3:19 to calculate the energy-flux density (energy density E′ per unit time) as follows:

(2.3:45) $$\frac{E'}{\Delta t} = \frac{1}{2}\cdot\rho\cdot v_x^3 .$$ [kg·m/s³]

The main difficulty in performing such calculations lies in determining the eddy kinematic viscosity. This can be accomplished by measuring the pressure along the axis of the jet. From the pressure, we can determine the velocity, which in turn allows us to calculate ε from Eqn 2.3:44.

### 2.3.6.4 Shock waves

As mentioned in 2.3.3.3, a weak disturbance in a fluid travels at the speed of sound. If the source of the disturbance is moving at less than the speed of sound, the disturbances will precede it and the pressure differential can be equalized.

However, if the source moves faster than the speed of sound, then at those points where pressure is higher the isentropic densification will cause a temperature increase, leading to an increase in the speed of sound and in the velocity of propagation. The faster-moving particles at higher pressure impact those associated with lower pressure, creating a steep pressure front – a *compression shock*.

Depending on the frame of reference selected, one can adopt either of two ways of looking at sonic booms. Typical examples:

- a moving pressure shock in a stationary medium, also known as a *shock wave* (caused, for instance, by the sudden release of a large quantity of gas under high pressure when an explosive is detonated),
- a stationary pressure shock in a moving medium (e.g. in the case of a model aircraft in a wind tunnel).

The plane on which the pressure shock lies is known as the *shock front*. This layer is extremely thin (with an order of magnitude of $10^{-4}$ mm). If the shock front is perpendicular to the direction of the parallel flow, one speaks of a *normal shock*

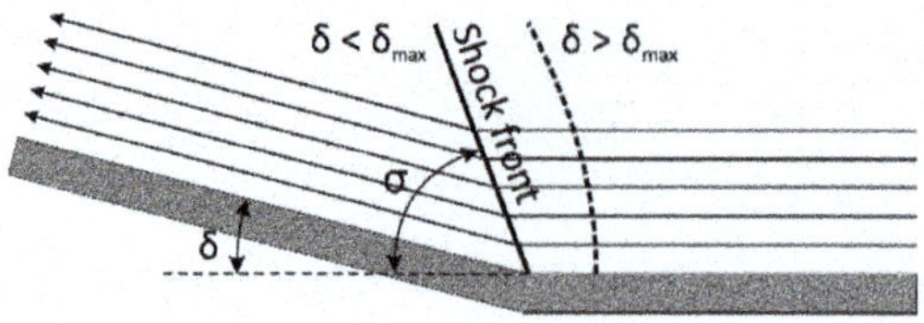


**Fig. 2.21.** Oblique shock. $\delta$ = corner angle, $\sigma$ = oblique shock angle. The dashed line represents the shock front when the maximum corner angle is exceeded.

*front*. Pressure, density and temperature all increase at the shock front, the velocity falls to below the speed of sound.

The values of flow characteristics following the shock can be calculated from the values of the same characteristics before it, using a system of non-linear equations consisting of a continuity equation, the law of conservation of momentum and an energy equation. As the phenomenon is of short duration, it may be assumed to be isentropic.

If a parallel flow travelling faster than the speed of sound is deflected by an oblique surface, an *oblique shock* occurs (see Fig. 2.21). In such a case, there is a sudden change not only in pressure, density, temperature and speed, but also in the direction of flow. Unlike in the case of a normal pressure shock, the flow may continue to move at supersonic speeds following the shock front, but with a lower Mach number than before the shock. The angle between the flow direction and the shock front is called the oblique shock angle $\sigma$. This angle is always greater than the Mach angle $\alpha$ introduced in 2.3.3.4.

If the corner angle $\delta$ changes, the oblique shock angle will also change. At a maximum corner angle $\delta_{max}$, which will depend on the Mach number, the shock front will start to separate from the break point of the deflection. Any of the following two cases may occur:

- $\delta$ very small. The oblique shock angle approaches the Mach angle $\alpha$. The supersonic flow does not generate a shock front.
- $\delta < \delta_{max}$. As the corner angle increases, so does the oblique shock angle, but its origin still lies at the break point of the deflection.
- $\delta > \delta_{max}$. The shock front has separated from the deflecting surface (Fig. 2.21, dashed line). It forms a vertical shock upstream of the wedge. As distance increases, this becomes an oblique shock (see Fig. 2.22).

The further the shock front lies from the source of the disturbance, the more it conforms to a line of which the gradient corresponds to the Mach angle. At a large distance from the centre of the disturbance, it is therefore possible to calculate flow velocity using (2.3:18), at least approximately.

### 2.3.6.5 De Laval nozzles

The main purpose of a nozzle is to turn gas escaping from a vessel into a focused jet that is capable of exerting a force and doing work. This is achieved by giving

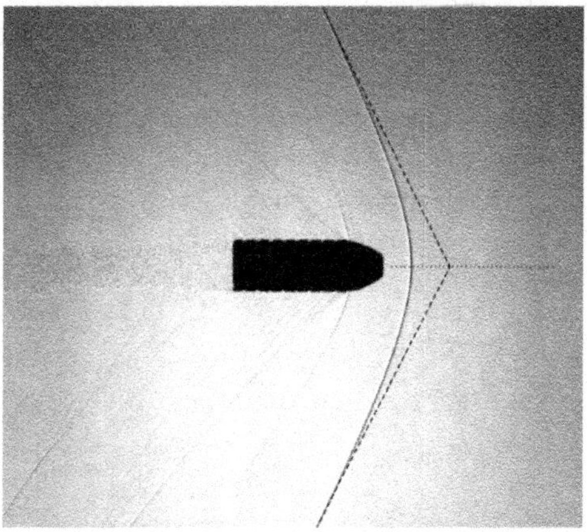

**Fig. 2.22.** Supersonic flow over a blunt object (a lead bullet for a muzzle loader). The shock has become separated. The dashed line indicates the associated Mach cone, from the angle of which it is possible to calculate the Mach number of the flow (1.11 in this instance).

the nozzle a suitable variable cross-sectional profile.

From the continuity equation and BERNOULLI's equation, we can now show that there is a close relationship between the Mach number of the flow and the form of the nozzle.

From the equation of motion, Eqn 2.3:36, with $\partial x/\partial t = 0$ und $\nu = 0$, we obtain BERNOUILLI's equation in differential form for a stationary, frictionless flow (the precondition for a jet):

$$(2.3{:}46) \qquad \frac{1}{\rho}\cdot\frac{\partial p}{\partial x} + v\cdot\frac{\partial v}{\partial x} = 0\,. \qquad [\mathrm{m/s^2}]$$

Multiplying the equation by $\partial x$ and inserting $\partial \rho$ in the first term and v in the second, we obtain:

$$(2.3{:}47) \qquad \left(\frac{\partial p}{\partial \rho}\right)\cdot\frac{\partial \rho}{\rho} + v^2\cdot\frac{\partial v}{v} = 0\,. \qquad [\mathrm{m^2/s^2}]$$

However, from Eqn 2.3:16a, the first factor in the first summand is the speed of sound squared, which means that we can insert the Mach number Ma in Eqn 2.3:47:

$$(2.3{:}48) \qquad \frac{\partial \rho}{\rho} + \mathrm{Ma}^2\cdot\frac{\partial v}{v} = 0\,. \qquad [-]$$

In the case of stationary flow, the mass flow rate remains constant over time ($dm/dt = 0$). The first summand in (2.3:48) can therefore be replaced with the aid of the continuity equation (2.3:19a). Calculation then gives us the following equation (where A = nozzle cross-sectional area):

$$(2.3{:}49) \qquad \frac{\partial v}{v} = -\frac{1}{1-\mathrm{Ma}^2}\cdot\frac{\partial A}{A}\,. \qquad [-]$$

From this equation, it is apparent that acceleration (i.e. $\partial v/v > 0$) only occurs if the cross-sectional area of the nozzle decreases ($\partial A/A < 0$) at subsonic velocities ($\mathrm{Ma} < 1$) and increases ($\partial A/A > 0$) at supersonic velocities ($\mathrm{Ma} > 1$).

In other words, subsonic flows will be accelerated by a convergent tube or channel cross-section, supersonic flows by a divergent nozzle. With a combination of these two types of nozzle, it is possible to achieve gas velocities well over the speed of sound. Such nozzles are called *de Laval* or *convergent-divergent* nozzles.

Equations 2.3:40 and 2.3:43 provide direct expressions for the geometric form of a de Laval nozzle. The first gives the mass flow rate for any given cross section. The second applies when the pressure in a given cross section is equal to the Laval pressure. This will be at the narrowest point of the de Laval nozzle, at a cross sectional area $A_m$. Suitably rearranged, these equations become:

$$\text{(2.3:50)} \quad \frac{\dot{m}}{A} = \sqrt{\frac{2\cdot\gamma}{\gamma-1}\cdot p_i\cdot\rho_i\cdot\left[\left(\frac{p}{p_i}\right)^{\frac{2}{\gamma}}-\left(\frac{p}{p_i}\right)^{\frac{\gamma+1}{\gamma}}\right]}, \qquad [\mathrm{kg/(m^2\cdot s)}]$$

$$\text{(2.3:51)} \quad \frac{\dot{m}}{A_m} = \sqrt{\gamma\cdot p_i\cdot\rho_i\cdot\left(\frac{2}{\gamma+1}\right)^{\frac{\gamma+1}{\gamma-1}}}. \qquad [\mathrm{kg/(m^2\cdot s)}]$$

If we divide the second equation by the first and substitute the equation that is valid in the isentropic case for the pressure ratio

$$\text{(2.3:52)} \quad \frac{p}{p_i} = \left(1+\frac{\gamma-1}{2}\cdot Ma^2\right)^{-\frac{\gamma}{\gamma-1}}, \qquad [-]$$

we obtain the following expression for the cross-section ratio:

$$\text{(2.3:53)} \quad \frac{A}{A_m} = \frac{1}{Ma}\cdot\left[\frac{2}{\gamma+1}\cdot\left(1+\frac{\gamma-1}{2}\cdot Ma^2\right)\right]^{\frac{\gamma+1}{2\cdot(\gamma-1)}}, \qquad [-]$$

where Ma is the Mach number at cross section A.

When designing a nozzle, the first step is to establish the required mass flow rate and pressure in the container $p_i$. These two parameters depend on the power required. Eqn 2.3:51 gives the cross section at the narrowest point $\dot{A}_m$, and Eqn 2.3:53 gives the cross-sectional profile as a function of Mach number. Fig. 2.23 shows the example of a nozzle designed for $\gamma = 1.25$ (the value for combustion gases) at exhaust velocities of up to Mach 2.

The lower part of Fig. 2.23 shows the conditions that can occur in a de Laval nozzle. These conditions depend on external pressure, and form four zones, delimited by three specific pressure levels, $p_1$, $p_2$ and $p_3$.

If external pressure $p_a$ is equal to internal pressure $p_i$, then clearly no flow will occur. If the external pressure falls, then initially subsonic flow will occur. In the convergent part of the nozzle, velocity will increase and pressure will decrease, while in the divergent part the pressure will increase again and the velocity will decrease correspondingly (shown by the dashed line in Fig. 2.23). If $p_a$ decreases further, the first pressure boundary $p_1$ will be reached. At this pressure the flow will just attain the speed of sound at the narrowest point in the nozzle (point Z in Fig. 2.23).

In accordance with Eqn 2.3:52, at Ma = 1 this point depends exclusively on the heat capacity ratio. At $\gamma = 1.25$, the ratio to the internal pressure is 0.5549.

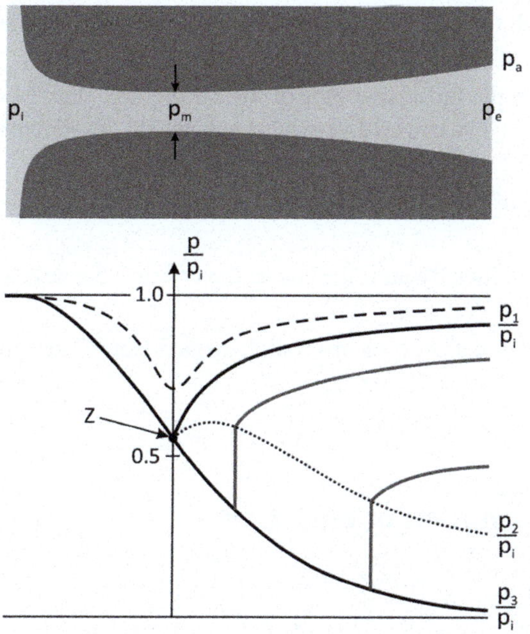


**Fig. 2.23.** A de Laval nozzle designed for $\gamma = 1.25$ at flow velocities of up to Mach 2. The narrowest point is indicated by arrows. Curves of pressure against distance at various levels of external pressure are shown below. See text for details.

However, the difference between $p_m$ and $p_1$ is too small to accelerate the gas further. Pressure increases again and flow velocity decreases.

If external pressure falls below $p_1$, the velocity of the gas will increase to supersonic levels after the narrowest point in the nozzle. However, the counter-pressure is still so high that after a certain distance a vertical shock occurs, pressure rises virtually instantaneously and the flow becomes subsonic. If pressure falls still further, this shock will occur closer to the end of the nozzle, and will occur at the point where pressure reaches $p_2$. This corresponds to the second pressure boundary; the gas will travel at supersonic speeds throughout the divergent part of the nozzle.

Because of the shock, pressure $p_e$ at the end of the nozzle is somewhat less than $p_2$. If pressure falls still further, the supersonic jet will therefore exit against a slight overpressure and will initially be compressed to a pressure in excess of $p_a$, creating oblique shocks before expanding to $p_a$.

The third pressure boundary is reached when $p_a = p_3$. At this point, the external pressure and the pressure at the end of the nozzle are equal ($p_e = p_a$). The gas passes through the entire nozzle at steadily increasing speed. The nozzle is then described as being *balanced*.

If external pressure falls below $p_3$, the pressure at the opening of the nozzle will be slightly higher; the jet will exit against a slight negative pressure and will oscillate as it expands to external pressure.

In summary, the following flow states can exist in a de Laval nozzle, depending on external pressure:

1 $p_r$–$p_1$. The gas passes through the entire nozzle at subsonic speeds. Speed decreases in the divergent part of the nozzle.

2 $p_1$–$p_2$. After passing through the narrowest part of the nozzle, the flow becomes supersonic. However, it also causes a vertical shock, which decelerates the flow to below the speed of sound.

3 $p_2$–$p_3$. In the divergent part of the nozzle, the flow is supersonic, and exits against a slight overpressure.

4 $p_3 = p_a$. Speed increases continuously until the end of the nozzle. The nozzle is balanced.

5 $p_3$–$0$. The supersonic jet exits the nozzle against a slight negative pressure and oscillates as it expands.

Nozzles in states 3 and 5 are said to be *unbalanced.*

## 2.4 The physical environment of ballistics

### 2.4.1 Orders of magnitude

Before we turn to the various sub-areas of ballistics, we need to take a look at the orders of magnitude that one encounters in this field.

Table 2.8 summarizes the ballistic data for a number of typical projectiles. The forces and accelerations in intermediate and terminal ballistics are huge. Furthermore, they arise in extremely short time periods and are generally associated with very high temperatures. It is difficult for us to visualize these phenomena using our usual quasistatic understanding. Furthermore, (highly complex) experiments have shown that under these highly dynamic, extreme thermal conditions, material characteristics such as strength and strain can attain values very different from those normally encountered.

Ballistics – especially intermediate and terminal ballistics – is very much an experimental science, and this is unlikely to change. Even experienced ballisticians regularly obtain experimental results that differ drastically from what they expected – but which can generally be explained.

Because the terminal ballistics of large-calibre weapons are so diverse, Table 2.8 does not include figures for such weapons.

Exterior ballistics is governed by the "normal" laws of physics. Nevertheless, it is very difficult to predict the complete movements of a projectile in all its degrees of freedom, as a projectile is subject to forces and moments that it is difficult or impossible to measure even during shooting tests or in a wind tunnel. As a result, one generally has to make do with relatively simple models. By using well-chosen parameters and correction factors, it is possible to obtain results that reproduce and predict real-life situations astonishingly accurately.

**Table 2.8.** Typical maximum values of pressure, force and acceleration, and average durations, rotation speeds and temperatures for a number of phenomena.

| Small-calibre weapons | | Interior ballistics | Exterior ballistics | Terminal ballistics |
|---|---|---|---|---|
| Pressure | [bar] | 2000–5000 | 0.5–1.0 | 10,000 |
| Force | [kN] | 10–15 | 0.001 | 100 |
| Acceleration | [m/s²] | 500,000–2,000,000 | –200 to –400 | –10,000,000 |
| Duration | [ms] | 0.5–1.5 | 100–500 | 0.1–1 |
| Speed of rotation | [1/s] | 3000–5000 | 3000–5000 | – |
| Temperature | [°C] | 2000–3000 | –20 to 200 | A few thousand |
| **Artillery and mortars** | | **Interior ballistics** | **Exterior ballistics** | **Terminal ballistics** |
| Pressure | [bar] | 500–3500 | 0.45–4.0 | |
| Force | [kN] | 750–6600 | 0.3–7.5 | Not |
| Acceleration | [m/s²] | 50,000–150,000 | –7 to –180 | possible |
| Duration | [s] | 0.008–0.025 | 20–60 | to state figures |
| Speed of rotation | [1/s] | 90–300 | 90–300 | |
| Temperature | [°C] | 2000–3000 | ? | |

### 2.4.2 Calibre and sectional density

Calibre and mass are always the first two geometrical/physical parameters used to describe a projectile. However, one should bear in mind that the term "calibre" can mean any of three different things:

- The diameter of the barrel (measured between the lands in the case of a rifled barrel).
- The diameter of the projectile (measured across the body of the projectile, if it has fins or driving bands).
- Nominal calibre (i.e. the official designation, usually based on the barrel diameter).

  *Exception*: In the case of a shotgun, the "bore" number is determined by the number of balls of the same diameter as the inside of the barrel that could be cast from a pound of lead. A larger bore number therefore corresponds to a small barrel diameter.

Interestingly, the physical/ballistics characteristics of a projectile do not depend exclusively on its calibre and mass, but also on the quotient of its mass and cross-sectional area (the nominal calibre is generally used). This is termed the cross sectional density (q) of a projectile.

$$(2.4{:}1) \qquad q = \frac{m}{\frac{1}{4}\cdot d^2\cdot\pi} \;. \qquad [g/mm^2],\ [kg/m^2]$$

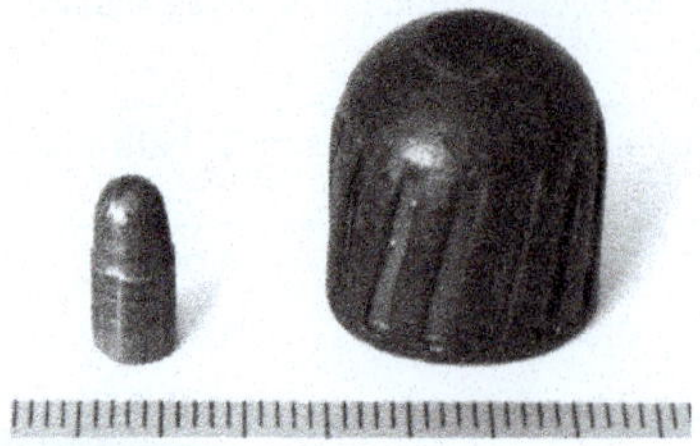

**Fig. 2.24.** Two projectiles with similar sectional densities and hence very similar ballistic characteristics, despite their different calibres and masses.

This is especially clear in the case of small-calibre weapons. A 12-bore shotgun slug, with a diameter of 18.2 mm and a mass of 28.3 g, and a 22 LR bullet with a diameter of 5.6 mm and a mass of 2.6 g (see Fig. 2.24), with the same muzzle velocity, have very similar ballistic characteristics: almost identical velocity plots, same windage and similar penetration.

In certain cases (e.g. when there is a large angle between the axis of the projectile and its trajectory), the sectional density used is that of the projectile's face area.

# 3 Interior ballistics

## 3.1 Introduction. History.

The process of firing a gun is one of the most extreme in classical physics. It is of very short duration and involves forces and accelerations of enormous magnitude, together with high temperatures (see Table 2.8). Recording and processing such events electronically poses problems that are difficult or even impossible to resolve, even with the most modern measuring and data-processing technology. Determining maximum gas pressure, the gas-pressure curve and the velocity of the projectile – especially muzzle velocity – is often considered the central problem of interior ballistics.

Ballistics was divided into interior and exterior ballistics in the mid-18$^{th}$ Century. From that point on, maximum pressure and muzzle velocity took centre-stage. The advances in physics made around this time were accompanied by the first theoretical studies on flow in a tube (LAGRANGE, 1793, published by S. POISSON in 1832). However, measuring techniques were not as advanced as the new theories, making it impossible to verify the relationships posited.

To measure maximum pressure and muzzle velocity, it is necessary to record parameters over extremely short time periods – milliseconds or fractions thereof. This was not possible with the time-measurement equipment available (chronometers and hour glasses).

In 1871, M. DEPREZ built a device capable of measuring maximum pressure fairly accurately (B. P. KNEUBUEHL 2013). By installing several such devices along a barrel, it was even possible to roughly measure change in pressure along its length. In 1864, Belgian officer P. E. LE BOULENGÉ (see also C. KUHN, 1866) invented a device with which it was possible to measure the velocity of a projectile along its trajectory. Acceleration due to gravity provided an indirect measure of time elapsed; two marks were made on a rod as it fell and the time was calculated from the distance between them.

The ability to measure pressure and velocity led to decisive progress in interior ballistics at the turn of the 18$^{th}$/19$^{th}$ Centuries. However, this knowledge also brought the realization that it would probably never be possible to fully understand the processes that occur in a weapon during firing.



B. P. Kneubuehl, *Ballistics*,
https://doi.org/10.1007/978-3-662-69237-0_3

## 3.2 Explosive materials

### 3.2.1 General

#### *3.2.1.1 Definitions and nomenclature*

Explosive materials are solid, liquid or gaseous materials that can burn rapidly without an external source of oxygen, yielding energy. In other words, they can react explosively. This reaction can be triggered mechanically (by shock or friction), thermally (by a spark, a flame or heat) or by a shock wave. Explosive materials are divided into three categories, according to their purpose:

- *Primary explosives* are triggered by mechanical or thermal effects. They serve to initiate other explosives.
- *Secondary explosives* (often referred to simply as *explosives*) are those that actually carry out the task in hand. They are less sensitive to shock, friction and electrostatic discharge. This makes them safer to handle than primary explosives, but at the same time they deliver more energy. They generally require a shock wave from a primary explosive in order to react.

Primary and secondary explosives are often both referred to simply as explosives.

- *Propellants* (powders) are explosives intended for use with projectiles (including rockets).
- *Pyrotechnic mixtures* are mixtures of oxygen carriers (oxidants) and fuels, and are designed to produce light, heat, smoke and/or sound.

One of the most important characteristics of an explosive material, which can also be used to define it, is its burning rate.

#### *3.2.1.2 Burning rate*

In an explosive, the reaction front at which chemical transformation (oxidation) occurs always moves at a given velocity perpendicular to the instantaneous surface. This velocity is termed the *burning rate.*

If the burning rate is of the order of millimetres or centimetres per second, we speak of *combustion.* This is primarily the case for propellants at normal pressure. If the reaction front moves at a significantly higher velocity, but at less than the speed of sound in the medium concerned, the process is called *deflagration.* The burning process is physically unstable; it progresses by means of the thermal energy that the reaction releases, at velocities of between a few tens of m/s and a few hundred m/s. If the burning rate exceeds the speed of sound, shock waves occur that are closely related to the chemical conversion. High pressures and temperatures occur in a very narrow, sharply-delimited zone, and it is these that are responsible for the progress of the reaction. This burning process is termed *detonation.* In such cases, the reaction front moves at velocities of between approximately 2 km/s and 10 km/s.

Explosives can burn, deflagrate or detonate. Which of these three occurs depends largely on the chemical composition of the material, the way in which it is ignited and the pressure to which it is subjected.

The differing characteristics of propellants and explosives mean that there are different ways of determining their burning rates. EN 13938-4 describes a process for measuring the burning rate of a propellant at normal pressure. However, the results are of little use to interior ballistics, so the method is not widely used. The burning rates of propellants are more usually determined using closed vessels, such as the manometric bomb or the Crawford bomb, to which a later section is dedicated (Section 3.5).

In theory, it is necessary to conduct measurements over a very short period of time in order to determine burning rates, but this can be avoided using the method invented by Henri J. DAUTRICHE in 1906. However, this approach does require the use of a length of detonating cord with a known detonation velocity.

Two detonators are inserted in a steel tube filled with the explosive under investigation, at a distance x apart. The two detonators are linked by a sufficient length of detonating cord of known detonation velocity $v_s$ (see Fig. 3.1). Because the two ends of the detonating cord are ignited at different times, the resulting two shock fronts meet at a certain distance y from the mid-point of the detonating cord. A clear mark is left on the lead block at that point. Basic physical/geometric calculations yield the following equation for the burning rate $v_{Sp}$ of the explosive material under investigation:

(3.2:1) $$v_{Sp} = \frac{x}{2 \cdot y} \cdot v_s \,. \qquad \text{[m/s]}$$

Modern short-time measurement technology makes it possible to determine the transit time of the detonation front between the two measuring points, which is probably more accurate.

## 3.2.2 Black powder

### *3.2.2.1 History*

Various types of saltpetre (nitrates) occur naturally. Nitre, Chile saltpetre and ammonium nitrate are often used as fertilizer. If ammonium nitrate is heated or initiated very quickly, it detonates, decomposing into nitrogen, oxygen and water; it is

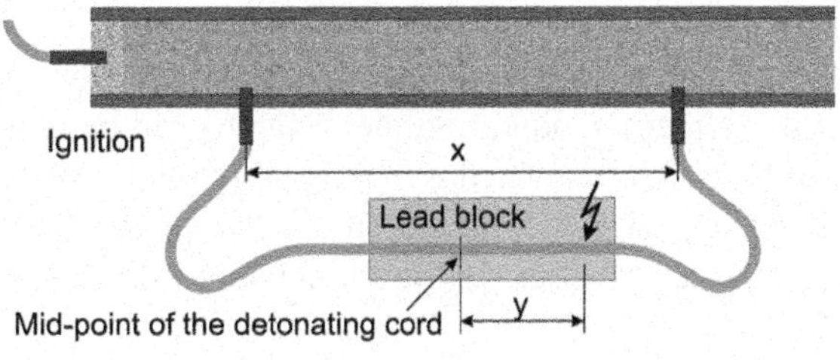


**Fig. 3.1.** Determining detonation velocity using the method devised by H. J. DAUTRICHE. See text for details.

an explosive. Nitre and Chile saltpetre are excellent sources of oxygen for combustion processes. If they are mixed with highly flammable materials, very rapid combustion or detonation can occur.

As these materials occur naturally, it is reasonable to assume that their burning characteristics were discovered a long time ago. It is therefore pointless to talk about the "invention" of black powder as having occurred at a particular time. The ancient Greeks and Romans appear to have been familiar with this type of highly combustible material, and to have used it for military purposes. Examples include the Spartans' allies, the Boeotians, who used such substances in the Peloponnesian War (431–404 BC). The famous mixture of saltpetre, carbon and sulphur that we now call black powder appears to have originated in China, where it is said to have been developed and first used between the 8$^{th}$ and 10$^{th}$ Centuries, for fireworks and rockets. How this know-how reached Europe, we can only guess. In the 13$^{th}$ Century, GENGHIS KHAN, his military commanders and his successors launched a series of military campaigns that took them as far as Europe (e.g. the Battle of Liegnitz, 1241) and Asia Minor. It is entirely possible that black powder – or at least information on how to produce it – travelled with him.

In Europe, two names are associated with this propellant. The first is that of English philosopher, theologian and natural scientist Roger BACON (1215–1290), who lived in Paris from 1241 to 1246. In a manuscript dating from around 1250, he describes a mixture of saltpetre, sulphur and carbon, mentioning it not as his invention but rather as a known material. Bacon did not talk about what this concoction could be used for, unlike the second person associated with it. The mythical Franciscan monk Berthold SCHWARZ, who lived in the monastery of Freiburg im Breisgau, Germany, and experimented with saltpetre, sulphur and carbon, appears to have used this mixture to propel objects. Schwarz lived from 1300 to 1390, so he could have heard of black powder. However, it is also possible that he discovered the mixture himself.

It is therefore fair to say that the origins of gun powder are lost in the mists of time. It may have been developed in any of several ways. There are two reasons for this uncertainty: such materials were always surrounded in secrecy, and *how* gun powder came to be invented is relatively unimportant from a historical point of view. However, the *use* of gun powder has had a huge influence on the history of humanity in the intervening thousand years.

#### 3.2.2.2 Components

Black powder consists of "ordinary" saltpetre, or nitre (potassium nitrate, $KNO_3$); sulphur (S) and carbon (C). Chile saltpetre (sodium nitrate, $NaNO_3$) can be used instead of nitre, to produce explosive saltpetre, which is mainly used in mining.

Saltpetre is found in small deposits and in salt deserts (nitre in Asia and Chile saltpetre in the Atacama Desert in South America). Nitre for black powder is also produced in saltpetre works, by mixing farmyard manure with lime ($CaCO_3$) and potash (potassium carbonate, $K_2CO_3$), and keeping the mixture moist with farm slurry.

Sulphur has been known of since antiquity. It occurs in pure form in volcanic regions and is embedded in many types of ore and mineral; obtaining it was never a problem.

Carbon is less simple to produce. It is usually manufactured as part of the process of making black powder, to keep quality and composition as consistent as possible. Carbon is obtained from wood by a process of pyrogenic distillation. Non-resinous, fine-grained types of wood such as alder buckthorn and grey alder are best suited, although hazel and willow are also used. Charcoal consists of carbon, hydrogen and oxygen. Hydrogen provides the necessary ignitability, oxygen the combustibility. Carbon with a high hydrogen and oxygen content (approx. 70% carbon) was used mainly in powder intended for hunting and target shooting. Carbon with a lower hydrogen and oxygen content (approx. 85% carbon) was used primarily for military purposes.

#### *3.2.2.3 Manufacturing process*

For centuries, the three components of black powder were blended as thoroughly as possible, using fine grains that formed a powder – hence the name. However, it was not possible to prevent the mixture segregating with time. This resulted in irregular burning and significant variation in the initial velocity of the projectile. In the early days of firearms this was of little significance, as they were in any case subject to considerable dispersion. As time went on, however – and especially as rifles entered widespread use – the quality requirements for powder became more stringent. It became necessary to develop processes that both prevented segregation and permitted production on an industrial scale. Over the years, a manufacturing process emerged that resulted in a much more coarse-grained end product.

The main steps in this process are as follows:

- The components are crushed, ground and blended.
- The powder is moistened and then compressed into cake, bonding the components together in such a way that they can no longer segregate.
- The cake is broken up into grains, which are then pre-graded using screens and polished in a polishing drum; the grains rub against each other, which deburrs them and polishes their surface, rendering them less hygroscopic.
- The powder is dried, reducing its moisture content from approx. 4%–5% to 0.25%–0.5%, and the grains graded by size.

Typical grain sizes for gun powder range from 0.2 mm to 1.6 mm (see Fig. 3.2). The type of powder used depends on a number of factors, including calibre, barrel length and projectile mass. Finer powders have a larger total surface area, and are hence more combustible than coarser powders.

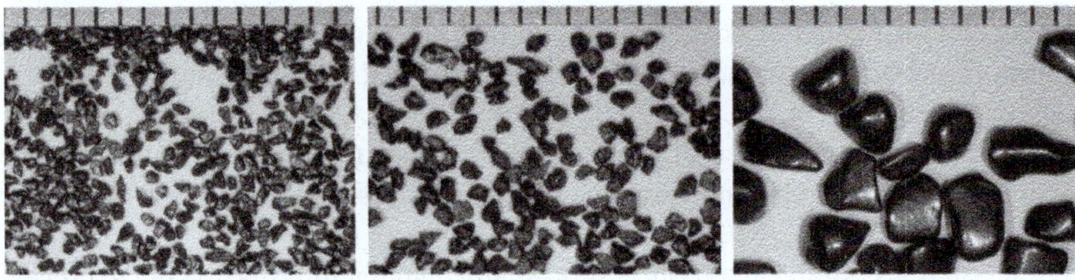

**Fig. 3.2.** Black powder grain sizes. *Left*: Grain Size 1, for pistols. *Centre*: Grain Size 2, for rifles. *Right*: Grain Size 5, for saluting guns.

#### *3.2.2.4 Composition and products of combustion*

The earliest black powders consisted of six parts nitre, one part sulphur and one part carbon. This equates to 75% nitre, 12.5% sulphur and 12.5% carbon, which corresponds fairly closely to today's standard composition (75% $KNO_3$, 10% S and 15% C).

Systematic study of the chemical processes and hence refinements to composition only became possible following advances in chemistry, especially those achieved by LAVOISIER in the second half of the 18th Century. As early as the 19th Century, it was discovered that the three components burned up completely at 74.65% saltpetre, 11.90% sulphur and 13.45% carbon.

The combustion of black powder proceeds as follows: The carbon is ignited and burns, producing carbon dioxide ($CO_2$). The resulting heat causes the sulphur to react with the saltpetre producing nitrogen ($N_2$), potassium sulphate ($K_2SO_4$), potassium carbonate ($K_2CO_3$) and potassium sulphate ($K_2S$), together with a few other substances. The nitrogen and carbon dioxide are responsible for the pressure created in the weapon. The remaining three substances (together with traces of others) are left as solid residue once combustion is complete. They account for 60% of the mass of the powder and are responsible for the smoke that emerges from the muzzle and for fouling of the 56. As these residues are water soluble, then can easily be removed.

### 3.2.3 Nitrocellulose propellants

#### *3.2.3.1 Nitrocellulose*

Black powder was not only used as a propellant for projectiles but also, from the second half of the 17th Century, as an explosive for mining. By the start of the 19th Century, however, it appeared that the maximum explosive power of black powder had been reached, and the search began for more powerful substances.

Experiments with the nitration of various substances were carried out in the 1830s, and in 1846 Christian F. SCHÖNBEIN, a German chemist lecturing at Basel

University, discovered *nitrocellulose* (also known as *gun cotton*, with the chemical name of cellulose nitrate). Rudolf C. BOETTGER discovered the same explosive in the same year, in Frankfurt am Main, Germany.

Nitrocellulose deflagrates rapidly on ignition, leaves no residue and produces no visible smoke. It therefore seemed to be a suitable replacement for black powder, at least as a propellant, but its burning rate was much too high.

The following year (1847), Italian chemist Ascanio SOBRERO created a highly explosive and shock-sensitive liquid explosive, *nitroglycerine* (NG, chemical name glyceryl trinitrate). However, SOBRERO considered this explosive too dangerous for practical use. Twenty years elapsed before Swedish chemist Alfred NOBEL succeeded in taming this substance (after several failures), by mixing it with diatomaceous earth, resulting in a usable explosive under the name Dynamite. However, nitroglycerine also plays an important role in propellants.

Nitrocellulose is made from short cotton fibres, known as linters. Cotton is an ideal raw material, as its fibres consist of almost pure cellulose. The cellulose is nitrated using nitrating acids, its fibrous structure remaining intact.

Nitrating acid is a mixture of concentrated nitric and sulphuric acids, in various ratios. For the nitration of cellulose, a ratio of approximately 25% $HNO_3$, 60% $H_2SO_4$ and 15% $H_2O$ is used. The degree of nitration can be controlled by altering the percentage of water used.

The nitric acid performs the actual nitration, while the sulphuric acid absorbs the water of reaction.

The nitrogen content of the nitrocellulose depends on the degree of nitration. If the product contains less than 12.75% nitrogen (10.6–12.7%) it is termed collodion cotton (cellulose dinitrate). Products with concentrations in excess of 12.75% are called gun cotton (cellulose trinitrate). The theoretical maximum nitrogen content – which can almost be achieved using special nitration processes – is 14.15%. However, commercial products manufactured using nitrating acids contain only approximately 13.5% nitrogen.

Propellants usually contain a mixture of gun cotton and collodion cotton. Collodion cotton used on its own forms the basis of celluloid, one of the first (thermosetting) plastics, which was used for many products including photographic film. It is also used as a binder in cellulose paints and adhesives.

During the second half of the 19th Century, there were several attempts to reduce the combustibility of nitrocellulose so that it could be used in a practical propellant. In 1884, some 40 years after the invention of nitroglycerine, P. VIEILLE in France and M. DUTTENHOFER in Germany finally developed a process that made it possible to handle this material, and hence also to store it. A few years later, A. NOBEL in Sweden (1887) and F. A. ABEL and J. DEWAR in England (1889) found ways of manufacturing propellants on the basis of nitrocellulose.

However, the NOBEL/DEWAR method differed from that of VIEILLE and DUTTENHOFER. As a result, there are two quite different ways of producing this category of propellant:

- The nitrocellulose is gelatinized using a solvent and mechanical energy (kneading and pressing). The solvent is then removed. This propellant is known as *solvent propellant.*

- A plasticizer is added to the nitrocellulose and a thermosetting plastic is produced by a combination of thermal and mechanical energy (heating and kneading). The resulting propellant is called *solventless propellant.*

If the nitrocellulose is the sole source of energy, this type of propellant is also called *single-base propellant.* If one or two further explosive substances are added (e.g. as energy-carrying plasticizers in the case of solventless propellants), the propellant is termed “double-base” or “multiple-base” propellant.

#### 3.2.3.2 Single-base powders

The basics of VIEILLE’s approach to taming nitrocellulose are still relevant today. The fibrous nitrocellulose is gelled, i.e. transformed into an amorphous, plastic mass, by adding alcohol and ether and then subjecting the material to prolonged kneading and pressing.

Unlike VIEILLE, DUTTENHOFER used ethyl acetate, which is no longer in common use.

It is also possible to use acetone instead of alcohol and ether. While alcohol and ether can only gel moderately nitrated collodion cotton, acetone can also gel highly nitrated gun cotton.

Once the material is capable of being kneaded, it can be formed into cakes and grains of any shape by extrusion and chopping. To maintain the chosen shape, the solvent must then be removed by drying.

These types of propellant are known as nitrocellulose (NC) or single-base (SB) propellants, as they contain only one substance that supplies energy.

NC propellants are mainly used for small- and medium-calibre weapons, but they are also found in medium artillery charges. They are modified to suit their intended role by the addition of various substances and mainly consist of 80–98% nitrocellulose and 1–4% stabilizers, surface treatment agents, plasticizers and other substances with functions such as reducing muzzle flash and protecting the barrel of the weapon. The stabilizers are particularly important, as they significantly extend the life of nitrocellulose, which is highly sensitive to temperature.

The performance of NC propellants can be enhanced by further processing (impregnation) with *nitroglycerine* (NG) e.g. to produce spherical grain or EI propellant, also known as semi-double-base propellant, or by adding crystalline energy vectors such as hexogen or nitroguanidine to the nitrocellulose before kneading it, to produce ECL or semi-nitramine propellant.

#### 3.2.3.3 Double-base and multiple-base powders

Double-base (DB) and multiple-base (MB) propellants can be manufactured with or without the aid of solvents.

A. NOBEL developed a non-solvent process in 1887, followed by F. A. ABEL and J. DEWAR in 1889. They discovered that it was possible to gel nitrocellulose using just nitroglycerine – a second source of energy. In modern methods of producing

DB and MB propellants that use this approach, the first step is to produce a propellant paste, which involves slowly adding nitroglycerine or diethylene glycol dinitrate (DEGDN) – or both – to the nitrocellulose after it has been slurried in water, while stirring vigorously. These two explosive substances may account for up to 50% of the propellant paste.

Nitroglycerine and DEGDN are liquid nitric acid esters, often referred to as explosive oils on account of their oily consistency.

In this process, the nitrocellulose binds almost all of the explosive oils and gels. The propellant paste is kept in conditioning storage for a certain time, to consolidate the bond between the two components. It is then mixed with the appropriate additives (stabilizers, muzzle flash suppressants, etc.) to produce the actual propellant. In the case of multiple-base propellant, the third energy source (usually nitroguanidine) is also added at this stage. The paste undergoes multiple passes through sets of rollers, producing a solid plastic mass that can be rolled up and sent to the press. It is heated to 70–80 °C, extruded in the required profile and chopped as appropriate.

The manufacture of double-base and multiple-base propellants using solvents involves first creating an intermediate product, a *powder preconcentrate* consisting of nitrocellulose and one or more explosive oils. The proportion of explosive oil (80%) is substantially higher than in the case of a propellant paste.

To produce the propellant itself, the powder preconcentrate, additives and solvent (generally a mixture of acetone and ethanol) are added to the kneader in succession and then kneaded for a number of hours. In the case of MB propellant, nitroguanidine is also added. Once mixing is complete, the propellant is extruded, chopped and dried.

Depending on the gelling agent used, double-base propellants are known as *nitroglycerine* (NG) or *glycol* propellants. They consist of 20–70% nitrocellulose, 20–50% explosive oil and 2–10% plasticizers, stabilizers and other additives. They deliver more energy that single-base propellants, but they also burn at a higher temperature, which places a greater load on the weapon. This disadvantage is mitigated by adding the third source of energy, nitroguanidine, which is a very insensitive, low-energy explosive. Multiple-base nitroguanidine propellant is a particularly low-energy, "cold" propellant, and is frequently used in weapons with a high rate of fire and for higher artillery charges. MB propellants contain approximately 30–55% nitroguanidine.

### 3.2.4 Propellants not based on nitrocellulose

#### *3.2.4.1 Polymer-bonded propellants*

Nitrocellulose-based propellants have two serious disadvantages: they ignite spontaneously at temperatures of less than 200 °C and they are relatively sensitive to mechanical effects. These disadvantages prompted the development of propellants that were less sensitive, especially to heat. For many years, they were sold as LOVA (low-vulnerability) propellants. The idea was that if they came under fire or were subjected to heat they would only burn, rather than deflagrating or detonating.

Secondary explosive materials such as hexogen, octogen, nitroguanidine and ammonium perchlorate are used as energy carriers. These are integrated into a binder in such a manner as to ensure even combustion. Depending on the manufacturing process, thermosetting resins, thermosetting plastics or gel bonding agents can be used. The binder may also supply energy or gas, in which case it is described as “energetic”. If this is not the case, it is termed “inert”.

LOVA propellants are mainly used for large-calibre weapons and solid-propellant rockets. However, they are also used for caseless artillery munitions, where they are also known as high-ignition-temperature propellants (HITP). They have not become widely used for this application, as they are highly sensitive to the effects of incoming fire.

#### *3.2.4.2 Composite propellants*

*Composite propellants* consist primarily of two compounds – an oxidant and a binder, the latter providing the fuel. They are used in rocket motors and gas generators.

They contrast with homogenous propellants, in which oxidant and fuel form a single chemical compound.

Ammonium nitrates and perchlorates are suitable oxidants, as are the nitrates and perchlorates of alkali metals. *Ammonium perchlorate* is often used, particularly on account of its high energy content and relatively pressure-independent burning rate. However, this oxidant has the disadvantage that it produces hydrogen chloride, which in turn forms smoke and hydrochloric acid fumes on contact with atmospheric humidity. *Ammonium nitrate* delivers less energy, but produces virtually no smoke and no corrosive by-products. *Potassium perchlorate* also delivers relatively little energy. Burning is strongly dependent on pressure, burning rate is high and this oxidant results in large quantities of smoke.

The binders used are thermosetting plastics (asphalt, PVC and polyisobutylene) and elastomers (polysulphide, polyurethane and polybutadiene). It is possible to increase the amount of energy using metallic additives such as aluminium powder. Burning rate is regulated by adding iron or copper oxides. The names of the propellants indicate the oxygen carrier used: APCP (ammonium perchlorate composite propellant) or ANCP (ammonium nitrate composite propellant).

### 3.2.5 Physical characteristics of propellants

In order to carry out interior ballistics calculations, we must know certain characteristics of the propellant used. These are as follows:

- The *explosion heat* $Q_{ex}$, which is the heat energy (in J/g) obtained by burning 1 g of propellant in an $N_2$ environment; if the propellant is burned in an $O_2$ environment, we speak of *heat of combustion.*

The total energy available for the firing of the projectile is found by multiplying the load mass by the explosion heat.

- The *specific gas volume* (in litres or $dm^3$): the quantity of gas produced by the combustion of 1 g of a propellant.
- The flame temperature $T_{ex}$ (formerly known also as the *explosion temperature*): the temperature at which the propellant burns in a space of fixed volume.
- The *specific energy* $F_{se}$ (formerly known also as the *propellant force*): a measure of the mechanical energy.

Specific energy is determined from the pressure increase in a closed bomb (of constant volume) that results from the burning of 1 g of a propellant, and corresponds to the resulting pressure energy.

- The *auto-ignition temperature*: the temperature at which the propellant ignites spontaneously.
- The percentage of *solid residues* (deposits in the barrel, plus smoke).

Table 3.1 lists typical values for the four most important types of propellant.

**Table 3.1.** Propellant characteristics

| | | BP | SB | DB | MB |
|---|---|---|---|---|---|
| Heat of combustion, $Q_{ex}$ | [J/g] | 2650 | 3000–4000 | 3500–5500 | 3000–4100 |
| Specific gas volume | [lt/g] | 0.28 | 0.90–0.97 | 0.80–0.86 | 0.90–1.00 |
| Flame temperature, $T_{ex}$ | [°C] | Approx. 2400 | 2400–2600 | 3000–3800 | 2500–3000 |
| Specific energy, $F_{se}$ | [J/g] | | 900–960 | 1100–1170 | 960–1050 |
| Auto-ignition temperature | [°C] | 300 | 170 | 160 | 150–170 |
| Solid residues | [%] | Approx. 40 | Approx. 1 | Approx. 1 | Approx. 1 |

*Key*: BP: Black powder; SB: Single-base propellant
DB: Double-base propellant; MB: Multiple-base propellant

### 3.2.6 Igniters and detonating agents

Propellants and explosives must be insensitive to such external influences as vibration, shock and heat, so that they can be stored, transported, handled and loaded safely. It is therefore not possible to initiate them simply, using heat or mechanical effects. Intermediate materials are required, and these fall into one of two categories:

- *Igniters*, which cause combustion or deflagration. Initiating the explosive material causes it to burn at a rate lower than the speed of sound in the material.
- *Detonating agents*, which cause detonation. Igniting the explosive material produces a shock wave that moves with the flame front, at a speed higher than the speed of sound in the material.

Both have the task of converting mechanical or thermal action into sufficient ignition of the explosive material. They therefore include an *initiator* in the form of a certain quantity of an initiating explosive that is sensitive to shock, friction or heat. Energy and oxygen carriers are also added, to ensure that energy is released rapidly and in sufficient quantity.

Originally, front-loading weapons were initiated using very finely-ground black powder – priming powder – which was placed in the pan and the flash hole in such a way that it could ignite the load. Priming powder was highly combustible and could be ignited using a match, or a spark from a flint in the case of a flintlock.

The first igniter for black powder, the *percussion cap*, was invented in 1818 by the Swiss J. EGG, who lived in England and also developed the percussion lock for front-loading rifles. He used *mercury fulminate* for the initiator and saltpetre to provide the energy and oxygen.

Other, more weapon- and environment-friendly primers have been developed over the last century. The first modification was to replace mercury with *lead styphnate* (lead trinitroresorcinate) and *tetrazene* as initiators and the highly corrosive saltpetre with *barium nitrate* (Sinoxid primers). With the introduction of lead-free ammunition, lead styphnate was replaced by *diazol* (diazodinitrophenol) and barium nitrate by other oxygen carriers (Sintox primers). Table 3.2 shows the compositions of the most common primers and Table 3.3. compares the physical characteristics of the most common primary explosives.

**Table 3.2.** Primer compositions for various types of propellant

| | Black powder | Conventional nitrocellulose | Lead-free nitrocellulose |
|---|---|---|---|
| Initiator | Mercury fulminate | Lead styphnate | Diazodinitrophenol |
| Sensitizer | | Tetrazene | Tetrazene |
| Energy carrier | Black powder | Antimony sulphide | NC propellant |
| Oxygen carrier | Potassium chlorate | Barium nitrate | Strontium nitrate |
| Various | Glass powder | Glass powder | |

**Table 3.3.** Physical characteristics of the main primary explosives

| Primary explosive | Density [g/cm³] | $Q_{ex}$ [J/g] | $v_{exp}$ [m/s] | GV [lt/g] | DP [°C] | SS [N·m] | FS [N] |
|---|---|---|---|---|---|---|---|
| Acetone peroxide | 1.22 | | 5300 | | | 0.3 | 0.1 |
| Lead azide | 4.80 | 1536 | 5180 | 0.31 | 315–360 | 3–6.5 | 0.1–1 |
| Lead styphnate | 3.00 | 1454 | 5200 | 0.47 | 275–280 | 2.5–5 | 1.5 |
| Diazol | 1.63 | 4000 | 6600 | 0.86 | 180–200 | 1.5 | |
| Nitroglycerine | 1.60 | 6680 | 7600 | 0.78 | 223–225 | 0.2 | > 350 |
| Mercury fulminate | 4.42 | 1788 | 5000 | | 165–170 | 1–2 | 3 |
| Tetrazene | 1.70 | | 1500 | | 140 | 1 | 8 |

*Key:* $Q_{ex}$: Heat of combustion; $v_{exp}$: Detonation rate; GV: Gas volume; DP: Deflagration point; SS: Shock sensitivity; FS: Friction sensitivity

Percussion caps are sufficient to ignite the quantities of propellant used in handguns and shoulder weapons. The loads used in medium- and large-calibre munitions require more powerful means of ignition. Depending on the mass of the load, a primer and an additional load consisting of black powder or porous NC propellant are combined into one component. In cartridge munitions and mortar bombs, this takes the form of a *priming screw*. In the case of separate-loading ammunition (where the propellant charges are generally sewn into cotton bags), this additional propellant (in the form of an *igniter pad*) is placed behind the main charge and ignited through the breech, using an *initiating cartridge*.

Primers are not capable of igniting secondary explosives. Smaller charges are triggered using *blasting caps* or electric initiators. Blasting caps are copper or aluminium cylinders containing an initiating explosive (lead azide or lead styphnate) and a high explosive such as penthrite or hexogen. They can be triggered mechanically (via a primer) or by heat (electrically). Larger charges and insensitive explosives are initiated using *detonators*, which are of basically the same construction with the addition of a high explosive booster.

## 3.3 Elements of interior ballistics

### 3.3.1 Means of accelerating a projectile

#### *3.3.1.1 General*

It is only possible to achieve an effect at a distance if one succeeds in transporting a sufficient quantity of energy to the desired location. This implies that two packages of energy must be available at the point of departure: one for the transport and one to do the work at the other end.

If we leave aside such energy carriers as lasers, sound waves, etc. and consider only conventional projectile weapons, the sole form of energy available for transport is kinetic energy (a mass provided with velocity) and the only form of energy available for achieving an effect at the target is also kinetic energy, or chemical energy in the form of an explosive that must be converted into kinetic energy at the target.

In all cases, therefore, a certain quantity of kinetic energy must be available at the point of departure. This can only be achieved by performing work on the mass that is to transport the energy (the projectile). There are many ways of storing the (potential) energy required to perform this work in the weapon.

#### *3.3.1.2 Spring energy*

The first throwing weapon designed to achieve longer ranges must have been the bow, in which work done by muscles is converted into the potential energy of a spring.

The oldest confirmed finding of a bow dates back to 6000 BC (i.e. the Middle Stone Age).

Weapons that accelerate projectiles using spring energy are still in use today. The best-known examples are the bow, crossbow and catapult.

The slingshot that David used so successfully against Goliath operates on a different principle, that of rotational energy, which is converted into translational energy when the centripetal force is removed.

The "interior ballistics" of spring weapons is based on the linear spring law:

$$F_{tf}(s) = D \cdot s \,. \qquad (3.3:1) \quad [N]$$

Where $F_{tf}$ = tension force, D = spring constant and s = draw length.

The total energy stored at a draw length s is

$$E = \int F_{sp}(s) \cdot ds = \tfrac{1}{2} \cdot D \cdot s^2 \,. \qquad (3.3:2) \quad [J]$$

Let $F_{tf,max}$ be maximum tension force and let $s_{max}$ be the corresponding draw length. The equations of velocity and acceleration of the projectile as it is being accelerated, as a function of displacement x, are:

(3.3:3) $$v(x) = \sqrt{\frac{F_{tf,max} \cdot x}{m} \cdot \left(2 - \frac{x}{s_{max}}\right)} \,, \qquad [m/s]$$

(3.3:4) $$v'(x) = \sqrt{\frac{F_{tf,max}}{m}} \cdot \frac{s_{max} - x}{\sqrt{s_{max} \cdot x \cdot (2 \cdot s_{max} - x)}} \,, \quad x > 0 \,. \qquad [1/s]$$

In these two equations, mass m consists not only of the mass of the projectile $m_p$ but also that of those parts of the weapon that also undergo acceleration (helical spring, bow arm and bow string, or rubber band). Mass m can therefore be divided into $m_p$ and a (fictitious) "carrier mass" $m_{ca}$. This is derived from the energy balance:

$$E_{tot} = \tfrac{1}{2} \cdot F_{tf,max} \cdot s_{max} = \tfrac{1}{2} \cdot \left(m_p + m_{ca}\right) \cdot v^2 \,, \qquad [J]$$

(3.3:5) $$m_{ca} = \frac{F_{tf,max} \cdot s_{max}}{v^2} - m_p \,. \qquad [kg]$$

The efficiency of the weapon is

(3.3:6) $$\eta = \frac{E_0}{E_{tot}} = \frac{m_p}{m_p + m_{ca}} \,. \qquad [-]$$

where $E_0$ is the initial energy of the projectile. For a given weapon, therefore, efficiency increases with projectile mass.

This is true for all projectile weapons, from a human hand throwing an object to a firearm.

The duration of the discharge process, $t_s$, which is often of interest in interior ballistics, is given by

(3.3:7) $$t(x) = \frac{s_{max}}{v^2} \cdot \arccos\left(1 - \frac{x}{s_{max}}\right) , \qquad [s]$$

(3.3:8) $$t_s = \frac{\pi}{2} \cdot \frac{s_{max}}{v^2} \,. \qquad [s]$$

Where $x = s_{max}$. This set of equations is valid for linear springs and is also reasonably valid for systems that behave in a manner similar to that of a linear spring, such as spring pistols and longbows. However, $F_{tf,max}$ must be so chosen that $\frac{1}{2} \cdot F_{tf,max} \cdot s$ corresponds to the effective spring work. Where the force-displacement curve deviates substantially from a straight line (e.g. in the case of a compound bow), the actual force-displacement relationship must be used.

(3.3:9) $$v(x) = \sqrt{\frac{2}{m} \cdot \int_0^x F_{tf}(s_{max} - \xi) \cdot d\xi} \,.$$

The maximum possible velocity will be obtained when $x = s_{max}$.

#### 3.3.1.3 Compressed gas

Energy can be stored not only by compressing or extending a spring, but also by compressing a gas. Several types of weapon or device use this source of energy. A gas is stored at a certain pressure in a chamber behind the projectile (pressure energy). When the weapon is fired, this pressure is released via a valve that opens very rapidly and applies energy to the base of the projectile, causing it to accelerate.

There are two ways of generating pressure:

- mechanical compression (in the case of an airgun)
- an air reservoir or gas cartridge (e.g. in the case of a $CO_2$ pistol or large-calibre weapon).

If we make certain simplifying assumptions, it is possible to calculate the displacement-velocity curve for a projectile accelerated in this fashion. If we assume that the valve opens instantaneously and that the expansion process is isentropic (i.e. no heat is exchanged with the surroundings) and hence

$$p \cdot V = \text{const} , \qquad T = \text{const.} \qquad \text{[J]}$$

then the pressure applied to the base of the shot after it has moved a distance x is given by

$$\text{(3.3:10)} \qquad p(x) = p_0 \cdot \left( \frac{V_0}{V_0 + A \cdot x} \right) - \frac{m_{Gas} \cdot v^2}{2 \cdot (V_0 + A \cdot x)} . \qquad [\text{N/m}^2]$$

Where: $V_0$ = compression chamber volume, $p_0$ = compression chamber pressure, A = barrel cross-sectional area, $m_{Gas}$ = mass of gas and v = projectile velocity. The gas mass is calculated from the compression chamber pressure and volume using Eqn 2.3:2. The second addend on the right-hand side is the reduction in pressure, calculated using BERNOULLI's equation.

In accordance with D'ALEMBERT's principle (Eqn 2.2:31), velocity and displacement are given by the following system of differential equations:

$$\text{(3.3:11)} \qquad \frac{dv}{dt} = \frac{p(x) \cdot A - F_b}{m_p + \frac{1}{2} \cdot m_{Gas}} , \qquad [\text{m/s}^2]$$

$$\text{(3.3:12)} \qquad \frac{dx}{dt} = v , \qquad [\text{m/s}]$$

where $m_p$ = projectile mass and $F_b$ = friction.

The three equations 3.3:10–12 describe the movement of the projectile and the pressure in the barrel of a compressed-gas weapon. If the corresponding weapon data are available, they yield plausible results.

Today, compressed gas is used mainly in small-calibre target weapons. In the first half of the 20$^{th}$ Century, however, guns using compressed gas were also produced, as they were much quieter than those using propellants. Minethrowers with calibres of 105 to 200 mm saw service during the First World War. These fired mines weighing 6.5 to 22.6 kg over distances of up to 800 m or 1250 m.

#### 3.3.1.4 *Explosive materials*

Section 3.2 described the characteristics of explosive materials, notably propellants. These burn without an external source of oxygen, produce large quantities of gas and, with the exception of black powder, leave virtually no solid residue.

It is scarcely surprising, therefore, that despite intensive efforts to develop other methods of propulsion, explosive-based propellants are still the most commonly used means of accelerating projectiles from guns and of powering smaller rockets.

### 3.3.2 Combustion of propellant

#### 3.3.2.1 *Burning rate*

As mentioned in 3.2.1.2, propellants always burn perpendicularly to their surface, at a specific burning rate r. If a layer of thickness de burns in time dt, then

$$(3.3{:}13)\qquad r = \frac{de}{dt}\,. \qquad [\mathrm{m/s}]$$

This *linear burning rate* is closely dependent on pressure and on the initial temperature of the propellant. Various attempts were made to determine burning rate during the 19th Century, both by means of experiments and through theoretical studies – qualitative and quantitative.

On the basis of his results, French chemist Paul VIEILLE proposed the following equation in 1893:

$$(3.3{:}14)\qquad \frac{de}{dt} = \beta\cdot\left(\frac{p}{p_0}\right)^{\alpha}. \qquad [\mathrm{mm/s}],\ [\mathrm{m/s}]$$

Where $p_0$ is constant standard pressure and both $\alpha$ and $\beta$ are constant for a given type of propellant.

VIEILLE took a value of 2/3 for $\alpha$ using his propellant. German chemist W. WOLFF later discovered a relationship with *loading density* (the ratio of propellant volume to the total volume available to the powder), prompting him to take values of between 0.46 and 0.66. Extensive tests by O. SCHMITZ (1913) led to values of between 0.9 and 1, while J. CORNER (1950) gives a range of 800 to 4000 bar. This was convenient in the era before computers, as it simplified interior ballistics calculations. It is possible to obtain the constant $\beta$ by measuring burning rate at standard pressure.

In the open air (at a standard pressure of 1 bar), gun powder burns very slowly – at a rate of approximately 0.1 mm/s. As pressure rises, burning rate increases rapidly. At 4000 bar, it may burn at around 60 cm/s. Table 3.4 lists some typical burning rates and their relationship to pressure.

French ballisticians H. MURAOUR and C. AUNIS (1949) also made a major contribution to studies on the relationship between pressure and burning rate:

**Table 3.4.** Burning rates of various propellants

| Propellant | Pressure [bar] | Burning rate [mm/s] |
|---|---|---|
| Black powder | 1 | 1.8 |
| Nitrocellulose propellant | 1 | 0.06–0.09 |
| Nitroglycerine propellant | 1 | 0.07–0.25 |
| " | 10 | 1.5 |
| " | 500 | 75 |
| " | 4000 | 600 |

$$(3.3{:}15) \qquad \frac{de}{dt} = \delta + \xi \cdot \frac{p}{p_0}, \qquad [\text{mm/s}], [\text{m/s}]$$

where δ is usually a constant and ξ is temperature-dependent. Although this equation was also found to be valid over a wide pressure range (200-4400 bar according to R. E. KUTTERER, 1959), the approach proposed by P. VIEILLE is the most commonly used today.

When calculating interior ballistic processes one must know not only the linear burning rate but also the way in which the overall propellant mass is converted into gas over time. Let e be the thickness of the propellant that has already burned, A(e) the total propellant surface burning at this point in time and V(e) the total remaining propellant volume. We can calculate the volume converted to gas in time dt, together with the burning rate, as follows:

$$(3.3{:}16) \qquad \frac{dV(e)}{dt} = -A(e) \cdot \frac{de}{dt}. \qquad [\text{m}^3/\text{s}]$$

The propellant volume V(e) that remains after a propellant thickness of e has burned is usually expressed in relation to the initial volume $V_0$.

$$(3.3{:}17) \qquad z = 1 - \frac{V(e)}{V_0}. \qquad [-]$$

Deriving Eqn 3.3:17 for time and substituting into (3.3:16), we obtain:

$$(3.3{:}18) \qquad \frac{dz}{dt} = \frac{A(e)}{V_0} \cdot \frac{de}{dt}. \qquad [1/\text{s}]$$

The equation is expanded by inclusion of the initial surface area $A_0$ and the thickness of the propellant $\rho_c$, giving us the relative (volumetric) burning rate:

$$(3.3{:}19) \qquad \frac{dz}{dt} = \frac{A(e)}{A_0} \cdot B \cdot \frac{de}{dt}, \quad \text{where} \qquad [1/\text{s}]$$

$$(3.3.19a) \qquad B = \frac{A_0 \cdot \rho_c}{m_c} \left( = \frac{A_0}{V_0} \right). \qquad [1/\text{m}]$$

B is the burn rate coefficient of the propellant, and depends exclusively on its geometry. The rate at which propellant is converted into gas depends on the quotient of the surface burning at a given moment in time and the initial surface, on the burn rate coefficient and on the burning rate. This quotient is represented as φ(z) and termed the (geometric) *form function* of the propellant.

$$(3.3{:}20) \qquad \varphi(z) = \frac{A(e)}{A_0} . \qquad [-]$$

The form function of a propellant therefore plays a decisive role in its burning, because the quantity of gas produced can be controlled by varying the shape of the propellant.

### 3.3.2.2 Form function

It may seem surprising that z does not appear on the right-hand side of Eqn 3.3:20. However, inspection of Eqn 3.3:17 reveals that the form function is defined by a parametric representation, with thickness of the burnt propellant e as its parameter. This means that φ(z) depends exclusively on the geometry of the individual propellant grains. Indeed, simple grain forms yield closed functions. Spherical and cuboid grains have the following simple function:

$$(3.3{:}21) \qquad \varphi(z) = (1-z)^{\frac{2}{3}} , \qquad [-]$$

for a cylinder with insulated end faces, the function is

$$(3.3{:}22) \qquad \varphi(z) = (1-z)^{\frac{1}{2}} . \qquad [-]$$

Table 3.5 shows the form functions for other typical, conventional grain shapes in parametric form. Figure 3.3 shows the functions in graphical form.

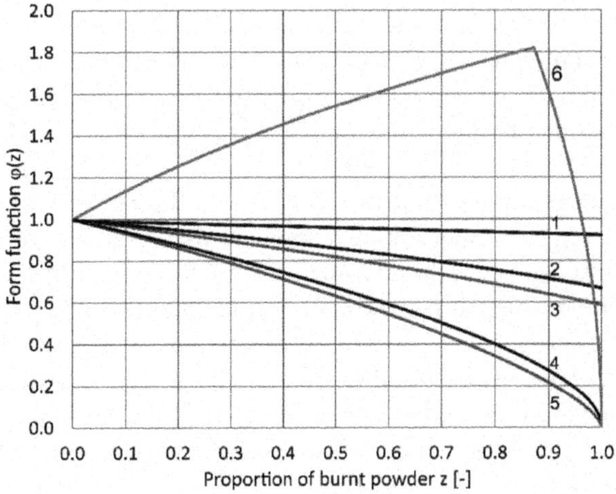


**Fig. 3.3.** Form functions for various grain shapes:
1: Cruciform
2: Strip
3: Short tube (L/D = 5)
4: Cylindrical
5: Spherical
6: Seven-hole

**Table 3.5.** Form functions for a number of typical grain shapes

| | |
|---|---|
| Tubular  | $z = 1-\left(1-\frac{2\cdot e}{r_a - r_i}\right)\cdot\left(1-\frac{2\cdot e}{c}\right)$ <br> $\varphi(z) = 1-\frac{4\cdot e}{r_a - r_i + c}$ |
| Seven-hole  | $z = 1-\left(1-\frac{2\cdot e}{c}\right)\cdot\left(1-\frac{2\cdot e\cdot(r_a+7\cdot r_i+3\cdot e)}{r_a^2-7\cdot r_i^2}\right)$ <br> $\varphi(z) = 1-2\cdot e\cdot\frac{r_a+7\cdot r_i-3\cdot c+6\cdot e}{c\cdot(r_a+7\cdot r_i)}$ |
| Cylindrical  | $z = 1-\left(1-\frac{e}{r}\right)^2\cdot\left(1-\frac{2\cdot e}{c}\right)$ <br> $\varphi(z) = \left(1-\frac{e}{r}\right)\cdot\left(1-\frac{e}{c}\right)$ |
| Strip  | $z = 2\cdot e\cdot\frac{(a\cdot b+a\cdot c+b\cdot c)-2\cdot e\cdot(a+b+c)}{a\cdot b\cdot c}$ <br> $\varphi(z) = 1-\frac{4\cdot e\cdot(a+b+c-3\cdot e)}{a\cdot b+a\cdot c+b\cdot c}$ |
| Cruciform  | $z = 1-\frac{(b-2\cdot e)\cdot(4\cdot a+b-2\cdot e)\cdot(c-2\cdot e)}{b\cdot c\cdot(4\cdot a+b)}$ <br> $\varphi(z) = \frac{2\cdot(c-2\cdot e)\cdot(2\cdot a+b-2\cdot e)+(b-2\cdot e)\cdot(4\cdot a+b-2\cdot e)}{2\cdot c\cdot(2\cdot a+b)+b\cdot(4\cdot a+b)}$ |

$r_a$ = cylinder radius, $r_i$ = radius of holes, a = larger dimension, b = size of smaller end, c = length. The thickness of propellant already burnt is represented by e.

If the surface area of the powder reduces as it burns, i.e. $\varphi(z) < 1$, the propellant is described as *degressive* (or sometimes, and inaccurately, as "offensive"). If the surface area increases, then $\varphi(z) > 1$ and the propellant is termed *progressive*. If the surface area remains constant during burning, the propellant is described as *neutral* ($\varphi(z) = 1$).

Convex grains always burn degressively. Spherical (Fig. 3.4 a), flake (Fig. 3.4 b) and short tubes (Fig. 3.4 c) are typical shapes for the propellants used in small-arms ammunition. Neutral propellants can be produced using grains in the form of cylinders with an isolated (non-combustible) surface shell or tubes with insulated ends. Long tube, strip and cruciform propellants, as used in conventional artillery, are virtually neutral. See Fig. 3.5. Progressive burning is achieved using cylindrical grains with multiple holes. For reasons of symmetry, grains are produced with 7, 19 and 43 holes (see Fig. 3.6 on p. 72 for examples,). These are mainly used for

**Fig. 3.4.** Grain shapes for small-arms ammunition: **a**: Spherical. **b**: Flake. **c**: Short tube.

medium-calibre weapons with long barrels, such as medium anti-aircraft guns and tank guns.

After they have burnt for a certain length of time, multiple-hole grains disintegrate, burning strongly degressively thereafter, as the particles are virtually triangular. This makes it necessary to show the form function in stages (see Fig. 3.3).

Because the form functions of real propellant grains are relatively complex, it is usual to consider one or more partial curves, using the following function (based on Eqns 3.3:21 and 3.3:22):

(3.3:23) $$\varphi(z) = \mu \cdot (1 + \lambda \cdot z)^n \,. \qquad [-]$$

This approximates the form functions found in practice, which have generally been determined empirically.

This approach was popular before computers became available, as taking $\lambda = 1$ and $n = ½$ permitted closed integration of the energy equation.

### 3.3.2.3 *Influencing burning rate by chemical means*

Eqn 3.3:19 introduced the burn rate coefficient B, in connection with the relative conversion rate of the propellant, dz/dt. The unit of measurement for B is [1/m], which means that B decreases as the size of the propellant grains increases. For geometrical reasons, therefore, larger grains will burn more slowly than small, leading to a slower increase in pressure. Small grains are advantageous in small-arms ammunition, however, as they can be packed more tightly.
Most short-grained propellants (see Fig. 3.4) therefore undergo a process that makes it possible to control burning rate by means other than changes to their geometrical form. This consists of applying a deterrent to the surface of the grains that

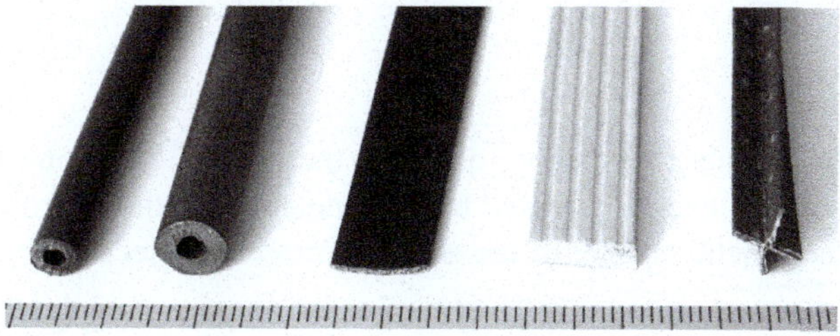

**Fig. 3.5.** Artillery propellants. *Left to right:* Two types of tubular propellant, thin strip, thick strip and cruciform. The length of the rods generally depends on the length of the chamber.

**Fig. 3.6.** *Left*: Seven-hole propellant for medium-calibre munitions. *Right*: Nineteen-hole propellant for tank guns.

penetrates to a certain depth and creates a less combustible outer layer. It is possible to increase the progressivity and hence the power of progressive multiple-hole propellants (see Fig. 3.6) by the same method. This provides a further means of influencing the burning behaviour of a propellant.

#### 3.3.2.4 The burning law

If we now substitute Eqns 3.3:14 and 3.3:20 into the equation of relative burning rate (3.3:19), we obtain the *burning law* named after P. J. CHARBONNIER (1908), which relates the conversion rate of the propellant to pressure.

(3.3:24) $$\frac{dz}{dt} = \phi(z)\cdot B\cdot\beta\cdot\left(\frac{p}{p_0}\right)^{\alpha} \quad [1/s]$$

Where $\varphi(z)$ = form function (generally from Eqn 3.3:23), B = burn rate coefficient (from Eqn 3.3:19a) and $\beta$ = linear burning rate at standard pressure $p_0$. The last two parameters (which are constants) are often combined and termed the *vivacity* of the propellant, $B_A$:

(3.3:25) $$B_A = B\cdot\beta = \frac{A_0\cdot\rho_c\cdot\beta}{m_c} \quad [1/s]$$

The vivacity determines the rate at which the propellant initially produces gas.

The names used are not entirely consistent. $B_A$ is sometimes termed the burn rate coefficient, with B unnamed. *Brisance* is sometimes used for one or other of the variables.

Vivacity is a propellant constant. Generally speaking, it can only be determined empirically, especially if the grains have undergone surface treatment to modify their burning rate. The manometric bomb (see 3.5.2) is used to measure this parameter.

### 3.3.3 The energy equation

To derive the interior ballistic energy equation, one uses the general equation of energy (Eqn 2.3:20) and sets the energy released by burning equal to the internal energy of the gases and the kinetic energy of the moving masses (projectile plus gases), $m^+$. Let z be the proportion of propellant mass that has been burned at a given point in time (see Eqn 3.3:17). We can write:

$$(3.3:26) \qquad Q_{ex} \cdot z \cdot m_c = C_v \cdot T \cdot z \cdot m_c + \frac{1}{2} \cdot m^+ \cdot v^2 + \Psi \ . \qquad [J]$$

If $$C_v = \frac{R}{\gamma - 1} \quad \text{and} \quad p \cdot V = R \cdot T \cdot z \cdot m_c \ ,$$

then $$Q_{ex} \cdot z \cdot m_c = \frac{p \cdot V}{\gamma - 1} + \frac{1}{2} \cdot m^+ \cdot v^2 + \Psi \ . \qquad [J]$$

$\Psi$ is the unknown "lost" energy. This consists of the thermal energy transferred to the barrel and the internal energy of the gases, together with energy lost elsewhere (e.g. friction). In practice, this is taken into account using an empirically-verified heat capacity ratio $\gamma^*$.

We must now determine the volume available to the gas. This consists of the volume of the chamber, plus the volume of that part of the barrel that has been made available by the movement of the projectile, minus the volume of the unburnt propellant and the covolume of the gases.

$$V = V_{ch} + A_b \cdot x - \frac{(1-z) \cdot m_c}{\rho_c} - \eta \cdot z \cdot m_c \ . \qquad [m^3]$$

Where $V_{ch}$ = chamber volume, $A_b$ = barrel cross-sectional area, x = displacement of projectile, $m_c$ = mass of propellant, $\rho_c$ = propellant density and $\eta$ = covolume of propellant gases.

Rearranging this equation and setting initial chamber volume $V_{ch} - m_c/\rho_c = V_{ch}^*$, we obtain the energy equation

$$(3.3:27) \qquad Q_{ex} \cdot z \cdot m_c = \frac{p}{\gamma^* - 1} \cdot \left[ V_{ch}^* - z \cdot m_c \cdot \left( \eta - \frac{1}{\rho_c} \right) + A_b \cdot x \right] + \frac{1}{2} \cdot m^+ \cdot v^2 \qquad [J]$$

In interior ballistics, this is generally known as the *RÉSAL's equation*, and is the primary equation of interior ballistics.

It now remains only to calculate the mass accelerated. This is generally given by

$$(3.3:28) \qquad m^+ = m_p + \varepsilon \cdot m_c \ . \qquad [g], [kg]$$

$\varepsilon$ is *SÉBERT's factor*. In the first half of the 20th Century, many experiments were conducted to find a value for this factor. Those experiments involved recoil measurements and measurements of pressure difference, and yielded values of between 0.25 and 0.5. A value of 0.5 seems to produce the best results (C. CRANZ, 1926, J. CORNER, 1950 and R. GERMERSHAUSEN, 1980).

### 3.3.4 Motion of the projectile in the barrel

The motion of the projectile in the barrel is calculated in accordance with d'ALEMBERT's principle (see Eqn 2.2:31). Let a = dv/dt. We obtain

$$(3.3:29) \qquad m \cdot \frac{dv}{dt} = \sum F_i \,. \qquad [N]$$

The expression on the right-hand side of this equation is the sum of the forces acting on the projectile as it passes through the barrel. The force $p \cdot A_b$ resulting from the pressure of the propellant gases applied to the base of the projectile (where $A_b$ = barrel cross-sectional area) accelerates the projectile, while friction $F_b$ between the projectile and the barrel opposes the motion of the projectile. The projectile also has to push the matter in the barrel forwards.

If that "matter" is air, then its mass will be negligible compared to that of the projectile. However, if the weapon is discharged under water, for instance, that will have a marked influence on the interior ballistics.

Two other forces will oppose the motion of the projectile at the beginning of the interior ballistics process. In order for the propellant to start burning as uniformly as possible across its entire surface, pressure must initially rise very rapidly. This means that the volume available for combustion must initially remain constant. In turn, this implies that the projectile must be prevented from moving for a certain time. The force to achieve this comes from the *engraving resistance* that the projectile must overcome in order to engage the rifling. In the case of cartridge munitions, this is supplemented by the *initial resistance* that opposes extraction of the projectile from the cartridge case. The equation of motion therefore becomes valid only when starting pressure $p_s$ has risen to the level required to overcome these two forces. We can therefore write:

$$(3.3:30) \qquad m \cdot \frac{dv}{dt} = p \cdot A_b - F_b \qquad \text{where } p > p_s \,. \qquad [N]$$

Furthermore, the projectile usually advances without rotating, possibly in free flight, between the moment at which it overcomes initial resistance and that at which it encounters engraving resistance. However, this is generally ignored for calculation purposes.

### 3.3.5 Pressure as a function of time and displacement

Once chamber pressure exceeds that required to extract the projectile from the cartridge case and/or to force it into the rifling, it will begin to move. Chamber pressure will continue to rise, increasing the burning rate and hence increasing the rate at which pressure rises. The projectile now accelerates very rapidly, making an increasingly large volume available to the gases. As long as the amount of gas produced continues to exceed the increase in volume, gas pressure will continue to rise. Gas pressure attains a maximum when the quantity of newly-produced gas becomes equal to the increase in volume at that moment. From that point onwards, pressure drops steadily. Once the propellant is fully burnt, pressure will still be high and the gas will expand (virtually isentropically), further accelerating the projectile. The pressure-time curve often shows "all burnt" and the start of expansion quite clearly.

With a lightly loaded cartridge and a long barrel, it is possible for the pressure-induced accelerating force to be smaller than the friction between the projectile and the barrel. In such cases, the muzzle velocity of the projectile will be lower than its maximum velocity in the barrel. This generally leads to greater vertical spread and less accuracy. In such cases, shortening the barrel increases muzzle velocity and produces better results.

Figs 3.7 to 3.14 show typical curves of pressure and velocity against time and (approximately) against displacement. Multiple curves are plotted together, to illustrate the disparate characteristics of propellants used with different calibres and projectile masses.

Figs 3.7 and 3.8 illustrate the difference between propellants for handguns and rifles. Because the pistol barrel is shorter, the propellant must burn considerably faster, accelerating the bullet more rapidly. Figs 3.9. and 3.10 compare the propel-

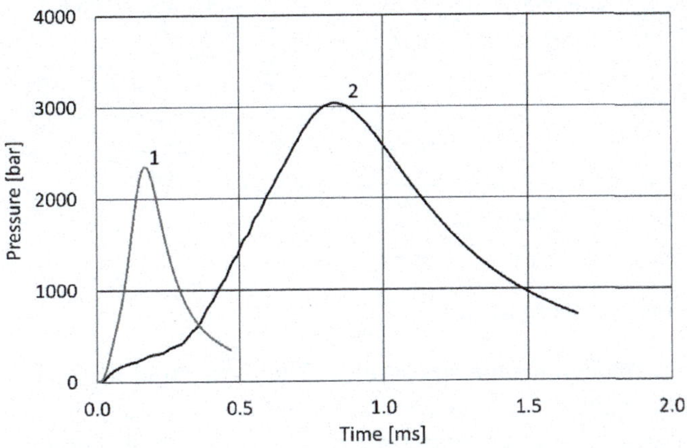


**Fig. 3.7.** Pressure-time curves for a pistol (9 mm Luger) and a rifle (7.5 × 55). Because the barrel of the pistol is shorter, the propellant must burn considerably faster.

1 Pistol
2 Rifle

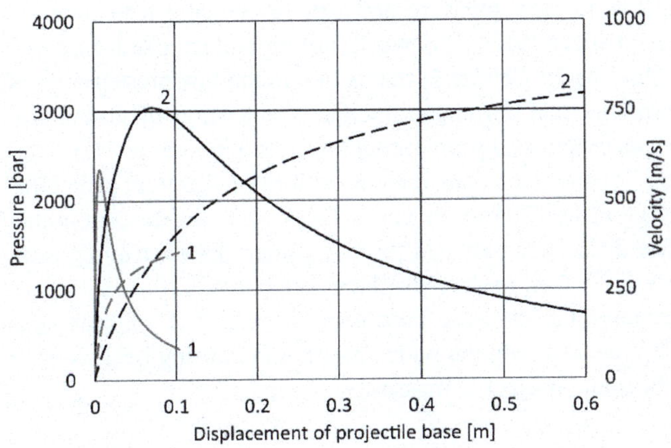


**Fig. 3.8.** Pressure-displacement (solid line) and pressure-velocity (dashed line) curves for the pistol and rifle from Fig. 3.7. These also show the difference in acceleration between the two bullets.

1 Pistol
2 Rifle

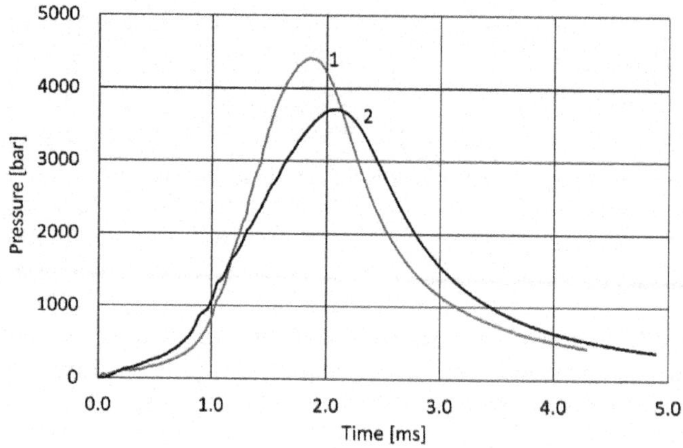


**Fig. 3.9.** Pressure-time curves for 30 × 173 mm rounds: a light sub-calibre APDS and a heavy solid shell. The light shell requires a faster-burning propellant.
1 APDS
2 Solid

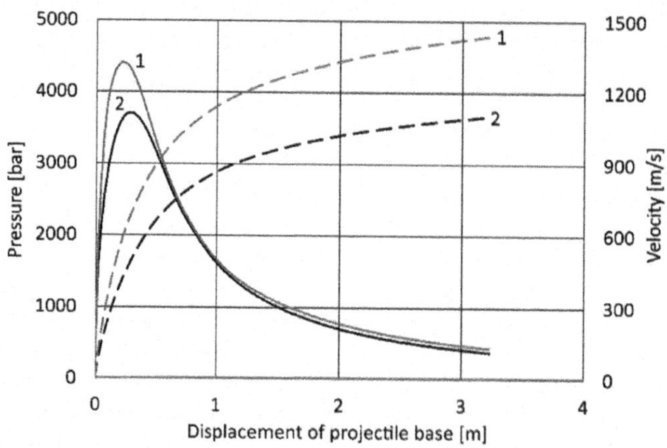


**Fig. 3.10.** Pressure-displacement (solid line) and pressure-velocity (dashed line) curves for the two cartridges in Fig. 3.9.
Note the higher acceleration in the case of the lighter projectile.
1 APDS
2 Solid

lants for light and heavy medium-calibre projectiles. The more rapid increase in pressure in the case of the lighter projectile shows that this is a degressive propellant. As in the case of the pistol with its shorter barrel, this leads to higher acceleration. Figs 3.11 and 3.12 show how pressure and velocity change with time and displacement for various artillery loads. It is worth noting the extended time over which pressure rises in the case of Charge 3; this is due to the low pressure. Small charges of this nature require fast-burning propellants consisting of grains with large surface areas (such as cruciform-grain propellants, see Fig. 3.5, p. 71)).

Figs 3.13 and 3.14 (p. 78) show the pressure curves for two different projectiles fired from a high-power gun (the 120 mm Pz Kan). The kink in the descent branch indicates "all burnt". The difference between the two projectiles is striking, especially as regards the area under the pressure-displacement curve.

Fig. 3.15 (p. 78) compares the pressure-time curves for all the calibres mentioned in the previous figures, showing the wide variation in burning behaviour required for different calibres and projectile masses.

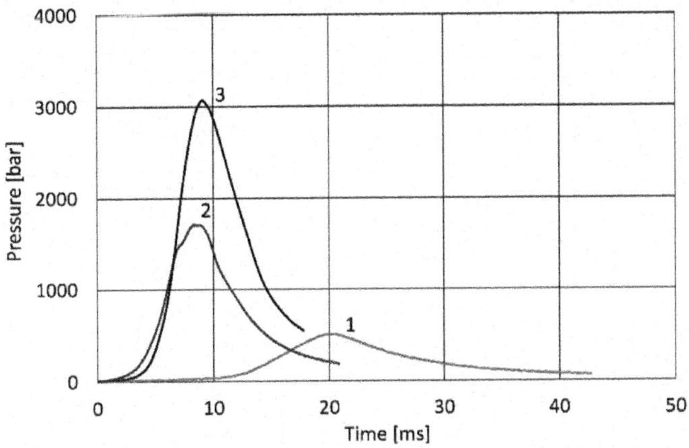


**Fig. 3.11.** Pressure-time curves for three 155 mm artillery loads. Note the longer time required to reach maximum pressure in the case of Charge 3.

1 Charge 3
2 Charge 7
3 Charge 9

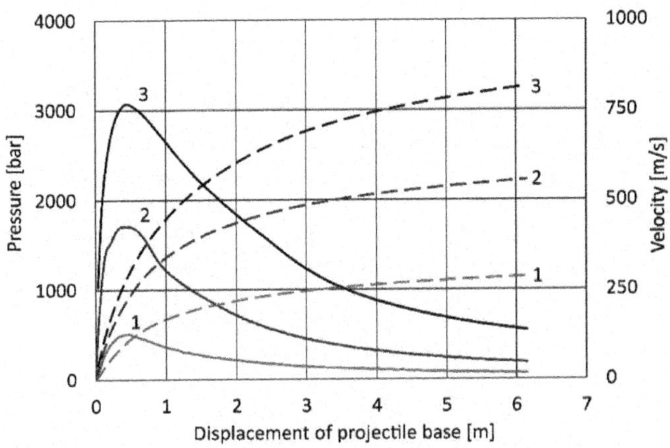


**Fig. 3.12.** Pressure-displacement (solid line) and pressure-velocity (dashed line) curves for the three artillery charges in Fig. 3:11.

1 Charge 3
2 Charge 7
3 Charge 9

### 3.3.6 Efficiency and maximum velocity

As mentioned in 2.3.2.5, firearms are heat engines, i.e. devices that convert thermal energy into mechanical energy. This being so, it is sometimes useful to known their mechanical efficiency, i.e. the proportion of total energy that can be used mechanically.

The total energy available to a projectile depends on the mass of the propellant and its combustion heat. The mechanical energy transferred to the projectile corresponds to the muzzle energy of the projectile and the kinetic energy of the gases (see Eqn 3.3:27). The efficiency of a weapon is therefore given by:

$$(3.3:31) \qquad \eta^{+} = \frac{m^{+} \cdot v_0^2}{2 \cdot m_c \cdot Q_{ex}} \qquad [-]$$

Where $m^{+} = m_p + \varepsilon \cdot m_c$ is total mass accelerated (Eqn 3.3:28), in which $m_p$ = mass of projectile and $m_c$ = load mass. $\varepsilon$ is SÉBERT's factor. For the actual effect of the weapon, $\eta$, only the muzzle energy of the projectile is relevant. In this instance, therefore, $m_p$ is used instead of $m^{+}$.

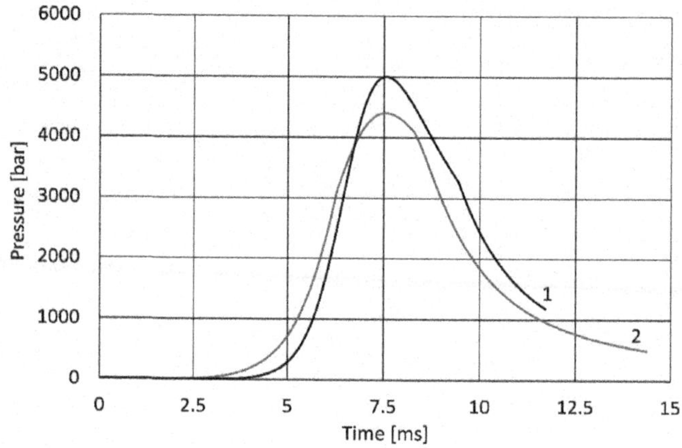


**Fig. 3.13.** Pressure-time curves for 120 mm tank rounds.

1 KE
2 HE

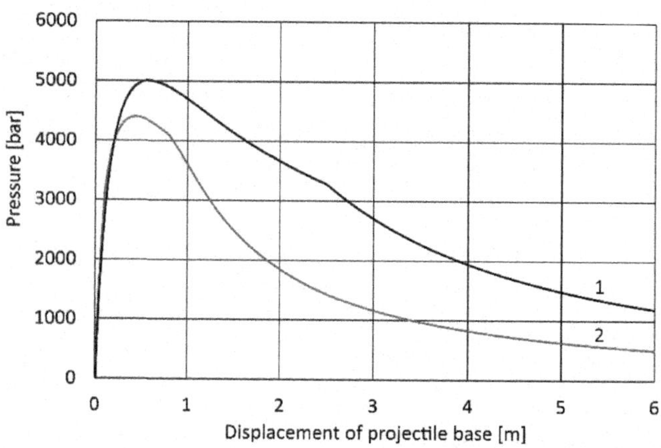


**Fig. 3.14.** Pressure-displacement curves for the two types of ammunition in Fig. 3:13. The kink in the descent branch indicates "all burnt".

1 KE
2 HE

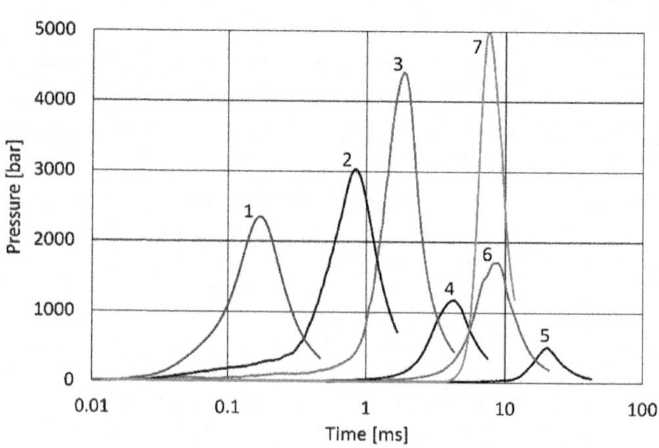


**Fig. 3.15.** Pressure-time curves for all calibres, plotted on a logarithmic scale.

1 9 mm Luger
2 7.5 × 55
3 30 × 173 mm APDS
4 120 mm mortar
5 155 mm, Charge 3
6 155 mm, Charge 7
7 120 mm Pz Kan, KE

The efficiency of a firearm generally lies between 0.25 and 0.45 and increases with projectile mass and calibre.

Knowing the mechanical efficiency allows us to calculate the *maximum possible muzzle velocity* of a firearm. One assumes that half of the propellant mass is accelerated along with the projectile (SÉBERT's factor, Eqn 3.3:28) and that the velocity of the propellant and gases $v_c$ increases linearly from the breech face ($v_c = 0$) to the base of the shot ($v_c = v_0$). For the energy of the propellant accelerated with the projectile, we can therefore write

$$E_c = \frac{1}{6} \cdot m_c \cdot v_0^2 \ . \qquad (3.3{:}32) \quad [J]$$

The energy equation is:

$$\frac{1}{2} \cdot \left( m_p + \frac{1}{3} \cdot m_c \right) \cdot v_0^2 = \eta^+ \cdot m_c \cdot Q_{ex} \ . \quad [J]$$

If we divide this equation by $m_c$ and assume

$$m_c \rightarrow \infty \ ,$$

we obtain the following expression for muzzle velocity:

$$v_0 = \sqrt{6 \cdot \eta^+ \cdot Q_{ex}} \ . \qquad (3.3{:}33) \quad [J]$$

So for instance, a propellant with an efficiency of 0.35 that generates 5000 kJ/kg combustion heat can accelerate a projectile to no more than approximately 3240 m/s.

## 3.4 Interior ballistics calculations

### 3.4.1 The thermodynamic equations

Sub-sections 3.3.2 to 3.3.4 present the physical laws that govern interior ballistics. If we summarize the equations derived in those sub-sections, we obtain the thermodynamic system of equations for interior ballistics ($p_s$ = starting pressure):

$$\frac{dz}{dt} = \phi(z) \cdot B_A \cdot \left( \frac{p}{p_0} \right)^{\alpha} , \qquad (3.4{:}1a) \quad [1/s]$$

$$\frac{dv}{dt} = \frac{p \cdot A_b - F_b}{m_p} \quad \text{if } p > p_s \ , \qquad (3.4{:}1b) \quad [m/s^2]$$

$$\frac{dx}{dt} = v \ . \qquad (3.4{:}1c) \quad [m/s]$$

In RÉSAL's equation (Eqn 3.3:27), the variable to be calculated is pressure, so we rearrange the equation accordingly:

(3.4:2) $$p = \frac{(\gamma^* - 1)\cdot\left[Q_{ex}\cdot z\cdot m_c - \frac{1}{2}\cdot(m_p + \varepsilon\cdot m_c)\cdot v^2\right]}{V^*_{ch} - z\cdot m_c\cdot\left(\eta - \frac{1}{\rho_c}\right) + A_b\cdot x}\,. \qquad [\mathrm{N/m^2}]$$

Equations 3.4:1a-c form a system of differential equations. If we take account of Eqn 3.4:2, those equations can be solved using a numerical method such as Runge-Kutta. This yields pressure (maximum pressure in particular), projectile velocity and the displacement of the projectile along the barrel, together with the characteristics of the process over time. First, however, we must collect a large number of values (see Table 3.6):

- *Propellant data*: Mass, density, thickness, initial surface area, form function, covolume, burning rate as a function of pressure, heat of combustion and specific energy.
- *Weapon data*: Barrel cross-section and length, rifling and chamber volume (only for cartridge ammunition where projectile and charge are separate).
- *Ammunition data*: Projectile mass, engraving force, friction, cartridge case volume (in the case of cartridge ammunition, and taking into account the portion of the projectile that protrudes into the cartridge case) and initial resistance.

The simplest solution is to request the necessary documents from the manufacturers. For small arms, there are two international standards for chamber and cartridge case dimensions: those of the CIP ("Commission Internationale Permanente pour

**Table 3.6.** Data required for interior ballistics calculations

| | | | | | |
|---|---|---|---|---|---|
| $A_0$ | [m²] | Initial surface area of propellant | $p_0$ | [N/m²] | Reference pressure |
| $A_b$ | [m²] | Barrel cross-section | v | [m/s] | Velocity |
| $B_A$ | [1/s] | Dynamic vivacity | x | [m] | Displacement of projectile |
| $F_b$ | [N] | Friction between projectile and barrel | z | [-] | Proportion of propellant burnt |
| $Q_{ex}$ | [J/g] | Heat of combustion | $\alpha$ | [-] | Pressure exponent for burning rate |
| $V_{ch}$ | [m³] | Chamber volume | $\beta$ | [m/s] | Burning rate at standard pressure $p_0$ |
| $F_{se}$ | [J/g] | Specific energy | $\varepsilon$ | [-] | SÉBERT's factor |
| $\mu, \lambda$ | [-] | Form function constants (3.3:23) | $\gamma^*$ | [-] | Adjusted isentropic exponent |
| $m_c$ | [kg] | Mass of propellant | $\eta$ | [m³/kg] | Covolume |
| $m_p$ | [kg] | Mass of projectile | $\varphi(z)$ | [-] | Form function |
| p | [N/m²] | Pressure | $\rho_c$ | [kg/m³] | Density of propellant |

l'Epreuve des Armes à Feu portatives", Permanent International Commission for the Proof of Small Arms) in Brussels and of the Sporting Arms and Ammunition Manufacturers' Institute (SAAMI) in the USA. Interior ballistics software is also available for small-arms ammunition (up to 15 mm) and certain medium-calibre munitions: *QuickLoad* by H. BRÖMEL, first released in 2009.

While the interior ballistics calculations are highly reliable, one should always check the results against the latest loading data from bullet and propellant manufacturers and respect the warnings in the software.

### 3.4.2 The expansion ratio

When designing the interior ballistics of a weapon one must comply with an important rule regarding chamber volume and barrel volume.

For cartridge munitions, chamber volume corresponds to the volume of the cartridge case, minus the volume of that portion of the projectile that protrudes into it.

Let $V_{ch}$ be chamber volume and let $V_b$ be barrel volume. The following rule of thumb applies:

(3.4:3) $$\frac{V_{ch} + V_b}{V_{ch}} = 1 + \frac{V_b}{V_{ch}} > 5...6 \ . \qquad [-]$$

This ratio of the total internal volume of a weapon to the chamber volume is known as the *expansion ratio* (German: "Entspannungsverhältnis", French: "rapport de détente").

Strictly speaking, the covolume of the gases should be subtracted from both the numerator and the denominator (J. CORNER, 1950). However, that would deprive the equation of its value as a "rule of thumb", as it would no longer be possible to conduct calculations easily. Furthermore, this simplification has little effect on the result.

The rule of Eqn 3.4:3 can be linked to the equation of state for gases (Eqn 2.3:2). As the projectile has only moved a short distance when gas pressure reaches a maximum (depending on the calibre, this will be somewhere between a few millimetres and a number of centimetres, but in any case a short distance compared to the length of the barrel), we can say, in approximate terms, that maximum gas pressure $p_{max}$ occurs within a volume corresponding to the chamber volume. By contrast, muzzle pressure $p_\mu$ occurs when the volume is that of the entire internal volume of the weapon. We can therefore write:

$$p_{max} \cdot V_{ch} = p_\mu \cdot (V_{ch} + V_b) \ , \qquad [J]$$

(3.4:4) $$\frac{p_{max}}{p_\mu} = 1 + \frac{V_b}{V_{ch}} \ . \qquad [-]$$

The expansion ratio is therefore approximately equal to the ratio of maximum pressure to muzzle pressure. The muzzle pressure of a weapon should hence be approximately five to six times lower than the maximum gas pressure. Failure to observe this rule will result in poor thermal efficiency, pronounced muzzle flash and considerable interference with the projectile by gas emerging from the muzzle.

Long-barrelled weapons almost always comply with the rule of Eqn 3.4:3. However, weapons with a small barrel volume (short-barrelled handguns and weapons with a very small calibre) often operate in a highly unsatisfactory manner from an interior ballistics point of view. The above rule also means that there are limits to chamber volume and hence to propellant quantity.

### 3.4.3 Twist and rifling groove forces

Long projectiles are stabilized in flight using the gyroscope principle. To make this happen, they have to be made to spin as they accelerate along the barrel. That is the role of the spiral lands and grooves in the barrel.

The angular velocity required to achieve stability depends solely on the projectile and its design (see 5.6). The first weapon parameter of relevance is the *twist length*, comparable to the pitch of a screw, which is the distance along the barrel that the projectile moves as it makes one rotation. The second is the *angle of twist*, which is the angle between the barrel axis and a tangent to the grooves.

The relationship between twist length $\Lambda$, calibre d und angle of twist $\Gamma$ is given by:

$$(3.4{:}5) \qquad \Gamma = \arctan\frac{d\cdot\pi}{\Lambda}\,. \qquad [-]$$

The angular velocity $\omega$ attained with a twist length of $\Lambda$ is related to muzzle velocity $v_0$ and is given by the following equation:

$$(3.4{:}6) \qquad \omega = 2\cdot\pi\cdot\frac{v_0}{\Lambda} = \frac{2}{d}\cdot v_0\cdot\tan\Gamma\,. \qquad [1/s]$$

When the projectile starts to move, its driving bands or jacket are forced into the lands and grooves of the barrel. As these are arranged at the angle of twist with respect to the axis of the barrel, forces arise between the sides of the grooves in the barrel and the imprint of the rifling in the part of the projectile that engages with it, and those forces cause the projectile to rotate. The forces are known as *rifling groove forces* $F_L$. The total force in the x direction is determined by D'ALEMBERT's principle (Eqn 2.2:31) and the moment equilibrium about the x-axis (see Fig. 3.16):

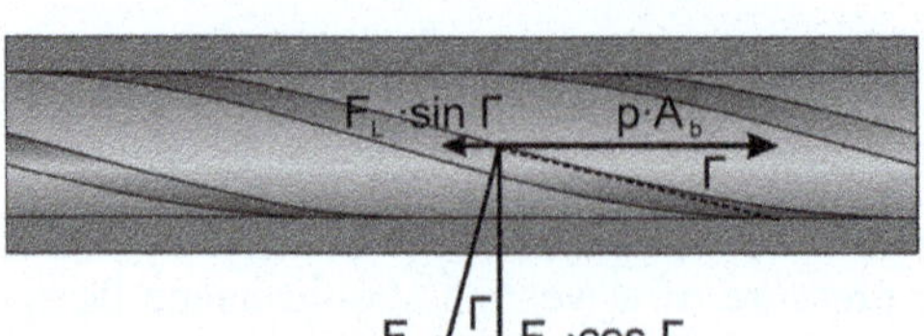


**Fig. 3.16.** Calculating the rifling groove forces. Angle of twist shown exaggerated for clarity.

$$p \cdot A_b - F_L \cdot \sin\Gamma - m_p \cdot \frac{dv}{dt} = 0 \, , \qquad \text{[N]}$$

$$F_L \cdot \frac{d}{2} \cdot \cos\Gamma - J_a \cdot \frac{d\omega}{dt} = 0 \, . \qquad \text{[J]}$$

Using the approximations $\sin^2\Gamma \approx 0$ and $\cos\Gamma \approx 1$, which are permissible in view of the small angle $\Gamma$, and taking into account Eqn 3.4:6, we obtain the following equation for total rifling groove force:

$$\text{(3.4:7)} \qquad F_L = \frac{\pi \cdot J_a}{m_p} \cdot p \cdot \tan\Gamma \, . \qquad \text{[N]}$$

Where $F_L$ = total rifling groove force, $J_a$ = axial moment of inertia of the projectile, $m_p$ = projectile mass, p = instantaneous gas pressure and $\Gamma$ = angle of twist.

The total rifling groove force is divided over the land/groove pairs (see Fig. 3.17). The individual forces generate shear stresses in the guiding surface of the projectile, and those stresses increase with gas pressure. They also increase if the lands are forced less into the guide bands or the cylindrical part of the projectile. If the stresses exceed the shear strength of the material in the projectile that engages with the barrel, the lands and grooves pressed into that material will break off. As a result, no spin will be imparted to the projectile (or too little) and it will be unstable when it leaves the barrel, causing it to topple.

Worn barrels that fire normally when cold may cause projectiles to topple once they warm up. The increase in calibre due to the increase in temperature means that the shear stresses exceed the strength of the projectile guiding surface. Once the barrel cools down, projectiles behave normally again.

If the barrel has eroded asymmetrically (in an oval pattern), the rifling groove force may be distributed over too few lands in the guiding band, which may also cause its shear strength to be exceeded. The shell will then fly insufficiently stabilized.

Increases in the interior ballistic performance of weapons and the need for longer projectiles with larger volumes (and hence greater useful loads) meant that the shear strength limits of projectile guidance surfaces were being exceeded, especially at maximum pressure shortly after the projectile began to move. This led to the devel-

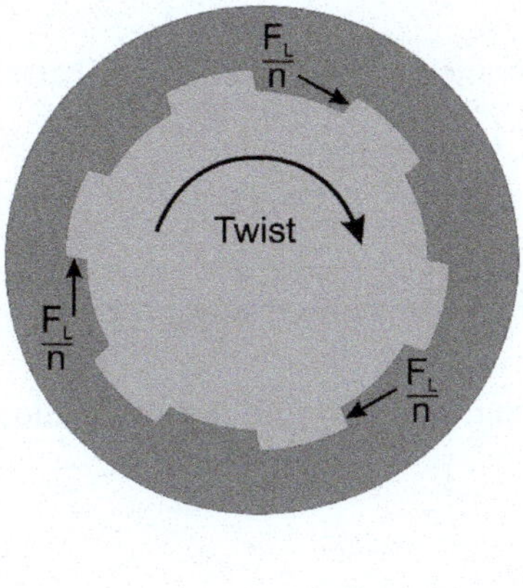


**Fig. 3.17.** Rifling groove forces. The force $F_L$ calculated by means of Eqn 3.4:7 is distributed over n grooves (for clarity, only some of the component forces are shown).

opment of barrels with *progressive twist*, in which the angle of twist increases along the length of the barrel in accordance with a certain rule until the final angle of twist needed to ensure the stability of the projectile is attained. Progressive twist is used primarily in high-power weapons, in which the projectile is subjected to high acceleration.

### 3.4.4 Pressure gradient in the barrel

The pressure curves in 3.3.5 (Figs 3.7–3.15) are always measured in the chamber. However, as projectile velocity increases, a pressure gradient is created between the breech face and the base of the projectile, which can be estimated fairly easily using BERNOULLI's equation (Eqn 2.3:22a).

The gases at the breech face are at rest, whereas those at the base of the projectile are travelling at the same velocity as the projectile. From BERNOULLI's equation, at gas velocity $v_{br} = 0$:

$$p_{br} = p_p + \frac{1}{2} \cdot \rho_c \cdot v_c^2 \,. \qquad \text{[Pa]}$$

Index c represents values for the gases, $p_{br}$ = pressure at the breech face and $p_p$ = pressure at the base of the projectile.

If we assume that the velocity of the gases increases linearly, giving a mean velocity $v_c = 0.5 \cdot v_p$, and if we also assume a mean density of $\rho_c$, then we can write

$$(3.4{:}8) \qquad p_{br} - p_p = \Delta p = \frac{1}{8} \cdot \rho_c \cdot v_p^2 \,. \qquad \text{[Pa]}$$

This means that the pressure on the base of the moving projectile is always lower than that in the chamber. The mean density of the gas after "all burnt" is determined by the mass of the propellant (which is equal to the mass of the gas) and the instantaneous barrel volume.

Eqn 3.4:8 is particularly useful if we wish to derive an approximate value for the effective muzzle pressure from the change in pressure measured in the chamber. The calculated pressure corresponds quite closely to the pressure measured at the muzzle.

Another approach is to apply the law of conservation of momentum to the barrel and projectile as a closed system.

$$(3.4{:}9) \qquad \left(m_r + \varepsilon' \cdot m_c\right) \cdot v_r = \left(m_p + \varepsilon \cdot m_c\right) \cdot v_p \,. \qquad \text{[N·s]}$$

The index r represents the values for the barrel, p and c the projectile and load as usual. ε is that proportion of the load that is accelerated with the projectile. ε′ is the proportion accelerated with the barrel.

Differentiating Eqn 3.4:9 for time and making the justified approximation $\varepsilon' \cdot m_c << m_r$, we obtain:

$$m_r \cdot \dot{v}_r = \left(m_p + \varepsilon \cdot m_c\right) \cdot \dot{v}_p \, . \qquad \text{[N]}$$

If we include the two equations of motion for the barrel and the projectile,

$$m_b \cdot \dot{v}_b = A_b \cdot p_{br} \, , \qquad \text{[N]}$$

$$m_p \cdot \dot{v}_p = A_b \cdot p_p \, , \qquad \text{[N]}$$

we obtain

$$\text{(3.4:10)} \qquad p_{br} - p_p = \Delta p = \frac{\varepsilon \cdot m_c}{m_p + \varepsilon \cdot m_c} \cdot p_{br} \, . \qquad \text{[Pa]}$$

If we make the right-hand sides of Eqns 3.4:8 and 3.4:10 equal, we can derive SÉBERT's factor ε from the muzzle pressure $p_{br}$ (measured at the breech face) and the muzzle velocity $v_p$:

$$\text{(3.4:11)} \qquad \varepsilon = \frac{m_p \cdot v_p^2}{8 \cdot p_{br} \cdot V_b - m_c \cdot v_p^2} \, . \qquad \text{[Pa]}$$

## 3.4.5 The gas dynamics model

### *3.4.5.1 History*

The first scientist to study the pressure, density and velocity of gas flow behind an accelerating projectile in a barrel was the Italian J. L. LAGRANGE (1736–1813). LAGRANGE lived in Paris from 1787 until his death, and it would appear that the government commissioned him to study this problem in 1793. However, his paper on the topic was only discovered after his death. He seems to have been dissatisfied with the results of his work and to have therefore decided not to publish. However, French mathematician and physicist S. D. POISSON (1832) saw new insights in this work and felt it should be published. Since that time, this question has been known as LAGRANGE's ballistic problem.

LAGRANGE could only approach this problem analytically, as no suitable numerical methods were available at the time. He was therefore obliged to simplify significant parts of the question. Even though his approach has long since been superseded, it does give us an initial insight into the thinking behind the gas dynamics model. We shall therefore outline it here.

LAGRANGE worked on the assumption that at the start of motion, the density of both gas and propellant is uniform (which effectively is the same as assuming that the propellant is fully burnt before the projectile begins to move). He also assumed that velocity, density and pressure are always constant within an infinitesimally thin segment of the barrel cross section – a one-dimensional model. To describe the movement of the gas, we consider a segment of thickness dx, which is located at position x at time t = 0. After time t, this segment has moved to a position at a

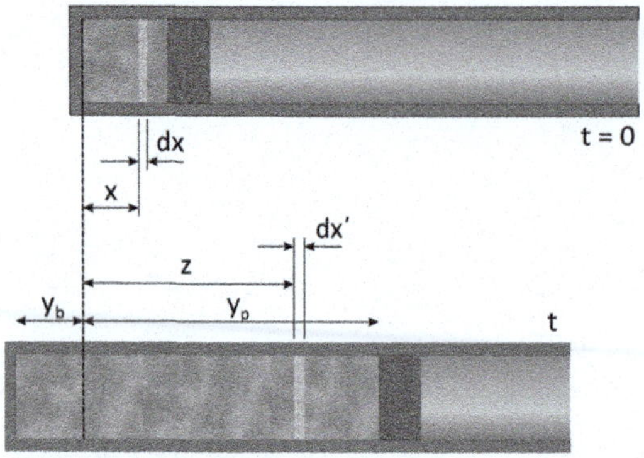


**Fig. 3.18.** LAGRANGE's approach to the gas dynamic model.
*Top*: Before start of motion (t = 0).
*Bottom*: At time t. The infinitesimal segment that was initially at x has now moved to z. t and x are the independent variables.

distance z from a fixed transverse surface (see Fig. 3.18). It is required to find the velocity v, density ρ, pressure p and position z of this segment as a function of the two independent variables x and t. For velocity we can write:

$$v = \frac{dz}{dt}, \qquad [\text{m/s}]$$

and for the instantaneous thickness of the segment

$$dx' = \frac{dz}{dx} \cdot dx \, . \qquad [\text{m}]$$

Because of the conservation of mass regarding the segment, its density ρ is given by

$$(3.4{:}12) \qquad \rho = \rho_0 \cdot \left(\frac{dz}{dx}\right)^{-1}, \qquad [\text{kg/m}^3]$$

where $\rho_0$ = density at time t = 0. LAGRANGE used the following expression for pressure p:

$$(3.4{:}13) \qquad p = \Psi_0 \cdot \left(\frac{dz}{dx}\right)^{-n}, \qquad [\text{N/m}^2]$$

where $\Psi_0$ = the volumetric energy density of the gas ("force élastique du gaz") before the projectile begins to move. He also extended BOYLE's law; rather than suppose that barrel pressure is directly proportional to density, he added an exponent n to render the equation more generally valid.

By so doing, he to some degree anticipated the isentropic relationship between pressure and density.

Applying D'ALEMBERT's principle to the segment gives us its equation of motion:

$$\frac{d^2z}{dt^2} \cdot \rho_0 \cdot A_b \cdot dx = A_b \cdot p - A_b \cdot \left(p + \frac{dp}{dx} \cdot dx\right), \qquad [\text{N}]$$

$$(3.4{:}14) \qquad \frac{d^2z}{dt^2} + \frac{1}{\rho_0} \cdot \frac{dp}{dx} = 0 \, . \qquad [\text{m/s}^2]$$

Substituting the derivative of Eqn 3.4:13 and combining all constants in $k^2$ yields the following differential equation:

$$(3.4:15)\qquad \frac{d^2z}{dt^2} = k^2 \cdot \frac{d^2z}{dx^2} \cdot \left(\frac{dz}{dx}\right)^{-(n+1)} . \qquad [m/s^2]$$

Fully describing the process in the barrel also requires the two equations of motion for the projectile and the barrel. If in addition to the variables in Fig. 3.18 we consider the mass of the projectile $m_p$ and of the barrel $m_b$, together with the pressure at the base of the projectile $p_p$ and that at the breech face $p_b$, we obtain

$$m_p \cdot \frac{d^2y_p}{dt^2} = A_b \cdot p_p , \qquad [m/s^2]$$

and

$$m_b \cdot \frac{d^2y_b}{dt^2} = -A_b \cdot p_b . \qquad [m/s^2]$$

$p_p$ and $p_b$ are given by Eqn 3.4:13, replacing z by $y_p$ and $y_b$:

$$(3.4:16)\qquad m_p \cdot \frac{d^2y_p}{dt^2} = A_b \cdot \Psi_0 \cdot \left(\frac{dy_p}{dx}\right)^{-n} , \qquad [m/s^2]$$

$$(3.4:17)\qquad m_b \cdot \frac{d^2y_b}{dt^2} = A_b \cdot \Psi_0 \cdot \left(\frac{dy_b}{dx}\right)^{-n} . \qquad [m/s^2]$$

The initial conditions for Eqns 3.4:15–17 were $t = 0$ and $z = x$, with velocity $= 0$ for gas, projectile and barrel.

By formulating Eqns 3.4:15–17, LAGRANGE had produced a system of equations that simultaneously described the motion of the gas, the projectile and the barrel, thereby establishing the basic principle of the gas dynamic model of interior ballistics. In attempting to solve this system he made an assumption that has proved very useful since; he postulated that the velocity of the gas in the barrel would increase linearly with distance (the "Lagrange approximation").

A number of other famous scientists studied LAGRANGE's ballistic problem during the course of the 19th Century, in addition to POISSON. German mathematician B. RIEMANN made an important contribution from a mathematical point of view, with his research into the differential equations of the motion of gas and their integration (B. RIEMANN, 1860).

RIEMANN did not address the interior ballistics problem in this work, rather describing the one-dimensional motion of a gas in a geometrical cylinder with no piston and no exchange of heat. From his point of view, the value of his publication lay in its potential applications to acoustics and in its theoretical treatment of non-linear partial differential equations.

The system of differential equations underlying this work consisted of the equation of the conservation of mass and the equation of motion, and took the following form (using the original notation, with u representing velocity):

$$\text{(3.4:18)} \qquad \frac{d\rho}{dt} = -\frac{d(\rho \cdot u)}{dx}, \qquad [\text{kg}/(\text{m}^3\cdot\text{s})]$$

$$\text{(3.4:19)} \qquad \rho \cdot \left(\frac{du}{dt} + u \cdot \frac{du}{dx}\right) = -\frac{dp(\rho)}{d\rho} \cdot \frac{d\rho}{dx}. \qquad [\text{N}/\text{m}^3]$$

RIEMANN succeeded in demonstrating that this system of equations could be fully integrated, on condition that the initial motion of all parts occurred in the same direction and that pressure and velocity were constant in any plane perpendicular to that direction. In his discussion of solutions, it transpired that expansion and compression waves of finite amplitude occur, causing pressure differences. This result was decisive as regards the interior ballistics aspects of this study and has formed an integral part of the gas dynamics aspect of interior ballistics ever since.

Over his brief lifetime (he died at the age of only 36) French physicist and ballistician P.-H. HUGONIOT focused on gas diffusion in tubes. He conducted a fundamental study on shock wave theory and applied the knowledge gained to LAGRANGE's ballistic problem. It is therefore not surprising that the general wave equation (P.-H. HUGONIOT, 1889) formed the precursor to the study of this problem, which was published posthumously. He used this to derive the following equation for the motion of the projectile:

$$\text{(3.4:20)} \qquad \frac{\partial^2 u}{\partial t^2} = \frac{\gamma \cdot p_0}{\rho_0} \cdot \left(1 + \frac{\partial u}{\partial x}\right)^{-\gamma-1} \cdot \frac{\partial^2 u}{\partial x^2}. \qquad [\text{m}/\text{s}^3]$$

Where $p_0$ = pressure before the start of motion, $\rho_0$ = density before the start of motion and $\gamma$ = heat capacity ratio.

HUGONIOT started by examining the initial motion of the projectile and the expansion wave as it starts to move towards the breech face. His approach, which to some extent already used the method of characteristics, was expanded upon significantly by the work of F.-F. GOSSOT and R. LIOUVILLE (1914) and then in particular that of A. E. H. LOVE and F. B. PIDDUCK (1922). Consideration was now also given to the waves reflected from the breech face to the base of the projectile, generating compound waves.

All of these calculations took place under the following assumptions:

- a The tube is smooth and cylindrical, without a wider chamber.
- b The propellant is completely burnt before the projectile starts to move.
- c Initial pressure, density and temperature are uniform.
- d The gases expand isentropically.
- e Covolume remains constant throughout the process.
- f Friction between projectile and barrel is zero.

Furthermore, the heat capacity ratio must be of the form

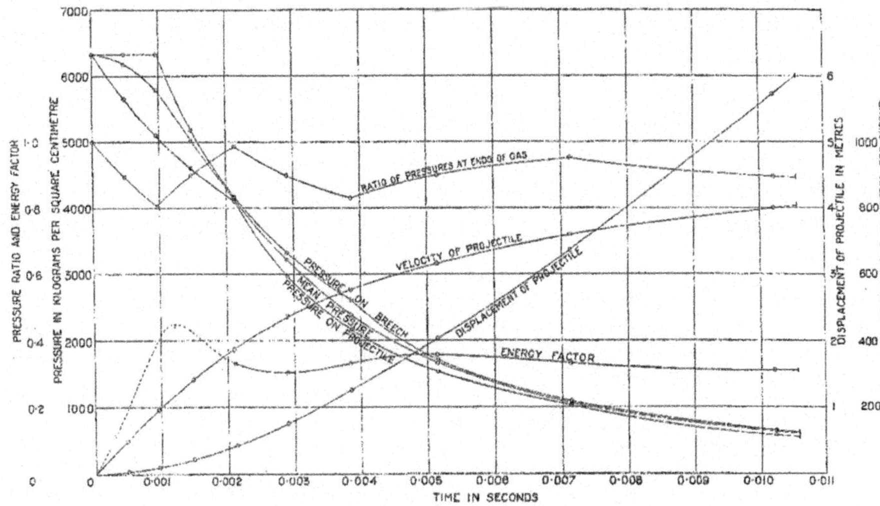


**Fig. 3.19.** Pressure, velocity and displacement curves as a function of time, from calculations performed by LOVE (LOVE and PIDDUCK, 1922).

$$(3.4{:}21) \qquad \gamma = \frac{2\cdot n+1}{2\cdot n-1}, \qquad \text{where n is a positive integer;} \qquad [\text{-}]$$

an approximation that is only valid for certain gases.

In the paper cited above, LOVE derives a generally-applicable equation which – under these conditions – provides solutions for pressure and velocity, if the necessary data are supplied. Such calculations were extremely labour intensive, however. Attempts were therefore made to find solutions using the method of characteristics. The graphical variant of this method made it possible to obtain results more quickly (see also P. DE HALLER, 1948).

Even at the time, the question was raised as to whether it was worth investing so much effort in calculations that required the use of such restrictive assumptions that the results were far removed from reality. This was most apparent as regards pressure at the beginning of the process; the requirement that the propellant be fully burnt before the projectile begins to move was completely at odds with what was known to occur (see Fig. 3.19). The work of LOVE and PIDDUCK has since been superseded.

HUGONIOT had already pointed out (H. HUGONIOT, 1889) that this assumption would lead to large errors, especially with slow-burning propellants, and PIDDUCK maintained, in LOVE and PIDDUCK (1922), Article 41, that "it appears hopeless to attempt to allow for the gradual burning of the propellant which occurs in actual guns."

Nonetheless, their approach was of value to the study of interior ballistics. Their results revealed the existence of hitherto unknown phenomena such as expansion waves and pressure gradients between breech face and projectile, plus the influence of the mass of gas that is also set in motion. However, they were of little use in the practical design of artillery.

If one makes the assumptions listed at a–f above, one always ends up with equations that can only be solved by means of approximation. LAGRANGE had already noted this in his original text. A breakthrough, in the sense of realistic results and physical questions, only became possible in the second half of the 20th Century, with the arrival of efficient numerical methods and the computing power that they required.

It is interesting to note that a similar development was underway at the same period in exterior ballistics, the second major sub-domain of ballistics. Here too, progress in numerical calculation methods made it possible to shift the focus from the mathematical processes for solving equations to the physical questions themselves.

#### 3.4.5.2 Current situation

*This section is based on documents supplied by Dr Rudi HEISER, Weil am Rhein, Germany.*

The opportunities presented today by numerical calculation methods, coupled with the increased power of computers, have made it possible to extend the gas dynamic models of interior ballistics considerably, in particular by eliminating some of the restrictions mentioned in the previous section. The most important step was being able to include the burning phase, rather than making assumption b (see p. 88). Not only gases but also solid grains of propellant influence the physical process in the space between the breech face and the projectile, and these must be taken into account. This led to the idea of a two-phase model.

To create such a model, one needs a new variable that corresponds to the ratio of the gas volume to the total volume, the latter consisting of the gas volume plus the volume of the load. This is known as the *porosity* $\zeta$ (generally indicated by $\alpha$ or $\varepsilon$).

$$\text{(3.4:22)} \qquad \zeta = \frac{V_g}{V_c + V_g} \,. \qquad [-]$$

This makes it possible to formulate a complete three-dimensional set of partial differential equations for the interior ballistics process, using the basic laws of conservation of mass, momentum and energy. In the following equations, those with a number ending in a apply to the (compressible) gas phase, while those ending in b apply to the (incompressible) propellant phase. For symbols, please see Table 3.7.

Equation of gas production:

$$\text{(3.4:23)} \qquad \dot{m}_g = (1-\zeta)\cdot\frac{A_c}{V_c}\cdot\rho_c\cdot\dot{r} \,. \qquad [\text{kg}/(\text{m}^3\cdot\text{s})]$$

Continuity equation:

(3.4:24a) $$\frac{\partial(\zeta\cdot\rho_g)}{\partial t}+\vec{\nabla}(\zeta\cdot\rho_g\cdot\vec{u}) = \dot{m}_g+\dot{m}_i\ ,$$ [kg/(m³·s)]

(3.4:24b) $$\rho_c\cdot\frac{\partial}{\partial t}(1-\zeta)+\rho_c\cdot\vec{\nabla}\left[(1-\zeta)\cdot\vec{u}_c\right] = -\dot{m}_g\ .$$ [kg/(m³·s)]

Conservation of momentum:

(3.4:25a) $$\frac{\partial(\zeta\cdot\rho_g\cdot\vec{u}_g)}{\partial t}+\vec{\nabla}(\zeta\cdot\rho_g\cdot\vec{u}_g\cdot\vec{u}_g)+\zeta\cdot\vec{\nabla}p =$$

$$= -D+\dot{m}_g\cdot(\vec{u}_c-\vec{u}_g)+\dot{m}_i\cdot\vec{u}_i\ ,$$ [N]

(3.4:25b) $$\rho_c\cdot\frac{\partial}{\partial t}\left[(1-\zeta)\cdot\vec{u}_c\right]+\rho_c\cdot\vec{\nabla}\left[(1-\zeta)\cdot\vec{u}_c\cdot\vec{u}_c\right]+(1-\zeta)\cdot\vec{\nabla}p =$$

$$= -D+\dot{m}_g\cdot\vec{u}_c-\vec{F}_c(\zeta)\ .$$ [N]

Conservation of energy:

(3.4:26) $$\frac{\partial(\zeta\cdot\rho_g\cdot e_g)}{\partial t}+\vec{\nabla}(\zeta\cdot\rho_g\cdot e\cdot\vec{u}_c)+\zeta\cdot p\cdot\vec{\nabla}\vec{u}_c =$$

$$= \dot{m}_c\cdot\left(Q+\tfrac{1}{2}\cdot\vec{u}_g\cdot\vec{u}_g\right)+\dot{m}_i\cdot Q_i+\vec{D}\cdot(\vec{u}_c-\vec{u}_g)\ ,$$ [N]

where $$Q = Q_{ex}+e_c+\tfrac{1}{2}\cdot\vec{u}_c\cdot\vec{u}_c-\vec{u}_c\cdot\vec{u}_g\ .$$ [J/kg]

A number of elements are required to fully describe the complete interior ballistics process. The most important of these are as follows:

- The burning law
- The equation of state for gases
- The equation of motion for the projectile
- The friction between projectile and barrel
- Heat transfer

**Table 3.7.** Variables used in the general gas dynamics model

| | | | | | |
|---|---|---|---|---|---|
| $\zeta$ | [–] | Porosity | $A_c$ | [m²] | Propellant grain surface area |
| c | – | Index for load | V | [m³] | Volume |
| g | – | Index for gas | $\dot{r}$ | [m/s] | Burning rate |
| $\rho$ | [kg/m³] | Density | D | [N] | Drag of propellant grains |
| $\vec{u}$ | [m/s] | Velocity | Q | [J/kg] | Energy per unit mass |
| $\dot{m}_g$ | [kg/(m³·s)] | Gas production rate | $\vec{F}_c$ | [N/m²] | Intergranular pressure |
| e | [J/kg] | Internal energy | i | – | Index for primer composition |

The calculation must also take into account the drag of the propellant grains and the interactions between them, the laws of conservation of momentum and energy for the detonating agent and the energy criterion for ignition of the main charge.

Working on this general basis, one can create a multitude of models, of various degrees of complexity, by reducing the dimensions and by (justifiably) disregarding some of the physical conditions listed above. For example, two-dimensional models (symmetrical about the axis) for two phases, with variable internal geometry (chamber and rear of projectile) and thermal conduction are already under development and in some cases have been successfully completed.

The gas dynamics models provide a more realistic pressure distribution and include the formation of pressure waves, unlike the thermodynamic models, in which gas pressure at the breech face is always lower than at the base of the projectile. They make it possible to simulate phenomena related to ignition and the structure of the load, which in turn makes it possible to assess the safety of a system. Such models are used primarily for the interior ballistics aspects of high-power gun design and for applications involving large propellant masses.

### 3.4.6 HEYDENREICH's method

Given today's numerical methods, determining pressure and velocity as a function of time would present no problems, were it not for the need to supply a large quantity of data. These are often difficult to obtain and virtually impossible to measure using simple equipment. This probably accounts for the fact that an empirical method – that of W. HEYDENREICH (1908) – is still in use.

HEYDENREICH's method is based on an analytical function, with which E. VALLIER (1899) first attempted to present pressure as a function of time, using measurements of maximum pressure and of muzzle velocity.

(3.4:27) $$p(t) = K_A \cdot t \cdot e^{-K_B \cdot t} \,. \qquad \text{[bar]}$$

While VALLIER made additional assumptions regarding the shape of the gas pressure curve, and attempted to obtain the constants $K_A$ and $K_B$ by that means, HEYDENREICH used a large quantity of empirical data, which he obtained from recoil measurements. He formulated the supposition that the ratio of $p_x$ at a given point x along the tube to maximum pressure $p_{max}$ was always the same function of the variables $x/x_{pmax}$, where $x_{pmax}$ = position of maximum pressure. For the first step of the calculation, he used the pressure ratio $\chi$:

(3.4:28) $$\chi = \frac{p_m}{p_{max}} \,. \qquad \text{[-]}$$

$p_m$ is the *mean pressure*, which here means the constant pressure that would be needed to accelerate the projectile and half the propellant mass (bearing in mind

SÉBERT's factor) to the same muzzle velocity as that resulting from the true pressure curve.

If maximum gas pressure is unknown, and cannot be measured, the necessary values for small arms cartridges can be found in weapon and reloading manuals or in the C.I.P. standards (2007). The precondition is that the cartridge utilizes the maximum gas pressure, at least approximately.

The structure of HEYDENREICH's equations is very simple and consistent. The desired value is obtained by multiplying the known value by the appropriate factor, which depends on the pressure ratio $\chi$. The values of these factors can be obtained from the empirically-derived functions $F_1$–$F_5$. See Table 3.8.

The equations are divided into two groups: those for the position of maximum pressure and those for the muzzle. The following equations give the displacement of the projectile $x_{pmax}$, time $t_{pmax}$ and projectile velocity $v_{pmax}$ at the time of maximum pressure (where $\ell_r$ = barrel length and $v_0$ = muzzle velocity:

(3.4:29a) $$x_{pmax} = \ell_r \cdot F_1(\chi) \quad [m]$$

(3.4:29b) $$t_{pmax} = \frac{2 \cdot \ell_r}{v_0} \cdot F_2(\chi) \quad [ms]$$

(3.4:29c) $$v_{pmax} = v_0 \cdot F_3(\chi) \quad [m/s]$$

The following equations give muzzle pressure $p_\mu$ and the time taken for the projectile to pass through the barrel $t_\ell$:

**Table 3.8.** Functions $F_1$–$F_5$ for HEYDENREICH's equations. The table below uses values obtained from testing, rather than those originally proposed by HEYDENREICH, as they correspond more closely to the characteristics of today's propellants (see OERLIKON Taschenbuch, 1981).

| $\chi$ | $F_1$ | $F_2$ | $F_3$ | $F_4$ | $F_5$ |
|---|---|---|---|---|---|
| 0.250 | 0.0313 | 0.139 | 0.324 | 0.216 | 0.725 |
| 0.275 | 0.0357 | 0.155 | 0.328 | 0.228 | 0.743 |
| 0.300 | 0.0402 | 0.172 | 0.333 | 0.242 | 0.762 |
| 0.325 | 0.0450 | 0.189 | 0.338 | 0.260 | 0.781 |
| 0.350 | 0.0500 | 0.207 | 0.343 | 0.278 | 0.800 |
| 0.375 | 0.0552 | 0.225 | 0.348 | 0.291 | 0.818 |
| 0.400 | 0.0608 | 0.244 | 0.354 | 0.304 | 0.836 |
| 0.425 | 0.0666 | 0.264 | 0.360 | 0.321 | 0.854 |
| 0.450 | 0.0729 | 0.284 | 0.366 | 0.340 | 0.873 |
| 0.475 | 0.0799 | 0.305 | 0.373 | 0.359 | 0.891 |
| 0.500 | 0.0877 | 0.326 | 0.380 | 0.382 | 0.910 |
| 0.525 | 0.0966 | 0.348 | 0.388 | 0.415 | 0.930 |
| 0.550 | 0.1059 | 0.370 | 0.396 | 0.454 | 0.950 |

(3.4:30a) $p_\mu = p_m \cdot F_4(\chi)$ [bar]

(3.4:30b) $t_\ell = \frac{2 \cdot \ell_r}{v_0} \cdot F_5(\chi)$ . [ms]

A further set of HEYDENREICH's empirical functions makes it possible to calculate pressure, velocity and time as a function of the displacement of the projectile. The parameter used here is the ratio of instantaneous displacement to displacement at the time of maximum pressure (from Eqn 3.4:29a):

(3.4:31) $\lambda = \frac{x}{x_{pmax}}$ . [-]

Coefficients $v_{pmax}$ and $t_{pmax}$ are calculated using Eqns 3.4:29c and 3.4:29b. The values of the empirical functions $F_6$–$F_8$ are to be found in Table 3.9.

(3.4:32a) $p_x = p_{max} \cdot F_6(\lambda)$ [bar]

(3.4:32b) $v_x = v_{pmax} \cdot F_7(\lambda)$ [m/s]

(3.4:32c) $t_x = t_{pmax} \cdot F_8(\lambda)$ . [ms]

If one knows the maximum pressure and the muzzle velocity for a given cartridge at a given barrel length, one can use HEYDENREICH's method to plot the curves of pressure and velocity along the entire travel of the projectile. This makes it possible to estimate muzzle velocity and muzzle pressure for any other barrel length.

**Table 3.9.** Functions $F_6$–$F_8$ for HEYDENREICH's equations.

| $\lambda$ | $F_6$ | $F_7$ | $F_8$ | $\lambda$ | $F_6$ | $F_7$ | $F_8$ |
|---|---|---|---|---|---|---|---|
| 0.25 | 0.741 | 0.392 | 0.610 | 8.00 | 0.284 | 2.445 | 2.309 |
| 0.50 | 0.912 | 0.635 | 0.780 | 9.00 | 0.248 | 2.509 | 2.451 |
| 0.75 | 0.980 | 0.834 | 0.903 | 10.00 | 0.220 | 2.566 | 2.589 |
| 1.00 | 1.000 | 1.000 | 1.000 | 11.00 | 0.199 | 2.615 | 2.725 |
| 1.25 | 0.989 | 1.140 | 1.081 | 12.00 | 0.181 | 2.659 | 2.858 |
| 1.50 | 0.965 | 1.262 | 1.154 | 13.00 | 0.164 | 2.702 | 2.988 |
| 1.75 | 0.932 | 1.366 | 1.219 | 14.00 | 0.150 | 2.740 | 3.116 |
| 2.00 | 0.898 | 1.468 | 1.282 | 15.00 | 0.137 | 2.777 | 3.253 |
| 2.50 | 0.823 | 1.632 | 1.394 | 16.00 | 0.125 | 2.811 | 3.390 |
| 3.00 | 0.747 | 1.763 | 1.495 | 17.00 | 0.117 | 2.837 | 3.502 |
| 3.50 | 0.675 | 1.875 | 1.589 | 18.00 | 0.109 | 2.862 | 3.618 |
| 4.00 | 0.604 | 1.983 | 1.682 | 19.00 | 0.102 | 2.887 | 3.740 |
| 4.50 | 0.546 | 2.068 | 1.769 | 20.00 | 0.096 | 2.910 | 3.816 |
| 5.00 | 0.495 | 2.140 | 1.851 | 25.00 | 0.073 | 3.003 | 4.455 |
| 6.00 | 0.403 | 2.269 | 2.012 | 30.00 | 0.058 | 3.075 | 5.031 |
| 7.00 | 0.338 | 2.363 | 2.163 | 35.00 | 0.048 | 3.162 | 5.657 |

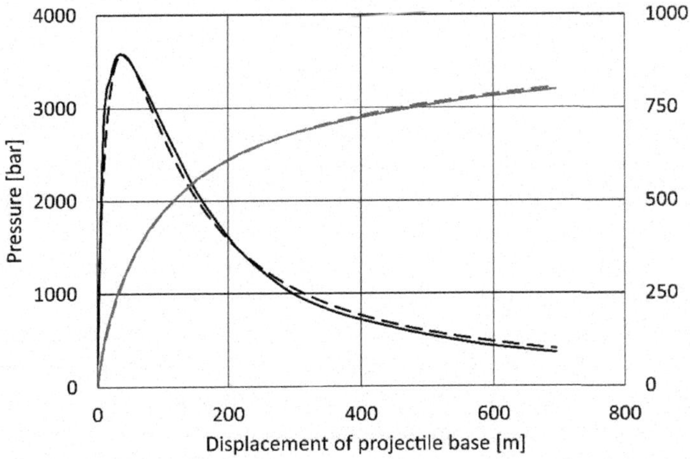


**Fig. 3.20.** Comparison between pressure-displacement and velocity-displacement curves calculated using HEYDENREICH's equations (solid lines) and those obtained by measurement (dashed lines). Cartridge: 7.5 × 55.

The basis of HEYDENREICH's method is purely empirical, and the method uses a fixed form of pressure curve, so one should not expect too much in terms of accuracy. Nevertheless, examples show that if one carries out calculations for known systems using the method, the results can correspond surprisingly closely to the measured values, especially if maximum pressure and muzzle velocity correspond to the true values (see Fig. 3.20, 7.5 × 55). For a smaller calibre (5.6 × 45), accuracy is poorer (see Fig. 3.21).

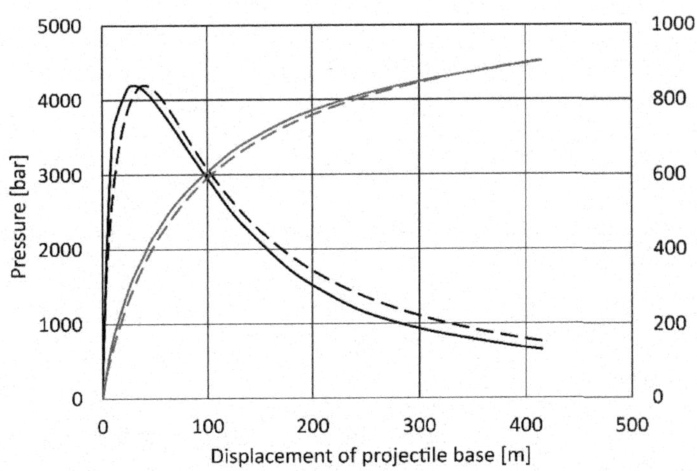


**Fig. 3.21.** Comparison between pressure-displacement and velocity-displacement curves calculated using HEYDENREICH's equations (solid lines) and those obtained by measurement (dashed lines). Cartridge: 5.6 × 45.

### 3.4.7 The open gun tube

#### 3.4.7.1 General

The interior ballistics of gas and alarm pistols, and of saluting guns, differs from that of conventional firearms most of all in that no projectile is involved. This means that there is no delimited combustion chamber. Furthermore, gas escapes via the muzzle as soon as barrel pressure starts to rise.

In order for propellant to ignite properly, it needs pressure to rise rapidly at the start of the process, which implies that it needs to be enclosed. In the case of blank cartridges, this is generally achieved by sealing or crimping the opening of the cartridge case. Saluting cartridges are generally closed off with a light material that does not burn easily, such as cork. Once the pressure in the cartridge case is sufficiently high for the cover to burst, or for the wadding to begin to move, both unburnt and partly-burnt powder is ejected into the barrel, from which point onwards the space available for combustion equates to the volume of the whole barrel. As a result, maximum gas pressure may occur anywhere in the barrel, and not necessarily in or near the cartridge.

Cartridges for gas and alarm pistols generally use nitrocellulose powder, whereas saluting cartridges almost exclusively use black powder.

#### 3.4.7.2 The extended system of equations

Taking as a basis the thermodynamic model of interior ballistics for tube weapons, we can derive the equations below for the open tube. These equations can be used for gas and alarm pistols and for saluting guns, as long as they use nitrocellulose powder.

We can use the burning law from Eqn 3.4:1a. The volume of the combustion chamber is basically the entire volume of the barrel, minus the volume of unburnt propellant, the covolume of the propellant gas and the volume of any baffles $V_{bf}$.

(3.4:33) $$V = A \cdot \ell_r - \frac{m_c}{\rho_c} \cdot (1 - z) - m_i \cdot \eta - V_{bf} \,. \qquad [\mathrm{m}^3]$$

Where $\ell_r$ = barrel length, $m_c$ = propellant mass, $\rho_c$ = propellant density, $\eta$ = covolume of propellant gases and $V_{bf}$ = volume of baffles (if present).

The mass of the gas in the barrel $m_i$ is the mass of propellant that has been burnt, minus the mass of the gas that has already left the muzzle $m_a$:

(3.4:34) $$m_i = m_c \cdot z - m_a \,. \qquad [\mathrm{kg}]$$

As the pressure in the barrel is almost always significantly higher than the pressure outside, one can always assume maximum mass flow rate (at Laval pressure):

$$\dot{m}_a = A \cdot \sqrt{\gamma \cdot p_i \cdot \frac{m_i}{V} \cdot \left(\frac{2}{\gamma+1}\right)^{\frac{\gamma+1}{\gamma-1}}} \; . \tag{3.4:35} \quad \text{[kg/s]}$$

The energy balance must take account of the pressure energy, the kinetic energy of the unburnt propellant particles, the kinetic energy of the gas flow in the barrel and the kinetic energy of the gas leaving the muzzle:

$$m_c \cdot z \cdot Q_{ex} = \frac{p_i \cdot V}{\gamma - 1} + \tfrac{1}{2} \cdot (m_c - m_a) \cdot v_i^2 + \tfrac{1}{2} \cdot m_a \cdot \frac{2 \cdot \gamma}{\gamma + 1} \cdot \frac{p_i \cdot V}{m_i} \; . \tag{3.4:36} \quad \text{[J]}$$

$\gamma$ = heat capacity ratio, $Q_{ex}$ = combustion heat of the propellant, $m_a$ = mass of gas that has already left the muzzle and $v_i$ = velocity of the gas in the barrel.

Both the unburnt propellant particles and the gas remaining in the barrel are in motion:

$$\dot{v}_i = \frac{\Delta p_i \cdot A}{m_i} \tag{3.4:37} \quad \text{[m/s}^2\text{]}$$

where $\Delta p_i$ = pressure gradient within the barrel.

The burning law, Eqns 3.4:33–36, the equation of motion (3.4:37) and that for displacement together form the following set of differential equations (where $p_m$ = Laval pressure):

$$\dot{z} = \phi(z) \cdot B_A \cdot \left(\frac{p}{p_0}\right)^n , \tag{3.4:38a} \quad \text{[1/s]}$$

$$\dot{m}_a = A \cdot \sqrt{\gamma \cdot p_i \cdot \frac{m_c \cdot z - m_a}{V} \cdot \left(\frac{2}{\gamma+1}\right)^{\frac{\gamma+1}{\gamma-1}}} , \tag{3.4:38b} \quad \text{[kg/s]}$$

$$\dot{v}_i = \frac{(p_i - p_m) \cdot A}{m_c \cdot z - m_a} , \tag{3.4:38c} \quad \text{[m/s}^2\text{]}$$

$$\dot{x} = v , \tag{3.4:38d} \quad \text{[m/s]}$$

$$p_i = \frac{m_c \cdot z \cdot Q_{ex} - \tfrac{1}{2} \cdot (m_c - m_a) \cdot v_i^2}{V \cdot \left(\frac{1}{\gamma - 1} + \frac{\gamma}{\gamma + 1} \cdot \frac{m_a}{m_c \cdot z - m_a}\right)} , \tag{3.4:39} \quad \text{[N/m}^2\text{]}$$

Eqn 3.4:39 is used to determine the internal pressure, which must be calculated after each integration step. Gas volume is given by Eqn 3.4:33.

This system of equations can be simplified by assuming that the velocity of the propellant particles and of the gases in the barrel is low, and can therefore be ignored. Under those conditions, one does not need Eqns 3.4:38c and 3.4:38d, nor the second addend in the numerator of Eqn 3.4:39. This assumption stems from measurement results, which show that maximum pressure occurs between the mouth of the cartridge and the middle of the barrel (depending on the design of the cartridge) and that the pressure differences within the barrel are relatively small.

The system of differential equations (3.4:38a–d) and Eqn 3.4:39 provide only a partial description of the phenomena that occur in an open tube. Inaccuracies arise particularly as a result of the following assumptions:

- The system of equations for the interior ballistics of guns assumes a delimited combustion chamber, closed off by the projectile. In the case of an open tube, the volume of the space that determines the rise in pressure is not known exactly. As an approximation, one initially uses the volume of half the length of the barrel (as burning starts at the rear). As the gases accelerate, this volume increases, eventually corresponding to the volume of the entire barrel.
- It is only possible to plot pressure for the entire volume available at any given moment. It is not possible to plot pressure as a function of displacement.
- The equations for the loss of mass due to outflow are only valid for a large vessel in which pressure and density are constant. For sufficiently short time steps, this is an acceptable approximation.

Despite the above, typical propellant data yield values that correspond approximately to those found in practice, with the exception of the time from primer firing to muzzle exit. The model is therefore suitable for parameter studies.

### 3.4.7.3 Gas pressure in an open gun tube

Gas pressure in an open tube as a function of time can be determined using simple meter tubes and blank cartridges, recording the pressure simultaneously using piezo-electric sensors at the mouth of the cartridge, the middle of the tube and just before the mouth of the tube. Figures 3.22–25 show typical examples of such measurements.

A number of phenomena stand out in the results of such experiments, particularly as regards the time of maximum pressure and the pressure just before the

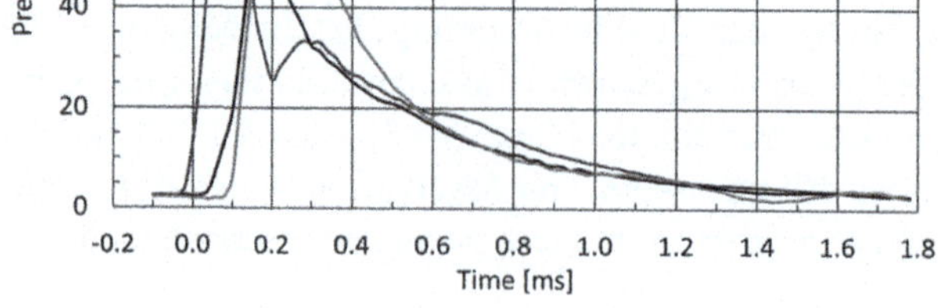


**Fig. 3.22.** Pressure curves for an 8 mm blank cartridge.
1: At the mouth of the cartridge.
2: In the middle of the tube.
3: 1 cm before the mouth of the tube.
Length of tube: 20 cm.
Pressure in the middle of the tube is lower than at the mouth of the cartridge and the mouth of the tube.

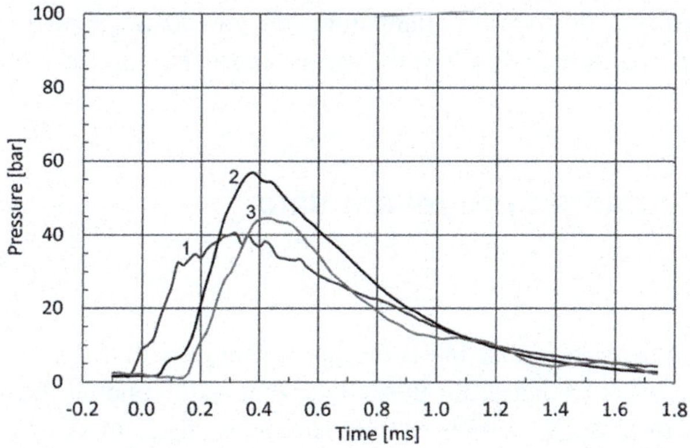


**Fig. 3.23.** Pressure curves for a 9 mm PA blank cartridge.
1: At the mouth of the cartridge.
2: In the middle of the tube.
3: 1 cm before the mouth of the tube.
Length of tube: 20 cm. Maximum pressure occurs in the middle of the tube. Pressure is higher at the mouth of the tube than at the mouth of the cartridge.

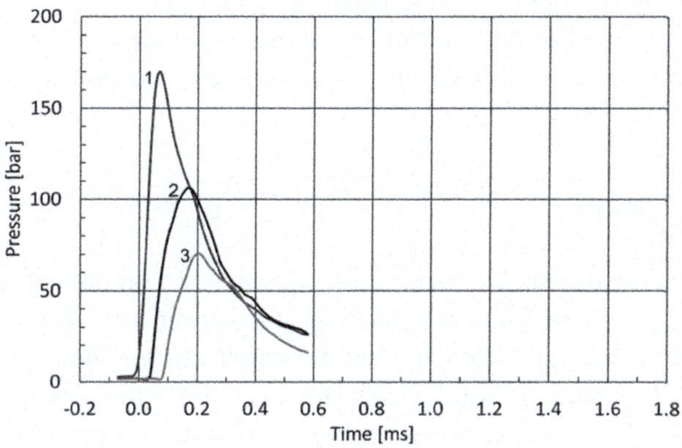


**Fig. 3.24.** Pressure curves for a 380 R (nitrocellulose powder) blank cartridge
1: At the mouth of the cartridge.
2: In the middle of the tube.
3: 1 cm before the mouth of the tube.
Length of tube: 20 cm. Maximum pressure occurs at the mouth of the cartridge, decreasing towards the mouth of the tube. Note the different scale used for the y axis.

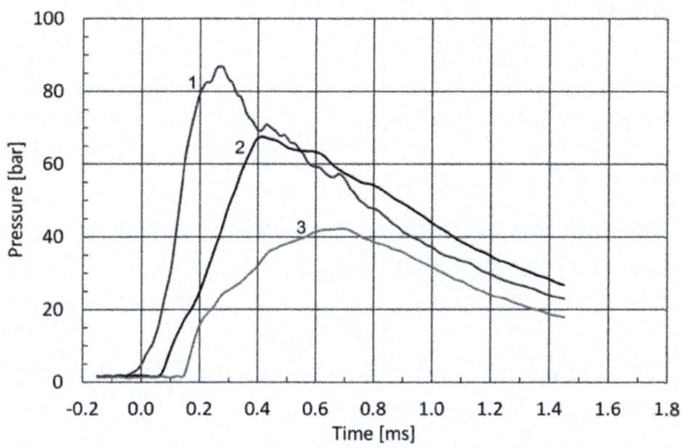


**Fig. 3.25.** Pressure curves for a 380 R (black powder) blank cartridge.
1: At the mouth of the cartridge.
2: In the middle of the tube.
3: 1 cm before the mouth of the tube.
Length of tube: 20 cm. Maximum pressure occurs at the mouth of the cartridge. The process takes significantly longer than in the case of nitrocellulose propellant.

mouth of the tube. In two instances, the latter is as high as – or higher than – the pressure at the mouth of the cartridge. Fluctuations indicate that in all cases expansion and compression waves are passing through the tube, as postulated by RIEMANN (see 3.4.5.1). The pressure becomes uniform over time, except in the case of the black powder cartridge.

The differences between nitrocellulose propellant and black powder are apparent from Figs 3.24 and 3.25; the two cartridges are of the same calibre, but loaded with different propellants.

## 3.5 Obtaining the required propellant data

### 3.5.1 General

Interior ballistics calculations are only possible if the propellant data in Table 3.6 are available. There are two ways of acquiring these data. One is to conduct thermodynamic calculations of the transformation of explosive substances (J. AKHAVAN, 2008, J. KÖHLER et al. 2008). The other is to carry out experiments.

Most propellant characteristics can be determined using a *manometric bomb* (also known as a *ballistic bomb*). A *Crawford bomb* can be used to measure the burning rate.

### 3.5.2 The manometric bomb

The basic principle of a manometric bomb is that a certain quantity of propellant is burned in a fixed volume $V_{mb}$, with maximum pressure $p_{mb}$ occurring at the end of burning. Because of the size of the gas molecules, and the pressure, one should in theory use the VAN DER WAALS equation (Eqn 2.3:2a). However, given the pressure levels involved, one can ignore the internal pressure. We can therefore express the pressure in a manometric bomb as

(3.5:1) $$p_{mb} \cdot (V_{mb} - \eta \cdot m_c) = m_c \cdot R \cdot T_{ex} \, . \qquad [J]$$

Where $m_c$ = propellant mass, $\eta$ = covolume, R = gas constant and $T_{ex}$ = temperature of the gases of combustion (*flame temperature* or *explosion temperature*), which can be determined empirically (but only with considerable effort) or calculated thermodynamically.

The product of gas constant and flame temperature is a propellant constant, the *specific energy* $F_{se}$ (previously known as the propellant force).

(3.5:2) $$F_{se} = R \cdot T_{ex} \, . \qquad [J/kg]$$

This is the pressure energy that is generated by the burning of 1 g of propellant and is available for transformation into mechanical energy.

If we divide Eqn 3.5:1 by the mass, include the *loading density* $\Delta$ ($m_c/V_{mb}$) and take account of Eqn 3.5:2 we obtain

(3.5:3) $$p_{mb} = \frac{\Delta \cdot F_{se}}{1 - \Delta \cdot \eta} \, . \qquad [Pa]$$

This equation was developed by English chemist F. A. ABEL and published in 1874, a year after the van der Waals equation. It is known in ballistics as *ABEL's equation*.

A manometric bomb consists of a thick-walled steel vessel with a defined internal volume, closed by a threaded plug (see Fig. 3.26). A certain quantity of the propellant under investigation is placed in the vessel. The vessel is then sealed and the propellant is ignited using an electrical igniter. The pressure indicated by the piezo-electric pressure sensor is recorded over time.

Eqn 3.5:3 contains two unknown variables, $F_{sp}$ and $\eta$, in addition to the two that can already be determined, $p_{mb}$ and $\Delta$. If pressure is measured in the manometric bomb using two different loading densities $\Delta_1$ and $\Delta_2$, giving us pressures $p_{mb1}$ and $p_{mb2}$, we obtain two equations that enable us to calculate the two unknown variables. These equations are as follows:

$$(3.5:4) \qquad F_{se} = \frac{(\Delta_2 - \Delta_1) \cdot p_{mb1} \cdot p_{mb2}}{\Delta_1 \cdot \Delta_2 \cdot (p_{mb1} - p_{mb2})}, \qquad [J/kg]$$

$$(3.5:5) \qquad \eta = \frac{\Delta_2 \cdot p_{mb1} - \Delta_1 \cdot p_{mb2}}{\Delta_1 \cdot \Delta_2 \cdot (p_{mb1} - p_{mb2})}. \qquad [m^3/kg]$$

A degree of scatter in the measurement results is inevitable, so measurements are carried out at a number of loading densities. Specific energy and covolume are then calculated using linear regression, on the basis of the following equation, which is derived from Eqn 3.5:3.

$$(3.5:6) \qquad \frac{1}{p_{mb}} = \frac{1}{F_{se}} \cdot \frac{1}{\Delta} - \frac{\eta}{F_{se}}. \qquad [1/Pa]$$

The reciprocal values from the measurements ($\Delta$, $p_{mb}$) form pairs of values, the gradients of the regression curves give the reciprocal value of the specific energy and this is used to calculate the covolume from the constant term.

Recording pressure as a function of time enables us to determine other propellant characteristics. Let z(t) be the proportion of the load that has been transformed. We can then find gas pressure at time t from

$$(3.5:7) \qquad p(t) = \frac{z(t) \cdot \Delta \cdot F_{se}}{1 - \eta \cdot z(t) \cdot \Delta - (1 - z(t)) \cdot \Delta / \rho_c}. \qquad [Pa]$$

The derivative with respect to time is then

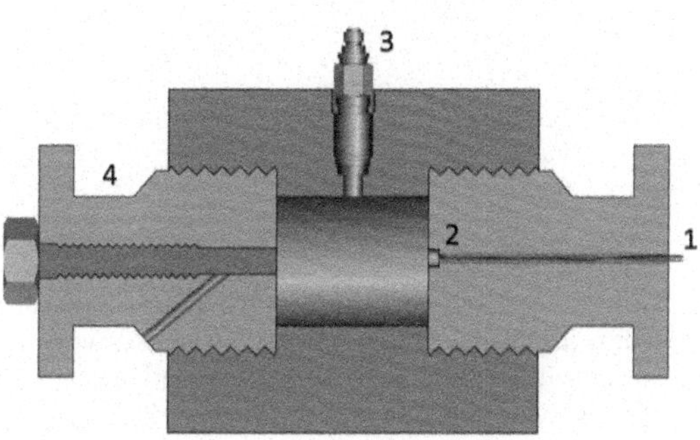


**Fig. 3.26.** A manometric bomb.
1: Electrical connection to igniter.
2: Igniter.
3: Piezo-electric pressure sensor.
4: Blow-off valve.

$$(3.5{:}8)\qquad \dot{p}(t) = \frac{\dot{z}(t)\cdot\Delta\cdot F_{se}\cdot\left(V_{mb}-\Delta/\rho_c\right)}{\left[1-\eta\cdot z(t)\cdot\Delta-\left(1-z(t)\right)\cdot\Delta/\rho_c\right]^2}\,. \qquad \text{[Pa/s]}$$

Instead of $\dot{z}(t)$, we can use the burning law of Eqn 3.4:1a and solve for the product of dynamic vivacity and form function:

$$(3.5{:}9)\qquad \frac{B_A\cdot\phi(z)}{p_0} = \frac{\dot{p}(t)}{p(t)}\cdot\frac{\left[1-\eta\cdot z(t)\cdot\Delta-\left(1-z(t)\right)\cdot\Delta/\rho_c\right]^2}{\Delta\cdot F_{se}\cdot\left(1-\Delta/\rho_c\right)}\,. \qquad \text{[1/(Pa·s)]}$$

If we assume that $\eta = 1/\rho_c$ and take account of Eqn 3.5:3, Eqn 3.5:9 simplifies to

$$(3.5{:}10)\qquad \frac{B_A\cdot\phi(z)}{p_0} = \frac{\dot{p}(t)}{p_{mb}\cdot p(t)}\,. \qquad \text{[1/(Pa·s)]}$$

The burning behaviour of the propellant is therefore known.

### 3.5.3 The Crawford bomb

Burning rate and its relationship to pressure, which are of importance to interior ballistics calculations, are determined using a *Crawford bomb*. This device was developed at the University of Minnesota in Minneapolis and first described in B. L. CRAWFORD et al., 1947.

A sample of the propellant under investigation is placed in a pressure-resistant steel vessel. The sample is usually cylindrical and is generally produced specially for the purpose. See Fig. 3.27. It has a diameter of 3–6 mm and a length of 150–250 mm, depending on the size of the bomb. The chamber is filled with nitrogen and the pressure adjusted to that at which one wishes to measure the burning rate. To ensure that the propellant rod burns linearly, its surface and free end are

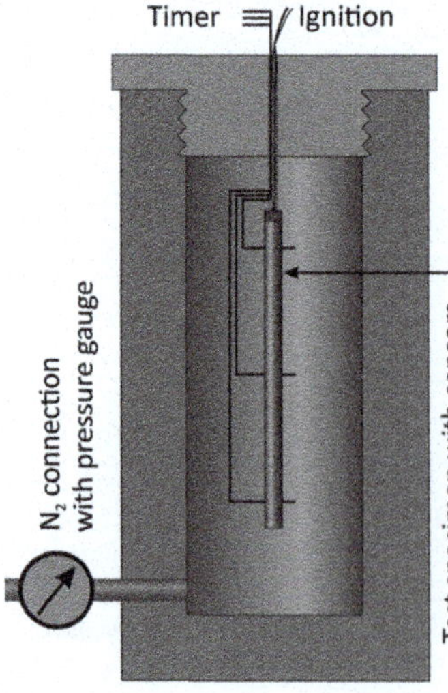


**Fig. 3.27.** A Crawford bomb. The pressure vessel is filled to a specified pressure with nitrogen in gaseous form. The surface layer and one end of the sample are insulated. The sample burns linearly, and the burning time between the probes is measured.

insulated with plastic. The rod is ignited electrically at the uninsulated end. Two or three sensors are inserted into holes in the rod. These are connected to a time-measuring instrument and send a signal when the flame front passes them. The burning rate at the chosen pressure is then calculated from the distance between the sensors and the times at which the flame front reaches each of them.

The chamber must be large enough that the gases produced as the propellant burns cause only an insignificant increase in pressure.

Given the manner in which measurements are made in the Crawford bomb, one cannot expect the results to directly reflect the values one would encounter in a real weapon; the sample is relatively small in diameter, and the insulated surface layer will influence burning. However, the method is well suited to comparisons between different propellants and to quality assessments.

### 3.5.4 Thermodynamics calculations

In addition to the empirical methods of acquiring the propellant data for interior ballistics calculations described above, there is also the option of calculating combustion heat $Q_{ex}$, flame temperature $T_{ex}$, specific energy $F_{se}$ and gas volume at standard temperature from the chemical molecular formula and the heat of formation of the components of the explosive material.

This is carried out at constant volume (as for the manometric bomb) in the case of propellant powder and at constant pressure in the case of rocket propellants.

However, this process belongs to the chemistry of explosives and lies outside the scope of the present book. For full information, see J. KÖHLER et al. (2008) and J. AKHAVAN (2008).

## 3.6 The interior ballistics of rockets

### 3.6.1 General

In this context, a rocket is a projectile that carries with it the propellant required for its propulsion. When this propellant is burned in a *combustion chamber*, gas is produced under pressure and is ejected via a de Laval nozzle at the rear, at high velocity, thereby generating momentum. Because of the law of conservation of momentum, that momentum is transferred to the projectile and the projectile

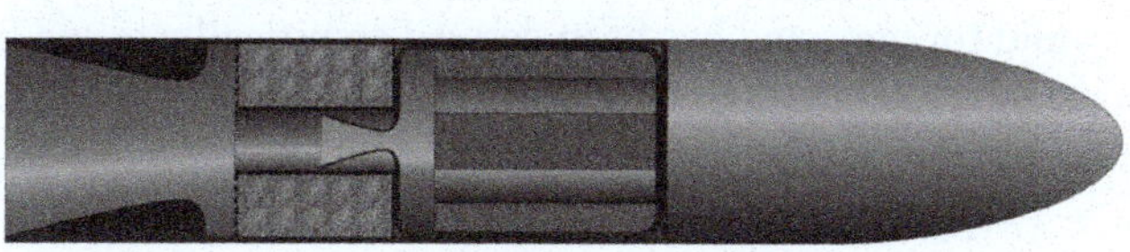

**Fig. 3.28.** Anti-tank rocket with booster (large nozzle) and sustain engine (small nozzle).

undergoes acceleration. The accelerating force is called *thrust*. It should be as constant as possible.

Because the gases are constantly exiting the nozzle, the pressure in the combustion chamber of a rocket is five to ten times lower than that in the chamber of a gun. This means that the projectile can be relatively light and, because pressure only builds up inside the projectile, launch systems can also be light.

Rockets often have two engines: a *booster*, which accelerates the rocket to a certain muzzle velocity (up to the end of the launcher) and a *sustain engine*, which causes the rocket to accelerate further during the initial phase of flight (see Fig. 3.28). Not surprisingly, these two engines must be designed very differently as regards both their propellant and their dimensions.

## 3.6.2 Propellant combustion and combustion chamber pressure

### *3.6.2.1 Propellant combustion*

From the nozzle theory equations in 2.3.6, it is clear that in order to achieve the highest possible exhaust velocity, and to keep that velocity constant for a certain length of time, the internal pressure in the combustion chamber needs to be as constant as possible. In conjunction with the lower pressure mentioned above, this requires the use of special rocket propellant with a low burning rate, and that propellant should exhibit little sensitivity to pressure. The candidates are:

- *Homogenous propellants*, in which oxidant and fuel form a single chemical compound. These are double-base propellants, with a structure similar to that of the artillery propellants mentioned in 3.2.3.3.
- *Composite propellants*, which consist of two components; an oxidant and an energy source, which also acts as a binder. These are described in greater detail in 3.2.4.2.

Liquid fuels are rarely used in rockets, so we shall not discuss them here.

The thermodynamic model of interior ballistics of a rocket motor assumes (as in the case of artillery) that the entire surface of the propellant starts to burn simultaneously. VIEILLE's law is generally used to calculate the burning rate (see Eqn 3.3:14):

$$\frac{de}{dt} = \beta \cdot \left(\frac{p_i}{p_0}\right)^{\alpha} . \qquad (3.6:1) \qquad [mm/s], [m/s]$$

Where e = thickness of propellant already burnt, $p_i$ = instantaneous pressure and $p_0$ = ambient pressure.

In this equation, which is valid for guns and has been derived empirically, $\alpha$ and $\beta$ are propellant-specific constants; the burning rate hence depends only on pressure. Given the short time for which the propellant burns (approximately 1–60 ms, de-

**Table 3.10.** Summary of the variables in 3.6.2.

| | | | | | |
|---|---|---|---|---|---|
| $e$ | [m] | Thickness of propellant burnt | $\rho_c$ | [kg/m³] | Density of propellant |
| $A_c$ | [m²] | Surface area of propellant | $\rho_i$ | [kg/m³] | Density of gases |
| $A_m$ | [m²] | Smallest nozzle cross-sectional area | $p_i$ | [Pa] | Pressure in the combustion chamber |
| $V_{ch}$ | [m³] | Volume of the combustion chamber | $p_0$ | [Pa] | Standard pressure |
| $V_i$ | [m³] | Free volume | $T_i$ | [K] | Combustion chamber temperature |
| $m_c$ | [kg] | Mass of propellant load | $c$ | [m/s] | Speed of sound |
| $m_a$ | [kg] | Mass of ejected gas | $\gamma$ | [–] | Heat capacity ratio |

pending on calibre) one can assume that temperature has no effect. Rocket motors operate at much lower pressures, but the duration of burning is between ten and a few hundred times greater. The hot combustion gases can therefore preheat the remaining propellant, increasing burning rate and hence increasing the value of coefficient $\beta$. This is compensated for in Eqn 3.6:1, by setting $\alpha < 1$. Using Eqn 3.3:17, we can write the relative burning rate as a function of volume in accordance with Eqn 3.3:18:

$$\frac{dz}{dt} = \frac{A(e)\cdot\rho_c}{m_c}\cdot\frac{de}{dt}, \qquad [1/s]$$

in which the volume of the original propellant load is replaced by its mass $m_c$ and its density $\rho_c$. Substituting Eqn 3.6:1, we obtain

$$(3.6:2) \qquad \frac{dz}{dt} = \frac{A(e)\cdot\rho_c}{m_c}\cdot\beta\cdot\left(\frac{p}{p_0}\right)^{\alpha}. \qquad [1/s]$$

### 3.6.2.2 Pressure in the combustion chamber

The gases generated by burning the propellant increase the pressure in the combustion chamber, but a proportion of those gases exits via the nozzle, owing to the pressure difference between the combustion chamber and the outside. The mass of propellant burned in time dt is:

$$(3.6:3) \qquad m_c\cdot\frac{dz}{dt} = \frac{d(\rho_i\cdot V_i)}{dt} + \frac{dm_a}{dt}. \qquad [kg/s]$$

If the nozzle is balanced, the narrowest part will be at Laval pressure (see 2.3.6.2) and, in accordance with Eqn 2.3:43, the mass exiting the nozzle per unit time will be

$$\frac{dm_a}{dt} = A_m \cdot \sqrt{\gamma \cdot p_i \cdot \rho_i \cdot \left(\frac{2}{\gamma+1}\right)^{\frac{\gamma+1}{\gamma-1}}} \quad . \qquad \text{[kg/s]}$$

See Table 3.10 for an explanation of the symbols.

To use this equation in Eqn 3.6:3, it is rearranged and the factors combined with the heat capacity ratio.

$$\frac{dm_a}{dt} = A_m \cdot p_i \cdot \sqrt{\frac{1}{\gamma \cdot R \cdot T_i}} \cdot \gamma \cdot \left(\frac{2}{\gamma+1}\right)^{\frac{\gamma+1}{2\cdot(\gamma-1)}} \quad . \qquad \text{[kg/s]}$$

Substituting the speed of sound

$$c = \sqrt{\gamma \cdot R \cdot T_i} \ , \qquad \text{[m/s]}$$

gives us

(3.6:4)
$$\frac{dm_a}{dt} = \frac{A_m \cdot p_i}{c} \cdot f(\gamma) \ . \qquad \text{[kg/s]}$$

where
$$f(\gamma) = \gamma \cdot \left(\frac{2}{\gamma+1}\right)^{\frac{\gamma+1}{2\cdot(\gamma-1)}} \quad . \qquad [–]$$

In order to derive the differential equation for pressure in the combustion chamber, we need expressions for the density of the gas in the space available for combustion

(3.6:5)
$$\rho_i = \frac{\gamma}{c^2} \cdot p_i \ , \qquad \text{[kg/m}^3\text{]}$$

for the volume available for combustion

(3.6:6)
$$V_i = V_{ch} - \frac{m_c}{\rho_c} \cdot (1-z) - \left(m_c \cdot z - \frac{dm_a}{dt}\right) \cdot \eta \ , \qquad \text{[m}^3\text{]}$$

and for the change in gas volume in time dt

(3.6:7)
$$\frac{dV_i}{dt} = \frac{m_c}{\rho_c} \cdot \frac{dz}{dt} = A(e) \cdot \beta \cdot \left(\frac{p}{p_0}\right)^{\alpha} \ . \qquad \text{[m}^3\text{/s]}$$

We use the product rule in the first addend on the right-hand side of Eqn 3.6:3 and then substitute the three equations 3.6:5–7.

Following a number of calculation steps and under the assumption that temperature in the combustion chamber is constant ($T_i = \text{const} \Rightarrow c = \text{const}$) we obtain

(3.6:8)
$$\frac{\gamma}{c^2} \cdot \frac{dp_i}{dt} \cdot \left[V_{ch} - \frac{m_c}{\rho_c} \cdot (1-z) - (m_c \cdot z - dm_a) \cdot \eta\right] +$$
$$+ \rho_c \cdot A_c(e) \cdot \beta \cdot \frac{p_i^{\alpha}}{p_0^{\alpha}} \cdot \left(\frac{\gamma}{c^2} \cdot \frac{p_i}{\rho_c} - 1\right) = -\frac{A_m \cdot p_i}{c} \cdot F(\gamma) \ . \qquad \text{[kg/s]}$$

This equation and Eqn 3.6:2 together form the system of differential equations for calculating pressure in the combustion chamber as a function of time. In this general form it can only be solved by numerical methods.

If we assume that the propellant burns with constant surface area, then pressure in the combustion chamber will also be constant, and can be calculated using Eqn 3.6:8. If $p_i$ = const then $dp_i/dt = 0$ and Eqn 3.6:8 can be simplified to

$$(3.6{:}9) \qquad \rho_c \cdot A_{c0} \cdot \beta \cdot \frac{p_i^\alpha}{p_0^\alpha} \cdot \left( \frac{\gamma}{c^2} \cdot \frac{p_i}{\rho_c} - 1 \right) = -\frac{A_m \cdot p_i}{c} \cdot F(\gamma) \,. \qquad \text{[kg/s]}$$

If we use Eqn 3.6:5 in the parentheses and then assume $\rho_i << \rho_c$ (which makes the expression in parentheses equal to –1), we can rearrange Eqn 3.6:9 and solve for $p_i$:

$$(3.6{:}10) \qquad p_i = \left( \frac{c \cdot \rho_c \cdot \beta}{F(\gamma) \cdot p_0^\alpha} \right)^{\frac{1}{1-\alpha}} \cdot \left( \frac{A_{c0}}{A_m} \right)^{\frac{1}{1-\alpha}} \,. \qquad \text{[Pa]}$$

This rearrangement focuses on the second parenthetical expression on the right-hand side, which represents the ratio of the (constant) propellant surface area $A_{c0}$ to the cross-sectional area of the narrowest part of the nozzle $A_m$. This ratio is identified, even in English, by the German term *Klemmung* (often referred to simply as Kn, but represented here by K) and is one of the most important parameters of a rocket motor:

$$(3.6{:}11) \qquad K = \frac{A_{c0}}{A_m} \,. \qquad [–]$$

We can therefore express Eqn 3.6:10 simply, thus:

$$(3.6{:}12) \qquad p_i = C \cdot K^{\frac{1}{1-\alpha}} \,, \qquad \text{where } C = \text{const.} \qquad \text{[Pa]}$$

From this equation it is immediately clear that $\alpha$ must be less than 1 for rocket propellant and should be as low as possible. The closer $\alpha$ gets to 1, the more sensitive pressure becomes to Kn (K in the notation used here).

For instance, if $\alpha = 0.8$, internal pressure becomes proportional to $K^5$. If the propellant surface area were to increase by just 10%, e.g. because of a crack, pressure would rise by over 60%.

### *3.6.2.3 Propellant geometry*

In the previous section, we saw how important it is that the surface area of a rocket propellant load remain constant. The most obvious shape with this characteristic is the *end burner*. This is a propellant cake insulated against burning on all sides, except at one end. The disadvantage of such a shape is that the walls of the combustion chamber adjacent to the burnt propellant become very hot during prolonged burning and have to be correspondingly thick and heavy. End burners are therefore rarely used today, except in boosters that are designed to burn only very briefly.

**Fig. 3.29.** Propellant cross sections for core-burning motors that burn with a constant surface area, at least for an extended period. The outer surfaces are insulated.

See Fig. 3.28 in 3.61 for an example. The propellant cake of the booster consists of a ring-shaped end burner surrounding the nozzle of the sustain motor. The sustain motor is a core-burning motor, of which the propellant also burns with constant surface area.

It is possible to use simple geometric forms to produce propellant cakes that burn with constant surface area, with the flame front reaching the walls of the combustion chamber last (a *core-burning motor*). Fig. 3.29 shows three such forms. More complex internal geometry, in conjunction with propellants of different burning rates, makes it possible to regulate combustion in such a way as to vary pressure over the duration of combustion. For instance, pressure can be adjusted in accordance with external conditions, or a booster and cruise engine can be combined in a single combustion chamber. See for instance W. WOLFF (2006).

## 3.6.3 Thrust

### *3.6.3.1 The rocket equation*

As mentioned in 3.6.1, a rocket is based on the principle of the conservation of momentum. Let $dm_p$ be the reduction in rocket mass, $dv_p$ the increase in rocket velocity, $dm_a$ the mass of the gas discharged via the nozzle and $v_a$ the velocity at which the gas leaves the nozzle. In the absence of other forces and given that $dm_p = dm_a$ we can write:

$$m_p \cdot dv_p = -dm_p \cdot v_a \,, \qquad [\text{kg·m/s}]$$

or

$$dv_p = -v_a \cdot \frac{dm_p}{m_p} \,. \qquad [\text{m/s}]$$

Integrating this equation from initial mass $m_{p0}$ to mass m gives us

$$v_p = -v_a \cdot \int_{m_{p0}}^{m} \frac{dm_p}{m_p} = -v_a \cdot \left(\ln m - \ln m_{p0}\right) , \qquad [\text{m/s}]$$

(3.6:13)
$$v_p = v_a \cdot \ln \frac{m_{p0}}{m} \,. \qquad [\text{m/s}]$$

This is the *general rocket equation*. If we replace m by the mass of the rocket at all-burnt, $m_{p0} - m_c$ (where $m_c$ = mass of the propellant load), Eqn 3.6:13 yields the maximum velocity at all-burnt, $v_{pe}$:

$$(3.6{:}14) \qquad v_{pe} = v_a \cdot \ln \frac{m_{p0}}{m_{p0} - m_c} = v_a \cdot \ln \mu \,. \qquad \text{[m/s]}$$

The quotient of initial mass $m_{p0}$ and the mass of the empty rocket $m_{p0} - m_c$ is known as its *mass ratio* $\mu$. To achieve a high final velocity, it is necessary to arrange for a high exhaust velocity and high mass ratio.

Part of the thrust propelling the rocket comes from the mass of exhaust gas per unit time multiplied by the exhaust velocity (from the law of conservation of momentum). A further portion of the thrust comes from the difference between the pressure at the nozzle exit and external pressure. This pressure difference exerts an additional force on the rear of the rocket. Thrust is therefore given by:

$$(3.6{:}15) \qquad F_T = \frac{dm_c}{dt} \cdot v_a + A_e \cdot (p_e - p_a) \,. \qquad \text{[N]}$$

Where $A_e$ = nozzle end diameter, $p_e$ = pressure at that point and $p_a$ = outside pressure.

If the rocket has a balanced de Laval nozzle (see 2.3.6.5), $p_e = p_a$ and the second term on the right-hand side disappears. In practice, it is most unlikely that a ballistic rocket would ever be balanced. Pressure at the rear of the rocket is determined largely by airflow over the body. In turn, that flow is determined by the velocity of the rocket, which is constantly changing during the acceleration phase. Furthermore, a rocket with a high apex (such as an artillery missile) will experience a reduction in outside pressure. This will affect both the design of the motor and the thrust used in the ballistic calculation for the boost phase of flight.

### 3.6.3.2 Calculating thrust. Nozzle design.

The equations in 2.3.6 can be used to calculate the thrust from a balanced nozzle. From Eqn 2.3:51, we obtain the following expression for the mass flow rate:

$$(3.6{:}16) \qquad \frac{dm_c}{dt} = A_m \cdot \sqrt{\gamma \cdot p_i \cdot \rho_i \cdot \left(\frac{2}{\gamma+1}\right)^{\frac{\gamma+1}{\gamma-1}}} \,, \qquad \text{[kg/s]}$$

and according to Eqn 2.3:39, fluid velocity $v_a$ at a pressure $p_e$ is

$$(3.6{:}17) \qquad v_a = \sqrt{\frac{2 \cdot \gamma}{\gamma - 1} \cdot \frac{p_i}{\rho_i} \cdot \left[1 - \left(\frac{p_e}{p_i}\right)^{\frac{\gamma-1}{\gamma}}\right]} \,. \qquad \text{[m/s]}$$

Substituting into 3.6:15, we obtain the following equation for thrust:

$$(3.6{:}18) \qquad F_T = A_m \cdot p_i \cdot Y \cdot \sqrt{1 - \left(\frac{p_e}{p_i}\right)^{\frac{\gamma-1}{\gamma}}} \,, \qquad \text{[N]}$$

in which Y is a pure function of the heat capacity ratio:

$$(3.6{:}19)\qquad Y = \gamma\cdot\sqrt{\frac{2}{\gamma-1}\cdot\left(\frac{2}{\gamma+1}\right)^{\frac{\gamma+1}{\gamma-1}}}\;. \qquad [\text{-}]$$

The quotient of thrust and the product of burning chamber pressure and narrowest nozzle cross section is termed the *thrust coefficient* $C_T$:

$$(3.6{:}20)\qquad C_T = \frac{F_T}{A_m\cdot p_i} = Y\cdot\sqrt{1-\left(\frac{p_e}{p_i}\right)^{\frac{\gamma-1}{\gamma}}}\;. \qquad [\text{N}]$$

This depends exclusively on the heat capacity ratio (and hence the propellant) and the ratio of nozzle exit pressure to combustion chamber pressure. The thrust coefficient is an important motor design parameter.

The geometrical form of a de Laval nozzle is calculated using Eqn 2.3:53, which governs the relationship between the nozzle cross-section and the Mach number of the flow at the relevant point. If we replace the Mach number by the ratio of combustion chamber pressure to pressure at the point under consideration, using Eqn 2.3:52, then following some calculation we obtain an equation with which we can determine exit cross-sectional area $A_e$ from the cross-sectional area of the narrowest part of the nozzle $A_m$, pressure at the nozzle exit $p_e$ and internal pressure $p_i$.

$$(3.6{:}21)\qquad \frac{A_e}{A_m} = \left(\frac{2}{\gamma+1}\right)^{\frac{1}{\gamma-1}}\cdot\left(\frac{\gamma-1}{\gamma+1}\right)^{\frac{1}{2}}\cdot\left(\frac{p_i}{p_e}\right)^{\frac{1}{\gamma}}\cdot\left[1-\left(\frac{p_i}{p_e}\right)^{\frac{1-\gamma}{\gamma}}\right]^{-\frac{1}{2}}\;. \qquad [\text{-}]$$

In the simplest case, this exit cross-sectional area is achieved through conical expansion. However, half the cone angle of the cone must not exceed 15° (W. WOLFF, 2006, P. VON BÖCKH, 2004, gives a value of 10°), otherwise the flow breaks away and the resulting turbulence leads to a sharp drop in the efficiency of the nozzle.

As the flow expands in the conical nozzle, radial velocity components will inevitably be present. As a result, velocity in the direction of travel will be somewhat lower. The mean value is given by

$$(3.6{:}22)\qquad v_{a\,eff} = \frac{v_a}{\alpha_0}\cdot\int_0^{\alpha_0}\cos\alpha\cdot d\alpha = v_a\cdot\frac{\sin\alpha_0}{\alpha_0}\;. \qquad [\text{m/s}]$$

If the conical nozzle has a half cone angle of 15°, velocity will be reduced by approximately 1.1%.

Nozzles for use at very high altitudes (with low outside pressures) would have to be very long, on account of the small cone angle. In such cases bell shapes are used, which allow for a larger bending angle, at which the flow is diverted by means of a Prandtl-Meyer expansion. The bell shape and its final angle can be so calculated as to create parallel flow (K. OSWATITSCH, 1976).

### 3.6.3.3 Determining thrust

There are two empirical methods of determining the thrust of a rocket for inclusion in trajectory calculations. The first is to use a rocket-engine test stand and record the time-force curve as the propellant burns, producing what is known as a *thrust diagram*. However, the thrust measured by this technique is not the same as that which occurs in flight, as a rocket on a test stand is not subjected to flow, which means that nozzle exit pressure differs from that which would occur in flight.

The other option is to record the velocity of the rocket in flight and calculate the accelerations from those measurements. Let $a_b(t)$ be acceleration during the boost phase and let $a_f(t)$ be deceleration during the midcourse phase following thrust decay. Thrust can be defined thus:

$$(3.6{:}23) \qquad F_T(t) = m_p(t)\cdot\left(a_b(t) - a_f(t)\right) \ . \qquad [\mathrm{N}]$$

However, even this thrust does not correspond to reality, as flow over the rocket in the midcourse phase is not the same as during acceleration. It is generally impossible to separate thrust and drag during the boost phase.

In practice, one generally uses the thrust diagram obtained on the test stand for the trajectory calculation and adapts drag to the results of the velocity measurements.

## 3.7 Special propulsion systems

### 3.7.1 General

Conventional guns have a number of major disadvantages: they are extremely heavy compared to the projectile (200 to 1000 times heavier), projectiles are subjected to high acceleration (and hence large forces) during firing and there are limits to muzzle velocity for weapons with conventional internal ballistics (see 3.3.6).

A number of systems have been developed in an effort to overcome one or other of these disadvantages. The main reason for the weight of a gun lies in the law of conservation of momentum,

$$m_b \cdot v_b = m_p \cdot v_p \ , \qquad [\mathrm{N{\cdot}s}]$$

which is what determines the recoil energy of the weapon

$$(3.7{:}1) \qquad E_b = \frac{m_p}{m_b}\cdot E_p \ . \qquad [\mathrm{J}]$$

Index w indicates the weapon, p the projectile. $E_p$ is therefore muzzle energy.

The energy transferred to the weapon on firing is equal to muzzle energy minus the ratio of projectile mass to weapon mass. High muzzle energies therefore require heavy weapons.

Even without resorting to rocket propulsion systems, it is possible to produce a weapon that generates a forward-directed momentum approximately equal to the rearward momentum. Such a weapon is subjected to virtually no net momentum and the recoil energy disappears. There are two basic ways of achieving this: expelling a mass similar to that of the projectile in the opposite direction or discharging a high-velocity gas jet from the rear of the weapon. Both methods have been used in practice. Other designs reduce the load on the projectile resulting from acceleration, or increase muzzle velocity. These will all be discussed in the following sections.

### 3.7.2 The Davis gun

A Davis gun consists of a tube with a firing mechanism. The round combines the projectile, the propellant and an inert countermass of which the mass is similar to that of the projectile (see Fig. 3.30). The propellant is placed between the projectile and the countermass. The pressure generated as it burns accelerates the projectile forwards and the countermass rearwards. If the countermass and the two acceleration distances are selected appropriately, the weapon will be recoilless.

Strictly speaking, recoillessness is achieved not by the weapon but by the munition, as the countermass forms part of it. In some cases the tube is disposable, which means that the "weapon" consists only of the firing mechanism.

We can readily extend the system of thermodynamic equations in 3.4.1 to cover the Davis gun. Let m be mass, v velocity, x displacement and F friction, with index 1 for the projectile and 2 for the countermass. We can derive a system of equations from Eqns 3.4:1a–c and 3.4:2, consisting of the burning law (a), the equations of motion (b–e) and the Résal equation (energy balance, 3.7:3):

(3.7:2a) $$\frac{dz}{dt} = \phi(z) \cdot B_A \cdot \left(\frac{p}{p_0}\right)^{\alpha} \qquad [1/s]$$

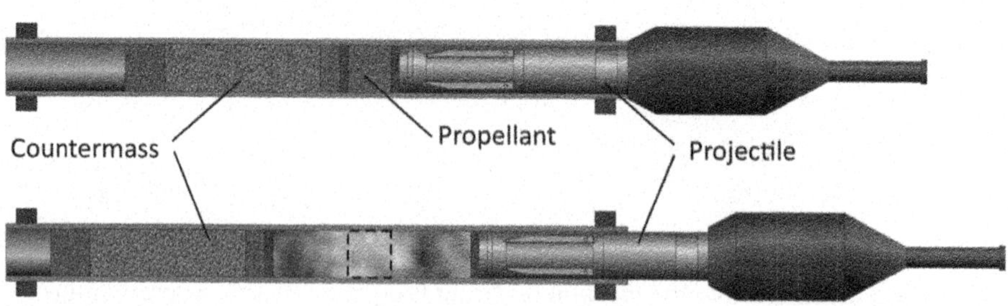


**Fig. 3.30.** A Davis gun. *Top*: Before firing. *Bottom*: After firing. The projectile is propelled forwards and the countermass rearwards. Displacement is measured from the rest position.

(3.7:2b) $$\frac{dv_1}{dt} = \frac{p \cdot A_b - F_1}{m_1} \quad [m/s^2]$$

(3.7:2c) $$\frac{dv_2}{dt} = \frac{p \cdot A_b - F_2}{m_2} \quad [m/s^2]$$

(3.7:2d) $$\frac{dx_1}{dt} = v_1 \quad [m/s]$$

(3.7:2e) $$\frac{dx_2}{dt} = v_2 \quad [m/s]$$

(3.7:3) $$p = \frac{(\gamma^* - 1) \cdot \left[Q_{ex} \cdot z \cdot m_c - \frac{1}{2} \cdot \left(m_1 \cdot v_1^2 + m_2 \cdot v_2^2\right)\right]}{V_b^* - z \cdot m_c \cdot \left(\eta - \frac{1}{\rho_c}\right) + A_b \cdot (x_1 + x_2)} \quad [N/m^2]$$

This system of equations has yielded good results for an existing weapon system and is also suitable for parameter studies.

The variable parameters (the mass, velocity and displacement of the countermass, plus the displacement of the projectile) mean that the Davis gun can be adjusted to provide suitable recoil compensation (in terms of time, displacement and force). Two disadvantages of the design are its weight (by comparison with that of the projectile) and the distances over which projectile and countermass must be accelerated, which mean that Davis guns have relatively long tubes by comparison with the muzzle velocities they achieve. These two factors impose limits on the power of a Davis gun. In addition, a safety zone is required behind the weapon.

### 3.7.3 Recoilless guns

Generating the counter momentum by ejecting a jet of gas from the rear of the weapon, rather than a solid mass, eliminates the Davis gun's shortcomings. However, the velocity of the gas must be sufficiently high to generate the required momentum. This is achieved by inserting a de Laval nozzle in the rear of the tube and using a charge that is sufficiently large to produce enough gas to both accelerate the projectile and generate the counter momentum. This is known as a *recoilless cannon* or *recoilless gun*. See Fig. 3.31.

To ensure that the propellant is ignited correctly, the nozzle must be closed off by a bursting disk or bursting foil. In some weapon systems, this forms part of the wall of the cartridge case, which is perforated. See Fig. 3.32.

Calculating the interior ballistics of a recoilless gun requires the system of thermodynamics equations from 3.4.1 and the equations for rocket thrust. For the first set of equations, one must take account of the fact that gas is being discharged via the nozzle. In the case of the rocket equations, the volume of the "combustion chamber" is constantly changing, as is chamber pressure. These two factors seriously

**Fig. 3.31**. How a recoilless gun works. The base of the round forms a bursting disk that allows pressure to rise to a certain level before it bursts. In some munitions, the casing performs this role. The gas generated as the propellant burns both generates the pressure needed to accelerate the projectile and creates a jet directed to the rear.

**Fig. 3.32**. Ammunition for a 106 mm recoilless gun. The nozzle is supplied with gas via the perforated shell case.

complicate the calculations associated with recoilless guns. For a comprehensive treatment, please see J. CORNER (1950).

CORNER applies his procedure not only to recoilless guns but to all interior ballistics processes with gas leakage, such as worn guns and mortars with baffles.

Guns that use gas discharge from a nozzle to achieve recoillessness are found in the form of portable systems and on light vehicles, aircraft (including helicopters) and ships. Calibres range from 30 mm to 120 mm.

However, recoilless guns also have their disadvantages: they often require a larger safety zone than a Davis gun, and the propellant charge is significantly larger than that of a conventional tube, which causes a louder report.

### 3.7.4 The high-low-pressure system

One major disadvantage of conventional guns is that pressure distribution is asymmetric over time. As a result, both projectile and barrel must be designed for a peak pressure that occurs only briefly, and only at the breech end of the barrel. The ideal would be for pressure to be constant from firing to muzzle exit, as this would considerably reduce the load on both barrel and projectile.

This led to the idea of designing the breech as a high-pressure chamber (the "main chamber") and dividing it from the barrel by means of a perforated disc, which could possibly be used to close off the end of the shell casing (see Fig. 3.33). The holes in the disc may be cylindrical or conical (Venturi nozzles). The pressure difference between the main chamber ($p_h$) and the barrel ($p_n$) generates a mass flow

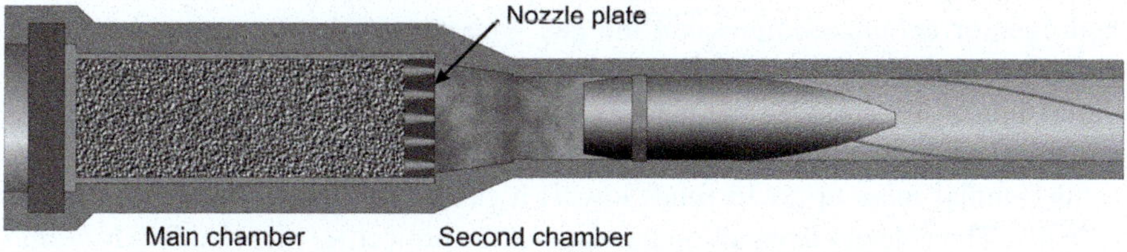


**Fig. 3.33.** Principle of a high-low (H/L) system. The propellant burns in the main (high-pressure) chamber. The resulting high pressure is expanded to a lower, more uniform level via the nozzle plate (with its plain or conical holes).

rate that initially depends solely on $p_h$. See 2.3.6.2. The flow rate causes pressure to fall (in accordance with BERNOULLI's equation).

By appropriate design of main chamber, nozzle plate and charge, this system can achieve a gas pressure curve that is considerably flatter than that of a conventional barrel. For an explanation as to how to calculate the internal ballistics of such a system (divided into "up to 'burnt'" and "after 'burnt'"), see J. CORNER (1950).

### 3.7.5 The light-gas gun

We saw in 3.3.6 that the muzzle velocity of a conventional propellant weapon cannot exceed a certain limit. One reason for this is the pressure difference between the breech face and the base of the projectile, which can only equalize at the speed of sound. According to Eqn 2.3:16, the speed of sound in a gas is

$$(3.7{:}4) \qquad c = \sqrt{\frac{\gamma \cdot p}{\rho}}\,, \qquad [\mathrm{m/s}]$$

and is hence inversely proportional to the density of the gas. The speed of sound is therefore higher in a lighter gas, so such a gas can equalize the pressure difference more rapidly, leading to a larger force being exerted on the base of the projectile.

A light-gas gun therefore consists of two stages (see Fig. 3.34). The first stage consists of a chamber filled with conventional propellant powder. When the propellant burns, a piston is fired into a second chamber (the pump tube) filled with

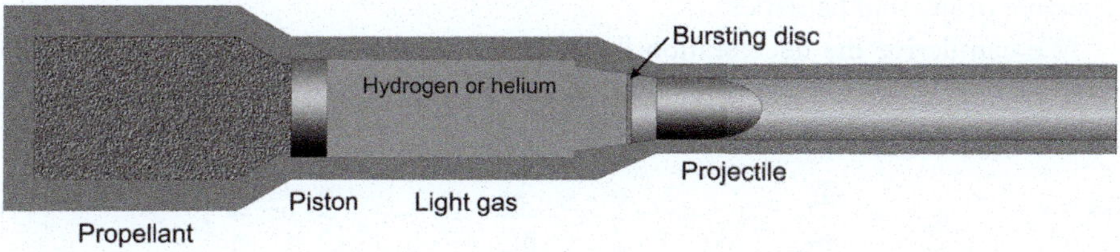


**Fig. 3.34.** How a light-gas gun works. When the propellant burns, the piston is fired into the pump tube, compressing the gas within. The disk bursts at a predefined pressure, allowing the gas to flow into the barrel at high velocity and accelerate the projectile.

hydrogen or helium, compressing the gas. The pump tube is separated from the barrel (and hence from the projectile) by a bursting disk. The disk bursts at a predefined pressure, allowing the highly compressed light gas to flow into the barrel and accelerate the projectile. Because the gas is much lighter than propellant gas (with a molar mass up to 15 times lower) it can be accelerated to a much higher velocity. The velocity depends on the mass of the projectile – approximately 7 km/s for projectiles of which the mass is measured in grams and up to 11 km/s for projectiles of which the mass is measured in milligrams.

Light-gas guns are mainly used for terminal ballistics studies of mini-meteorite impacts on spacecraft, and to study meteorite impacts on Earth.

## 3.7.6 The principle of sub-calibre projectiles

### *3.7.6.1 Basics*

We pointed out the significance of sectional density in 2.4.2. This factor is especially relevant at the boundaries of the sub-domains of ballistics, where it makes contradictory demands of projectile manufacturers. These contradictions can be presented in the form of simple laws.

In interior ballistics, the mechanical work that the propellant gas performs on the projectile is equal to the kinetic energy transferred to the projectile. We can therefore write:

$$\tfrac{1}{2}\cdot m_p\cdot v_0^2 \;=\; p_m\cdot A_p\cdot x\;. \qquad [\mathrm{J}]$$

Where $A_p$ = cross-sectional area of the projectile, $p_m$ = a constant, mean pressure that performs the same work as the actual pressure pattern and x = the distance covered by the projectile.

If we solve this equation for muzzle velocity and use the definition of sectional density $q = m_p/A_p$ (Eqn 2.4:1 in 2.4.2) we obtain

$$(3.7{:}5) \qquad v_0 \;=\; \sqrt{\frac{2\cdot p_m\cdot A_p\cdot x}{m_p}} \;=\; \sqrt{\frac{2\cdot p_m\cdot x}{q}}\;. \qquad [\mathrm{m/s}]$$

The muzzle velocity is therefore inversely proportional to the square root of q. Because a high value of $v_0$ is desirable, so is a low sectional density – from the perspective of internal ballistics.

We can derive the deceleration that a projectile experiences in flight from the drag (Eqn 2.3:25a). Once again taking account of sectional density, deceleration is therefore given by

$$(3.7{:}6) \qquad a_p \;=\; \frac{F_W}{m_p} \;=\; -c_W\cdot\frac{\rho}{2}\cdot v^2\cdot\frac{A_p}{m_p} \;=\; -c_W\cdot\frac{\rho}{2}\cdot v^2\cdot\frac{1}{q}\;. \qquad [\mathrm{J}]$$

This means that a projectile will decelerate less if it has a high sectional density, i.e. if it has a small diameter and is as heavy as possible.

A projectile therefore needs a low sectional density to maximize acceleration in the barrel and a high sectional density in flight (the First ballistic paradox; see B. P. KNEUBUEHL, 2013). The challenge for projectile manufacturers was therefore to fulfil these contradictory requirements by appropriate projectile design.

### 3.7.6.2 Tapered-bore guns

Because the mass of the projectile has to meet other requirements, it seemed logical to design it in such a way that its cross-sectional area was initially large, and then reduce that area before the projectile left the barrel. With this aim in mind, a patent was registered in 1903 that involved inserting a conical element in a barrel, or fitting such a device to the muzzle, in order to squeeze the projectile down to a smaller calibre. However, this would make it necessary to build a projectile with softer material on the outside that could be compressed (see Fig. 3.35). Spin is transferred to the projectile in the first, cylindrical part of the barrel.

Compressing the projectile in the smooth, conical section of the barrel reduces the projectile's axial moment of inertia and hence increases its angular velocity (on account of the law of conservation of angular momentum). This renders the projectile particularly stable.

Tapered-bore barrels first appeared in the Second World War, when they were mainly used on anti-tank guns, to achieve the high muzzle velocities required to penetrate armour.

We can use the system of thermodynamic differential equations from 3.4.1, as long as we take account of the varying barrel cross-section. We replace Eqn 3.4.1b by

$$(3.7{:}7) \qquad \frac{dv}{dt} = \frac{1}{m_p}\cdot\left(p\cdot A_r(x) - F_r\right) \qquad \text{with } p > p_s\,, \qquad [\mathrm{m/s^2}]$$

where $p_s$ = pressure on entering the conical section of the barrel. Eqn 3.4:2 for pressure can then be written:

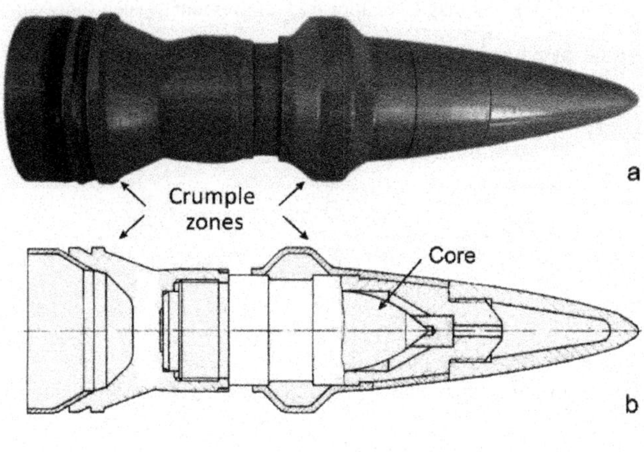


**Fig. 3.35** **a.** 47 mm anti-tank projectile, designed for a barrel tapering from 47 mm to 35 mm.
**b.** The cross-sectional drawing illustrates its complex internal structure.

$$(3.7{:}8)\qquad p = \frac{(\gamma^{*}-1)\cdot\left[Q_{ex}\cdot z\cdot m_c - \frac{1}{2}\cdot(m_p+\varepsilon\cdot m_c)\cdot v^2\right]}{V_b^{*} - z\cdot m_c\cdot\left(\eta - \frac{1}{\rho_c}\right) + \int_0^x A_r(\xi)\cdot d\xi}\ .\qquad [\mathrm{N/m^2}]$$

From experiments (R. GERMERSHAUSEN, 1980), the effect of deforming the projectile would appear to be minimal, and can apparently be ignored. On the other hand, the taper does affect gas flow in the barrel, and it is assumed that this reduces the pressure gradient (J. CORNER, 1950).

"Salvo Squeeze Bore" pistols do exist, however, incorporating barrels that compress the bullet from 9 mm Luger to 7.62 mm, or from .45 Auto to .38 (see B. P. KNEUBUEHL, 2013).

### 3.7.6.3 Sabot rounds

A design emerged around the middle of the 20th Century that may well have been inspired by the old muzzle loaders.

To facilitate loading, these used slightly sub-calibre shot. However, this made it necessary to insert some intermediate material between the powder and the ball, to seal the barrel and to ensure that the pressure was applied to the projectile.

The idea is to use a projectile of significantly smaller calibre than the barrel, enveloping it in a lightweight structure that guides it along the barrel and imparts spin if required. This additional structure is separated from the projectile after the round leaves the muzzle by the forces acting upon it (air drag or centrifugal force). Inside the barrel, the projectile will have the low sectional density that would be associated with the calibre, whereas the projectile in flight will have a higher sectional density by comparison with the calibre of the projectile. The first projectiles of this type were designed for rifled barrels and had a disc at the rear (generally made of aluminium) and a supporting structure for the projectile itself (see Fig. 3.36). These were known as *sabot* rounds (using cup or base sabots).

The word "sabot" is in fact the French word for a wooden shoe; the earliest devices used to stabilize sub-calibre projectiles were made of wood, and may have appeared reminiscent of such footwear.

Today, the most common sabot round is the fin stabilized (long-rod) type, which has an extremely high sectional density in flight. The multi-segmented aluminium or plastic sabot is located in front of the stabilizing fins (spindle sabot). See Figs 3.37

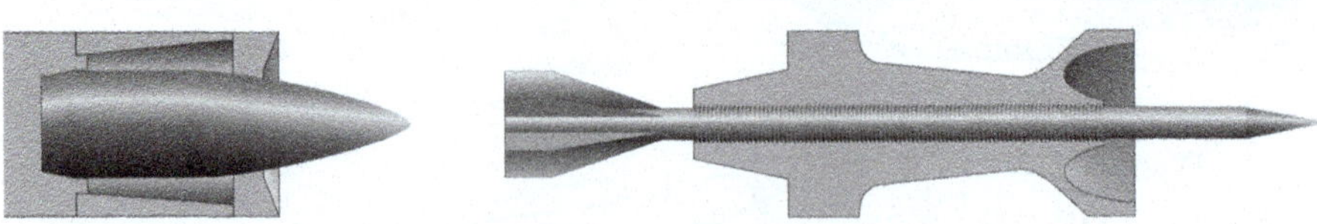

**Fig. 3.36.** (*Left*). Principle of a sabot round for use in a rifled barrel.

**Fig. 3.37.** (*Right*). Principle of a fin stabilized projectile with a spindle sabot for use in a smooth barrel.

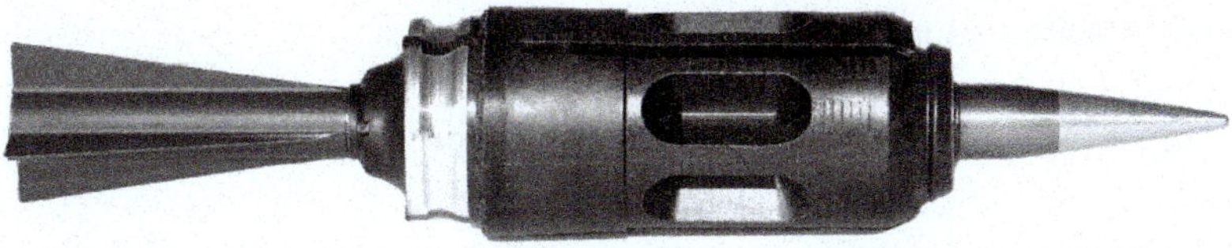

**Fig. 3.38.** Example of a fin stabilized projectile with sabot for use in a medium-calibre rifled gun.

and 3.38. Fin-stabilized projectiles for rifled barrels require a sabot that will slide through the barrel without engaging the rifling, as the stabilizing fins are not compatible with the high rotational speed of a spin-stabilized projectile.

It is apparently easier to find a way of discarding the sabot without disturbing the flight of the projectile than to resolve the problems associated with conical barrels (manufacturing such barrels and producing a projectile with a deformable outer layer). With a few rare exceptions, conical barrels therefore disappeared soon after the appearance of the first sabot rounds.

## 3.8 The acceleration of fragments

### 3.8.1 General

Fragments are generally accelerated using explosives. Their initial velocity (and therefore their initial energy) depends on the energy balance and hence on the law of conservation of energy. Fragments may also be created when a bullet chips pieces out of a hard material or when the projectile itself breaks up. In such cases the fragments acquire most of their energy from a transfer of momentum.

### 3.8.2 Fragments accelerated by explosives

#### *3.8.2.1 The Gurney formula*

During the Second World War, R. W. GURNEY (1943), a solid-state physicist from Princeton, working at the US Army Ballistics Research Laboratory in Aberdeen, Maryland, USA, developed a number of simple principles for estimating the initial velocity of fragments accelerated by explosives. See also J. A. ZUKAS et al. (1998). He made the following assumptions:

- The detonation of a given explosive produces a fixed amount of kinetic energy per unit mass, which is divided over the casing and the explosion gases.
- The explosion gases are of constant density.
- The velocity profile of the explosion gases is linear.

It is interesting to note that the work done to burst the casing can apparently be neglected.

From the law of conservation of energy:

$$(3.8{:}1) \qquad m_c \cdot \varepsilon = \tfrac{1}{2} \cdot m_f \cdot v^2 + \tfrac{1}{2} \cdot \int_0^y \rho_c(x) \cdot A(x) \cdot v_c^2(x) \cdot dx \; . \qquad [J]$$

Where $m_c$ = mass of the explosive charge, $\varepsilon$ = energy of the explosive per unit mass, $m_f$ = mass of the accelerated fragments, v = velocity of the accelerated fragments, $\rho_c$ = density of the explosion gases, $v_c$ = velocity of the explosion gases, A = cross-sectional area of the explosion gas channel and y = length of the explosion gas channel.

For a cylindrical casing of radius b and length y (see Fig. 3.39) we can write

$$(3.8{:}2a) \qquad \rho_c = \frac{m_c}{b^2 \cdot \pi \cdot y} \, , \qquad [kg/m^3]$$

$$(3.8{:}2b) \qquad A(x) = 2 \cdot x \cdot \pi \cdot y \, , \qquad [m^2]$$

$$(3.8{:}2c) \qquad v_c(x) = \frac{x}{b} \cdot v \; . \qquad [m/s]$$

Substituting Eqns 3.8:2a-c into 3.8:1, integrating and rearranging gives us

$$(3.8{:}3) \qquad \frac{v}{\sqrt{2 \cdot \varepsilon}} = \left( \frac{m_f}{m_c} + \frac{1}{2} \right)^{-\frac{1}{2}} . \qquad [-]$$

The following equations apply to various other geometries. Sphere:

$$(3.8{:}4) \qquad \frac{v}{\sqrt{2 \cdot \varepsilon}} = \left( \frac{m_f}{m_c} + \frac{3}{5} \right)^{-\frac{1}{2}} \qquad [-]$$

Symmetrical sandwich:

$$(3.8{:}5) \qquad \frac{v}{\sqrt{2 \cdot \varepsilon}} = \left( \frac{m_f}{m_c} + \frac{1}{3} \right)^{-\frac{1}{2}} \qquad [-]$$

Asymmetrical sandwich (with wadding):

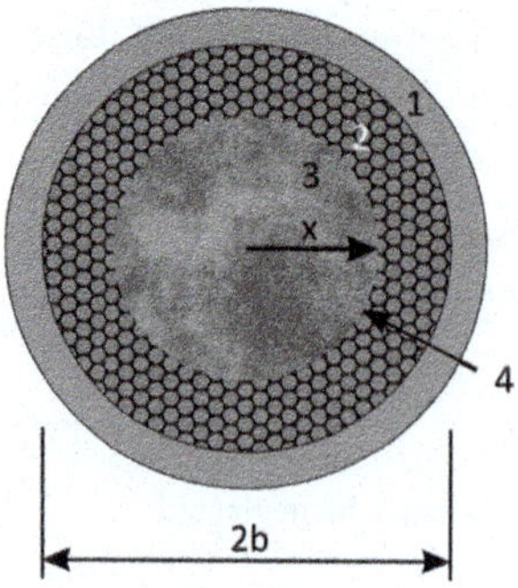


**Fig. 3.39.** Derivation of the Gurney equation.
1 Casing (material to be accelerated).
2 Explosive.
3 Explosion gases.
4 Flame front.

(3.8:6) $$\frac{v}{\sqrt{2\cdot\varepsilon}} = \left(\frac{1 + C^3}{3\cdot(1 + C)} + \frac{m_t}{m_c}\cdot C^2 + \frac{m_f}{m_c}\right)^{-\frac{1}{2}},$$ [-]

where $$C = \frac{1 + 2\cdot\frac{m_f}{m_c}}{1 + 2\cdot\frac{m_t}{m_c}}.$$ [-]

Where $m_t$ = mass of wadding.

If $m_t = \infty$, then from Eqn 3.8:6 we obtain an expression for a symmetrical sandwich. If $m_t = 0$, however, we obtain an expression for the open-faced sandwich configuration:

(3.8:7) $$\frac{v}{\sqrt{2\cdot\varepsilon}} = \left(\frac{\left(1+2\cdot\frac{m_f}{m_c}\right)^3 + 1}{6\cdot\left(1+\frac{m_f}{m_c}\right)} + \frac{m_f}{m_c}\right)^{-\frac{1}{2}}.$$ [-]

The variable with the dimensions of a velocity,

$$v_G = \sqrt{2\cdot\varepsilon}$$ [m/s]

is known as the *Gurney constant*, or *Gurney velocity*. This varies according to the explosive. Table 3.11 (next page) shows a number of typical values.

The dimensionless quotient of fragment velocity and Gurney constant is determined solely by geometry and the ratio of the accelerated mass to the mass of the explosive.

The reciprocal value of this mass ratio ($m_c/m_f$) is the quantity of explosive available to accelerate the fragment.

Fig. 3.40 (next page) shows this function for the most important situations. If the mass of the fragments is more than twice that of the explosive, the arrangement of the explosive ceases to be significant.

### 3.8.2.2 Local application of the Gurney formula

When the Gurney formula is used, the initial velocity of fragments accelerated by explosives is generally calculated from the ratio of the total mass of the munition to the mass of the explosive. This yields a constant value for all fragments. However, measurements of fragment velocity reveal that values can vary considerably over the surface of the projectile.

If one assumes that detonation starts at the fuze, then the amount of explosive available to accelerate the fragments will vary over the surface of the projectile.

**Table 3.11**. Densities and Gurney constants for a number of explosives and propellants

| Explosive | Designation/composition | Density [kg/m$^3$] | $v_G$ [m/s] |
|---|---|---|---|
| Composition A-3 | 91% RDX, 9% wax | 1.59 | 2630 |
| Composition B | 63% RDX, 36% TNT, 1% wax | 1.72 | 2710 |
| Composition C-3 | 71% RDX, 4% TNT, 10% DNT, ... | 1.60 | 2680 |
| Cyclotol | 75% RDX, 25% TNT | 1.754 | 2790 |
| Hexogen | RDX | 1.77 | 2830 |
| Hexogen | 97% RDX, 3% wax | 1.75 | 2800 |
| LX-14 | 95.5% HMX, 4.5% polymer binder | 1.83 | 2800 |
| Pentrit | PETN | 1.76 | 2930 |
| Octogen | HMX | 1.89 | 2970 |
| Octol | 75% HMX, 25% TNT | 1.82 | 2830 |
| PBX 9404 | 94% HMX, 3% NC, 3% CEF | 1.84 | 2900 |
| PBX 9502 | 95% TATB, 5% binder | 1.885 | 2377 |
| Tetryl | | 1.62 | 2500 |
| Trinitrotoluene | TNT | 1.63 | 2370 |
| Tritonal | 80% TNT, 20% A (moulded/pressed) | 1.69/1.55 | 2200 |
| NC and Ngl propellant | | 1.55 | 1680 |
| Black powder | | 1.30 | 950 |

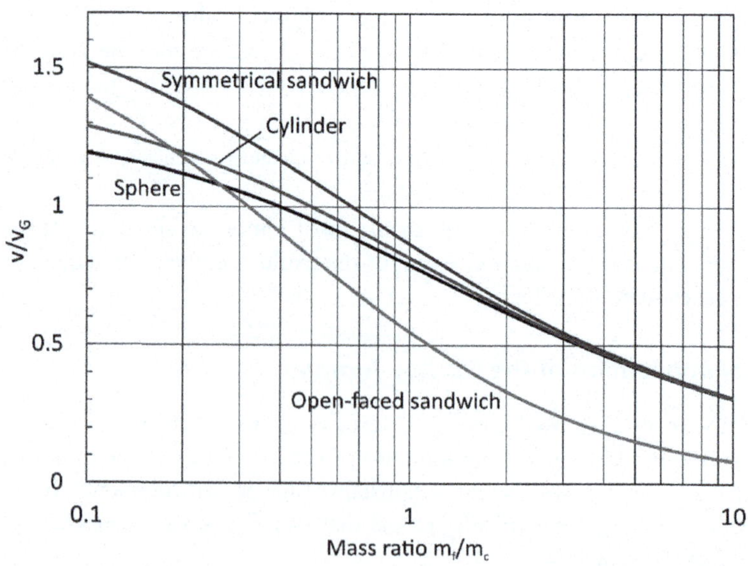


**Fig. 3.40**. Curves of Gurney functions from Eqns 3.8:3–3.8:5 and 3.8:7. If the mass of the fragments is more than approximately twice that of the explosive, the arrangement of the explosive ceases to be significant.

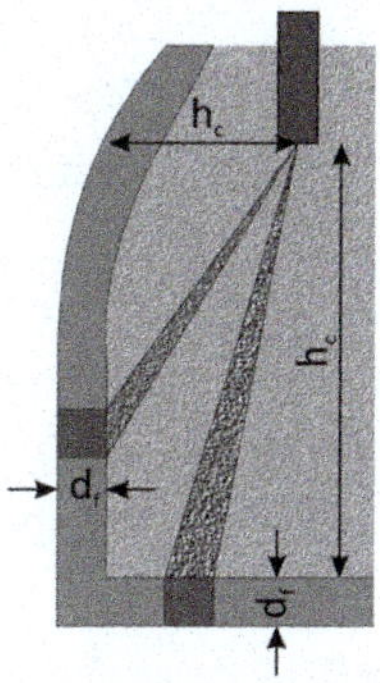


**Fig. 3.41**. (*Left*). Local calculations of fragment velocities using the amount of explosive between fragment and fuze.

**Fig. 3.42**. (*Right*) Comparison between measured and calculated fragment velocities. Upper figure: Measured. Lower figure: Calculated with "Gurney local".

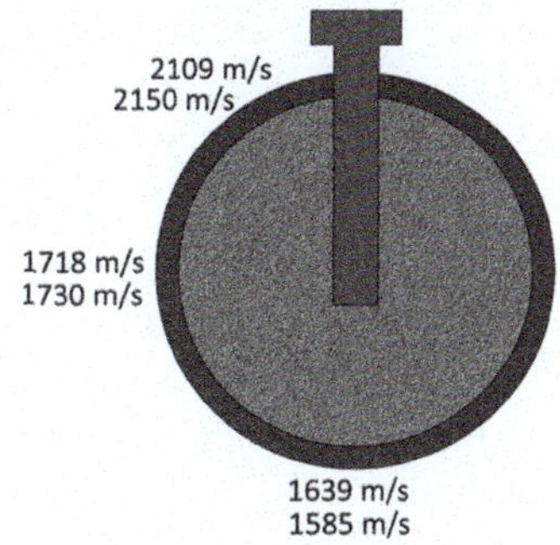


This amount can be calculated for a given segment of the projectile from the mass of the segment and the mass of the pyramid of explosive of which the tip lies at the point of detonation (the fuze). See Fig. 3.41. The mass ratio is

(3.8:8) $$\frac{m_f}{m_c} = \frac{3 \cdot d_f \cdot \rho_f}{h_c \cdot \rho_c} \; . \qquad [-]$$

Where $d_f$ = thickness of projectile casing, $h_c$ = height of pyramid of explosive (measured from its base), $\rho_f$ = density of casing and $\rho_c$ = density of explosive.

Using this in the Gurney formula for a sphere yields more precise initial velocity values for the fragments.

*Examples*: In the case of an almost spherical hand-grenade, detonation is initiated slightly away from the centre. This means that the quantity of explosive accelerating fragments varies quite significantly over the surface of the casing, leading to variations in initial velocity. Because measurements made using x-rays and aluminium foil targets were available, it was possible to verify the results of the locally-applied Gurney equation. The discrepancies were less than 5% in all cases. See Fig. 3.42, (B. P. KNEUBUEHL, 1999).
Results for mortar bombs were of similar quality. There was a difference of less than 3.5% between calculated values and x-ray measurements for a 60 mm mortar bomb.

## 3.8.3 Fragments accelerated by momentum

### *3.8.3.1 Origins. Possible velocities.*

Impulse-accelerated fragments may be created by two mechanisms:

- A projectile striking small masses (stones).
- A projectile shattering against a hard object.

In the first instance, one can calculate the initial velocity of the fragment using the law of impact (B. P. KNEUBUEHL, 1999). A fragment resulting from such an impact cannot be accelerated to a velocity greater than

$$(3.8:9) \qquad v_f = v_p \cdot \sqrt{\frac{2 \cdot \frac{m_p}{m_f}}{\frac{m_p}{m_f} + 1}}, \qquad [m/s]$$

where $v_f$ = fragment velocity, $v_p$ = velocity of projectile, $m_p$ = mass of projectile and $m_f$ = mass of fragment.

As Eqn 3.8:9 assumes a perfectly elastic collision, the calculated velocity is an upper limit.

*Example*: If an 11.3 g projectile strikes a 1 g stone at 650 m/s, the highest velocity the stone can attain is 650·1.36 = 880 m/s.

Taking the second situation: if a projectile strikes a hard surface at a sufficiently steep angle, but is unable to penetrate it, then all the energy it is carrying will be transferred to the target over a distance approximately equal to its own length. Very high forces occur, causing the projectile to break up and accelerating the resulting fragments.

A special type of fragment acceleration occurs when a soft core projectile carrying sufficient energy strikes a hard surface. The forces mentioned above cause compressive stresses in the lead core that burst the jacket of the projectile. The resulting fragments can attain quite high velocities.

Generally speaking, it is impossible to make an estimate of the fragment velocity in this second case of acceleration due to momentum. However, if such fragments cause damage, it is often possible to deduce their initial energy.

### 3.8.3.2 Case study

During a live-firing exercise involving the use of hand grenades and assault rifles, on an Alpine meadow littered with large rocks, a soldier throwing hand grenades was wounded in the eye by a fragment from the jacket of a bullet. The fragment penetrated the eye to a depth of approximately 25 mm. The mass of the fragment was 0.05 g. From its dimensions, its sectional density was 0.0087 g/mm$^2$. Wound ballistics literature (K. SELLIER, 1982, and B. P. KNEUBUEHL et al., 2022) indicates that a fragment of this nature would have to be travelling at a velocity of at least 115 m/s to penetrate the eye. A more slowly-moving fragment would not have done so.

This assumes that the resistance to the fragment of the sclera (the white of the eye) is of the same order of magnitude as that of the cornea.

As the fragment penetrated the eye to a depth of approximately 2.5 cm, its velocity must have been at least 10 m/s to 20 m/s higher than the minimum required for penetration, so a rough estimate puts the impact velocity at no less than 130 m/s. It is estimated that the victim was approximately 15 m from the point at which the fragment originated, so the initial velocity must have been approximately 900 m/s.

# 4 Intermediate ballistics

## 4.1 Introduction

The two primary areas of ballistics – interior and exterior – had been seen as separate since the mid-18th Century. During the course of the 20th Century, a new sub-discipline emerged, covering the transition between the two. The origins of this development probably lie in the work of C. CRANZ and H. SCHARDIN. In 1929, those two researchers developed a cinematograph that could project up to 24 optically separate images on a stationary film at up to 3 million images/second, with exposure times of 0.1 μs. See Fig. 4.1. For the first time, it was now possible to observe the rapid phenomena that occur as the projectile exits the barrel, together with a number of terminal ballistics phenomena.

24 spark gaps provided short-time exposure, and an optical system of one large lens and 24 small lenses projected each spark onto the corresponding area of the fixed film. For a detailed description, please see R. E. KUTTERER, 1959 and B. P. KNEUBUEHL, 2013.

This made it increasingly clear to what influences the projectile is exposed immediately after leaving the muzzle, how those influences affect the motion of the

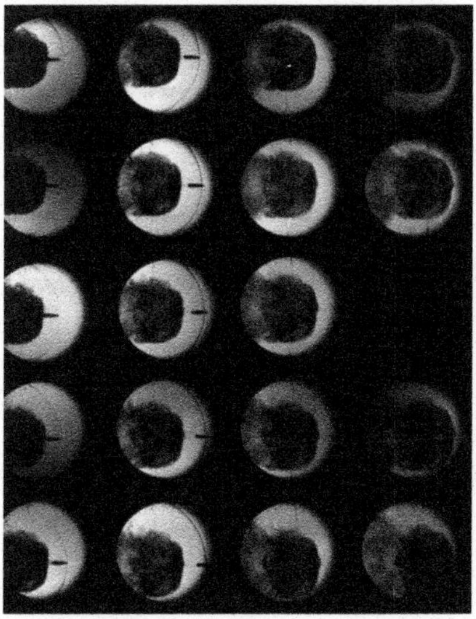

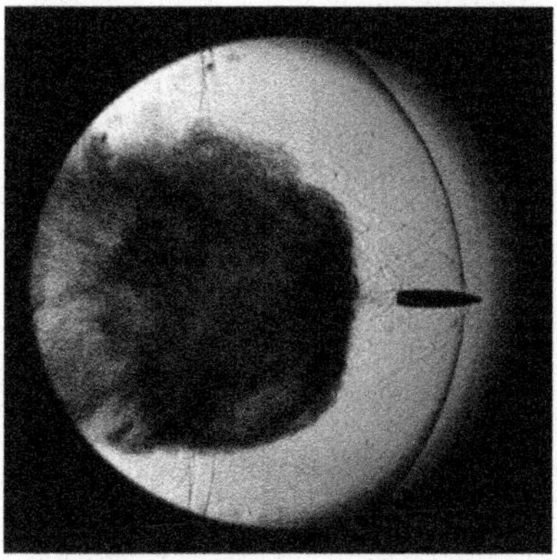

**Fig. 4.1.** *Left*: 20 individual images of a 7.5 × 55 Swiss (GP 11) bullet exiting the muzzle, produced by a Cranz-Schardin camera. Chronological order, earliest image at top. *Above*: Enlargement of image No. 8.



B. P. Kneubuehl, *Ballistics*,
https://doi.org/10.1007/978-3-662-69237-0_4

projectile and their effects on the firer or mounting. R. E. KUTTERER (1959) described the flow of gas from the muzzle and used the German term "Zwischenballistik", which literally means "between ballistics", i.e. "between" interior and exterior ballistics. R. GERMERSHAUSEN et al., (1980) used the German term "Abgangsballistik" (literally "discharge ballistics"), although that work examined only the angle of jump. "Abgangsballistik" is also the title of the present chapter in the original German. The book by C. L. FARRAR et al. (1983) devotes a whole chapter to this area of ballistics, under the title of "intermediate ballistics". That chapter covers not only gas flow but also flash, muzzle blast, silencers and muzzle brakes. Over time, intermediate ballistics expanded to also encompass the discarding of sabots from sub-calibre munitions.

Most processes related to the exit of the projectile from the muzzle do not lend themselves to a theoretical approach. One is therefore obliged to describe the phenomena and to focus on qualitative physical interpretation.

## 4.2 Muzzle phenomena

### 4.2.1 Gas flows

#### *4.2.1.1 Gas flows as the projectile leaves the barrel*

The pressure curves in 3.3.5 show that pressure reaches a maximum within the first few calibres of the projectile's displacement. The projectile is forced into the barrel and against the column of air under extremely high acceleration (see Table 2.8). This produces a shock wave in the air column. That shock wave precedes the projectile down the barrel and produces a precursor blast shock at the muzzle before

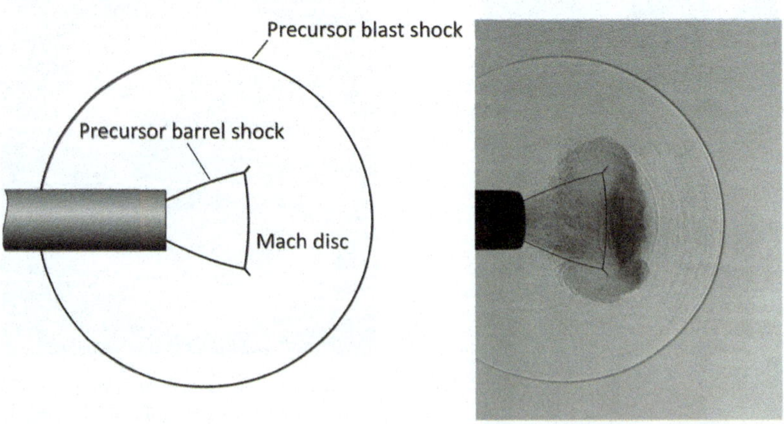


**Fig. 4.2.** **a**. Diagram and **b**. image of gas exiting the muzzle ahead of a projectile. The precursor blast shock is visible, but the ear cannot distinguish it from the muzzle blast.

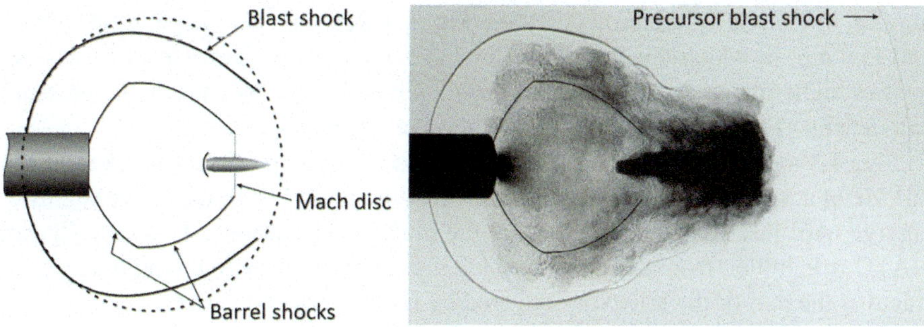


**Fig. 4.3.** **a**. Diagram and **b**. image of gas flow exiting the muzzle behind the projectile. Muzzle blast and shock waves are clearly visible, and the precursor blast shock is still present.

the projectile emerges (see Fig. 4.2). Because no projectile ever provides a completely gas-tight seal, the gases emerging from the muzzle also include propellant gas. The jet expands in accordance with the laws governing the discharge of gas from a muzzle against a negative pressure, producing small shock waves and a Mach disc, in which the flow slows from supersonic to subsonic.

The gases in the barrel are still under high pressure (muzzle pressure) when the projectile emerges from the muzzle. As a result of the large pressure difference relative to ambient pressure, expansion now produces a significantly more powerful shock wave, the *muzzle blast*. Disturbance due to the presence of the projectile and the gases behind it mean that the muzzle blast often expands in a pattern that deviates substantially from a sphere (see Figs 4.3 and 4.4). At the same time, high flow velocities occur in the muzzle, and gases overtake the projectile, disturbing its flight. Further shock waves are produced, known as barrel shocks, together with a Mach disc. As the outflowing gas velocity decreases, the Mach disc is sucked back through the muzzle, producing a rarefaction wave in the barrel.

In the case of artillery, this can mean that if the breech is opened too soon, propellant gas is ejected rearwards and can ignite on combining with oxygen in the atmosphere. A bore evacuator eliminates this backflow. See R. GERMERSHAUSEN, 1980.

The outflowing gas velocity is so high that a shock wave is always visible near the

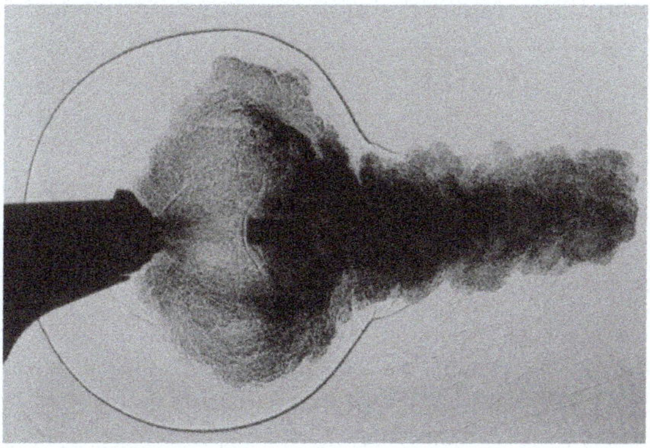

**Fig. 4.4.** Shadowgraph of a pistol muzzle (9 × 18 mm). Exposure time approximately 1 µs. Muzzle blast, shock waves and a shock wave behind the tail of the bullet are all clearly visible. The bullet is travelling at approximately 330 m/s.

muzzle, at the rear of the projectile. See Fig. 4.4, Fig. 4.5 (next page, third image) and Fig. 4.6, in which the first image includes a close-up. These point to gas velocities higher than the local speed of sound, which in turn depends on the pressure and density of the gases inside the "shock wave envelope".

Figs 4.5 and 4.6 show two series of shadowgraphs. The first was taken at the muzzle of a 357 Magnum revolver, and covers the period from when gas begins to emerge until just after the bullet clears the muzzle. The second shows a 7.62 mm NATO rifle bullet, from the time at which it exits the muzzle to the point at which it leaves the part of the trajectory covered by intermediate ballistics.

The images are all taken from the same position, but show different shots, with the camera triggered progressively later.

Because the gas is travelling at such high speed, its mass generates additional momentum, which combines with the projectile momentum to provide the total recoil momentum acting on the mounting or the firer. It is sometimes claimed that the gases exiting the muzzle further accelerate the projectile. However, it is clear that the resulting increase in velocity can be ignored, as the gas only acts on the projectile for a few tens of microseconds.

A very rough calculation shows that even at high gas densities and velocities, projectile velocity can only increase by less than 1 m/s.

#### 4.2.1.2 Outflowing gas velocity

The velocity of the gas as it leaves the muzzle plays an important role in determining recoil momentum and energy. Measuring this parameter is difficult, however, as the pressure, density and temperature of the gas all vary considerably in the vicinity of the muzzle.

For the sake of simplicity, let us assume that when the projectile passes through the muzzle, barrel pressure is uniform. Let us also assume that the gas is ejected from the barrel so rapidly that we can ignore the transfer of heat between the gas and its surroundings (i.e. we assume that the process is isentropic). The gases then leave the muzzle at Laval velocity (see Eqn 2.3:42) and with the mass flow rate given by Eqn 2.3:43. The momentum flow rate is therefore:

$$\dot{m} \cdot v_m = A_m \cdot \sqrt{\gamma \cdot p_i \cdot \rho_i \cdot \left(\frac{2}{\gamma+1}\right)^{\frac{\gamma+1}{\gamma-1}} \cdot \frac{2 \cdot \gamma}{\gamma+1} \cdot \frac{p_i}{\rho_i}} \; . \qquad [\mathrm{N}]$$

Simplifying, we obtain

$$\dot{m} \cdot v_m = A_m \cdot \gamma \cdot p_i \cdot \left(\frac{2}{\gamma+1}\right)^{\frac{\gamma}{\gamma-1}} , \qquad [\mathrm{N}] \tag{4.2:1}$$

where $v_m$ = mean outflowing gas velocity determining total momentum, $A_m$ = muzzle cross-sectional area, $p_i$ = internal pressure and $\gamma$ = heat capacity ratio.

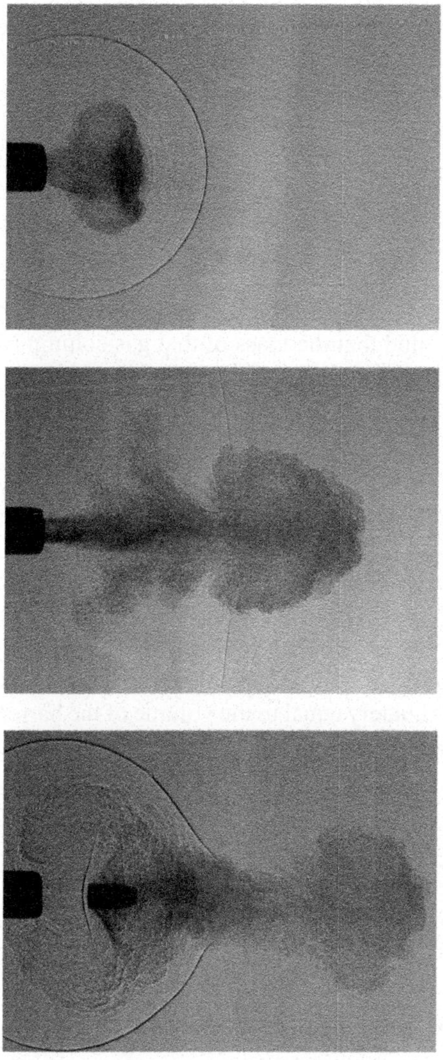

**Fig. 4.5.** *Above.* Shadowgraphs taken at the muzzle of a 357 Magnum revolver, from when gas begins to emerge until just after the bullet clears the muzzle.

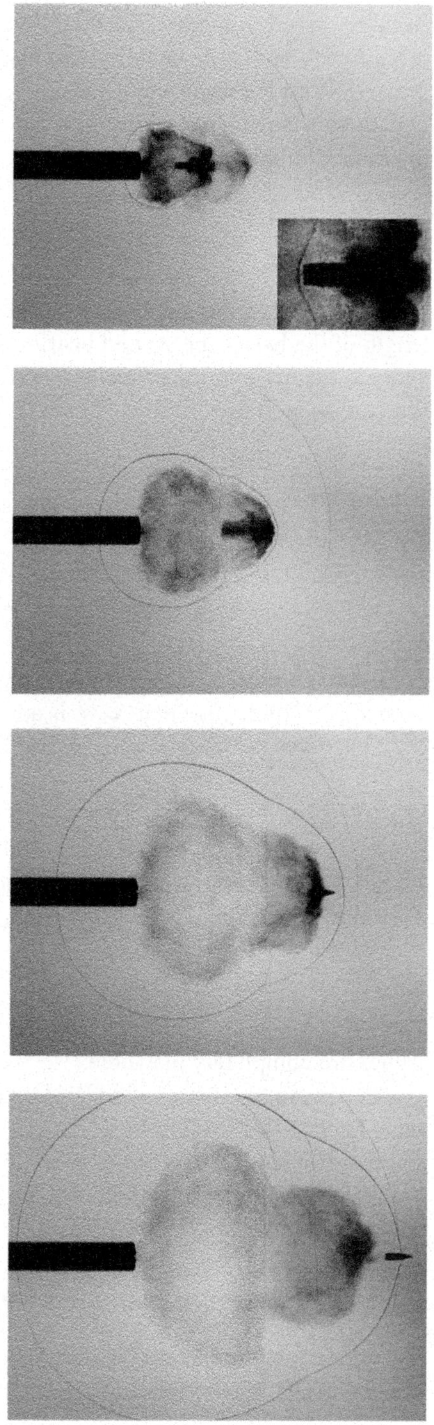

**Fig. 4.6.** *Right*: Shadowgraphs of a 7.62 mm NATO rifle bullet, from the time at which it exits the muzzle to the time at which it leaves the part of the trajectory covered by intermediate ballistics. The first frame includes a close-up showing more clearly how the propellant gases flow past the bullet.

Because these formulae assume negligible velocity inside the barrel, the calculation is relative to projectile velocity. We discretize Eqn 4.2:1 and introduce an arbitrary factor of ½ on the right-hand side, as we can be certain that mass flow rate does not remain constant for the entire time during which gas escapes through the muzzle (bearing in mind SÉBERT's factor):

$$\frac{\Delta m}{\Delta t} \cdot v_m = \frac{1}{2} \cdot A_m \cdot \gamma \cdot p_i \cdot \left( \frac{2}{\gamma+1} \right)^{\frac{\gamma}{\gamma-1}} . \qquad \text{[N]}$$

Let the length of the gas column in the barrel (which is roughly the same as the length of the barrel) be $\ell$, and bearing in mind that the mass of that gas column is equal to the mass of the propellant $m_c$ and that internal pressure is equal to muzzle pressure $p_m$ we can write

$$m_c \cdot v_m = \frac{1}{2} \cdot A_m \cdot \gamma \cdot p_m \cdot \left( \frac{2}{\gamma+1} \right)^{\frac{\gamma}{\gamma-1}} \cdot \frac{\ell}{v_m} , \qquad \text{[N·s]}$$

(4.2:2) $$v_m = \sqrt{\frac{p_m \cdot A_m \cdot \ell}{2 \cdot m_c} \cdot \gamma \cdot \left( \frac{2}{\gamma+1} \right)^{\frac{\gamma}{\gamma-1}}} . \qquad \text{[m/s]}$$

We use the typical value for the heat capacity ratio of propellant gas ($\gamma = 1.25$) and take account of the fact that $A_m \cdot \ell$ is approximately equal to the volume of the barrel $V_b$. This gives us the following approximate equation:

(4.2:3) $$v_m = 0.59 \cdot \sqrt{\frac{p_m \cdot V_b}{m_c}} . \qquad \text{[m/s]}$$

As this velocity is relative to that of the projectile $v_0$, we obtain the following expression for outflowing gas velocity $v_{ex}$ (exhaust velocity):

(4.2:4) $$v_{ex} = \sqrt{v_0^2 + v_m^2} . \qquad \text{[m/s]}$$

Table 4.1 shows a number of examples based on measured pressure curves. The values are completely plausible.

The literature contains very little information of this type. The OERLIKON Taschenbuch (1981) refers to a theory proposed by H. KRATZ, but cites no other sources. R. GERMERSHAUSEN et al. (1980) give an estimated value of $v_a = 1000$ m/s. Somewhat surprisingly, D. E. CARLUCCI et al. (2008) use the normal speed of sound in air for $v_m$.

#### 4.2.1.3 Gas flow from an open barrel

In saluting guns and alarm pistols, the load is only held in place lightly by wadding, and the barrel is otherwise open, leaving aside any baffles. The resulting pressure

**Table 4.1.** Approximate outflowing gas velocities from Eqns 4.2:3 and 4.2:4

| | | 9 mm Luger | 7.5 × 55 Swiss | 35 × 228 |
|---|---|---|---|---|
| Calibre | [mm] | 9 | 7.5 | 35 |
| Displacement | [mm] | 104 | 603 | 2930 |
| Muzzle pressure | [bar] | 342 | 721 | 560 |
| Mass of propellant | [g] | 0.305 | 2.92 | 322 |
| $v_0$ projectile | [m/s] | 356 | 795 | 1175 |
| $v_m$ | [m/s] | 507 | 479 | 412 |
| $v_{ex}$ | [m/s] | 620 | 928 | 1245 |

The values in the first two columns are taken from pressure measurements. Those in the third column are taken from OERLIKON Taschenbuch (1981).

distribution in the barrel is discussed in 3.4.7. The previous section uses gas dynamic equations to estimate outflowing gas velocity. In studying alarm pistols, gas velocity at the muzzle was measured directly, using three independent methods (B. P.KNEUBUEHL, 1998).

It is possible to measure the propagation velocity of the gas jet using multiple shadowgraphs with very short exposure times (of the order of 1 μs). Distortion due to perspective means that it is only possible to measure the velocity of the tip of the jet along its own axis. Velocity is calculated by measuring the distance covered between two images and dividing this by the time elapsed.

The second method involves measuring pressure along the axis of the jet at intervals of 30, 50 and 100 mm. Velocity is then calculated from pressure using BERNOULLI's equation plus Eqn 2.3:44. Setting $r = 0$ gives the velocity field of a free jet along its axis:

$$(4.2:5) \qquad v_x(x,0) = \frac{3}{8 \cdot \pi} \cdot \frac{I}{\rho \cdot \varepsilon \cdot (x + x_0)} \,. \qquad [\text{m/s}]$$

According to Eqn 2.3:44, velocity at the muzzle ($x = 0$) would be undefined. To make it possible to use the measured values, the starting point is displaced by $x_0$.

By performing a best-fit calculation over the three measuring intervals, it is possible to determine the factor $I/\varepsilon$ and $x_0$, and hence the velocity at the muzzle.

The third approach to calculating mean outflowing gas velocity involves determining the total muzzle momentum of the ejected gas. This can be achieved using a ballistic pendulum with a pendulum weight made of a material (such as gelatine) that can be fully penetrated by the gas jet. The maximum velocity $v_{pe}$ of the pendulum can be calculated from its deflection, using the conservation of momentum and of energy. By performing measurements using the same cartridge and different pendulum masses $m_{pe}$, we can use linear regression with the target function

$$(4.2:6) \qquad v_{pe} = i \cdot \frac{1}{m_{pe}} \,, \qquad [\text{m/s}]$$

**Table 4.2.** Outflowing gas velocity: Comparison between methods of measurement. 8 mm blank cartridge, NC propellant. (PTB = Physikalisch-Technische Bundesanstalt, the (German) National Metrology Institute).

| Weapon | Barrel length | Measured values [m/s] | | |
|---|---|---|---|---|
| PTB No. | [mm] | Pressure | Optical | Pendulum |
| 176/2 | 53 | | 3278<br>3057 | |
| 100/2 | 56 | 3334 | 2856<br>2968 | 2722 |
| 78 | 57 | | 3251<br>3300 | |
| 388 | 57 | | 3282 | |
| 437 | 125 | 2196 | 2288<br>2814 | 2079 |

to determine i, the total momentum of the gas emerging from the muzzle. The mass of the gas is known (as it is equal to the mass of the propellant), so we can calculate the (mean) outflowing gas velocity.

Table 4.2 shows a number of measured values of outflowing gas velocity. If one bears in mind the relatively large differences between individual blank cartridges and the difficult measuring conditions, the results are surprisingly comparable (with the exception of one measurement using the pressure method).

Pressure measurements made at the side of the gas flow field were used to determine kinematic viscosity ε, which in turn was used to approximately calculate the velocity field immediately after the muzzle using Eqn 2.3:44. Figures 4.7 and

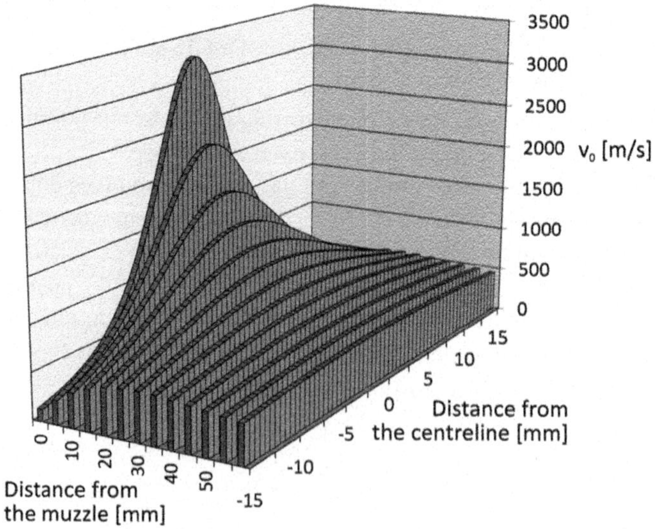


**Fig. 4.7.** Distribution of gas velocity at the muzzle of an alarm pistol (PTB No. 78).

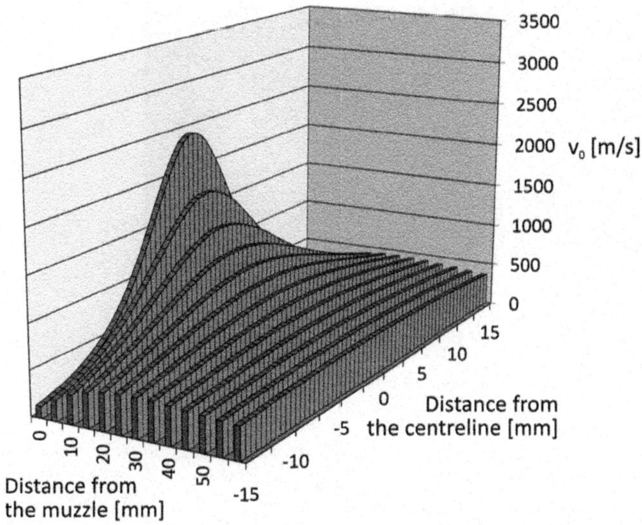


**Fig. 4.8.** Distribution of gas velocity at the muzzle of an alarm rifle (PTB No. 437).

4.8 show such velocity distributions for a pistol (PTB No. 78) and a rifle (PTB No. 437). See also Table 4.2. The very high outflowing gas velocity at the muzzle is worth noting, as is the rapid drop in velocity.

The free outflow of gas from an open barrel is obviously not comparable with the flow of gas behind a projectile as it exits the muzzle. The projectile prevents the gas from accelerating freely over the first few millimetres, which means that it does not attain the high velocities of a gas emerging from an open barrel.

### 4.2.2 Flash

Flash at the muzzle comes from two sources. In the case of a short-barrelled weapon, or when using progressive propellants, the projectile often leaves the barrel before the propellant has finished burning. The burning powder carried by the gases causes *primary flash*. In addition, the gases behind the projectile can ignite in front of the muzzle in combination with the oxygen contained in the atmosphere. This is known as *secondary flash*, and always occurs at a distance from the muzzle. Both are visible in Fig. 4.9 (next page), which is taken from a video. Secondary flash often takes the form of a ring, as the flammable gas/oxygen mix forms at a distance from the line of fire. Figs 4.10 and 4.11 show further examples, for very different calibres.

**Fig. 4.9.** Primary and secondary flash from a pistol chambered in 9 mm Luger. The secondary flash takes the form of a ring.

**Fig. 4.10.** Secondary flash from a 155 mm self-propelled howitzer. Gas emerging rearwards from the muzzle brake is also visible.

**Fig. 4.11.** Ring-shaped secondary flash from an 84 mm Carl Gustaf anti-tank weapon. The fireball from the recoil compensation jet is visible to the rear.

# 4.3 Barrel exit

## 4.3.1 Barrel oscillation and angle of jump

### *4.3.1.1 Origin and effects*

After the propellant is ignited, pressure rises very rapidly, reaching a maximum when the projectile has moved only a few calibres (see 3.3.5). At that point, the pressure in the barrel lies between a few hundred bar (in the case of small artillery loads) and several thousand (in the case of high-power guns and small arms). This results in a very large dynamic force, which acts not only on the base of the shot but also on the barrel, causing it to oscillate longitudinally and transversely. Oscillations of this type propagate very rapidly in steel: at approximately 5900 m/s longitudinally and 3200 m/s transversely. As a result, they affect the whole barrel a long time before the projectile reaches the muzzle. The transverse oscillations continually change the direction in which the axis of the barrel is pointing, which means that the barrel is not usually pointing in the intended direction when the projectile passes through the muzzle. In turn, this means that when the projectile leaves the barrel, its angle of departure does not correspond to the elevation set. The difference in angle is known as the *angle of jump*. Regardless of calibre, it will be of the order of 1–3 mils (approx. 3–10 minutes of arc).

In the case of a weapon on a mounting, the angle of jump is fairly constant. It is generally measured for each weapon and must merely be verified periodically. However, this angle can vary considerably for a shoulder-fired weapon. How the firer holds the weapon will directly affect barrel oscillation.

A group of precision shooters was asked first to fire normally, and then with the weapon pressed harder into the shoulder than usual. Scatter was similar in both cases, but for all participants the mean point of impact moved by 10–15 cm at a range of 100 m, which corresponds to an angle of 1–1.5 mils.

If two shooters use the same weapon without adjusting the sights, the angle of departure can easily vary by 1–2 mils, which will shift the mean point of impact by 30–60 cm at 300 m.

Given the extreme accuracy that precision and sport shooters can achieve, one can safely assume that barrel oscillation remains very similar if the weapon is always held in an identical fashion.

Occasionally, one encounters situations that clearly show the effect of barrel oscillation on mean point of impact.

Pistol ammunition was undergoing test firing as part of acceptance testing, using a test gun. It was noted that bullets with a lower muzzle velocity were hitting the target measurably higher than faster bullets. Initially, this appeared inexplicable. However, if one bears in mind that the faster bullets will spend less time in the barrel, it is clear that they could be leaving the muzzle just when the barrel is at the "up" end of its cycle. Fitting an additional weight to the barrel changed its moment of inertia and hence the frequency of the oscillations and hence the relationship between velocity and point of impact.

The angle of jump has less effect on artillery than on direct-fire weapons, corresponding to some 0.5–1% of range at lower elevations and decreasing sharply with increasing elevation.

Angle of jump can only be determined empirically. Various methods are available, with various degrees of accuracy.

#### *4.3.1.2 Determining angle of jump*

The simpler methods of determining angle of jump work on the assumption that barrel oscillation does not displace the muzzle transversely. In other words, the bore axis continues to pass through the origin of the frame of reference. This means that only the change in angle is measured.

A cardboard disc is placed at a certain distance from the muzzle (a few tens of metres in the case of artillery). The projection of the bore axis (T) is marked on the disc using a laser or optical boresight. See Fig. 4.12. The shot passes through the disc at point S, at a time $t_S$ after leaving the muzzle. During that time, the projectile is subject to gravity. If its trajectory had been a straight line, it would have hit the disc at a height

(4.3:1) $$y_A = y_S + \frac{1}{2} \cdot g \cdot t_S^2 \qquad [\text{m}]$$

above the height of the muzzle. On the basis of Fig. 4.12, we obtain the following expression for the angle of jump $\Delta\theta_0$:

(4.3:2) $$\Delta\theta_0 = \arctan\left(\frac{y_A}{x_S}\right) - \arctan\left(\frac{y_T}{x_S}\right). \qquad [\text{mil}]$$

If we measure the horizontal component of muzzle velocity, rather than flight time, and if we approximate the trajectory of the projectile by a parabola (which is acceptable in view of the short distance involved) then from Eqn 2.2:34 we obtain

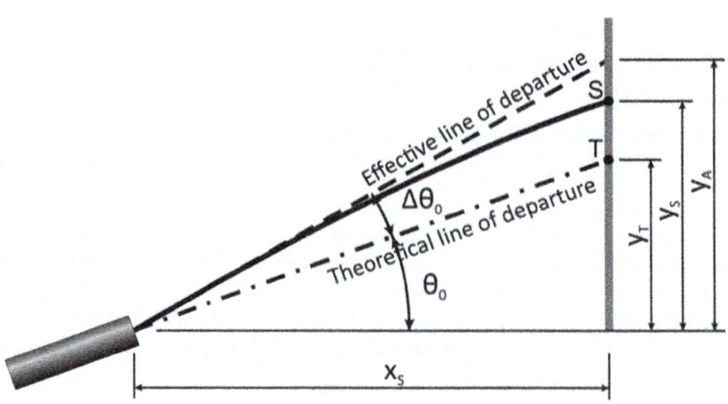


**Fig. 4.12.** Calculating the angle of jump (simple method). See text for details.

(4.3:3) $$y_S = x_S \cdot \tan(\theta_0 + \Delta\theta_0) - \frac{g \cdot x_S^2}{2 \cdot v_{0x}^2},$$ [m]

which in turn gives the following expression for the effective angle of departure:

(4.3:4) $$\theta_0 + \Delta\theta_0 = \arctan\left(\frac{y_S}{x_S} + \frac{g \cdot x_S}{2 \cdot v_{0x}^2}\right),$$ [mil]

where $\theta_0$ is the elevation of the barrel.

As for the previous method, we suppose that the muzzle lies on a line passing through the origin.

If we do not make this assumption, we need additional information, which we can obtain by using an additional cardboard disc (see Fig. 4.13). The effective angle of departure and the deflection of the muzzle from its position of rest are given by the system of equations below:

$$\left| \begin{array}{lcl} y_1 & = & y_0 + x_1 \cdot \tan(\theta_0 + \Delta\theta_0) - \dfrac{g \cdot x_1^2}{2 \cdot v_{0x}^2} \\ y_2 & = & y_0 + x_2 \cdot \tan(\theta_0 + \Delta\theta_0) - \dfrac{g \cdot x_2^2}{2 \cdot v_{0x}^2} \end{array} \right| .$$ [m]

This system of equations has the solutions

(4.3:5) $$\theta_0 + \Delta\theta_0 = \arctan\left(\frac{y_2 - y_1}{x_2 - x_1} + \frac{g \cdot (x_2 + x_1)}{2 \cdot v_{0x}^2}\right)$$ and [mil]

(4.3:6) $$y_0 = \frac{y_2 \cdot x_1 - y_1 \cdot x_2}{x_1 - x_2} + \frac{g \cdot x_1 \cdot x_2}{2 \cdot v_{0x}^2} .$$ [m]

We can go a step further; instead of assuming a parabolic trajectory, we can use an approximation formula such as that of PITON-BRESSANT (see 5.4.6.3). To do so, we

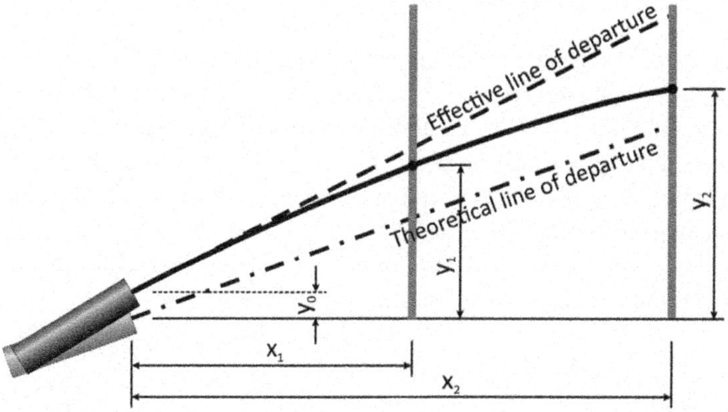


**Fig. 4.13.** Calculating the angle of jump and the deflection of the muzzle. See text for details.

require additional information – either the measured time of flight to one of the points along the trajectory or a further measuring point in the form of a third disc. This method is useful when it is necessary to place the discs at a greater distance from the muzzle, e.g. in the case of high muzzle pressure.

If the projectile is subject to pronounced yaw, it is important not to confuse yaw amplitude with angle of jump. The linear method is sufficient if one wishes to measure any transverse angle of jump.

The effects of angle of jump $\Delta\theta_0$ on the point of impact can be estimated using approximation formulae. For the lower group of angles, the vertical error at a given range is given by

$$(4.3{:}7) \qquad \Delta y = x_e \cdot \frac{1-\tan^2\theta}{\cos\theta} \cdot \Delta\theta_0 \qquad [\mathrm{m}]$$

and for the upper group of angles the range error at a given target height is given by

$$(4.3{:}8) \qquad \Delta x_e = x_e \cdot \left(1 - \frac{\sin 2\cdot(\theta + \Delta\theta)}{\sin 2\cdot\theta}\right), \qquad [\mathrm{m}]$$

where $x_e$ = range that would have been achieved with zero jump angle.

## 4.3.2 Recoil

### *4.3.2.1 Causes of recoil*

When the weapon is fired, the projectile and the propellant gas behind it are subjected to forward momentum. This produces rearward momentum of the same magnitude, the *recoil momentum*. That momentum is the direct cause of the mechanical forces that act on the mounting or the firer when the weapon is discharged.

### *4.3.2.2 Determining recoil momentum*

Theoretically, one should take account of the motion and friction of the projectile and of the propellant gas in the barrel when calculating total momentum. However, this would mean undertaking complex calculations involving data that are difficult to obtain and would not significantly improve accuracy. It is much simpler, and sufficiently accurate, if we calculate the momentum of the projectile and of the propellant gas as they pass through the muzzle.

In accordance with Eqn 2.2:12, *projectile momentum* $I_p$ is muzzle velocity $v_0$ multiplied by projectile mass $m_p$:

$$(4.3{:}9) \qquad I_p = m_p \cdot v_0 \,. \qquad [\mathrm{N{\cdot}s}]$$

If we disregard the shock and rarefaction waves in the barrel between the breech face and the base of the projectile, we can suppose that the gas at the breech face is stationary while that at the base of the projectile is moving at the same velocity as the projectile. As a linear approximation, we therefore suppose that the gas column is moving towards the muzzle at half the speed of the projectile. As the mass of the gas is equal to that of the propellant $m_c$, the momentum of the gas in the barrel $I_n$ is given by

(4.3:10) $$I_n = m_c \cdot \tfrac{1}{2} \cdot v_0 \; .$$ [N·s]

The two momentum components in Eqns 4.3:9 and 4.3:10 depend on $v_0$ and together make up the *muzzle momentum* $I_M$. Combining those two equations, we can write

(4.3:11) $$I_m = \left(m_p + \tfrac{1}{2} \cdot m_c\right) \cdot v_0 \; .$$ [N·s]

Once the projectile has exited the muzzle, the gases escape at velocity $v_{ex}$ as explained in 4.2.1.2. If we know the muzzle pressure, we can calculate that velocity approximately, using Eqns 4.2:3 and 4.2:4. Muzzle pressure can be determined using interior ballistics software such as that of H. BRÖMEL (2009) or the Heydenreich method. In some instances, measured values may be available.

If we do not have access to the data needed to calculate outflowing gas velocity $v_{ex}$, we can use the following estimates:

- shoulder weapons: 850–1050 m/s,
- handguns: 500–650 m/s.

The lower values are appropriate at low muzzle pressure (in the case of a long barrel or degressive propellant) and the higher values at higher pressures (short barrel or progressive propellant).

The momentum of the gases that expand after the projectile has exited the barrel is the *additional momentum* $I_N$.

**Table 4.3.** Data and momentum values for the bullets in Table 4.1

| | | 9 mm Luger | 7.5 × 55 Swiss | 35 × 228 |
|---|---|---|---|---|
| Calibre | [mm] | 9 | 7.5 | 35 |
| Mass of bullet | [g] | 8 | 11.3 | 550 |
| Mass of propellant | [g] | 0.305 | 2.92 | 322 |
| $v_0$ projectile | [m/s] | 356 | 795 | 1175 |
| $v_a$ from Eqn 4.2:4 | [m/s] | 620 | 928 | 1245 |
| Momentum of bullet (from Eqn 4.3:9) | [N·s] | 2.85 | 8.98 | 646 |
| Muzzle momentum (from Eqn 4.3:11) | [N·s] | 2.90 | 10.14 | 835 |
| Additional momentum acting on bullet (from Eqn 4.3:12) | [N·s] | 0.19 | 2.71 | 401 |
| Recoil momentum (from Eqn 4.3:13) | [N·s] | 3.09 | 12.85 | 1236 |

(4.3:12) $I_N = m_c \cdot v_{ex}$ [N·s]

*Recoil momentum* $I_R$ is equal to the sum of muzzle momentum and additional momentum. Its magnitude is the same as that of the total momentum.

(4.3:13) $I_R = \left(m_p + \frac{1}{2} \cdot m_c\right) \cdot v_0 + m_c \cdot v_{ex}$ . [N·s]

Table 4.3 gives the momentum values for the bullets in Table 4.1.

#### 4.3.2.3 Recoil velocity and energy

As a result of the recoil momentum, the weapon moves rearwards at a certain velocity $v_R$. From the equation for momentum and with weapon mass $m_W$, we obtain

(4.3:14) $v_R = \frac{I_R}{m_w}$ . [m/s]

However, the recoil velocity has no value other than to provide a degree of clarity. It cannot be used to evaluate recoil.

A barrel or weapon that moves backwards at a certain velocity possesses kinetic energy, *recoil energy* $E_R$. As recoil energy is a measure of the capacity of the moving weapon to do work, it is the only measurable parameter with which one can objectively assess recoil. From Equations 2.2:19 and 4.3:14 we obtain

(4.3:15) $E_R = \frac{I_R^2}{2 \cdot m_w}$ . [J]

A given quantity of recoil energy is absorbed by performing work in the form of braking force, which acts over a certain stopping distance (Eqn 2.2:16). In the case of artillery, this is achieved by tensioning springs or compressing a fluid.

For a person firing a weapon from the shoulder, this braking force (the *recoil force*) accounts for a large part of the perceived recoil. For a given recoil energy, stopping distance and recoil force are inversely proportional to one another. Increasing the stopping distance (e.g. by wearing padded clothing between rifle butt and shoulder) therefore reduces recoil force.

If a handgun is gripped firmly or a rifle pressed firmly into the shoulder, it becomes one with the corresponding part of the firer's body. When using Eqn 4.3:15 to calculate recoil energy, one must add a percentage of the firer's body mass to the mass of the weapon. This significantly reduces recoil energy.

In the case of a smaller weapon, one can measure recoil energy by suspending the weapon as a pendulum and measuring the deflection of the pendulum when the weapon is fired. The weapon can also be mounted on a low-friction cart and measurements taken of the velocity of the cart under the influence of recoil. The axis of the barrel must be so arranged that no rotation moment is produced when the weapon is fired (see Fig. 4.14).

**Fig. 4.14.** Measuring recoil energy. The barrel of the weapon is exactly in the plane of the wheel axles, so that no rotation moment is produced when the weapon is fired.

#### *4.3.2.4 Torque during firing*

For various technical reasons, virtually all weapons are designed with the barrel above the point at which the weapon is supported. As a result, recoil momentum acts at a certain distance from the point at which the required opposing force is applied (see Fig. 4.15). This generates a torque that causes the barrel to rotate upwards about a fulcrum. In the case of a mounted weapon, this torque must be countered by the mounting. A handgun or shoulder weapon will “kick” upwards.

Rotation begins as soon as the projectile starts its journey down the barrel. By the time a handgun bullet leaves the barrel, the muzzle will have risen by a number of millimetres. Because this process is readily reproducible, the firer does not notice it.

Rotation continues after the projectile has left the barrel. In the case of high recoil energies, the weapon may rise significantly above the original line of departure. The fulcrum lies approximately mid-way between the axis of the barrel and the point at which the weapon is supported. Let the distance between barrel axis and supporting point be r, and let $I_{rot}$ be the rotational component of the recoil momentum. Angular momentum L is then

(4.3:16) $$L = I_{rot} \cdot r \,. \qquad [\mathrm{kg \cdot m^2/s}]$$

From Eqn 4.3:15, energy of rotation is

(4.3:17) $$E_{rot} = \frac{L^2}{2 \cdot J_w} = \frac{I_{rot}^2 \cdot r^2}{2 \cdot J_w}\,, \qquad [\mathrm{J}]$$

where $J_w$ = moment of inertia of the weapon with respect to the fulcrum.

**Fig. 4.15.** Rotation moment, taking a pistol as an example. The weapon is supported below the line of action of the recoil momentum. $F_a$ = supporting force. Other symbols as in 4.3.2.2.

Let $I_t$ be the linear (translational) component of the recoil momentum and $E_t$ the corresponding energy. We can then use Eqns 4.3:15 and 4.3:17 to obtain the following energy equation:

$$(4.3{:}18)\qquad E_R = E_t + E_{rot} = \frac{I_t^2}{2\cdot m_w} + \frac{I_{rot}^2\cdot r^2}{2\cdot J_w}\,. \qquad [J]$$

The moment of inertia of the weapon $J_w$ can be estimated from $\beta\cdot m_w\cdot r^2$, where $\beta$ is a constant and $m_w$ is the mass of the weapon.

$$\frac{I_R^2}{2\cdot m_w} = \frac{I_t^2}{2\cdot m_w} + \frac{I_{rot}^2\cdot r^2}{2\cdot\beta\cdot m_w\cdot r^2}\,, \qquad [J]$$

$$(4.3{:}19)\qquad I_R^2 = I_t^2 + \frac{I_{rot}^2}{\beta}\,. \qquad [(N\cdot s)^2]$$

The total recoil energy is divided into rearwards recoil energy and rotational energy. The more effectively one counteracts rotation, the more strongly one feels the rearwards recoil. The more one allows the weapon to rotate, the less one feels the rearwards recoil. As the sum of these two energy components is always equal to total recoil energy, this forms a suitable parameter for evaluating recoil.

### *4.3.2.5 Means of influencing recoil*

The gases behind the projectile exit the muzzle at high speed (see 4.2.1.2). In conjunction with their mass, this results in a large quantity of energy. It therefore seems logical to use this energy to reduce recoil and rotation moment. All that is required is to direct the gases that expand roughly spherically after leaving the muzzle (see Fig. 4.3) in the opposite direction to the projectile by means of a *muzzle brake* installed on the barrel.

There are several designs using baffles installed in front of the muzzle (active muzzle brakes) or apertures in the side of the barrel (reactive muzzle brakes). The degree to which recoil is counterbalanced is determined by the angle of deflection. An angle of deflection greater than 90° is required to achieve a satisfactory effect

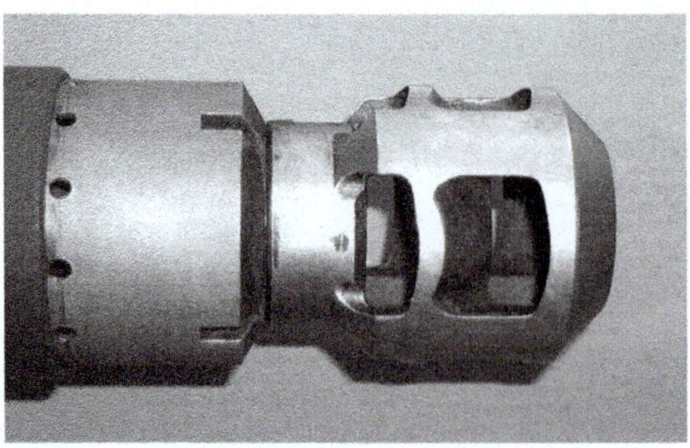

**Fig. 4.16.** Muzzle brake of a medium-calibre gun, with two chambers (two baffles mounted one behind the other). Gases are directed not only rearwards to the side, but also upwards and downwards.

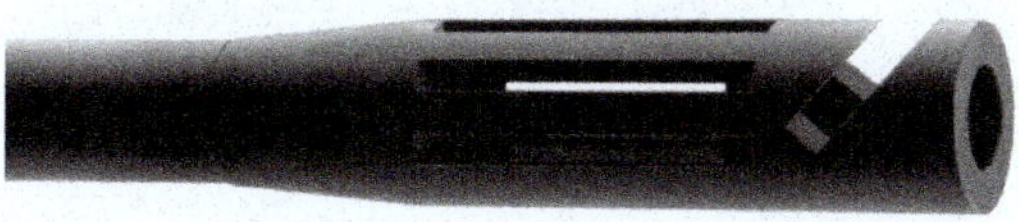

**Fig. 4.17.** Muzzle brake with compensator on a SIG Sauer 2000 precision rifle.

(see Fig. 4.16). However, such angles expose the gun crew to high blast pressures and also to propellant gas, so muzzle brakes are usually fitted to guns only if the gun crew is protected.

For small arms, reactive devices are generally used, with round or slit-shaped openings at the sides. To minimize the effect on the firer, these openings are directed perpendicularly to the direction of fire or no more than slightly rearwards. One occasionally encounters devices with openings directed upwards and forwards. These are designed to counteract the upwards kick mentioned in 4.3.2.4 (see Fig. 4.17) and are sometimes referred to as *compensators*.

However, lateral apertures have an additional purpose; they ensure better dispersion of the gases behind the projectile and hence reduce secondary flash, even if they do not reduce recoil. Devices of this type are generally known as *flash hiders* (see Fig. 4.18).

### 4.3.3 The sabot-discarding process

In 3.7.6, we encountered the first ballistic paradox, for which conical barrels were proposed as a possible solution. The second, more common way of ensuring that sectional density is low inside the barrel and high outside is to use a sub-calibre projectile of the type mentioned in the same section. Such projectiles must discard their guiding components after leaving the barrel of the weapon.

Those guiding components are known as *sabots*. In the case of spin-stabilized projectiles, they are discarded by means of centrifugal force, whereas fin-stabilized projectiles use aerodynamic forces. Verification that a projectile discards its sabot without suffering disturbance is generally performed using imaging methods, such as shadowgraphy or incident light photography (see Figs 4.19 and 4.20).

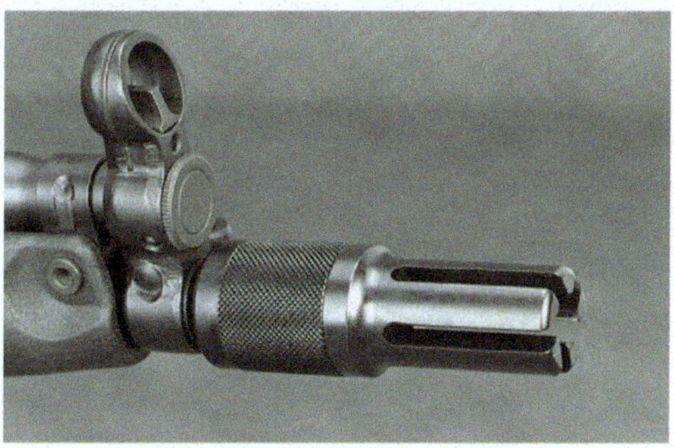

**Fig. 4.18.** Simple flash hider fitted to the muzzle of a small-calibre weapon. This device is not designed to reduce recoil.

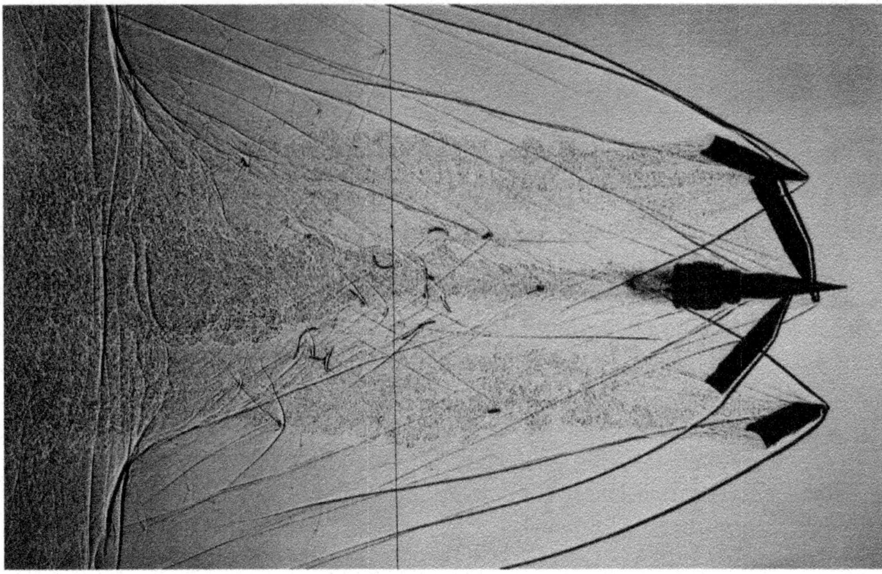

**Fig. 4.19.** Shadowgraph of a spin-stabilized 20 mm sub-calibre projectile discarding the front part of the sabot. Its rear part has not yet separated from the tail of the projectile.

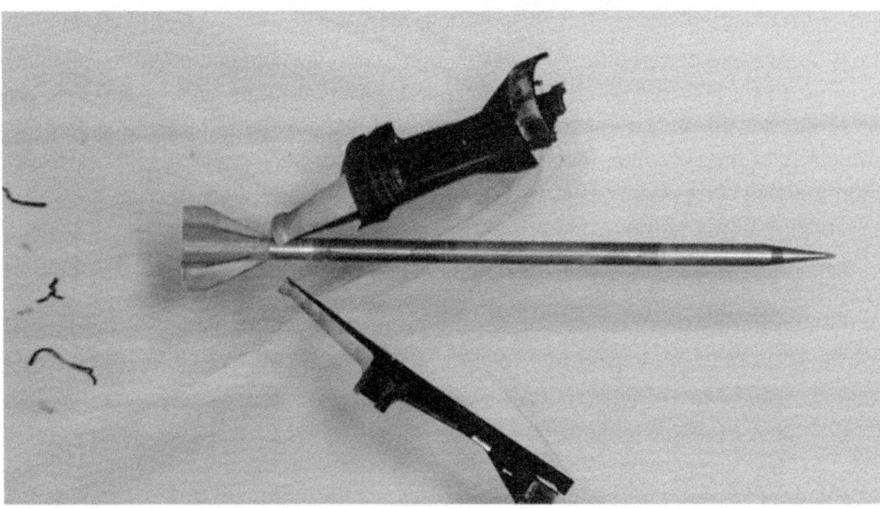

**Fig. 4.20.** Incident light photograph of a sabot separating from a long-rod projectile. Airflow engages with the leading edges of the sabot components, tipping them backwards.

# 5 Exterior ballistics

## 5.1 History

Exterior ballistics is the most fundamental branch of ballistics. The flight or movement of an arrow or a ball attracted the attention of ancient Greek thinkers. Examples include ZENON's arrow paradox (ZENON lived from approximately 490 to 430 BC). The first description of a projectile trajectory appears as part of kinematics in the philosophical works of ARISTOTLE (384–322 BC), who devoted considerable attention to motion and repose in "Physics, Lectures on Nature". In discussing the change of position of a body, he distinguished between natural and unnatural (or violent) movements. He classed continued movement in a straight line as "natural" and movements involving a change in direction as "unnatural", completely excluding the possibility that there could be any combination of the two.

This resulted in a three-part trajectory model (see Fig. 5.1). The first part consisted of a "natural" movement in a straight line from A to B, the second part was a "violent" curved section (B to C) and the third and final part was once again "natural" free fall from C to D. This Aristotelian trajectory model was accepted for almost 2000 years.

It was not until the 16th Century that another scientist investigated the flight of projectiles: the Italian Nicolò TARTAGLIA (1500–1557).

TARTAGLIA owes his fame primarily to a procedure for solving the general cubic equation, and the related court case that set him against doctor and mathematician Girolamo CARDANO (1501–1576).

In 1537, TARTAGLIA published a work in Italian titled "Nova Scientia", which was translated into English, with a commentary in Matteo VALLERIANI, 2013. This was

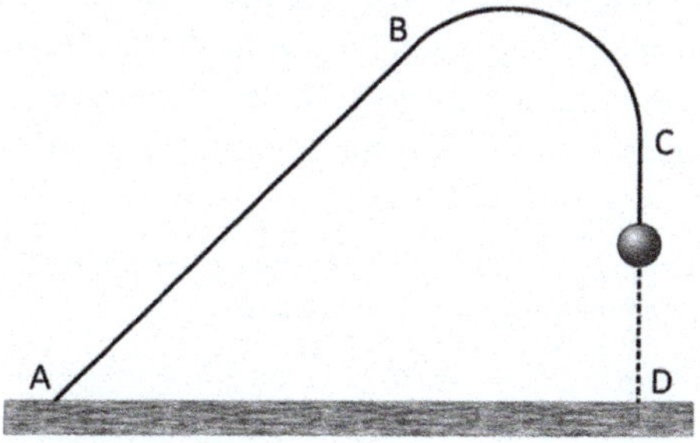


**Fig. 5.1.** Aristotelian trajectory. The trajectory starts with a "natural" straight line, followed by a "violent" curved section and ending with "natural" free fall.



B. P. Kneubuehl, *Ballistics*,
https://doi.org/10.1007/978-3-662-69237-0_5

the first attempt to build a theory of exterior ballistics on a scientific (geometrical) basis. TARTAGLIA based his work in part on that of Euclid, in that he attempted from a small number of basic rules (axioms) to derive a limited number of equations that would exactly describe the trajectory of a projectile. However, he lacked the fundamental concept of acceleration. As a result, his trajectories were still completely under the influence of Aristotle, with the geometric form of Fig. 5.1, although he had postulated the existence of curved trajectory segments in both the ascending and descending branches. He was the first to answer the question regarding the elevation required to achieve maximum range, concluding from the fact that range is zero at elevations of 0° and 90° that a projectile will attain maximum range when fired at an elevation of 45°.

A further century elapsed before the next, decisive step towards understanding the ballistic trajectory. It was no less a person that Galileo GALILEI (1564–1641), who in 1638 presented a comprehensive treatment of kinematics in "Discorsi", his principal work on the subject of physics.

GALILEO wrote his "Discorsi" in the form of a conversation between three people that took place over four days. Because of his difficulties with the Roman Catholic church, the book was published not in Italy, but in Leiden, by Elsevier. Trajectories are among the topics discussed on the fourth day.

On the basis of his discoveries concerning gravitational acceleration, GALILEO proved in the fourth part of his "Discorsi" that a trajectory consists of two independent movements: horizontal motion, of which the speed remains constant because of inertia, and vertical motion, with constant downwards acceleration. The resulting trajectory takes the form of a quadratic parabola. However, he had discovered during his study of falling objects that drag increases with speed, which means there must exist some height at which drag equals weight and motion becomes uniform. If a trajectory were to extend to a great height, this law would mean that it could not take the form of a parabola.

However, if the projectile is not too light and the height not too great, deviation from the quadratic law of falling bodies remains small. Given this empirical result, GALILEO retained the parabola as the form of a trajectory in such cases. He did, however, point out that this approximation was not valid for the trajectories of projectiles discharged from firearms, on account of their "unnatural momentum".

Despite GALILEO's discoveries, ballisticians treated all trajectories as parabolas until Isaac NEWTON (1643–1727) created an entirely new basis for mechanics in his "Principia" (1687 and 1713). In drop tests using large, hollow glass spheres, NEWTON discovered that drag must be considerable, and assumed that it was proportional to velocity squared. However, he said nothing about the form of the resulting trajectory.

It is interesting to note that NEWTON had already recognized that this assumption regarding drag meant that the meridian curve of a projectile of minimum drag must be transcendental.

A number of other famous scientists addressed the calculation of trajectories in the second half of the 17th Century and during the 18th, including Christiaan HUYGENS (1629–1695) and Johann BERNOULLI (1667–1748). A breakthrough occurred when English scientist Benjamin ROBINS (1707–1751) invented the ballistic pendulum, making it possible to measure the velocity of a projectile. He described this in "New Principles of Gunnery" (1742), in which he addressed not only the problem of the "resisting power of the air" and the form of a trajectory, but also – probably for the first time – the "force" of gunpowder, i.e. interior ballistics. This work contained so many new ideas that the great Leonhard EULER (1707–1783) translated it into German, despite its mathematical errors and shortcomings. He enhanced the work with additional material and explanations, thereby creating the first complete theory of the motion of a body in flight (see Fig. 5.2).

EULER continued to study the problem of modelling a trajectory. In what was probably his most important work on the subject (L. EULER, 1755) he wrote:

*Although this question can readily be reduced to a purely analytical problem, the great NEWTON has searched in vain for a solution, despite his inspired investigations. The finding of a solution under the assumption that resistance is proportional to velocity having been so successful, it is difficult to imagine that one should*

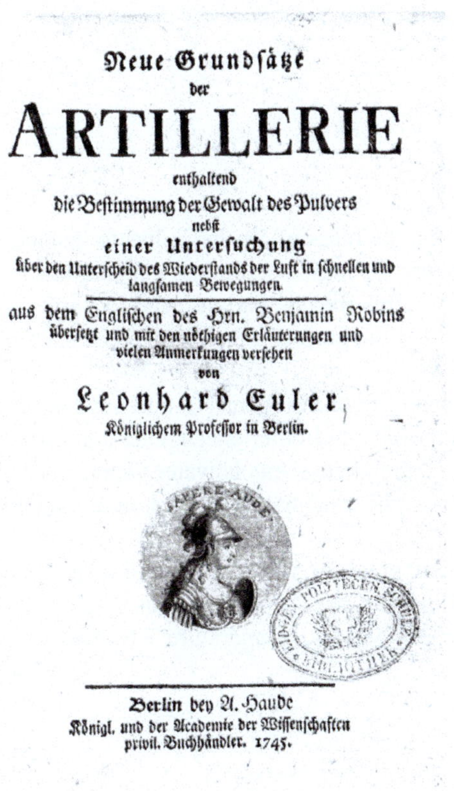
Neue Grundsätze
der
ARTILLERIE
enthaltend
die Bestimmung der Gewalt des Pulvers
nebst
einer Untersuchung
über den Unterscheid des Wiederstands der Luft in schnellen und langsamen Bewegungen
aus dem Englischen des Hrn. Benjamin Robins
übersetzt und mit den nöthigen Erläuterungen und vielen Anmerkungen versehen
von
Leonhard Euler,
Königlichem Professor in Berlin.

Berlin bey A. Haude
Königl. und der Academie der Wissenschaften privil. Buchhändler. 1745.

**Fig. 5.2.** The title page of the German version of Benjamin ROBINS' work, translated and amended by Leonhard EULER. The book was published in Berlin in 1745. It became so well known that 40 years later it was translated into French and back into English. Title page of a copy held by the Eidgenössischen Polytechnischen Schule (Federal Polytechnical College), today's ETH Zürich.

*not achieve the aim if one assumed resistance to be proportional to the square of velocity, in view of the far more difficult problems that have already been solved. It was Johann BERNOULLI who first solved this problem and described the form of the curve."*

However, he went on to add that one could derive no practical use from this solution, however good it might be in theory. He stated that the main reason was that in practice the parameters would include not only initial velocity and angle of departure, but also the mass and diameter of the projectile. In place of an analytical solution, he proposed the compiling of tables, an idea that has remained in use in modern times (e.g. in SIACCI's methods).

Two parameters make it impossible to solve the system of general exterior ballistic equations of motion analytically: the form- and velocity-dependent drag coefficient of the projectile (which is also influenced by the Mach number via temperature) and the decrease in atmospheric density with altitude.

Until the second half of the 20th Century, one was therefore confronted in exterior ballistics with a problem very similar to that encountered in interior ballistics: the system of differential equations describing a trajectory could not be solved analytically. As a result, ballisticians devoted considerable effort to finding the best possible approximations. Only when efficient numerical methods and the computers to apply them became available did physical ballistic questions take centre stage.

## 5.2 The atmosphere

### 5.2.1 Physical characteristics

The drag on a projectile depends in part on the mass of air that it has to displace. As is generally the case for gases, we use not the mass itself in our calculations but the density, i.e. the mass per unit volume. This means that we need to know the atmospheric density at the current position of the projectile in order to calculate its trajectory.

Air is a mixture of gases, consisting primarily of nitrogen ($N_2$) and oxygen ($O_2$) together with carbon dioxide ($CO_2$), argon (Ar) and other gases. One component that has a significant influence on atmospheric phenomena is water vapour ($H_2O$). The percentage of water vapour in the atmosphere is highly variable, unlike that of the other gases.

The state of the atmosphere is defined by pressure, temperature and relative humidity. For the purposes of ballistics, it may be seen as an ideal gas. This means that the relationship between pressure p, density ρ and temperature T (in Kelvin) is given by Eqn 2.3:3, i.e.

$$(5.2:1) \qquad \frac{p}{\rho} = R \cdot T \,. \qquad [\mathrm{J/kg}]$$

We can measure atmospheric pressure using a barometer and temperature with a thermometer, and can then calculate the atmospheric density needed to determine aerodynamic forces using Eqn 5.2:1. With atmospheric pressure p in millibar (mbar) or hectopascal (hPa) and temperature $T_C$ in Celsius, we obtain the following equation for atmospheric density (in kg/m³):

$$(5.2:2) \qquad \rho = \frac{100 \cdot p}{R_A \cdot (T_C + 273.15)} . \qquad [\text{kg/m}^3]$$

where $R_A$ is the special gas constant of air. The value for dry air, $R_{A,d} = 287.06$ J/(kg·K), can be taken as a good approximation.

However, the gas constant increases with atmospheric humidity, which means that atmospheric density decreases. If we require greater precision, we can take this factor into account using the following equation for the gas constant of moist air:

$$(5.2:3) \qquad R_{A,m} = \frac{R_{A,d}}{1 - r_h \cdot \frac{p_{wa}}{p} \cdot \left(1 - \frac{R_{A,d}}{R_{wa}}\right)} . \qquad [\text{J/(kg·K)}]$$

Where $R_{A,m}$ = the gas constant of moist air, $R_{A,d}$ = the gas constant of dry air = 287.06 J/(kg·K), $R_{wa}$ = the gas constant of water vapour = 461.52 J/(kg·K), p = current atmospheric pressure, $p_{wa}$ = the saturation vapour pressure of water (both in mbar) und $r_h$ = relative humidity (relative to 1).

Saturation vapour pressure depends on temperature. We can obtain an approximate value for this parameter using the Magnus equation (D. SONNTAG 1990):

$$(5.2:3a) \qquad p_{wa} = 6.112 \cdot \exp\left(\frac{17.62 \cdot T_C}{243.12 + T_C}\right) . \qquad [\text{mbar}]$$

As humidity has only a minor influence on atmospheric density (generally less than 1%), we usually ignore it in the case of flat-trajectory weapons with short projectile flight times.

Atmospheric pressure results from the mass of the atmosphere and therefore decreases as altitude increases. Temperature also plays a decisive role here; it likewise decreases as pressure decreases (see Eqn 5.2:1). The mean change in temperature in the various layers that make up our atmosphere is linear. This means that the increase or decrease in temperature is constant for a given change in altitude. The change in temperature per unit altitude is known as the *temperature gradient* $\tau$. If temperature change is linear, this gradient is constant:

$$(5.2:4) \qquad dT = \tau \cdot dh . \qquad [\text{K}]$$

Pressure drop in layer dh is given by:

$$(5.2:4a) \qquad dp = -\rho \cdot g \cdot dh . \qquad [\text{mbar}]$$

Replacing $\rho$ by Eqn 5.2:1, taking dh from Eqn 5.2:4, separating the variables and integrating between height $h_1$ (with $p_1$ and $T_1$) and height $h_2$ (with $p_2$ and $T_2$), we obtain:

(5.2:5) $$p_2 = p_1 \cdot \left(1 + \frac{\tau \cdot \Delta h}{T_1}\right)^{-\frac{g}{R_A \cdot \tau}},$$ [mbar]

where $\Delta h = h_2 - h_1$, $g$ = acceleration due to gravity and $R_A$ = the special gas constant for air. ($T_1$ in Kelvin).

At greater altitudes (over 10 km), temperature may be constant within certain altitude ranges. In such cases, we cannot calculate pressure drop using Eqn 5.2:5 (as $\tau = 0$) but we can obtain it directly, by integrating Eqn 5.2:4a (after replacing $\rho$ by Eqn 5.2:1).

(5.2:6) $$p_2 = p_1 \cdot e^{-\frac{g \cdot \Delta h}{R_A \cdot T}}.$$ [mbar]

We can calculate the corresponding air densities by substituting the pressure values into Eqn 5.2:2.

In the case of flat trajectories, in which the highest point of the trajectory (the apex) is only slightly higher than the points of departure and impact, no significant changes in pressure are to be expected. Temperature will also remain virtually constant over the entire flight of the projectile, given the short time of flight. We can therefore assume constant atmospheric density.

Atmospheric pressure and temperature not only change with altitude but are also strongly influenced by weather. The question therefore arises as to which data one should use in creating tools such as range tables. Standard atmospheres have been created for this purpose, in which change in pressure and temperature as a function of altitude is precisely specified.

## 5.2.2 Standard atmospheres

### *5.2.2.1 General*

We can calculate change in pressure as a function of altitude using Eqn 5.2:5, as long as we know the pressure and temperature at a specific altitude, the *reference altitude*, and as long as the temperature gradient is constant. If we assign a subscript 0 to values at reference altitude $y_0$, pressure at a altitude y is given by:

(5.2:7) $$p(y) = p_0 \cdot \left(1 + \frac{\tau}{T_0} \cdot (y - y_0)\right)^{-\frac{g}{R_A \cdot \tau}},$$ [mbar]

where $g$ = acceleration due to gravity and $R_A$ = is the special gas constant of air. $T_0$ in Kelvin.

If temperature is constant within an atmospheric layer, we can use Eqn 5.2:6:

(5.2:8) $$p(y) = p_0 \cdot e^{-\frac{g \cdot (y - y_0)}{R_A \cdot T_0}}.$$ [mbar]

To define a standard atmosphere we therefore need to establish a reference altitude, together with the pressure and temperature at that altitude. Such atmospheres

**Table 5.1.** Geopotential altitudes in the ICAO standard atmosphere.

| Lower boundary [m] | Upper boundary [m] | Effective altitude [m] | Temperature gradient τ [K/m] | Level name |
|---|---|---|---|---|
| 0 | 11,000 | 11,019 | -0.0065 | Troposphere |
| 11,000 | 20000 | 20,063 | 0.0 | Tropopause |
| 20,000 | 32000 | 32,162 | 0.001 | Stratosphere |
| 32,000 | 47,000 | 47,350 | 0.0028 | Stratosphere. |
| 47,000 | 51,000 | 51,413 | 0.0 | Stratopause |
| 51,000 | 71,000 | 71,802 | -0.0028 | Mesosphere |
| 71,000 | 84,850 | 86,000 | -0.002 | Mesopause |

assume that temperature changes in steps, remaining constant within each atmospheric layer. Acceleration due to gravity is usually assumed to also be constant. As a result, atmospheric pressure and density are determined solely by temperature.

#### 5.2.2.2 The ICAO standard atmosphere

The best-known and most widely-used standard atmosphere is that published by the International Civil Aviation Organisation (ICAO). Sea level was chosen as reference altitude, with the following standardized values:

- Atmospheric pressure: 1013.25 mbar
- Temperature: 288.15 K, (= 15.0 °C)
- Acceleration due to gravity: 9.80665 $m/s^2$

This gives an atmospheric density of 1.225 $kg/m^3$.

Because of the assumption that acceleration due to gravity is constant, geopotential altitudes are used to define the standard atmosphere by means of temperature gradients (see Table 5.1).

Geopotential altitude is the altitude at which a body possesses the same potential energy at constant acceleration due to gravity as it does at the corresponding geometric altitude when subjected to the true acceleration due to gravity, which decreases with altitude. Geopotential altitude is therefore always slightly lower than geometric altitude.

The standard is not arbitrary, of course. It is based on observations and measurements conducted over many years at many locations, and is so designed that the standard atmosphere corresponds approximately to the mean state of the actual atmosphere. Table 5.2 shows temperature, pressure and density for the ICAO standard atmosphere, up to and including the stratosphere.

**Table 5.2.** ICAO standard atmosphere

| Geopototential altitude [m] | Atmospheric pressure [mb] | Temperature [°C] | Temperature [K] | Atmospheric density [kg/m³] | g(y) [m/s²] |
|---|---|---|---|---|---|
| 0 | 1013.25 | 15.00 | 288.15 | 1.225 | 9.8067 |
| 250 | 983.6 | 13.40 | 286.53 | 1.196 | 9.8059 |
| 500 | 954.6 | 11.80 | 284.90 | 1.167 | 9.8051 |
| 750 | 926.3 | 10.10 | 283.28 | 1.139 | 9.8043 |
| 1000 | 898.7 | 8.50 | 281.65 | 1.112 | 9.8036 |
| 1250 | 871.8 | 6.90 | 280.03 | 1.085 | 9.8028 |
| 1500 | 845.6 | 5.30 | 278.40 | 1.058 | 9.8020 |
| 1750 | 819.9 | 3.60 | 276.78 | 1.032 | 9.8013 |
| 2000 | 795.0 | 2.00 | 275.15 | 1.006 | 9.8005 |
| 2500 | 746.8 | -1.25 | 271.90 | 0.957 | 9.7990 |
| 3000 | 701.1 | -4.50 | 268.65 | 0.909 | 9.7974 |
| 3500 | 657.6 | -7.75 | 265.40 | 0.863 | 9.7959 |
| 4000 | 616.4 | -11.00 | 262.15 | 0.819 | 9.7943 |
| 4500 | 577.3 | -14.25 | 258.90 | 0.777 | 9.7928 |
| 5000 | 540.2 | -17.50 | 255.65 | 0.736 | 9.7913 |
| 5500 | 505.1 | -20.75 | 252.40 | 0.697 | 9.7897 |
| 6000 | 471.8 | -24.00 | 249.15 | 0.660 | 9.7882 |
| 6500 | 440.3 | -27.25 | 245.90 | 0.624 | 9.7867 |
| 7000 | 410.6 | -30.50 | 242.65 | 0.590 | 9.7851 |
| 7500 | 382.5 | -33.75 | 239.40 | 0.557 | 9.7836 |
| 8000 | 356.0 | -37.00 | 236.15 | 0.525 | 9.7821 |
| 8500 | 331.0 | -40.25 | 232.90 | 0.495 | 9.7805 |
| 9000 | 307.4 | -43.50 | 229.65 | 0.466 | 9.7790 |
| 9500 | 285.2 | -46.75 | 226.40 | 0.439 | 9.7775 |
| 10,000 | 264.4 | -50.00 | 223.15 | 0.413 | 9.7759 |
| 10,500 | 244.7 | -53.25 | 219.90 | 0.388 | 9.7744 |
| 11,000 | 226.3 | -56.50 | 216.65 | 0.364 | 9.7729 |
| 20,000 | 54.75 | -56.50 | 216.65 | 0.088 | 9.7454 |
| 32,000 | 8.68 | -44.50 | 228.65 | 0.0132 | 9.7089 |
| 47,000 | 1.11 | -2.50 | 270.65 | 0.0014 | 9.6635 |

#### 5.2.2.3 *Other standard atmospheres*

The atmospheric profile departs from the norm in certain areas such as the Alps, deserts and northern regions. In such areas it makes sense to use a different standard atmosphere. Measurements conducted in the mid-20th Century revealed the temperature gradient over Switzerland to be 1 °C/1000 m less than that of the ICAO standard. A specific standard atmosphere has therefore been defined for military purposes in Switzerland, with a standard pressure at sea level of 1018.7 mbar, a standard temperature of 283.95 K (= 10.8 °C) and a temperature gradient of –0.0055 K/m. This has made it possible to significantly increase the accuracy of calculations for trajectories of which the apex height corresponds to Alpine altitudes.

Certain other countries or groups of countries have their own standard atmospheres. In some cases this is for historical reasons (e.g. in eastern countries), while in others (e.g. Switzerland) such atmospheres are designed to take account of the geographical situation. The standard atmosphere in the countries of the former Soviet Union, for instance, had a standard temperature at sea level of 288.9 K and a temperature gradient of –0.006328 K/m. France also has its own standard atmosphere.

### 5.2.3 True conditions

#### *5.2.3.1 Pressure and temperature*

In most cases, one does not know the actual makeup of the atmosphere at the time of firing. Ballistic calculations are therefore normally carried out using a standard atmosphere. How accurately one can predict the trajectory depends on the degree to which the actual atmosphere diverges from the standard. The temperature profile is the most important factor in this.

For long-range weapons such as artillery and heavy mortars, the influence of the atmosphere is so significant that it is necessary to measure the current temperature profile before firing. Using this profile and the atmospheric pressure measured at a reference altitude (generally that of the muzzle), one can determine the current atmospheric density at a given altitude. This results in significantly more accurate predictions of trajectory and point of impact.

Conditions may depart significantly from those of the standard in winter, for instance, if inversion of the temperature gradient occurs. Under such conditions, air temperature at higher altitudes (generally above a layer of cloud) is greater than at lower altitudes. In other words, temperature initially rises with altitude and then falls, hence the term "inversion". Another exceptional situation may occur in hot summer weather, when solar radiation raises ground temperatures to much higher levels than normal. Under such conditions, temperature drops significantly more rapidly in the lower layers of the atmosphere than predicted by the standard atmosphere.

Such temperature variations have little effect on flat-trajectory weapons. In most cases, one need only measure pressure and temperature at muzzle altitude and use the normal temperature gradient for calculations.

#### *5.2.3.2 Wind*

Wind is the result of pressure differentials in the atmosphere. The direction and speed of the moving air depend on the positions of the high and low-pressure centres, and on the difference in pressure between them. The direction is also influenced by the Coriolis force of the rotating Earth.

In meteorology, one always labels a wind according to the direction from which it comes. In ballistics, however, one uses a velocity vector, i.e. the direction in which the wind is blowing.

At ground level, wind flow is turbulent because of relief and the roughness of the surface. As altitude increases, the reduction in friction causes velocity to increase, and the Coriolis force causes the wind to turn to the right in the northern hemisphere and the left in the southern. We therefore have to distinguish between wind at ground level and wind in the unobstructed upper atmosphere.

In the case of neutral thermal layering (where rising air masses have the same temperature profile as the atmosphere), wind speed increases almost logarithmically at altitudes of up to approximately 100 m .

At greater altitude, one encounters layers of air with approximately constant wind speed and direction. These layers can be from several hundred metres to a few kilometres in height. The boundaries between layers appear to be narrow, even sharp.

From a ballistics point of view, we should remember that for flat trajectories it is only the wind near to ground level that counts, and that the time of flight – i.e. the time for which the projectile is exposed to the wind – is very short; a matter of seconds. Steep trajectories often involve flight times of a few tens of seconds, with the projectile passing through the upper layers of the atmosphere for most of its flight. It will pass through the turbulent layers near the ground so rapidly that they will scarcely affect its trajectory.

## 5.3 Forces and moments acting on a projectile

### 5.3.1 Earth-related accelerations

#### *5.3.1.1 Centrifugal acceleration*

The ballistic coordinate system used to describe a trajectory is always linked to Earth, and is hence an accelerated system. It is therefore necessary to take the related accelerations into account. Centrifugal acceleration acts in the opposite direction to Earth's centripetal acceleration (see Eqn 2.2:10) and may be written as $a_f = r \cdot \omega^2$. It always acts perpendicular to Earth's axis of rotation. The defining radius r is that of Earth at the latitude at which the movement commences (see Fig. 5.3). Let $\beta$ be latitude, $R_E$ Earth's radius and $\omega_E$ Earth's angular velocity, then if $R_E = 6{,}371{,}000$ m and $\omega_E = 7.292115 \cdot 10^{-5}$ 1/s we can write:

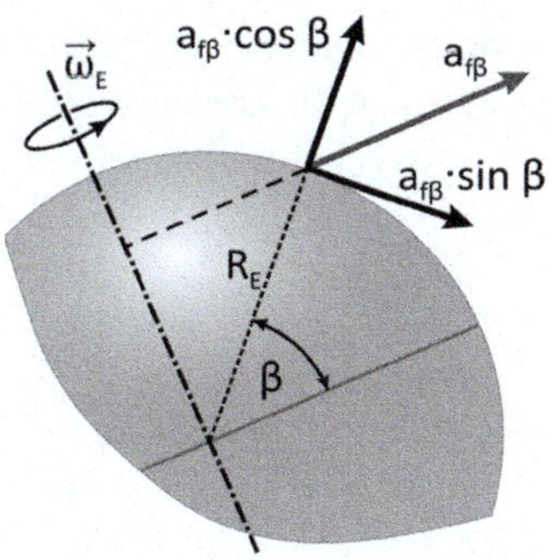


**Fig. 5.3.** The radial and tangential components of centrifugal acceleration. $\beta$ = latitude.

(5.3:1) $$a_{f\beta} = \omega_E^2 \cdot R_E \cdot \cos\beta \qquad [m/s^2]$$

In practice, we divide centrifugal acceleration into a radial component and a tangential component relative to the north-south direction. The radial component acts against acceleration due to gravity and is generally subsumed into it. However, the tangential component has to be transformed into the current ballistic coordinate system, and we must take account of its effect in our calculations. The two components are:

(5.3:2a) $$a_{fr} = a_{f\beta} \cdot \cos\beta = \omega_E^2 \cdot R_E \cdot \cos^2\beta\ , \qquad [m/s^2]$$

(5.3:2b) $$a_{ft} = a_{f\beta} \cdot \sin\beta = \omega_E^2 \cdot R_E \cdot \cos\beta \cdot \sin\beta\ . \qquad [m/s^2]$$

Centrifugal acceleration increases with height above the ground, on account of the increasing distance:

(5.3:3) $$a_f(h) = \omega_E^2 \cdot (R_E + h) \cdot \cos\beta = a_{f\beta} \cdot \left(1 + \frac{h}{R_E}\right)\ . \qquad [m/s^2]$$

For typical ballistic ranges, we can use the latitude of the point of departure for our calculations and can safely ignore any change in latitude along the trajectory. We can also ignore the increase in radius due to the height of the projectile, which we should theoretically add to the latitude radius. The error in centrifugal force is less than one per mille, even at great apex heights, and is of the same order of magnitude as the error caused by taking a mean value for Earth's radius.

### *5.3.1.2 Acceleration due to gravity*

Calculations at Earth's surface and at low altitudes are generally carried out using the standard value for acceleration due to gravity, which for the ICAO standard atmosphere is $g_n = 9.80665$ m/s$^2$. However, acceleration due to gravity consists not only of attraction (gravitation) but also of the radial component of the centrifugal acceleration due to Earth's rotation. These two components change with distance from Earth's surface in accordance with different laws. That applying to centrifugal acceleration is given in Eqn 5.3:3, and also applies to the radial component.

*Acceleration due to gravity* is governed by the law of gravitation. Let the mass of Earth be $m_E$, Earth's radius $R_E$ and the gravitational constant G. Then at Earth's surface (i.e. at sea level):

$$(5.3:4) \qquad g_0 = G \cdot \frac{m_E}{R_E^2}, \qquad [m/s^2]$$

and at an altitude h:

$$(5.3:5) \qquad g(h) = G \cdot \frac{m_E}{(R_E + h)^2}. \qquad [m/s^2]$$

Dividing Eqn 5.3:5 by Eqn 5.3:4 gives us

$$\frac{g(h)}{g_0} = \frac{R_E^2}{(R_E + h)^2} = \left(1 + \frac{h}{R_E}\right)^{-2}. \qquad [-]$$

Expanding the last expression into a series, considering only the linear term and solving for g(h), we obtain:

$$(5.3:6) \qquad g(h) = g_0 \cdot \left(1 - \frac{2 \cdot h}{R_E}\right). \qquad [m/s^2]$$

As acceleration due to gravity decreases from the equator towards the poles, $g_0$ is not constant, but varies with latitude thus:

$$(5.3:7) \qquad g_0(\beta) = 9.814 \cdot \left(1 + 1.834 \cdot 10^{-3} \cdot \sin^2 \beta - 1.981 \cdot 10^{-5} \cdot \sin^2 2 \cdot \beta\right). \qquad [m/s^2]$$

For latitudes of between 46° and 55°, we can take a constant value of $g_0 = 9.825$ m/s² with no noticeable loss of accuracy.

The same applies to acceleration due to gravity:

$$(5.3:8) \qquad g_e(\beta) = 9.7805 \cdot \left(1 + 5.266 \cdot 10^{-3} \cdot \sin^2 \beta - 5.9 \cdot 10^{-6} \cdot \sin^2 2 \cdot \beta\right). \qquad [m/s^2]$$

Between the latitudes mentioned above, we can safely use a constant value of $g_{e0} = 9.8111$ m/s².

We then obtain the change in acceleration due to gravity with increasing altitude by combining Eqns 5.3:6 and 5.3:3, rearranging the latter to solve for the radial component of centrifugal acceleration:

$$g_e(h) = g_0 \cdot \left(1 - \frac{2 \cdot h}{R_E}\right) - a_{fr0} \cdot \left(1 + \frac{h}{R_E}\right). \qquad [m/s^2]$$

However, at Earth's surface we can write $g_0 = g_{e0} + a_{fr0}$

$$g_e(h) = \left(g_{e0} + a_{fr0}\right) \cdot \left(1 - \frac{2 \cdot h}{R_E}\right) - a_{fr0} \cdot \left(1 + \frac{h}{R_E}\right), \qquad [m/s^2]$$

and by simplifying, we obtain:

(5.3:9) $$g_e(h) = g_{e0} \cdot \left(1 - \frac{2 \cdot h}{R_E}\right) - a_{fr0} \cdot \frac{3 \cdot h}{R_E} .$$ [m/s²]

The correction term on the right-hand side of this equation is very small and is only of relevance at high altitudes (over 10,000 m).

### 5.3.1.3 Coriolis acceleration

In 2.2.7.2, we introduced the Coriolis acceleration, to which a body is subjected when it moves outwards from the centre of a rotating disc in a radial direction. This acceleration – which acts only within the rotating system – acts perpendicular to the vector of angular velocity and to the velocity of the body itself, so in this two-dimensional case it acts in a tangential direction.

We cannot use Eqn 2.2:53 to calculate the Coriolis acceleration acting on a moving projectile, as its velocity vector changes as it travels along its trajectory. We need to use the general form of the equation of motion in a rotating coordinate system, of which the Coriolis component is as follows (W. GREINER, 1989):

(5.3:10) $$\vec{a}_c = 2 \cdot (\vec{v} \times \vec{\omega}) .$$ [m/s²]

Where $\vec{v}$ = the velocity of the projectile and $\vec{\omega}$ = Earth's angular velocity.

The Coriolis acceleration therefore acts perpendicularly to the plane formed by Earth's axis and the direction in which the body is moving. Its direction therefore changes continuously along the trajectory. We hence need to determine its vector components in a ballistic coordinate system of which the origin A lies at latitude $\beta$ and of which the line of departure and a line pointing north enclose the angle $\alpha$. $\alpha$ is the *azimuth*, the angular displacement from North, in a clockwise direction.

We can see from Fig. 5.4 that it is possible to decompose angular velocity at the origin of the ballistic coordinate system A into one component tangential to Earth's surface and one that is perpendicular to it. The tangential component is projected onto the x and z axes of the ballistic coordinate system, as shown in Fig. 5.5. For this system, we can write:

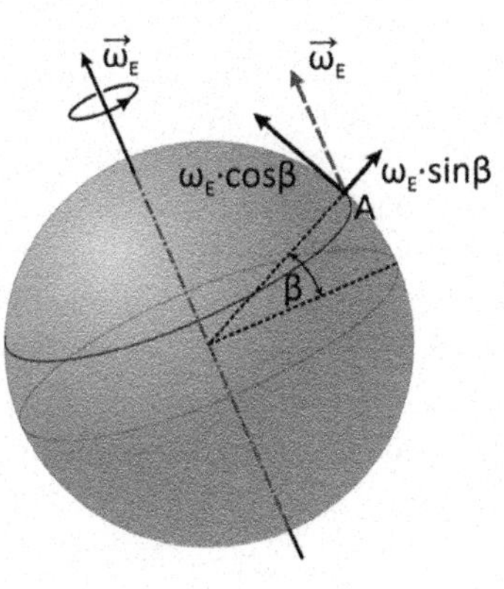


**Fig. 5.4.** (Left). Decomposing angular velocity into a tangential and a perpendicular direction.

**Fig. 5.5.** (Right). Decomposing the tangential components in accordance with the directions of the ballistic coordinate system.

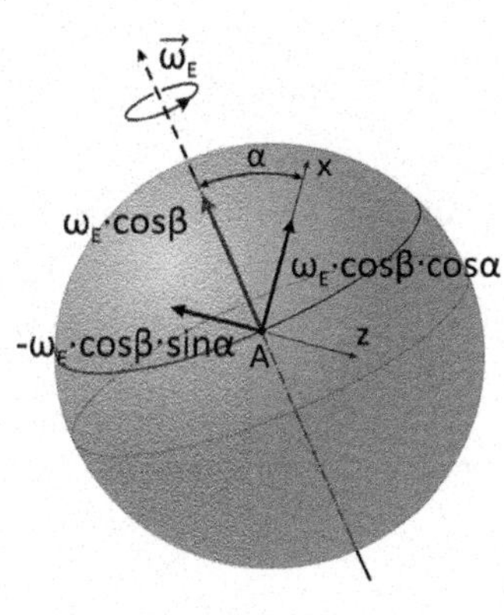

(5.3:11) $\vec{\omega}_E = \omega_E \cdot (\cos\beta \cdot \cos\alpha\,,\ \sin\beta\,,\ -\cos\beta \cdot \sin\alpha)\ .$ [1/s]

In the same system, the projectile velocity vector is:

(5.3:12) $\vec{v} = (v_x\,,\ v_y\,,\ v_z)\ .$ [m/s]

The vector product of these two (in accordance with Eqn 5.3:10) is the vector of the Coriolis acceleration in the ballistic coordinate system:

(5.3:13) $$\vec{a}_c = 2 \cdot \omega_E \cdot \begin{pmatrix} -v_y \cdot \cos\beta \cdot \sin\alpha - v_z \cdot \sin\beta \\ v_z \cdot \cos\beta \cdot \cos\alpha + v_x \cdot \cos\beta \cdot \sin\alpha \\ v_x \cdot \sin\beta - v_y \cdot \cos\beta \cdot \cos\alpha \end{pmatrix} .$$ [m/s²]

This vector can be decomposed into one direction-dependent component and one component that is independent of direction:

(5.3:14) $\vec{a}_c = \vec{a}_{cu} + \vec{a}_{c\alpha}\ ,$ [m/s²]

where $\vec{a}_{cu} = 2 \cdot \omega_E \cdot \sin\beta \cdot (-v_z, 0\,, v_x)\ ,$ [m/s²]

and $$\vec{a}_{c\alpha} = 2 \cdot \omega_E \cdot \cos\beta \cdot \begin{pmatrix} -v_y \cdot \sin\alpha \\ v_z \cdot \cos\alpha + v_x \cdot \sin\alpha \\ -v_y \cdot \cos\alpha \end{pmatrix} .$$ [m/s²]

We can readily include Coriolis acceleration in a trajectory calculation in the form of Eqn 5.3:14.

The question now is how significant the effect on trajectory of this acceleration due to Earth's rotation can be, and in which direction it acts. We shall simplify things by assuming that the z component of velocity is zero. This allows us to express the trajectory velocity of the projectile as a function of the angle between the tangent to the trajectory and the horizontal (the *angle of trajectory* $\theta$) thus:

(5.3:15) $\vec{v} = v \cdot (\cos\theta\,, \sin\theta\,, 0)\ .$ [m/s]

Substituting into Eqn 5.3:10, in conjunction with Eqn 5.3:11, this gives us

(5.3:16) $$\vec{a}_c = 2 \cdot v \cdot \omega_E \cdot \begin{pmatrix} -\sin\theta \cdot \cos\beta \cdot \sin\alpha \\ \cos\theta \cdot \cos\beta \cdot \sin\alpha \\ \cos\theta \cdot \sin\beta - \sin\theta \cdot \cos\beta \cdot \cos\alpha \end{pmatrix} .$$ [m/s²]

This allows us to examine the magnitude and the direction of Coriolis acceleration separately. Fig. 5.6 shows the change in direction of Coriolis acceleration along the trajectory (the spatial vector) for three azimuths, at a latitude of 47°N. A projectile fired northwards experiences only a lateral acceleration. If the angle of departure is greater than the angle of the latitude, acceleration acts to the left; if it is less, acceleration acts to the right. For all other directions, the acceleration vectors change

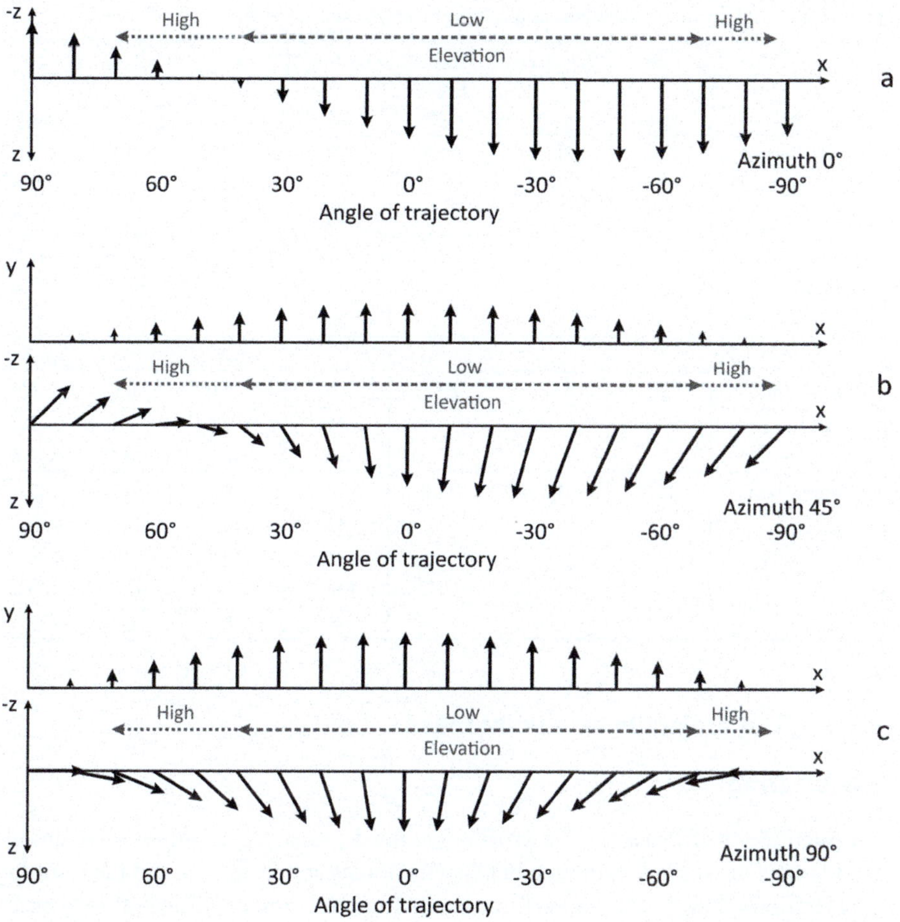


**Fig. 5.6.** The direction of Coriolis acceleration in the ballistic coordinate system as a function of line of departure (azimuth) and angle of trajectory (the angle of the tangent to the trajectory). **a**. Shot fired northwards. **b**. Shot fired north-eastwards. **c**. Shot fired eastwards. a has only a z component (x and y are both zero). b and c are shown in plan and side view. Trajectories from left to right. Angle of departure positive, angle of descent negative. Angle of trajectory for all flat trajectories around 0°.

along the trajectory. Projectiles fired in the opposite direction experience a change of sign. The effect on them is therefore inverted or acts in the opposite direction.

Table 5.3 summarizes deviation due to Coriolis acceleration for a number of typical projectiles. The time of flight of the projectile clearly has a major influence on the order of magnitude of these values.

**Table 5.3.** Trajectory deviations as a result of Coriolis acceleration.

| | Calibre [mm] | Distance [m] | Time of flight [s] | Azimuth [°] | x [mm] | y [mm] | z [mm] |
|---|---|---|---|---|---|---|---|
| Rifle bullet | 7.62 | 300 | 0.42 | 0 | 0.0 | 0.0 | 7.0 |
| | | | | 45 | 0.0 | 4.6 | 7.0 |
| | | | | 90 | -0.0 | 6.6 | 7.0 |
| Medium-calibre projectile | 20 | 1000 | 1.16 | 0 | 0.0 | 0.0 | 67.0 |
| | | | | 45 | -0.1 | 44.3 | 67.0 |
| | | | | 90 | -0.1 | 62.6 | 67.1 |
| | | | | | [m] | [m] | [m] |
| Mortar bomb | 81 | 2700 | 29.5 | 0 | 0.0 | 0.0 | 2.6 |
| | | | | 45 | -1.4 | 3.1 | 3.2 |
| | | | | 90 | -2.0 | 4.3 | 4.6 |
| Artillery shell | 155 | 30,000 | 77.0 | 0 | 0 | 0 | 102 |
| | | | | 45 | -26 | 91 | 113 |
| | | | | 90 | -36 | 129 | 138 |
| | | | | 135 | -26 | 91 | 164 |
| | | | | 180 | 0 | 0 | 175 |

## 5.3.2 The projectile as a point mass

### *5.3.2.1 Drag*

To calculate the trajectory of a profile, we need to know not only the conditions obtaining at the start of its motion but also the forces acting upon it, and the resulting accelerations. The most obvious force stems from the pressure distribution over the surface of the projectile, which is governed by BERNOULLI's Law. This corresponds to the surface integral of the product of pressure p and surface element $d\vec{A}$, and is known as *drag*

$$(5.3{:}17) \qquad \vec{F}_D = \int_O p \cdot d\vec{A} \; . \qquad [\mathrm{N}]$$

We saw in 2.3.4.3 that the pressure generated by flow is referenced to the stagnation pressure on a large plane surface. Because the surface areas of real bodies can vary widely, it is useful also to assign the reference pressure a fixed *reference area* $A_0$. Together with the stagnation pressure, this gives us a fixed reference force, the *dynamic pressure resistance* $F_{DP}$:

$$(5.3{:}18) \qquad F_{DP} = \frac{1}{2} \cdot \rho \cdot v^2 \cdot A_0 \; . \qquad [\mathrm{N}]$$

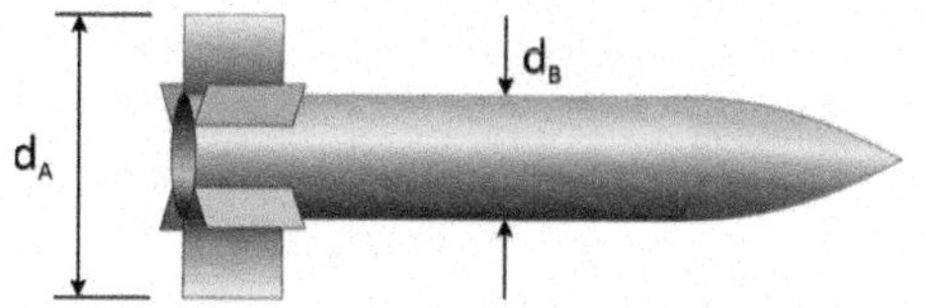


**Fig. 5.7.** Reference areas used to determine $C_D$.
- Cross-sectional area of largest diameter $d_A$; this is the value generally used in aerodynamics.
- Cross-sectional area of the nominal calibre; $d_B$: this is the value used in ballistics.

The true drag experienced by an object travelling through the air is always referenced to this dynamic pressure resistance. The ratio between drag $F_D$ and dynamic pressure resistance is called the *drag coefficient* $C_D$.

$$C_D = \frac{F_D}{F_{DP}}, \qquad (5.3{:}19) \quad [-]$$

$$F_D = C_D \cdot \frac{1}{2} \cdot \rho \cdot v^2 \cdot A_0 \,. \qquad (5.3{:}20) \quad [N]$$

Where $\rho$ = atmospheric density, $v_0$ = projectile velocity and $A_0$ = reference area.

In Eqn 5.3:20, drag, atmospheric density and velocity are all fixed physical quantities. The drag coefficient therefore depends on the reference area selected. Large reference areas result in small values of $C_D$ and vice versa. When looking at a $C_D$ curve for a body, one must always check the surface $A_0$ on which it is based. In ballistics, this is generally the cross-sectional area of the nominal calibre (which is usually different from the cross-sectional area of the projectile), whereas aerodynamicists often use the cross-sectional area of the largest diameter of the body. This can lead to very different $C_D$ values for the same drag in the case of a finned rocket (see Fig. 5.7).

Pressure distribution over the body subjected to flow and the area concerned are heavily affected by flow velocity and the form of the projectile. These relationships affect the drag coefficient, which will therefore depend on the shape of the body and on flow velocity. Furthermore, as we saw in 2.3.6.5, in the case of stationary, frictionless flow (which we can assume for projectiles) it is not the velocity that determines the type of flow but the Mach number. $C_D$ is therefore plotted against Mach number and not velocity. Figure 5.8 shows some typical examples.

#### *5.3.2.2 Weight*

After drag, the most basic force affecting a projectile in free flight is its weight.

$$F_W = m_p \cdot g \,. \qquad (5.3{:}21) \quad [N]$$

The usual decomposition of weight, into a tangential component and a component perpendicular to that, shows that in the ascending branch of the trajectory the tangential component reinforces drag and helps decelerate the projectile, whereas in the descending branch it counteracts the braking effect of the air. See Fig. 5.9 a and b.

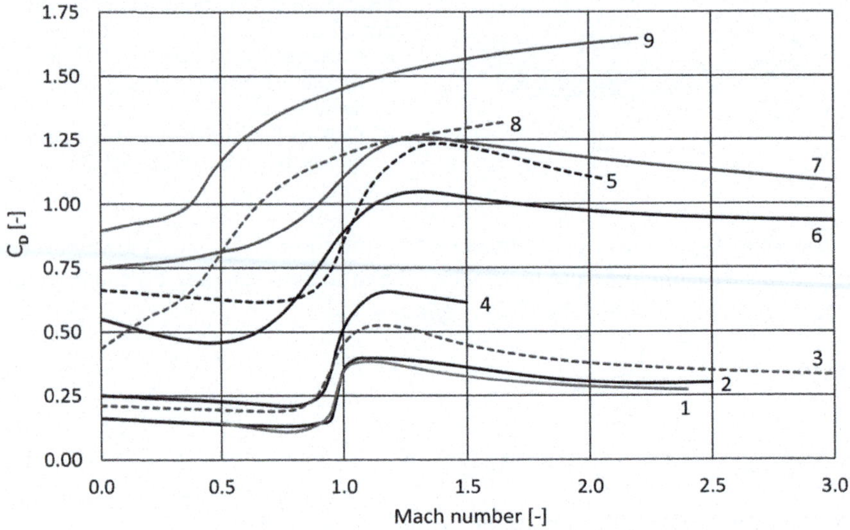


**Fig. 5.8.** Drag coefficient $C_D$ for a number of projectiles. *1*: 155 mm HE Shell, *2*: 7.5 mm rifle bullet, *3*: 5.56 × 45 (SS92) rifle bullet, *4*: Pistol bullet (9 mm Luger) *5*: Slug (12/70 calibre), *6*: Sphere, *7*: Cube, *8*: Cylinder, diameter/length ratio (d/l) = 1:2.6, *9*: Cylinder, d/l = 1:4.

The perpendicular component, however, as a central force, causes a continuous change in direction and is hence responsible for the curvature of the trajectory.

At greater ranges, Earth's curvature means that the gravitational acceleration vector in the ballistic coordinate system becomes less and less perpendicular to the x-z plane. As a result, the weight of the projectile acquires a component $F_{WR}$ that acts in the opposite direction to that in which the projectile is moving:

$$(5.3{:}22) \qquad F_{WR}(x) = m_p \cdot g \cdot \sin\left(\frac{x}{R_E}\right) \qquad [N]$$

where x = range and $R_E$ = Earth's radius.

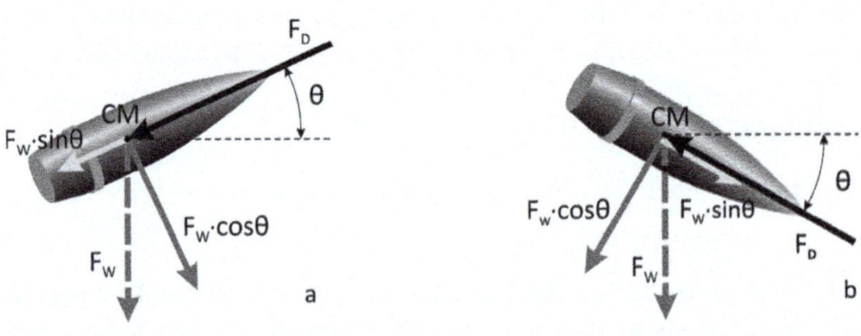


**Fig. 5.9.** Projectile weight $F_W$ and its components. **a**. In the ascending branch. **b**. In the descending branch. Where θ = angle of trajectory (angle between tangent to the trajectory and the horizontal) and $F_D$ = drag.

### 5.3.3 The projectile as a rigid body

#### *5.3.3.1 The motion of a projectile*

Most projectiles fired from tube weapons are stabilized by being caused to spin very rapidly about their longitudinal axis, through the action of spiral grooves in the barrel – the rifling. The projectile becomes a heavy, symmetrical gyroscope. Fig. 5.10 shows the analogy between a fixed, heavy gyroscope and a free-flying projectile. In the case of a gyroscope, the centre of rotation O (the tip of the gyroscope) is fixed, and a force (the weight of the gyroscope) acts at the *centre of mass* CM. The distance between CM and O causes this weight to exert a torque on the gyroscope. A projectile, however, rotates about its centre of mass CM, and an aerodynamic force acts at a point AF (the *point of application of the aerodynamic force*), located between the tip of the projectile and its centre of mass. Like the gyroscope, the projectile also experiences a torque, on account of the distance between AF and CM, the *lever arm of the aerodynamic force* $e_A$.

As a heavy, symmetrical gyroscope, the spin-stabilized projectile makes the precession movement described in 2.2.8.4. As a result, its longitudinal axis does not lie along the direction of travel, but instead rotates about the tangent to the trajectory, describing a conical surface of which the tip lies at the centre of mass of the projectile. The direction of travel and the axis of the projectile form an angle, known as the *angle of attack*. The plane created by the sides of the angle of attack is the *resistance plane*.

A projectile is usually also subjected to the nutation described in 2.2.8.3. This causes an additional movement of the projectile axis, which can be interpreted as a pitching motion about a continuously changing transverse axis and is superimposed on precession.

The projectile also rotates about a horizontal transverse axis, because its longitudinal axis follows a tangent to the trajectory (see Fig. 2.2). The total angle through which it rotates about this transverse axis corresponds to the difference between the angle of trajectory at the moment of firing (the elevation) and the (negative) angle

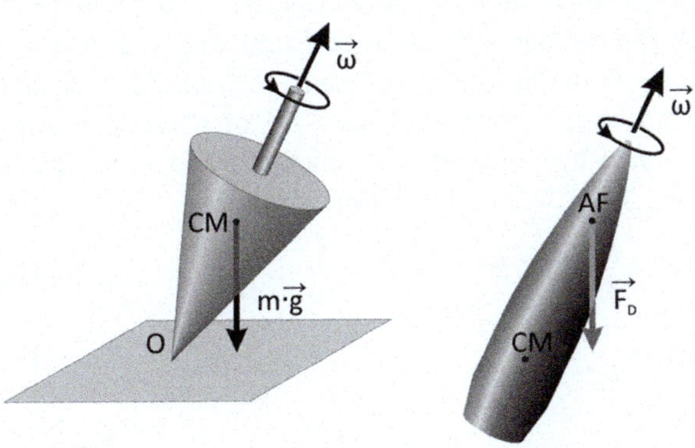


**Fig. 5.10.** Analogy between a symmetrical fixed gyroscope and a projectile in free flight.
O: Centre of rotation of gyroscope
AF: Point of application of the aerodynamic force $F_A$. See text for details.

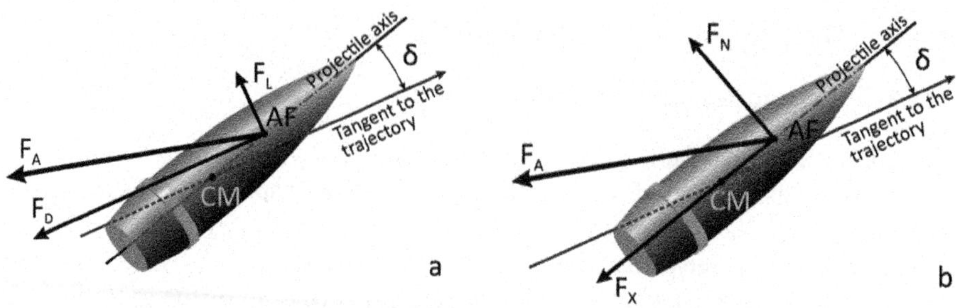


**Fig. 5.11.** Decomposition of the resultant aerodynamic force. **a**. Decomposition into components aligned with the tangent to the trajectory and perpendicular thereto gives drag and lift. **b**. Decomposition into components aligned with the axis of the projectile and perpendicular thereto gives the tangential and normal forces, the latter also being responsible for the torque.

of trajectory at the point of impact. The rapid rotation about the longitudinal axis that is required to stabilize the projectile will also influence the forces acting upon it.

### 5.3.3.2 Forces and moments resulting from non-axial flow

Because the angle of attack is continuously changing, the line of action of the resulting aerodynamic force $F_A$ acting on the projectile coincides neither with the axis of the projectile nor with its direction of travel. There are two useful ways of decomposing this force:

- Into the direction of movement and perpendicular to it (as per Fig. 5.11 a)
- Into the direction of the projectile axis and perpendicular to it (as per Fig. 5.11 b).

The most suitable approach depends on the frame of reference chosen. The force acting in the direction of movement is known as *drag* $F_D$ (as mentioned in 2.3.4.3), while the force acting perpendicular to this is the *lift force* $F_L$.

Lift force always acts in the resistance plane, which means it can act in any direction, including sideways or downwards.

The second approach to vector decomposition therefore results in an *axial force* $F_X$ and a *normal force* $F_N$. Because the contact point of the resulting aerodynamic force lies between the centre of mass and the tip of the projectile, the normal force also produces a rotation moment about the centre of mass. All of these forces are referenced to the stagnation pressure by means of a reference area and a coefficient. See the equation for drag (Eqn 5.3:20). We can therefore write:

(5.3:23) $$F_L = C_L \cdot \frac{\rho}{2} \cdot v^2 \cdot A_0 \,, \qquad [N]$$

(5.3:24) $$F_X = C_X \cdot \frac{\rho}{2} \cdot v^2 \cdot A_0 \,, \qquad [N]$$

(5.3:25) $$F_N = C_N \cdot \frac{\rho}{2} \cdot v^2 \cdot A_0 \ . \qquad [N]$$

The rotation moment (generally known as the *overturning moment*) is also linked to a reference distance, for which the nominal calibre d is used.

(5.3:26) $$M_M = C_M \cdot \frac{\rho}{2} \cdot v^2 \cdot A_0 \cdot d \ . \qquad [N \cdot m]$$

Let $e_A$ be the lever arm of the aerodynamic force. From Eqns 5.3:25 and 5.3:26 we obtain:

(5.3:27) $$e_A = \frac{M_M}{F_N} = \frac{C_M}{C_N} \cdot d \ . \qquad [m]$$

Aerodynamic force $F_A$ is given by the surface integral of the pressure distribution over the projectile (see Eqn 5.3:17). This means that it depends on the Mach number of the flow and on the angle of attack. This affects all the components mentioned above and, because they are referenced to stagnation pressure, it also affects the coefficients. If we know the coefficients, we also know the forces and moments themselves.

#### 5.3.3.3 Forces and moments resulting from rotation about a transverse axis

If a force acts on a body subjected to flow, the resulting acceleration is not as great as one would expect it to be from NEWTON's law.

(5.3:28) $$m \cdot a = F - m^* \cdot a \ . \qquad [N]$$

$m^*$ is proportional to the mass of the fluid displaced by the body. Acceleration in a flow therefore exerts forces. This becomes apparent in the pitch of a projectile about a non-stationary transverse axis. These forces stem from three factors: the angular velocity of pitch $\omega_P$, change in this value and change in angle of attack $\delta$. The forces generated by the two angular velocities $\omega_P$ and $\dot{\delta}$ exert a damping effect on the pitching motion. For that reason, they are often grouped together and referred to as the *pitch damping force* $F_E$:

(5.3:29) $$F_E = C_E \cdot \frac{\rho}{2} \cdot v^2 \cdot A_0 \cdot \frac{\omega_p \cdot d}{v} \ . \qquad [N]$$

Their point of application does not usually coincide with the centre of mass. The pitch damping force therefore creates a *pitch damping moment* $M_H$:

(5.3:30) $$M_H = C_H \cdot \frac{\rho}{2} \cdot v^2 \cdot A_0 \cdot d \cdot \frac{\omega_p \cdot d}{v} \ . \qquad [N \cdot m]$$

In 5.3.3.1 we mentioned the rotation about a transverse axis that occurs because the axis of the projectile follows a tangent to the trajectory. This generates a (small) crossflow across the forward part of the projectile (between the tip and the centre of mass) and an opposing flow across the rear part. The resulting force and its moment are generally integrated into the lift force and the rotation moment.

#### *5.3.3.4 Forces and moments resulting from rotation about the longitudinal axis*

The frictional force on the projectile surface resulting from axial flow contributes to drag, but for reasons of symmetry does not contribute to lateral force. However, it does create a moment that opposes rotation about the longitudinal axis and therefore reduces angular velocity. This moment is called the *roll damping moment* $M_I$:

$$(5.3{:}31) \qquad M_I = C_I \cdot \frac{\rho}{2} \cdot v^2 \cdot A_0 \cdot d \cdot \frac{\omega \cdot d}{v} . \qquad [\mathrm{N{\cdot}m}]$$

Because a projectile virtually always has a non-zero angle of attack, it is subjected to a continuous crossflow that is superimposed on that mentioned in 5.3.3.3. Because the projectile is rotating rapidly about its longitudinal axis, the Magnus effect leads to a force acting upon it, the *Magnus force* $F_K$. In the case of a normal, spin-stabilized projectile, the contact point of that force lies behind the centre of mass. This results in a rotation moment, the *Magnus moment*:

$$(5.3{:}32) \qquad F_K = C_K \cdot \frac{\rho}{2} \cdot v^2 \cdot A_0 \cdot \frac{\omega \cdot d}{v} . \qquad [\mathrm{N}]$$

$$(5.3{:}33) \qquad M_J = C_J \cdot \frac{\rho}{2} \cdot v^2 \cdot A_0 \cdot d \cdot \frac{\omega \cdot d}{v} . \qquad [\mathrm{N{\cdot}m}]$$

The contact point of the Magnus force is strongly influenced by the nature of the projectile surface (roughness, driving bands, etc.). If the projectile is poorly designed, the contact point may be located at a position forward of the centre of mass. In such cases, the Magnus moment will have a destabilizing effect.

Pitch about a transverse axis also causes crossflow. Because the projectile is rotating about its longitudinal axis, that crossflow exerts a force, on account of the Magnus effect. This force is known as the *pitch Magnus force* $F_{KP}$. There is also a corresponding *pitch Magnus moment* $M_{JP}$.

$$(5.3{:}34) \qquad F_{KP} = C_{KP} \cdot \frac{\rho}{2} \cdot v^2 \cdot A_0 \cdot \frac{\omega \cdot d}{v} \cdot \frac{\dot{\delta} \cdot d}{v} . \qquad [\mathrm{N}]$$

$$(5.3{:}35) \qquad M_{JP} = C_{JP} \cdot \frac{\rho}{2} \cdot v^2 \cdot A_0 \cdot d \cdot \frac{\omega \cdot d}{v} \cdot \frac{\dot{\delta} \cdot d}{v} . \qquad [\mathrm{N{\cdot}m}]$$

For further information on the origins of these equations, please see 5.5.1, "Trajectory calculations with six degrees of freedom". For comprehensive derivations based on the physical relationships, please see H. MOLITZ, R. STROBEL (1963).

#### *5.3.3.5 The cause of lateral deviation*

The superimposition of gyroscopic motion and rotation about a horizontal transverse axis means that the tip of a projectile spinning clockwise will on average lie slightly to the right of the tangent to the trajectory. As a result, the projectile experiences a continuous (but slight) flow from the left, which causes it to deviate to the right. For a trajectory with an elevation of less than approximately 60° (and especially for a flat trajectory), the effect of this crossflow will be greater than that of the Magnus force. Projectiles with a clockwise spin therefore deviate to the right, while those with an anticlockwise spin deviate to the left. At elevations above 60°, the Magnus force starts to predominate, and the point of impact moves gradually to the left (or to the right in the case of a projectile with anticlockwise spin).

### 5.3.4 Forces and moments acting on the projectile

In addition to weight and Coriolis force, a rigid-body projectile is subjected to the forces and moments shown in Table 5.4, which are all characterized by their coefficients.

**Table 5.4.** The forces and moments acting on a projectile, together with their coefficients

| Force | | | Moment | | |
|---|---|---|---|---|---|
| $F_D$ | Drag force | $C_D$ | $M_I$ | Roll damping moment | $C_I$ |
| $F_L$ | Lift force | $C_L$ | $M_M$ | Overturning moment | $C_M$ |
| $F_K$ | Magnus force | $C_K$ | $M_J$ | Magnus moment | $C_J$ |
| $F_E$ | Pitch damping force | $C_E$ | $M_H$ | Pitch damping moment | $C_H$ |
| $F_{KP}$ | Pitch Magnus force | $C_{KP}$ | $M_{JP}$ | Pitch Magnus moment | $C_{JP}$ |

## 5.4 Trajectory calculations using the point-mass model

### 5.4.1 Equations of motion

#### *5.4.1.1 Physical derivation*

If we ignore Coriolis acceleration and possible transverse components of acceleration due to gravity, calculating a trajectory is a two-dimensional problem. The system of equations of motion is therefore as follows:

$$(5.4{:}1) \qquad \ddot{\vec{r}}(t) = -\frac{F_D}{m_p}\cdot\frac{\dot{\vec{r}}(t)}{v} + \vec{g}\,. \qquad [\mathrm{m/s^2}]$$

Under this assumption, we have the following components:

$$(5.4{:}2)\qquad \ddot{x}(t) = -\frac{F_D}{m_p}\cdot\frac{\dot{x}(t)}{v}, \qquad [m/s^2]$$

$$\ddot{y}(t) = -\frac{F_D}{m_p}\cdot\frac{\dot{y}(t)}{v} - g\,. \qquad [m/s^2]$$

From Eqn 2.3:25a, the magnitude of drag $F_D$ is a function of velocity v, atmospheric density ρ(y) and the drag coefficient $C_D$, which in turn (owing to the relationship with Mach number) is dependent upon temperature and hence upon height y. We can therefore write:

$$(5.4{:}3)\qquad F_D = F_D(v, y, C_D)\,. \qquad [m/s^2]$$

Using Fig. 5.9, we can therefore write down a force equation for the tangential direction:

$$m\cdot\dot{v} = -F_D - F_W\cdot\sin\theta = -F_D - m\cdot g\cdot\sin\theta\,, \qquad [N]$$

with angle of trajectory θ. Dividing by the mass gives us an equation for speed as a function of time:

$$(5.4{:}4)\qquad \dot{v} = -\frac{F_D}{m} - g\cdot\sin\theta\,. \qquad [m/s^2]$$

Eliminating drag from Eqns 5.4:2 gives the equation for the angle of trajectory. This is achieved by multiplying the first equation by $\dot{y}$ and the second by $\dot{x}$, and subtracting the first from the second. This yields:

$$\ddot{y}\cdot\dot{x} - \dot{y}\cdot\ddot{x} = -g\cdot\dot{x}\,. \qquad [m^2/s^3]$$

Dividing by $\dot{x}^2$ we obtain:

$$\frac{d}{dt}\left(\frac{\dot{y}}{\dot{x}}\right) = -\frac{g}{\dot{x}}\,. \qquad [1/s]$$

As $\dot{y}/\dot{x} = \tan\theta$, we can write:

$$\frac{d}{dt}(\tan\theta) = \frac{1}{\cos^2\theta}\cdot\dot{\theta} = -\frac{g}{v\cdot\cos\theta} \qquad [1/s]$$

$$(5.4{:}5)\qquad \dot{\theta} = -\frac{g}{v}\cdot\cos\theta\,. \qquad [1/s]$$

This equation and Eqn 5.4:4 together form a system of equations describing the velocity vector (magnitude and direction) as a function of time.

#### 5.4.1.2 *Trajectory geometry derivation*

The equations of motion for the point mass can be derived from the differential geometry characteristics of the trajectory.

Let $\vec{r}(t)$ be any trajectory and

$$\vec{v}(t) = \dot{\vec{r}}(t) . \qquad [m/s]$$

$\vec{v}$ is always tangential to the trajectory. Taking the tangential direction vector $\vec{e}_t$ and the arc length s, we can write:

$$(5.4{:}6) \qquad \vec{v}(t) = v(t)\cdot\vec{e}_t\left[s(t)\right] . \qquad [m/s]$$

Deriving for time, we obtain:

$$(5.4{:}7) \qquad \vec{a}(t) = \dot{\vec{v}}(t) = \frac{d}{dt}\left[v\cdot\vec{e}_t(s)\right] = \dot{v}\cdot\vec{e}_t + v^2\cdot\frac{d\vec{e}_t}{ds} . \qquad [m/s^2]$$

FRENET's formulae for differential geometry give the following equation for the tangential vector:

$$\frac{d\vec{e}_t}{ds} = \frac{1}{R}\cdot\vec{e}_n , \qquad [m/s^2]$$

where R = radius of curvature of the trajectory at point s(t) and $\vec{e}_n$ = the normal vector. Acceleration (Eqn 5.4:7) is thus decomposed into tangential and normal components:

$$(5.4{:}8) \qquad \vec{a}(t) = \vec{a}_t + \vec{a}_n = \dot{v}\cdot\vec{e}_t + \frac{v^2}{R}\cdot\vec{e}_n . \qquad [m/s^2]$$

On the basis of Fig. 5.9 (p. 162), we obtain the following expression for the tangential and normal acceleration components:

$$(5.4{:}9a) \qquad \vec{a}_t = -\left(\frac{F_D}{m} + g\cdot\sin\theta\right)\cdot\vec{e}_t , \qquad [m/s^2]$$

$$(5.4{:}9b) \qquad \vec{a}_n = -g\cdot\cos\theta\cdot\vec{e}_n . \qquad [m/s^2]$$

Comparing coefficients with Eqn 5.4:8 and applying

$$ds = R\cdot d\theta = v\cdot dt , \qquad [m]$$

$$\frac{1}{R} = \frac{1}{v}\cdot\frac{d\theta}{dt} , \qquad [1/m]$$

yields the same system of equations as we already derived at 5.4.1.1, but this time based on trajectory geometry. To determine position, we need to add the two equations for the position coordinates:

$$(5.4{:}10) \qquad \left|\begin{array}{lll} \dot{v} = -\dfrac{F_D}{m} - g\cdot\sin\theta , & & [m/s^2] \\ \dot{\theta} = -\dfrac{g}{v}\cdot\cos\theta , & & [1/s] \\ \dot{x} = v\cdot\cos\theta , & & [m/s] \\ \dot{y} = v\cdot\sin\theta . & & [m/s] \end{array}\right.$$

This system is often described as a trajectory geometry system, because it represents the trajectory by plotting the velocity vector (magnitude and direction) as a function of time. For many years it was the most commonly-used system.

#### 5.4.1.3 The principal equation of exterior ballistics

If we arrange this equation thus:

(5.4:11) $$\frac{d}{dt} = \frac{d\theta}{dt}\cdot\frac{d}{d\theta} = -\frac{g\cdot\cos\theta}{v}\cdot\frac{d}{d\theta}\,, \qquad [1/s]$$

we can use the angle of trajectory as our integration variable in the system of equations at 5.4:10, instead of time of flight. For the first equation, we obtain:

$$\frac{dv}{dt} = -\frac{g\cdot\cos\theta}{v}\cdot\frac{dv}{d\theta} = -\frac{F_D}{m} - g\cdot\sin\theta\,. \qquad [m/s]$$

Arranging the equations in the usual manner and applying the procedure to the three other equations, we obtain the following system:

(5.4:12) $$\left|\begin{aligned} \frac{dv}{d\theta} &= \frac{F_D}{m}\cdot\frac{v}{g\cdot\cos\theta} + v\cdot\tan\theta\,, && [m/s] \\ \frac{dt}{d\theta} &= -\frac{v}{g\cdot\cos\theta}\,, && [s] \\ \frac{dx}{d\theta} &= -\frac{v^2}{g}\,, && [m] \\ \frac{dy}{d\theta} &= -\frac{v^2}{g}\cdot\tan\theta\,. && [m] \end{aligned}\right.$$

In this form, the first equation is of particular significance. If we multiply it by cos θ and then subtract v·sin θ from both sides, we obtain:

$$\frac{dv}{d\theta}\cdot\cos\theta - v\cdot\sin\theta = \frac{F_D(v)}{m\cdot g}\cdot v\,, \qquad \text{or} \qquad [m/s]$$

(5.4:13) $$\frac{d}{d\theta}\left(v\cdot\cos\theta\right) = \frac{F_D(v)}{m\cdot g}\cdot v\,. \qquad [m/s]$$

This equation is seen as the principal equation of external ballistics. If we assume that acceleration due to gravity g and atmospheric density ρ are both constant (and hence independent of altitude y), it is possible to integrate this equation analytically for special forms of the function $F_D(v)$ and hence to plot velocity as a function of the angle of trajectory (the *hodograph*). In the days before computers, a large number of drag laws were developed with the aim of making it possible to solve the principal equation.

## 5.4.2 Historical remarks on trajectory calculations

### 5.4.2.1 Representing drag

As mentioned in 5.1, even the leading figures of mathematics, such as NEWTON, BERNOULLI, d'ALEMBERT and EULER had only partly succeeded in modelling trajectories mathematically. Analytical integration of the principal equation of external ballistics depended (and still depends) on the nature of the drag function $F_D(v)$, and initially very few functions were discovered that allowed the differential equation to be solved.

NEWTON modelled trajectories using the linear function $F_D(v) = k \cdot v$, while BERNOULLI calculated trajectories not only for the supposed square law, but even for integer powers: $F_D(v) = k_n \cdot v^n$. As early as 1744, D'ALEMBERT showed that it was possible to solve the principal equation (Eqn 5.4:13) using the functions $F_D(v) = k_0 + k_n \cdot v^n$, $F_D(v) = a + b \cdot \log v$ and similar, by means of quadratures, i.e. by numerical integration.

As researchers acquired increasing knowledge and experience, it became apparent that the drag function would depend not only on velocity but also, to a significant degree, on atmospheric density (for which an average value had been used hitherto) and hence must be an empirical function.

Over time, two approaches to the exterior ballistic problem emerged. One was to continue the search for functions that would both describe drag as accurately as possible and permit integration of the principal equation (generally using quadratures). The other was to divide the trajectory into short steps, with drag constant within each step, and then assemble these to give the overall trajectory.

### 5.4.2.2 SIACCI's method

A third approach emerged towards the end of the 19th Century, thanks mainly to the efforts of Italian ballistician F. SIACCI. This approach involved so modifying the principal equation that it was possible to calculate any point along the trajectory (generally the end) using an empirical drag law: $F_{DS}(v) = k_S \cdot f_S(v)$. To achieve this, a number of less important variables in the principal equation were replaced by suitable mean values. This made it possible to separate the variables of the principal equation, allowing integration:

(5.4:14) $$d(\tan\theta) = \frac{g \cdot du}{u \cdot k_S \cdot f(u) \cdot \varepsilon}, \qquad [-]$$

with $$u = \frac{v \cdot \cos\theta}{\sigma}, \qquad [\mathrm{m/s}]$$

where $\sigma = \cos\theta_1$ and $\varepsilon = \cos\theta_2$ are the two variables for which suitable mean values are selected. To make the calculation possible, four "primary functions" were defined, which allowed direct integration of the principal equation.

$$(5.4{:}15)\quad \begin{aligned} T(u) &= -\int^{u} \frac{d\upsilon}{f(\upsilon)}\,, && [\text{s/m}] \\ D(u) &= -\int^{u} \frac{\upsilon\cdot d\upsilon}{f(\upsilon)}\,, && [\text{-}] \\ J(u) &= -2\cdot g\cdot\int^{u} \frac{d\upsilon}{\upsilon\cdot f(\upsilon)}\,, && [\text{1/m}] \\ A(u) &= -\int^{u} \frac{J(\upsilon)\cdot\upsilon\cdot d\upsilon}{f(\upsilon)}\,. && [\text{1/m}] \end{aligned}$$

Any lower limit could be selected for the integral, as calculation only involved differences between the values of these expressions.

In a further step, SIACCI refined this method by defining the two variables for which suitable mean values were required as follows:

$$\sigma = \cos\theta_0 \quad \text{and} \quad \varepsilon = \xi\cdot\cos^2\theta_0\,. \qquad [\text{m/s}]$$

Because $\theta_0$ is the initial value, it was possible to replace the two adjustment variables $\sigma$ and $\varepsilon$ with just one, $\xi$. Integrating 5.4:14 and using further equations from the Eqn 5.4:12 system yielded the following equations (known as "Siacci III").

$$(5.4{:}16)\quad \begin{aligned} \tan\theta &= \tan\theta_0 - \frac{1}{2\cdot k_S\cdot\xi\cdot\cos^2\theta_0}\cdot\left[J(u)-J(u_0)\right], && [\text{-}] \\ x &= \frac{1}{k_S\cdot\xi}\cdot\left[D(u)-D(u_0)\right], && [\text{m}] \\ t &= \frac{1}{k_S\cdot\xi\cdot\cos\theta_0}\cdot\left[T(u)-T(u_0)\right], && [\text{s}] \\ y &= x\cdot\tan\theta_0 - \frac{x}{2\cdot k_S\cdot\xi\cdot\cos^2\theta_0}\cdot\left[\frac{A(u)-A(u_0)}{D(u)-D(u_0)} - J(u_0)\right]. && [\text{m}] \end{aligned}$$

The primary functions could be tabulated for the usual drag laws known at the time, and the key trajectory values could then be calculated using the equations at 5.4:16.

In the pre-computer era, SIACCI's method was probably the most important method of modelling trajectories using drag functions and integration of the principal equation. For more comprehensive accounts of this method, please see R. L. MCCOY (1999), G. HAUCK (1972) and H. MOLITZ, R. STROBEL (1963). The first of these also includes a large number of tables and example.

#### *5.4.2.3 Drag laws and standard projectiles*

Attempts were made to use drag laws to establish a deceleration function that would be valid for all projectiles and could be used to integrate the principal equation. The approach was generally of the form

$$(5.4{:}17) \qquad F_D / m = a = k \cdot f(v) = k \cdot v^2 \cdot F(v) \,. \qquad [\mathrm{m/s^2}]$$

Three drag laws rose to prominence at the time, of which one has become so widely accepted that it is still used today in one particular area of ballistics.

One of these laws was formulated by SIACCI, who in 1896 attempted to combine results from firing tests in England, German and Russia, with different projectiles, into a single model. His law (with subscript S) was:

$$(5.4{:}18) \qquad f_S(v) = 0.2002 \cdot v - 48.05 + \sqrt{(0.1648 \cdot v - 47.95)^2 + 9.6} + B \,, \qquad [\mathrm{m^2/s^2}]$$

$$\text{with} \qquad B = \frac{0.0442 \cdot v \cdot (v - 300)}{371 + \left(\dfrac{v}{200}\right)^{10}} \,. \qquad [\mathrm{m^2/s^2}]$$

For constant $k_S$ he used

$$(5.4{:}19) \qquad k_S = \frac{1000 \cdot d^2 \cdot \rho \cdot i_S}{1.206 \cdot m} \,. \qquad [1/\mathrm{m}]$$

Where $d$ = calibre, $\rho$ = atmospheric density, $m$ = projectile mass and $i_S$ is a dimensionless form factor through which the equation could be adjusted to allow for different projectile shapes.

By substituting Eqn 5.3:20 into the left-hand side of Eqn 5.4:17 and Eqn 5.4:18 into the second form of the right-hand side, we obtain an equation linking the drag coefficient $C_D$ and SIACCI's drag law. After brief calculation, we obtain:

$$(5.4{:}20) \qquad c_{DS}(v) = \frac{8000 \cdot i_S \cdot F_{DS}(v)}{1.206 \cdot \pi} \quad \text{with} \quad F_{DS}(v) = \frac{f_S(v)}{v^2} \,. \qquad [-]$$

Krupp – the most important European artillery manufacturer in the second half of the 19th Century and at the beginning of the 20th – took another approach. The company designed a standard projectile (Fig. 5.12), with which a large number of firing tests were conducted. The result was the Krupp drag law. This drag law was structured in line with Eqn 5.4:17, with the constant k (subscript K) calculated as follows:

$$(5.4{:}21) \qquad k_K = \frac{10^4 \cdot d^2 \cdot \pi \cdot \rho \cdot i_K}{1.22 \cdot 4 \cdot m} \,. \qquad [1/\mathrm{m}]$$

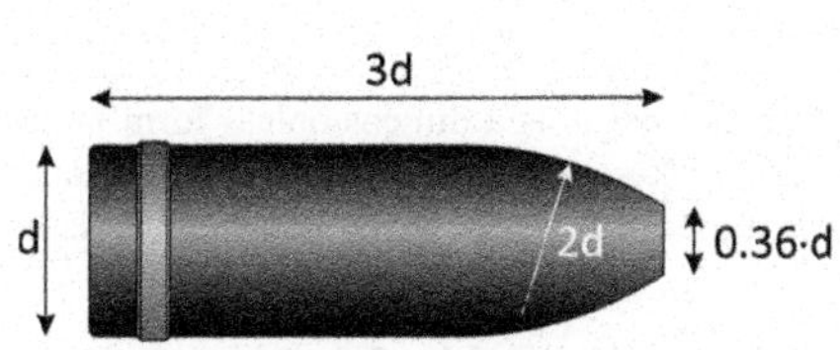


**Fig. 5.12.** Krupp standard projectile, 10 cm calibre.

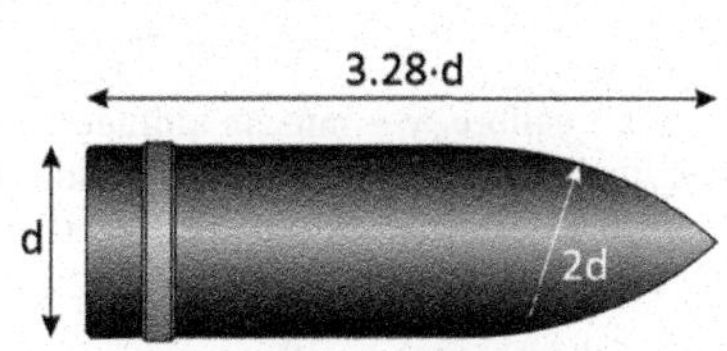


**Fig. 5.13.** The G1 standard projectile of the Gâvres Commission.

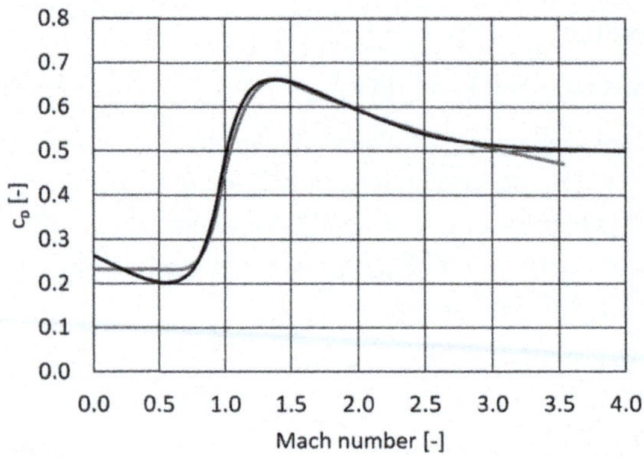


**Fig. 5.14.** The $C_D$ curves for the two standard projectiles.
*Grey line*: Krupp projectile.
*Black line*: Gâvres G1 projectile.

This drag law also used a dimensionless form factor ($i_K$). For the Krupp projectile, this was set to 1.

(5.4:22) $$c_{DK}(v) = 1.64\cdot 10^4 \cdot i_K \cdot F_{DK}(v) \quad \text{with} \quad F_{DK}(v) = \frac{f_K(v)}{v^2}\,. \qquad [-]$$

The functions $F_{DS}$ (Siacci) and $F_{DK}$ (Krupp) were both tabulated. The coefficient function for the Krupp projectile is shown in Fig. 5.14 (grey line).

In 1829, the French Navy set up the *Groupe d'études et de recherches en balistique, armes et munitions* (Ballistics, weapons and munitions study and research group) at Gâvres, on the Atlantic coast of France. This group, later known as the "Gâvres Commission", had a large gunnery range on the beach at Gâvres. Between 1873 and 1898, a very large number of tests were conducted, mainly using naval artillery firing the G1 standard projectile (see Fig. 5.13). The results formed the basis of the Gâvres Commission drag law (subscript G):

(5.4:23) $$f_G(v) = k_G \cdot v^2 \cdot \left[0.255 + B\cdot \arctan\left(\frac{v-330}{50}\right)\right]\cdot 10^{\frac{v-600}{10^6}}, \qquad [m^2/s^2]$$

where $$B = \frac{\sqrt[4]{1+0.0392\cdot\left(\frac{v-330}{500}\right)^8}}{27226+494\cdot\left(\frac{v-330}{50}\right)^2}, \qquad [-]$$

and $$k_G = \frac{d^2 \cdot i_G \cdot 10^{0.000045\cdot y}}{m}, \qquad [1/m]$$

where d = calibre, y = muzzle altitude, m = projectile mass and $i_G$ is a dimensionless form factor through which the equation could be adjusted to allow for different projectile shapes. The value of 10 raised to a power in the term $k_G$ is the atmospheric density at altitude y, and has the same dimension.

Fig. 5.14 also shows the curve of the coefficient function for the G1 projectile (as a black line).

In contrast to the other two laws mentioned above, the Gâvres law attempted to take account of the rapid increase in drag at around the speed of sound, using an arctangent function. Furthermore, an altitude-dependent factor was included in $k_G$, to take account of atmospheric density.

### 5.4.2.4 The Ingalls tables and the ballistic coefficient (BC)

In 1883, working on the basis of the Krupp test firings conducted between 1875 and 1881, Russian Major-General N. MAYEVSKI formulated his "zone laws of air resistance" that became widely known, especially in the USA, where they are still used. In addition to the Krupp results, he used those of experiments conducted in the UK (by F. BASHFORTH) and his own experiments.

MAYEVSKI published the first version in his work *Traité de balistique extérieure*, which appeared in Paris in 1872.

MAYEVSKI divided the velocity range from 0 to 900 m/s into five zones. For each zone, he used the approach from Eqn 5.4:17, expressing deceleration as velocity raised to a power:

$$(5.4:24) \qquad a_{M,i} = k_{M,i} \cdot \frac{d^2 \cdot \pi}{4} \cdot \frac{\rho}{1.206} \cdot v^{n_i}, \quad i = 1 \ldots 5 \; . \qquad [\mathrm{m/s^2}]$$

The constant pairs of values ($k_{M,i}$, $n_i$) were so chosen that the equation correctly predicted the deceleration of the Krupp standard shell. The fifth velocity zone was later extended out to 1100 m/s and split into three. The deceleration of any shell in the i-th velocity range can be referenced to the drag of the standard shell using a ratio factor C:

$$(5.4:25) \qquad a_i = \frac{a_{M,i}}{C} \; . \qquad [\mathrm{m/s^2}]$$

It was possible to compensate for the discontinuous form of MAYEVSKI's law of resistance using SIACCI's drag law (Eqn 5.4:18). By choosing a suitable value for $k_S$, one could achieve close correlation both with calculations conducted in accordance with MAYEVSKI's law and with experiments.

In the USA, ballistician and artillery colonel J. M. INGALLS took MAYEVSKI's method and SIACCI's drag law, and converted them into what he called "English units". He compiled a set of tables – the *Ingalls Tables* (J.M. INGALLS, 1918). These were first published in 1883 and included not only the primary Siacci functions (5.4:15) but also additional tables for specific ballistic problems, such as wind and changes in atmospheric density. In the introduction to his tables, Ingalls stated: "The fundamental hypothesis upon which they are based is that the ratio of the retardations of any two different projectiles moving at the same speed is independent of the velocity, assuming air conditions, as barometer, wind, etc., the same." C in Eqn 5.4:25 was therefore a constant, termed the *ballistic coefficient*, today commonly abbreviated to BC.

Using this hypothesis, the same shape of $C_D$ curve is used for every shell as for the standard shell. The use of BC merely displaces this curve, without changing its form. BC is calculated by adjusting the calculated values to match empirical values as closely as possible, under a range of conditions (atmospheric density, temperature, etc.).

In general, BC is defined as follows (J. M. INGALLS, 1918)

$$(5.4{:}26) \qquad BC = \frac{\rho_0}{\rho} \cdot f_W \cdot f_a \cdot \frac{m}{\beta \cdot i \cdot d^2}, \qquad [-]$$

where m = mass of shell, d = calibre, $\rho_0$ = standard atmospheric density, $\rho$ = actual atmospheric density at muzzle altitude, $f_W$ is the wind factor, which corrects for the influence of downrange wind and $f_a$ is the altitude factor, which corrects for the mean altitude of the trajectory. $\beta$ is the integrating factor, which is intended to compensate for the error introduced by the approximation resulting from the application of SIACCI's method to MAYEVSKI's zone laws of air resistance. i is the *form factor* (to use the term preferred by MCCOY and others; INGALLS calls i the "coefficient of form"). This allows for the influence on drag of the difference in form between the standard shell and the shell under consideration. The original of Ingalls' Tables gives weight in pounds, velocity in feet per second and calibre (diameter of projectile) in inches.

Under normal conditions, $\rho = \rho_0$ and the correction factors are equal to unity, with the exception of i. The equation for BC then becomes:

$$(5.4{:}27) \qquad BC = \frac{m}{i \cdot d^2}. \qquad [-]$$

Given its definition as the ratio between two decelerations (see Eqn 5.4:25) the ballistic coefficient is dimensionless. The form of Eqns 5.4:26 and 5.4:27 means that i cannot be dimensionless but has the somewhat unusual (in physical terms) unit of square inches per pound.

US literature uses the term *sectional density* to refer to the value of expression $m/d^2$ as shown in both formulae. This is equal to sectional density as understood in the present work, multiplied by $4/\pi$. However, this implies the existence of a direct link between form factor i and drag coefficient $C_D$. If we use subscripts of 0 for the standard shell and 1 for some other shell, then on the basis of Eqns 5.4:25 and 5.4:27 we can write:

$$(5.4{:}28) \qquad BC_1 = \frac{a_0}{a_1} = \frac{m_1}{i_1 \cdot d_1^2}. \qquad [-]$$

Furthermore:

$$BC_0 = 1 = \frac{m_0}{i_0 \cdot d_0^2}, \qquad [-]$$

and hence

$$(5.4{:}29) \qquad i_0 = \frac{m_0}{d_0^2}. \qquad [\text{lbs/in}^2]$$

The form factor of the standard shell is therefore equal to its sectional density in the US sense of the term. Setting $m_0$ = 1 lb and $d_0$ = 1 in (J. S. HATCHER, 1962) this becomes unity, as required.

If we then insert the classic aerodynamic form in Eqn 5.4:28 for both decelerations, we obtain

$$a_j = C_D \cdot \frac{\rho \cdot v^2}{2} \cdot \frac{d_j^2 \cdot \pi}{4 \cdot m_j}, \quad j = 0,1 \ . \qquad [-]$$

Following some calculation, and using Eqn 5.4:29:

$$(5.4:30) \qquad \frac{i_0}{i_1} = \frac{C_{D0}}{C_{D1}} \ . \qquad [-]$$

The form factor of any shell is related to that of the standard shell in the same way as the corresponding drag coefficients. We can calculate it from the $C_D$ function for the shell.

$$(5.4:31) \qquad i_1 = \frac{C_{D1}}{C_{D0}} \cdot \frac{m_0}{d_0^2} \ . \qquad [\text{lbs/in}^2]$$

Subscript 0 indicates the values for the standard shell, subscript 1 those for the shell under study.

A few comments regarding this approach to calculating trajectories – which is still used in a number of countries, especially the USA, primarily to calculate the trajectories of bullets for hunting and target weapons.

- The calculations use the Siacci method for flat trajectories, over which atmospheric density can be assumed constant.
- There are limits to the extent to which one can use a single standard projectile as the basis for calculating the trajectories of all the disparate forms that exist. This led to the introduction of additional standard shells, designated G2, G5, G6, G7, G8 and GL. G7 has a slender nose and boattail. However, calculations for modern small-calibre projectiles make almost exclusive use of G1 (Fig. 5.13), with occasional use of G7.
- Because calculations are referenced to a standard projectile and its $C_D$ curve, trajectories for other projectiles can only ever be approximations. This is sufficient for small-calibre projectiles and typical ranges, but can lead to large errors at maximum range.
- Alongside MAYEVSKI's and INGALLS' definitions of BC (see Eqn 5.4:25), there is also a definition that uses the principal equation of the exterior ballistics, Eqn 5.4:13 (see A. GLUCHOFF, 2011):

$$(5.4:32) \qquad \frac{d}{d\theta}\left(v \cdot \cos\theta\right) = \frac{F_D(v)}{g \cdot BC} \cdot v \ , \qquad [\text{m/s}]$$

retaining the form of Eqn 5.4:27. In this case, the BC can no longer represent the ratio of the decelerations to each other, as it cannot be dimensionless.

- The form factor i is proportional to $C_D$, but in the BC it appears in the denominator (Eqn 5.4:27). The BC values therefore behave in exactly the opposite manner to $C_D$. The less a projectile decelerates, the greater its BC.

#### 5.4.2.5 *Lateral deviation (derivation)*

Because of the work involved, both techniques for calculating trajectories – integrating the principal equation using drag laws and numerical integration in steps – were only used to model trajectories in two dimensions. It was therefore necessary to determine the lateral discrepancy between point of impact and initial line of departure (which CRANZ had already ascribed to Magnus force and projectile spin as early as 1896) separately from the trajectory. The approach was fairly obvious: an acceleration (*lateral acceleration* $d_z$) was introduced that acted perpendicular to the line of departure. Because the mean lateral velocity of the projectile is low (approximately 2–7 m/s, depending on distance), it was possible to ignore the air resistance opposing lateral movement and to treat lateral deviation simply as a quadratic function of time of flight:

$$(5.4{:}33) \qquad z(t) = d_z \cdot t^2 \,. \qquad [\mathrm{m}]$$

Initially, $d_z$ was assumed constant, and was determined from test firings. However, it became apparent that the increasing ranges of artillery (which implied longer times of flight) meant that Eqn 5.4:33 required modification.

$$(5.4{:}34) \qquad z(t) = d_z(\theta_0) \cdot t^n \,. \qquad [\mathrm{m}]$$

The constant was made a function of elevation $\theta_0$, and in some cases values greater than 2 were used for the exponent n.

Introducing trajectory calculations with three degrees of freedom, thereby taking account of the lateral component of drag, usually results in constant values of lateral acceleration, which makes sense from a physical point of view.

### 5.4.3 Equation systems for programming

#### 5.4.3.1 *Numeric calculations*

The second method of calculating a trajectory mentioned in 5.4.2.1, which involved breaking the trajectory down into short segments, became increasingly useful as numerical mathematics and electrical/electronic calculating systems progressed. The method developed by C. RUNGE and W. KUTTA for the numerical integration of systems of differential equations (see G. ENGELN-MÜLLGES et. al., 2011) contributed to ensuring the widespread adoption of this approach, at least in Europe.

Such calculations were extremely labour intensive, however. Compiling a set of two-dimensional range tables and diagrams for one artillery gun using electrical calculating machines occupied a dozen people for several months.

The power of modern computers has removed the limitations. This means that ballisticians scarcely need to think about calculation procedures and hence have considerably more time for ballistics questions, such as how to minimize drag or stabilize the projectile.

### 5.4.3.2 Systems of equations with two and three degrees of freedom

The point-mass model assumes that drag $F_D$ acts along the axis of the projectile and that this always coincides with the tangent to the trajectory. The general equation of motion for the point mass is therefore:

$$(5.4{:}35) \qquad \ddot{\vec{r}} = -\frac{F_D}{m}\cdot\frac{\dot{\vec{r}}}{v} + \vec{g} + \vec{d}_z + \vec{a}_c\,, \qquad [\text{m/s}^2]$$

where $\vec{r} = (x, y, z)$ the radius vector of the point on the trajectory, $F_D = F_D(v, y)$ the magnitude of resistance, m = projectile mass, v = the magnitude of velocity (i.e. speed), $\vec{g}$ is a vector that encompasses both acceleration due to gravity and centrifugal acceleration, $\vec{a}_c$ = Coriolis acceleration and $\vec{d}_z = (0, 0, d_z)$ the lateral acceleration responsible for lateral deviation. If we write this out in components and reduce $\ddot{\vec{r}} = \dot{\vec{v}}$ to the first order, we obtain the following system of differential equations for the motion of a projectile:

$$(5.4{:}36) \qquad \left|\begin{array}{lll} \dot{v}_x = -\frac{F_D}{m}\cdot\frac{v_x}{v} + g_x + a_{cx}\,, & & [\text{m/s}^2] \\ \dot{v}_y = -\frac{F_D}{m}\cdot\frac{v_y}{v} + g_y + a_{cy}\,, & & [\text{m/s}^2] \\ \dot{v}_z = -\frac{F_D}{m}\cdot\frac{v_z}{v} + g_z + d_z + a_{cz}\,, & & [\text{m/s}^2] \\ \dot{x} = x_x\,, & & [\text{m/s}] \\ \dot{y} = v_y\,, & & [\text{m/s}] \\ \dot{z} = v_z\,. & & [\text{m/s}] \end{array}\right.$$

Deceleration due to aerodynamic force:

$$(5.4{:}37) \qquad \frac{F_D}{m} = a = C_D\cdot\frac{1}{2}\cdot\rho\cdot v^2\cdot\frac{1}{q}\,, \qquad [\text{m/s}^2]$$

where q = sectional density of the projectile (see Eqn 2.4:1). We obtain the angle of trajectory at any given moment from the components of the velocity vector:

$$(5.4{:}38) \qquad \theta = \arctan\left(\frac{v_y}{v_x}\right). \qquad [-]$$

The system forming Eqn 5.4:36 is the classical physical form of the equations of motion for ballistics, in which time is the independent variable. In many cases, however, it is useful to rearrange the equations so that another variable becomes the independent variable. It is useful to calculate the flat trajectories of direct-fire weapons using a system in which integration is performed over the horizontal distance. If we multiply each individual equation of the system forming Eqn 5.4:36 by

$$\frac{1}{v_x} = \frac{dt}{dx} , \qquad (5.4{:}39) \quad [\text{s/m}]$$

we obtain the following system, which is often used in practice:

$$(5.4{:}40) \quad \left| \begin{array}{lll} v_x' = -\frac{F_D}{m} \cdot \frac{1}{v} + \frac{g_x}{v_x} + \frac{a_{cx}}{v_x} , & & [1/\text{s}] \\ v_y' = -\frac{F_D}{m} \cdot \frac{v_y}{v \cdot v_x} + \frac{g_y}{v_x} + \frac{a_{cy}}{v_x} , & & [1/\text{s}] \\ v_z' = -\frac{F_D}{m} \cdot \frac{v_z}{v \cdot v_x} + \frac{g_z}{v_x} + \frac{d_z}{v_x} + \frac{a_{cz}}{v_x} , & & [1/\text{s}] \\ t' = \frac{1}{v_x} , & & [\text{s/m}] \\ y' = \frac{v_y}{v_x} , & & [-] \\ z' = \frac{v_z}{v_x} . & & [-] \end{array} \right.$$

By substituting in a manner analogous to Eqn 5.4:39, we can also make slope distance, the length of the trajectory or the height of the projectile the independent variable. Substituting in accordance with Eqn 5.4:11 allows us to also select the angle of trajectory as the integration variable. There are often cases in which such systems can be useful.

For flat trajectories over short ranges (with times of flight of up to about 7 s), we can assume acceleration due to gravity as constant and ignore Coriolis acceleration (as its influence will be small by comparison with scatter). If we make these assumptions, we can construct the following simple, readily-programmed system on the basis of Eqn 5.4:36:

$$(5.4{:}41) \quad \left| \begin{array}{lll} \dot{v}_x = -\frac{F_D}{m} \cdot \frac{v_x}{v} , & & [\text{m/s}^2] \\ \dot{v}_y = -\frac{F_D}{m} \cdot \frac{v_y}{v} - g , & & [\text{m/s}^2] \\ \dot{x} = v_x , & & [\text{m/s}] \\ \dot{y} = v_y . & & [\text{m/s}] \end{array} \right.$$

With the horizontal distance as independent variable, this system is particularly suitable for producing range tables.

$$(5.4{:}42)\quad \left|\begin{array}{llr} v_x' = -\dfrac{F_D}{m}\cdot\dfrac{1}{v}\,, & & [1/s] \\ v_y' = -\dfrac{F_D}{m}\cdot\dfrac{v_y}{v_x\cdot v}-\dfrac{g}{v_x}\,, & & [1/s] \\ t' = \dfrac{1}{v_x}\,, & & [s/m] \\ y' = \dfrac{v_y}{v_x}\,. & & [-] \end{array}\right.$$

It is possible to calculate all the values usually found in a range table (time of flight, final velocity, energy, angle of shot and angle of trajectory) for all ranges, with a single integration operation.

#### *5.4.3.3 The initial value problem*

The usual form of a trajectory calculation is an initial value problem, in which the initial values are known, and we wish to find trajectory values that meet certain criteria. The initial values are determined by muzzle velocity $v_0$ and elevation $\theta_0$, and normally we can write:

$$(5.4{:}43)\quad \left|\begin{array}{llr} v_{x0} = v_0\cdot\cos\theta_0\,, & & [m/s] \\ v_{y0} = v_0\cdot\sin\theta_0\,, & & [m/s] \\ v_{z0} = 0\,, & & [m/s] \\ t_0 = 0\,, & & [s] \\ \vec{r}_0 = (x_0, y_0, z_0) = (0,0,0)\,. & & [m] \end{array}\right.$$

To determine drag $F_D$ at any given point, we need additional information. Atmospheric density is calculated from atmospheric pressure at muzzle altitude and the variation in pressure and temperature with altitude. We also need to know temperature in order to find the Mach number, which determines the instantaneous drag coefficient. Which of these initial and supplementary values are required, and to what degree of accuracy, will depend upon the type of calculation and its purpose.

The trajectory calculations yield the position and velocity of the projectile at discrete points (the interval between these points depending upon the integration step size) and other specific points, such as the highest point of the trajectory (the *apex*) and the range $x_E$ corresponding to a given target altitude $y_E$ or the target altitude at a given range (see Fig. 5.15). It is also possible to obtain further data using simple geometric functions, such as the angle of trajectory (using Eqn 5.4:38) and the angle of descent $\omega_H$ at the base of the trajectory, which is the vertically opposite angle to the angle of trajectory at that point. The angle of descent $\omega_T$ with relation

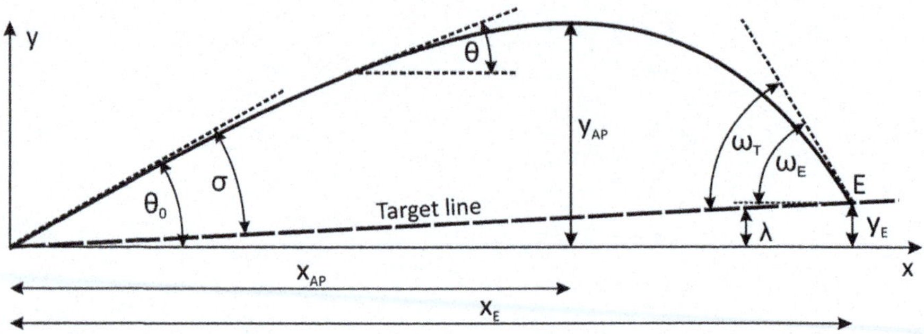


**Fig. 5.15.** Principal trajectory data. $x_E$: range, $y_E$: target height, $x_{AP}$: horizontal distance to apex, $y_{AP}$: apex height, λ: angle of sight, $\theta_0$: elevation, σ: angle of shot, θ: angle of trajectory, $\omega_T$: angle of descent with respect to the target line (the line connecting the muzzle and the target).

to the target line is then the sum of the angle of descent $\omega_E$ at the point of impact and the *angle of sight* λ. The angle between the target line and the direction of shot (the *angle of shot* σ) is then given by finding the difference between $\theta_0$ and λ.

These data can be used in the following applications of trajectory calculation using the point-mass model:

***Predictions***. These include constructing tools such as range tables, trajectory cards and sights, and predicting deviation of point of impact due to other influences, such as range wind and crosswind, or changes in altitude (and hence in atmospheric density). It is also possible to predict the ballistic characteristics of new projectiles.

One generally assumes standard conditions for such calculations, i.e. initial values as per Eqn 5.4:43, a standard atmosphere (generally ICAO) and muzzle altitude at sea level.

In countries with widely varying altitudes, it is common to use standard muzzle altitudes above sea level. In Germany, for instance, it is common to use a standard altitude of 100 m above sea level, in conjunction with the ICAO standard atmosphere. In Switzerland, one of four standard altitudes is used, depending on the anticipated primary deployment zone, along with a standard atmosphere adapted to Alpine regions (see 5.2.2.3).

In addition, it is essential to have as precise a drag function as possible. For new projectile types, one will have either to make an estimate on the basis of aerodynamic calculations or else choose the $C_D$ of a known projectile of similar form; the calibre of this other projectile can be completely different.

***Parameter studies*** In this context, parameter studies are calculations of which the purpose is to ascertain the influence of individual parameters on trajectory, such as the effect of a change in projectile mass on sensitivity to wind or on range, for a given muzzle energy. It is also possible to study the relationships between parameters, which can be useful for various purposes. For instance, it can be very useful to know the relationship between angle of descent and the elevation of a gun for the purposes of trajectory reconstruction (see 8.2.1.2). Such studies makes it possible to choose the optimum characteristics of a new projectile, such as mass and velocity.

In such cases, calculations are far less expensive and time-consuming than tests, so it is possible to perform calculations for many more situations and options than one could test empirically. This produces results that are more reliable and can be applied to a wider range of cases.

***Reconstruction***. There are all too many instances in which projectiles kill or endanger people, or cause damage. This may be the result of an accident, criminal activity or even a breach of international law, e.g. when projectiles hit the personnel or vehicles of international aid organizations. In such cases, it is important for legal reasons to find out where the projectile(s) came from.

The coordinates of the point of impact form the starting point for such investigations, together with the angle of descent and the direction of arrival. Velocity is estimated on the basis of the damage or injuries caused. We can integrate the systems of equations at 5.4.3.2 using negative time or distance steps, which makes it possible to determine the point of departure. It is important to establish the meteorological data for the time at which the projectile was fired (atmospheric density, pressure, etc.) and include these in the calculations. These can generally be obtained via the official meteorological service from the records of a nearby weather station. Wind speed and direction are particularly important, as these have a major influence on the trajectory.

If the drag parameters of the projectile are not known, it is possible to use those for projectiles of similar shape. As the value of $C_D$ depends only on form, calibre and mass are irrelevant as far as this is concerned.

For a more comprehensive treatment of forensic trajectory reconstruction, please see 8.2.1.

#### 5.4.3.4 Some tips on integration

The most suitable method of numerical integration is generally the classical fourth-order Runge-Kutta method. One elegant and quick variant, which also requires fewer registers, is the Runge-Kutta-Gill method (see A. RALSTON, H. S. WILF, 1967). This method was especially popular in the early days of programmable calculators, when they had limited computing power and memory, but is still used in contemporary mathematical software packages.

Flat trajectories and those with lower elevations can be integrated with time or distance as the independent variable. It is better to integrate trajectories with higher elevations over time; the steepness of the descending branch requires very small integration steps if one integrates over distance.

Calculations for specific points such as the apex and the end of the trajectory are carried out by repeatedly reducing the size of the integration step until the criterion for the point in question is fulfilled to the required degree of precision.

At the apex, for instance, the criterion is $v_y = 0$.

The same procedure is also necessary at the boundaries between atmospheric layers, e.g. when the temperature gradient or wind direction changes. Here, one must integrate up to the layer boundary, and then start again using the new data. It is best to so choose the first step in the new layer as to maintain the previous step pattern.

## 5.4.4 The influence of wind

### 5.4.4.1 Equations of motion taking account of wind

The drag on a projectile is caused by the stagnation pressure, which in turn is determined by the relative wind velocity. In other words, it is the velocity of the air relative to the projectile that matters, and not the velocity of the projectile relative to some fixed point. If the velocity of the wind is $\vec{v}_w$, the air strikes the projectile at $\vec{v}_{rel} = \vec{v} - \vec{v}_w$. The equation of motion Eqn 5.4:35 then becomes:

$$\text{(5.4:44)} \qquad \ddot{\vec{r}} = -\frac{F_D}{m} \cdot \frac{\left(\dot{\vec{r}} - \vec{v}_w\right)}{\left|v_{rel}\right|} + \vec{g} + \vec{d}_z + \vec{a}_c \, . \qquad [\text{m/s}^2]$$

This second-order differential equation can be transformed into a system of six first-order equations in a manner analogous to that described in 5.4.3.2.

$$\text{(5.4:45)} \quad \left|\begin{array}{lll} \dot{v}_x & = -\dfrac{F_D}{m} \cdot \dfrac{v_x - v_{wx}}{\left|v_{rel}\right|} + g_x + a_{cx} \, , & [\text{m/s}^2] \\ \dot{v}_y & = -\dfrac{F_D}{m} \cdot \dfrac{v_y - v_{wy}}{\left|v_{rel}\right|} + g_y + a_{cy} \, , & [\text{m/s}^2] \\ \dot{v}_z & = -\dfrac{F_D}{m} \cdot \dfrac{v_z - v_{wz}}{\left|v_{rel}\right|} + g_z + d_z + a_{cz} \, , & [\text{m/s}^2] \\ \dot{x} & = v_x \, , & [\text{m/s}] \\ \dot{y} & = v_y , & [\text{m/s}] \\ \dot{z} & = v_z \, , & [\text{m/s}] \end{array}\right.$$

with aerodynamic force

$$\text{(5.4:46)} \qquad F_D = C_D \cdot \frac{1}{2} \cdot \rho \cdot \left(\dot{\vec{r}} - \vec{v}_w\right)^2 \cdot A_0 \, . \qquad [\text{N}]$$

This system can be rearranged for other integration variables as described in 5.4.3.2 and can be integrated numerically as described in 5.4.3.4 (using the Runge-Kutta method), using the initial values of Eqn 5.4:43.

### 5.4.4.2 Transformation into a wind-axis system

In addition to numerical integration of the system of differential equations (Eqn 5.4:45) there is also an elegant, direct means of calculating the crosswind deflection of a projectile. This was discovered by French ballistician Isidor DIDION (1798–1878) (I. DIDION, 1852, B. P. KNEUBUEHL, 1976). The principle is simple: The equation of motion is transformed into a *wind-axis* system, i.e. a coordinate system that moves with the wind, in which one can assume the results of the trajectory calculation to be known. Transforming back into an earth-axis system then yields the deflection caused by the wind.

Differential equation 5.4:44 is subjected to the following transformation into the wind-axis star coordinate system:

(5.4:47) $\vec{r}^{\,*} = A\cdot(\vec{r}-\vec{v}_W\cdot t)$ , with [m]

(5.4:48) $\dot{\vec{r}}^{\,*} = A\cdot(\dot{\vec{r}}-\vec{v}_W)$ , and [m/s]

$\ddot{\vec{r}}^{\,*} = A\cdot(\ddot{\vec{r}}-\dot{\vec{v}}_W)$ . [m/s²]

Assuming a constant wind velocity $\dot{\vec{v}}_W = 0$, we obtain:

(5.4:49) $\ddot{\vec{r}}^{\,*} = A\cdot\ddot{\vec{r}}$ . [m/s²]

It is quite reasonable to assume wind velocity as constant, because both for flat trajectories close to the ground and for steep trajectories in free air within the atmospheric layers, times of flight are generally short by comparison with the time taken for wind velocity to change (B. KLOSE, 2016).

Eqns 5.4:48 and 49 are solved for $\dot{\vec{r}}$ or $\ddot{\vec{r}}$ and the results substituted into Eqn 5.4:44:

$$A^{-1}\cdot\ddot{\vec{r}}^{\,*} = -\frac{F_D}{m}\cdot\frac{A^{-1}\cdot\dot{\vec{r}}^{\,*}+\vec{v}_W-\vec{v}_W}{\left|A^{-1}\cdot\dot{\vec{r}}^{\,*}+\vec{v}_W-\vec{v}_W\right|} + \vec{g} + \vec{a}_c\,, \qquad \text{[m/s}^2\text{]}$$

(5.4:50) $$\ddot{\vec{r}}^{\,*} = -\frac{F_D}{m}\cdot\frac{\dot{\vec{r}}^{\,*}}{\left|\dot{\vec{r}}^{\,*}\right|} + A\cdot(\vec{g}+\vec{a}_c)\,. \qquad \text{[m/s}^2\text{]}$$

Regardless of the (constant) transformation matrix A chosen, this ensures that the wind vector in the star coordinate system is zero. Eqn 5.4:50 is therefore the equation for a trajectory in a wind-free system. At the same time, however, new initial conditions are created, with muzzle velocity $\dot{\vec{r}}^{\,*}(0) = \vec{v}_0^{\,*}$ and elevation $\theta_0^*$.

By choosing matrix A appropriately, one can ensure that vector $\vec{v}_0^{\,*}$ lies in the $x^*$ - $y^*$ plane of the star coordinate system. A is then a rotation matrix with an angle of ξ between the projections of $\vec{v}_0$ and $\vec{v}_0-\vec{v}_W$ in the x-z plane as angle of rotation (see Fig. 5.16). The y axis is the axis of rotation. Matrix A:

(5.4:51) $$A = \begin{pmatrix} \cos\xi & 0 & \sin\xi \\ 0 & 1 & 0 \\ -\sin\xi & 0 & \cos\xi \end{pmatrix}, \qquad \text{with} \qquad \text{[-]}$$

(5.4:52a) $$\cos\xi = \frac{v_{0x} - v_{wx}}{\sqrt{(v_{0x} - v_{wx})^2 + v_{wz}^2}} ,$$ [-]

(5.4:52b) $$\sin\xi = \frac{-v_{wz}}{\sqrt{(v_{0x} - v_{wx})^2 + v_{wz}^2}} .$$ [-]

At t = 0, then from Eqn 5.4:48 we obtain:

(5.4:53) $$\vec{v}_0^* = A \cdot (\vec{v}_0 - \vec{v}_w) ,$$ [m/s]

and from Fig. 5.16 we can write the following equation for the angle of departure in the system represented by dashed lines:

(5.4:54) $$\tan\theta_0^* = \frac{v_{0y} - v_{wy}}{\sqrt{(v_{0x} - v_{wx})^2 + v_{wz}^2}} .$$ [-]

We know the initial conditions for differential equation 5.4:50 from Eqns 5.4:53 and 54 in the star coordinate system, and the equation can be integrated using one of the usual methods. This gives us the points along the trajectory in the (wind-axis) star system, from which we can select any desired point, e.g. the point of impact in the plane of the muzzle $\vec{r}_E^* = \vec{r}_E^*(t_E^*)$. That point is then inserted in the transformation equation (Eqn 5.4:47) and the equation solved for vector $\vec{r}_E$ in the earth-axis system. We obtain:

(5.4:55) $$\vec{r}_E = A^{-1} \cdot \vec{r}_E^* + \vec{v}_w \cdot t_E^* ,$$ [m]

or, in component form:

(5.4:56a) $$x_E = x_E^* \cdot \cos\xi - z_E^* \cdot \sin\xi + v_{wx} \cdot t_E^* ,$$ [m]

(5.4:56b) $$y_E = y_E^* + v_{wy} \cdot t_E^* ,$$ [m]

(5.4.56c) $$z_E = x_E^* \cdot \sin\xi + z_E^* \cdot \cos\xi + v_{wz} \cdot t_E^* .$$ [m]

Substituting 5.4:52a and b into 5.4:56a and c and rearranging gives us the following equations for $x_E$ and $z_E$:

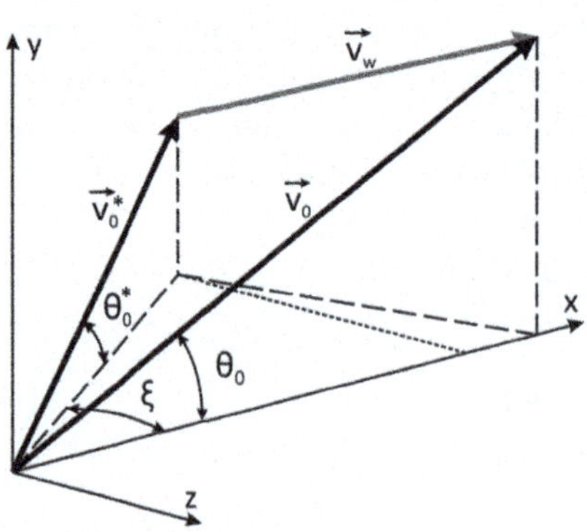


**Fig. 5.16.** Transformation between earth-axis and wind-axis coordinate systems.

$$(5.4{:}57)\qquad x_E = \frac{x_E^* + z_E^* \cdot \dfrac{v_{wz}}{v_{0x} - v_{wx}}}{\sqrt{1 + \left(\dfrac{v_{wz}}{v_{0x} - v_{wx}}\right)^2}} + v_{wx} \cdot t_E^*\,, \qquad [\mathrm{m}]$$

$$(5.4{:}58)\qquad z_E = \frac{z_E^* - x_E^* \cdot \dfrac{v_{wz}}{v_{0x} - v_{wx}}}{\sqrt{1 + \left(\dfrac{v_{wz}}{v_{0x} - v_{wx}}\right)^2}} + v_{wz} \cdot t_E^*\,. \qquad [\mathrm{m}]$$

As we have seen, any wind direction can be decomposed into one component in the direction of shot and one perpendicular to it. We can therefore discuss both equations separately for downrange wind and crosswind. Vertical winds are generally ignored in ballistics. If $v_{wy} = 0$, then from Eqn 5.4:56b we immediately see that $y_E = y_E^*$.

One reason for ignoring vertical wind is that it is virtually impossible to measure upward and downward air currents using the balloon sondes operated by artillery weather services, as it is impossible to differentiate between vertical wind velocity and the vertical velocity of the balloon. The latter depends on the velocity of the balloon relative to the surrounding air, and is influenced by any vertical wind.

Eqns 5.4:57 and 58 include range $x_E^*$ and lateral deviation $z_E^*$ in the wind-axis star system. Time of flight varies little – if at all – which means we can assume $t_E^* = t_E$. If one simplifies the star coordinate system by ignoring other sideways accelerations such as lateral and Coriolis acceleration (which will have little effect on pure wind deviation), one requires only two-dimensional integration in the wind-axis system. This means that we can also assume $z_E^* = 0$. However, that will limit the precision and scope of these equations

### 5.4.4.3 *DIDION's longitudinal wind equation*

From Eqns 5.4:57 and 58, and under the above-mentioned assumptions, we obtain the following deviations for a pure range wind ($v_{wz} = 0$):

$$(5.4{:}59)\qquad x_E = x_E^* + v_{wx} \cdot t_E^*\,, \qquad [\mathrm{m}]$$

$$\text{with}\qquad z_E = z_E^* = 0\,. \qquad [\mathrm{m}]$$

We can therefore calculate the influence of downrange wind from the range in a wind-free system, by adding the distance obtained from the wind velocity and time of flight in a wind-free system, observing the correct sign. This involves no lateral deviation.

The equation is not suitable for artillery, as it results in errors of 30% and higher. It is, however, suitable for small-calibre projectiles at an elevation of over 100 mils (approx. 5.6°); errors will be less than 10%. The error increases considerably at lower elevations, and can exceed 100% in the case of short, very flat trajectories.

#### 5.4.4.4 *DIDION's crosswind equation*

Crosswind deflection is generally of greater interest than any alteration in range, especially for flat trajectories. Under certain limiting conditions, DIDION's crosswind equation yields good results, which make the equation of practical value.

If only a crosswind is present, substituting $v_{wx} = 0$ in Eqns 5.4:57 and 58 yields:

(5.4:60) $$x_E = \frac{x_E^* + z_E^* \cdot \varepsilon}{\sqrt{1+\varepsilon^2}}\,, \qquad \text{[m]}$$

(5.4:61) $$z_E = \frac{z_E^* - x_E^* \cdot \varepsilon}{\sqrt{1+\varepsilon^2}} + v_{wz} \cdot t_E^*\,, \qquad \text{[m]}$$

where $$\varepsilon = \frac{v_{wz}}{v_{0x}}\,. \qquad \text{[-]}$$

If $\varepsilon \ll 1$ (i.e. $v_{wz} \ll v_{0x}$, which can almost always be assumed for flat trajectories), then we can linearize Eqns 5.4:60 and 5.4:61 and by substituting $z_E^* = 0$ and $t_E^* = t_E$ we obtain:

(5.4:62) $$x_E = x_E^*\,, \qquad \text{[m]}$$

(5.4:63) $$z_E = -x_E^* \cdot \frac{v_{wz}}{v_{0x}} + v_{wz} \cdot t_E\,. \qquad \text{[m]}$$

Bearing in mind Eqn 5.4:62, we can simply rearrange Eqn 5.4:63 to obtain Didion's crosswind equation:

(5.4:64) $$z_E = v_{wz} \cdot \left( t_E - \frac{x_E}{v_{0x}} \right)\,. \qquad \text{[m]}$$

The main factor limiting the range over which DIDION's equation remains valid is the use of a two-dimensional calculation in the wind-axis system. The deflection by the wind of the effective trajectory creates a component of drag that acts in the opposite direction to the wind (see Fig. 5.17). That component is neglected when integrating in the wind-axis system. This means that:

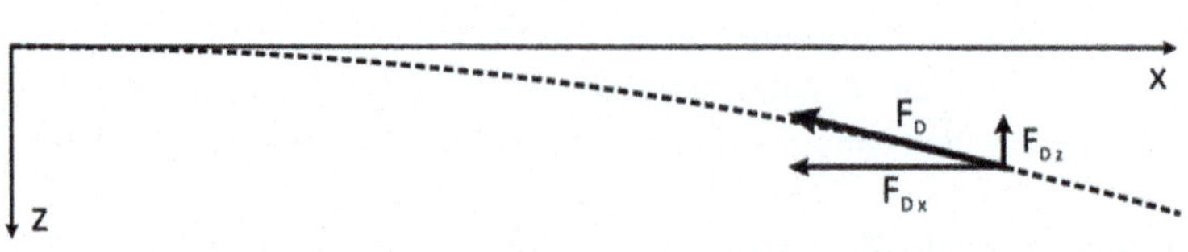


**Fig. 5.17.** Deflection due to a crosswind creates a component of drag that acts in the opposite direction to the wind.

- Eqn 5.4:64 produces values that are too high
- as time of flight increases, the values resulting from the equation deviate further from those obtained by integrating the system of Eqn 5.4:45.

For a comparison between the results of DIDION's equation and those of the system of equations, please see Fig. 5.18. The error resulting from use of DIDION's equation is less than 1% for times of flight of less than 7 s in the case of handguns (which corresponds to a range of approx. 1 km) and 15 s (a range of 2 km to 3 km, depending on calibre) in the case of shoulder weapons. This means that DIDION's crosswind equation produces very good results for all practical small-arms trajectories and is hence suitable for calculating crosswind deflection where it is not possible to integrate the system of differential equations.

However, Eqn 5.4:64 does require one to know time of flight, in addition to range and muzzle velocity, which are already known. If no range table with this information is available, one can proceed as follows:

If final velocity $v_E$ is known, one can calculate time of flight to a range $x_E$ from the mean velocity:

$$t_E = \frac{2 \cdot x_E}{v_0 + v_E} \; . \qquad \text{[s]} \tag{5.4:65}$$

If $v_E$ is also unknown, it can be determined from the total range using the estimated values below, by estimating the final velocity. This method can be used with rifles at ranges of up to approx. 300 m.

- Heavy, slender pointed bullets (with calibres of 7 mm and above) decelerate by approximately 0.5 m/s per metre distance. A bullet with a more rounded nose (such as a typical hunting bullet) will decelerate by approximately 0.7–0.8 m/s per metre.
- Light, slender pointed bullets (with calibres of 6 mm and below) decelerate by approximately 1 m/s per metre distance. A bullet with a more rounded nose will decelerate by approximately 1.2–1.3 m/s per metre.

Calculations carried out using this approach will of course be less accurate, but the results yield perfectly usable estimates.

When comparing the ballistic characteristics of two types of ammunition, one generally looks at the crosswind deflection that occurs with a crosswind of 1 m/s. This is known as *sensitivity to crosswind* and is given by the following equation, derived from Eqn 5.4:64.

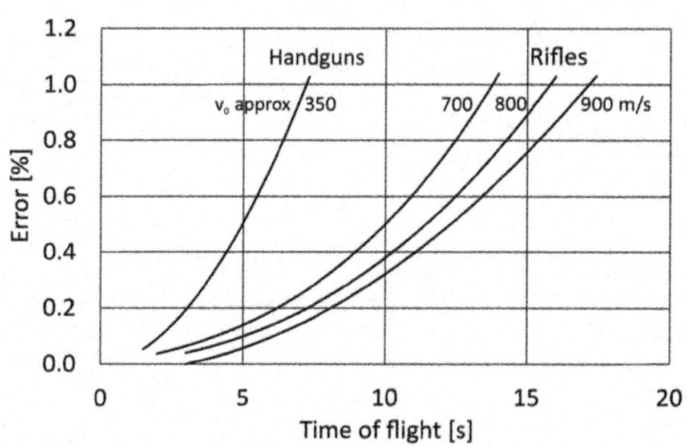


**Fig. 5.18.** Accuracy of DIDION's equation for calculating crosswind deviation. For the typical range of a light weapon (with a time of flight of less than 5 s), the error is very small.

$$z'_E = \frac{z_E}{v_{wz}} = \left(t_E - \frac{x_E}{v_{0x}}\right). \qquad (5.4:66) \quad [s]$$

### 5.4.4.5 Measuring sensitivity to crosswind

The concept of switching coordinate systems used to derive DIDION's equation can also be used to measure sensitivity to crosswind empirically. If, at the moment the weapon is fired, it is moved at a uniform velocity $v_q$ transverse to the direction of shot, the projectile will be subjected to a uniform crosswind of opposite direction but identical magnitude in the weapon-based coordinate system, (which moves with the weapon). In the absence of wind, the earth-axis coordinate system will also be the wind-axis system (see Fig. 5.19).

The lateral movement of the weapon at a velocity $v_q$ displaces the theoretical point of impact

$$z_g = v_q \cdot t_E \qquad [m]$$

In the target plane relative to the line of departure, where $t_E$ = time of flight. However, the actual point of impact will differ from that point. The difference is due to the displacement caused by the airstream $\Delta z$, which is related to lateral velocity $v_q$. The sensitivity to wind of the projectile is equal to $\Delta z$ multiplied by $v_q$.

When carrying out such measurements, one needs to accurately measure not only the lateral velocity of the weapon and the time of flight of the projectile, but also the time from the moment of firing to the moment at which the projectile exits the barrel and the velocity of any natural wind.

A suitable measuring system has been successfully tested for various projectiles, including rockets. See B. P. KNEUBUEHL (1979) for further details, including instructions regarding construction of such an installation and analysis of results.

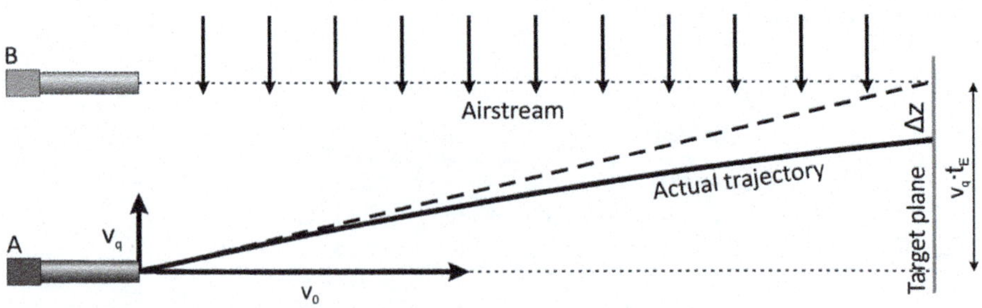


**Fig. 5.19.** Experimental determination of sensitivity to crosswind. A: Position of the weapon when fired. B: Position of the weapon when the projectile reaches the target plane. The difference $\Delta z$ between the actual point of impact and the geometrically predicted point of impact is equal to the crosswind deviation at $v_q$.

### 5.4.5 Trajectory characteristics

#### *5.4.5.1 General*

Trajectories calculated in accordance with the point-mass model have a number of characteristics that can be deduced from the equations of motion. While they have no direct significance, they do enhance understanding and give a feel for the geometrical/physical process involved. Most of them are obvious, but we shall briefly look at how they are derived, as words such as "obvious", "clear" and "self-explanatory" should be used with caution, especially in ballistics (as we saw in 2.4).

A trajectory consists of two main parts: the *ascending branch* and the *descending branch*. The boundary between these two regions is the highest point of the trajectory, at which the y component of the projectile's velocity falls to zero. This is known as the *apex*. It is sometimes referred to as the *vertex*, but this is incorrect; see 5.4.5.4.

The following comments all assume that the projectile is passive, i.e. that it does not have its own source of energy. The equations for the two-dimensional trajectory form the starting point for these remarks, but that is not much of a restriction, as the characteristics under consideration are not affected by the lateral deviation of the trajectory.

#### *5.4.5.2 Characteristics of trajectory velocity*

The trajectory velocity of a projectile possesses the following six characteristics, which all stem from the equations of motion of the two-dimensional model and the basic laws of physics:

1 The horizontal component of trajectory velocity continually decreases.

2 Trajectory velocity continually decreases during the ascending branch.

3 Trajectory velocity is greater at a given height in the ascending branch than at the same height in the descending branch.

4 The magnitude of vertical velocity is greater at a given height in the ascending branch than at the same height in the descending branch.

5 Trajectory velocity cannot exceed a certain maximum.

6 Relative extreme values of trajectory velocity can only occur in the descending branch.

Reasoning:

*1* Dividing Eqn 5.4:41/1 by $v_x$ and dt gives:

$$\frac{dv_x}{v_x} = -\frac{F_D}{m \cdot v} \cdot dt \, . \qquad [-]$$

Integrating between two points, 1 und 2, with $t_2 > t_1$:

$$\ln\frac{v_{x2}}{v_{x1}} = -\int_{t_1}^{t_2}\frac{F_D}{m\cdot v}\cdot dt \ , \qquad [-]$$

(5.4:67) $$v_{x2} = v_{x1}\cdot\exp\left(-\int_{t_1}^{t_2}\frac{F_D}{m\cdot v}\cdot dt\right) . \qquad [m/s]$$

$F_D$, m and v are always positive, so the exponential function must have a negative exponent, which means that $v_{x2} < v_{x1}$.

*2* This is a direct result of the law of conservation of energy, and the change in the kinetic energy of the projectile must be equal to the work done in opposition to drag and the effect of gravity:

$$\frac{1}{2}\cdot m\cdot d\left(v^2\right) = -F_D\cdot ds - m\cdot g\cdot\sin\theta\cdot ds \ . \qquad [J]$$

Following integration, with $s_2 > s_1$ and substituting $dy = \sin\theta\cdot ds$

$$\frac{1}{2}\cdot m\cdot\int_{s_1}^{s_2} d\left(v^2\right) = -\int_{s_1}^{s_2} F_D\cdot ds - m\cdot g\cdot\int_{y_1}^{y_2} dy \ , \qquad [J]$$

and after performing the integration

(5.4:68) $$\frac{1}{2}\cdot m\cdot\left(v_2^2 - v_1^2\right) = -\int_{s_1}^{s_2} F_D\cdot ds - m\cdot g\cdot\left(y_2 - y_1\right) . \qquad [J]$$

As $y_2 > y_1$, the right-hand side is negative, which means that the left-hand side must also be negative, so $v_2 < v_1$.

*3* Because $y_2 = y_1$, but $s_2 > s_1$ still applies, the second summand on the right-hand side of Eqn 5.4:68 disappears. The right-hand side remains negative, which means that the left-hand side must also be negative, so in accordance with the same reasoning as set out above, $v_2 < v_1$. This means that at the base of the trajectory in particular, $v_E$ must always be less than $v_0$.

*4* We rearrange Eqn 5.4:41/2 thus:

$$\frac{dt}{dy} = \frac{1}{v_y} \ , \qquad [s/m]$$

and make y the integration variable

$$v_y\cdot dv_y = -\frac{F_D}{m}\cdot\frac{v_y}{v}\cdot dy - g\cdot dy = -\frac{F_D}{m}\cdot\sin\theta\cdot dy - g\cdot dy \ , \qquad [m^2/s^2]$$

and with $dy = v\cdot\sin\theta\cdot dt$ we obtain

$$d\left(v_y^2\right) = -\frac{F_D}{m}\cdot\sin\theta\cdot v\cdot\sin\theta\cdot dt - g\cdot dy \ . \qquad [m^2/s^2]$$

We integrate from Point 1 in the ascending branch to Point 2 in the descending branch

$$(5.4{:}69)\qquad \int_1^2 d\left(v_y^2\right) = v_{y_2}^2 - v_{y_1}^2 = -\int_{t_1}^{t_2} \frac{F_D}{m}\cdot v\cdot \sin^2\theta\cdot dt - \int_{y_1}^{y_2} g\cdot dy\ . \qquad [m^2/s^2]$$

Because $y_2 = y_1$, the second summand on the right-hand side disappears, the right-hand side remains negative and the left-hand side must also therefore be negative. This proves Statement 4.

In preparation for proving Statements 5 and 6, we shall show that trajectory velocity reaches a minimum in the descending branch. Substituting

$$\frac{dt}{ds} = \frac{1}{v} \qquad [s/m]$$

transforms Eqn 5.4:10/1 into the following equation, which also represents the conservation of energy:

$$(5.4{:}70)\qquad v\cdot dv = -\left(\frac{F_D}{m} + g\cdot \sin\theta\right)\cdot ds\ . \qquad [m^2/s^2]$$

At the apex, $\sin\theta = 0$ and $F_D/m > 0$. This means that dv must be less than zero. Velocity falls still further at the apex, displacing the minimum into the descending branch.

*5* Having proved the above, we need only consider the descending branch. The projectile is subject to gravity. It therefore gains in velocity after passing the point of minimum trajectory velocity. However, as deceleration increases quadratically with velocity, and acceleration due to gravity increases only slightly as altitude decreases, a point in time is reached at which deceleration equals acceleration, and hence $dv = 0$. From Eqn 5.4:70 and using Eqn 5.4:37, we can derive the following:

$$\frac{F_D}{m} = C_D\cdot\frac{\rho}{2\cdot q}\cdot v^2 = -g\cdot\sin\theta\ . \qquad [m/s^2]$$

Because the angle of trajectory $\theta$ at this point in time is in the region of $-90°$, we can take $\sin\theta = -1$. Solving for velocity gives

$$(5.4{:}71)\qquad v = \sqrt{\frac{2\cdot q\cdot g}{C_D\cdot\rho}}\ . \qquad [m/s]$$

This is the *terminal velocity* of a projectile. It forms the upper limit for the velocity of a falling projectile, thereby proving Statement 5.

A relative maximum velocity is reached only in very steep descending branches and hence only in the case of a high elevation. For most trajectories, velocity continues to increase up to impact.

*6* We have already proven that the relative minimum occurs in the descending branch. As regards the relative maximum, we must distinguish between two cases:

- Terminal velocity is significantly lower than the speed of sound. This is the case with low sectional densities (small-calibre projectiles and lighter medium-calibre projectiles). In such cases, the value of $C_D$ remains approximately constant. Because atmospheric density rises with decreasing altitude to a greater degree than does acceleration due to gravity, velocity starts to decrease again once the projectile has reached its relative maximum. However, this is only the case once the projectile has descended below muzzle height. Until that point, the projectile is still accelerating.

– The terminal velocity of projectiles with a greater sectional density (heavy medium-calibre and artillery projectiles) is close to the speed of sound. Any further increase in velocity causes a very large rise in $C_D$, which causes the projectile to decelerate sharply. That causes a decrease in $C_D$ and the projectile accelerates again. A kind of oscillation occurs, between acceleration and deceleration, leading to a mean rate of fall. This rate of fall decreases as the projectile descends, again because of the rising atmospheric density. Here again, therefore, the projectile attains a relative maximum velocity, at a point above muzzle height.

This behaviour is clearly audible in the impact zone of artillery shells. The velocity of the descending shells oscillates between low supersonic and high subsonic velocity, producing a clear rattling sound caused by the interrupted shock waves.

Figures 5.20, 5.21 and 5.22 show velocity/range curves for a small-calibre, medium-calibre and artillery projectile at high elevation. The characteristics mentioned above are clearly visible. The velocity of the rifle bullet (Fig. 5.20) is still increasing at the end of its trajectory. That of the medium-calibre projectile (Fig. 5.21) is approaching a maximum and the velocity of the artillery shell (Fig. 5.22) has already reached its maximum. At lower elevations (under approx. 800 mils), the artillery shell also continues to accelerate until it reaches muzzle height.

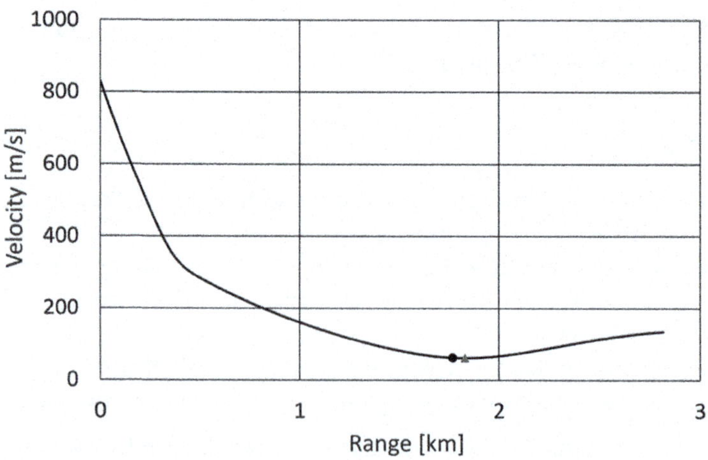


**Fig. 5.20.** Velocity of a rifle bullet, elevation 1100 mils. The black circle indicates the apex, the grey triangle the point of lowest velocity.

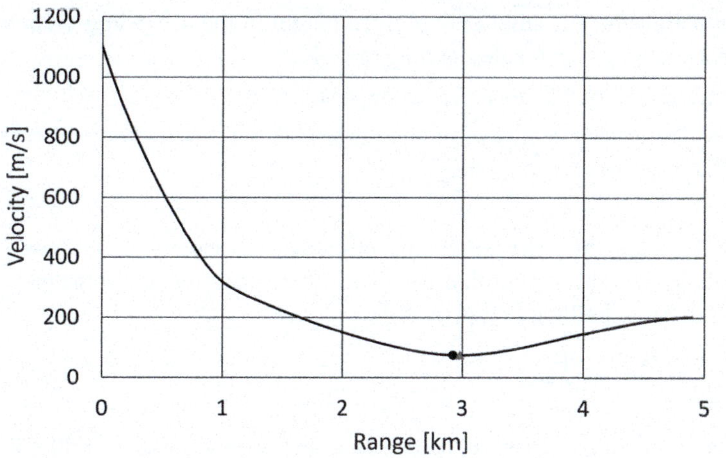


**Fig. 5.21**. Velocity of a 20 mm anti-aircraft shell, elevation 1200 mils. Symbols as in Fig. 5.20.

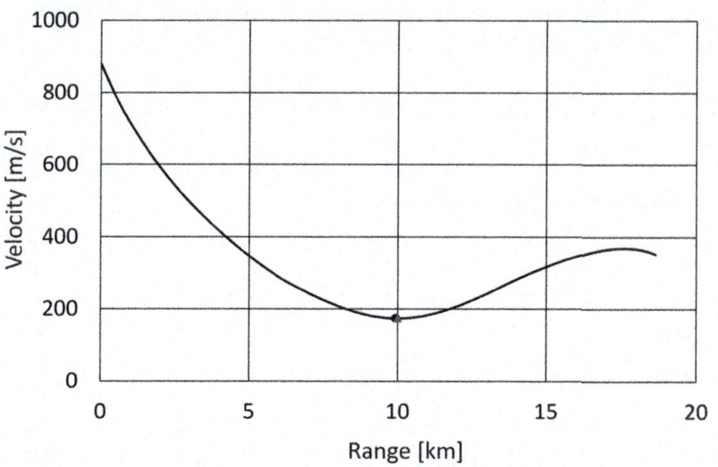


**Fig. 5.22**. Velocity of a 155 mm artillery shell, elevation 1200 mils. The black circle indicates the apex, the grey triangle the point of lowest velocity. Velocity attains a relative maximum before the projectile reaches muzzle height.

It is interesting to note the kink in the curve in the vicinity of the speed of sound. This is less pronounced for larger calibres.

### *5.4.5.3 Geometric characteristics of the trajectory*

In addition to the characteristics of trajectory velocity, we should note the following geometrical characteristics of the trajectory:

7 The angle of trajectory decreases continuously along the trajectory. It is positive in the ascending branch and negative in the descending branch.

8 The angle of trajectory is smaller at a given height in the ascending branch than at the same height in the descending branch.

9 The descending branch has a vertical asymptote.

10 The apex of a trajectory is always situated behind the mid-point between two points of the trajectory of equal height.

Reasoning:

*7* According to Statement 1, the horizontal component of velocity $v_x = v \cdot \cos\theta$ decreases monotonically. This means that $d(v \cdot \cos\theta) < 0$. Examination of the principal equation of external ballistics (Eqn 5.4:13)

$$\frac{d(v \cdot \cos\theta)}{d\theta} = \frac{F_D \cdot v}{m \cdot g}, \qquad \text{[m/s]}$$

of which the right-hand side is positive, reveals that this is only possible if $d\theta < 0$. This applies to the entire trajectory, as does the principal equation.

***8*** The characteristic to be demonstrated is a relationship between angle of trajectory and trajectory height. The second and fourth equations of the system at Eqn 5.4:10 include these two variables.

$$\dot{\theta} = -\frac{g}{v} \cdot \cos\theta, \qquad \text{[1/s]}$$

and
$$\dot{y} = v \cdot \sin\theta. \qquad \text{[m/s]}$$

Dividing the fourth equation by the second gives us

$$\frac{dy}{d\theta} = -\frac{v^2}{g} \cdot \frac{\sin\theta}{\cos\theta}, \qquad \text{[m]}$$

and
$$\frac{d(\cos\theta)}{\cos\theta} = \frac{g}{v^2} \cdot dy. \qquad \text{[-]}$$

Integrating from Point 1 $(y_1, \theta_1)$ to Point 2 $(y_2, \theta_2)$:

(5.4:72)
$$\ln(\cos\theta_2) - \ln(\cos\theta_1) = \int_{y_1}^{y_2} \frac{g}{v^2} \cdot dy. \qquad \text{[-]}$$

Integration from Point 1 in the ascending branch to Point 2 in the descending branch is interrupted at the apex (with $\theta_G = 0$ at the apex).

$$\ln(\cos\theta_1) = -\int_{y_1}^{y_G} \frac{g}{v^2} \cdot dy, \qquad \text{[-]}$$

$$\ln(\cos\theta_2) = -\int_{y_2}^{y_G} \frac{g}{v^2} \cdot dy. \qquad \text{[-]}$$

Because the velocity at a given height in the ascending branch is always greater than the velocity at the same height in the descending branch (Statement 3) and $\theta$

decreases monotonically, then from the last two equations:

$$\ln(\cos\theta_1) \geq \ln(\cos\theta_2) \quad \Rightarrow \quad \cos\theta_1 \geq \cos\theta_2 \ , \qquad [-]$$

and hence $|\theta_1| \leq |\theta_2|$ . [-]

This also means that the descending branch is always steeper than the ascending branch, and that the angle of descent at muzzle height is always greater than the elevation.

***9*** In the explanation for Statement 1, we derived Eqn 5.4:67, in which drag can be inserted using Eqn 5.4:37

$$v_{x2} = v_{x1} \cdot \exp\left( -\int_{t_1}^{t_2} \frac{C_D \cdot \rho \cdot v}{2 \cdot q} \cdot dt \right) \qquad [\mathrm{m/s}]$$

where q = sectional density.

The integrand in the exponent is positive and has a constant denominator, which means that for $t_2 \rightarrow \infty$ the integral also tends to infinity and the horizontal velocity therefore has a limit of 0. However, this also means that for every trajectory in the atmosphere there is a maximum range that cannot be exceeded, which confirms the existence of a vertical asymptote in the descending branch.

***10*** This statement is also based on Statement 8. Because the ascending branch is flatter than the descending branch, the apex is displaced away from the origin of the trajectory.

A trajectory in a vacuum and in a parallel gravitational field has the form of a parabola (see Eqn 2.2:34) and is hence symmetrical. In accordance with Statement 10, drag causes asymmetry in the trajectory, and this becomes increasingly pronounced the more the projectile decelerates. Projectiles with a low sectional density moving at high speed follow more asymmetrical trajectories, whereas the trajectories of heavy, slow-moving projectiles are almost parabolic. This is clearly

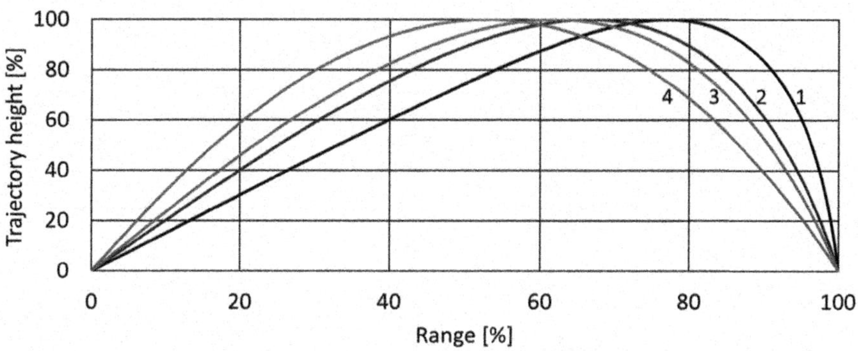


**Fig. 5.23.** Percentage presentation of four trajectories with different decelerations. Trajectory 1: light, fast fragment. Trajectory 2: light 5.56 × 45 military bullet. Trajectory 3: heavy 7.5 × 55 military bullet. Trajectory 4: artillery shell.

visible in Fig. 5.23, which shows four trajectories, ranging from light, very fast fragments to a heavy, slow artillery shell on a percentage scale.

The fast-moving fragment (Trajectory 1 in Fig. 5.23, mass 1.1 g, $v_0$ 2000 m/s) reaches its apex after covering 77% of its total distance, while the slow-moving artillery shell (Trajectory 4, mass 43 kg, $v_0$ 470 m/s) has already reached the highest point of its trajectory after covering 53% of its total distance. The light and heavy military bullets (Trajectories 2 and 3) reach the apex at 66% and 62% respectively. All four trajectories were calculated for the same elevation, 600 mils. Elevation is also a factor. At lower elevations (up to maximum range), asymmetry increases slightly, decreasing slightly at higher elevations.

#### *5.4.5.4 Specific points along the trajectory*

A number of points along the trajectory are of particular interest in terms of trajectory geometry. Each of these points constitutes an extreme of one sort or another, and they always appear in the same order along the descending branch. These points are as follows:

1 Apex (highest point of the trajectory).
2 Point of greatest trajectory angular velocity.
3 Point of smallest radius of trajectory curvature, also known as the vertex.
4 Point of lowest tractability (usually the point of greatest stability).
5 Point of lowest velocity.

Fig. 5.24 shows these points for the trajectory of a 5.56 × 45 bullet. The diagram is drawn to scale. As elevation increases, the point of lowest velocity moves towards the apex (see Figs 5.20–5.22), and with it the other points.

The vertex of the trajectory is the point at which the radius of curvature of the trajectory is at its smallest. It is always the next-but-one special point after the apex.

The point of lowest tractability is also of particular relevance to the motion of a projectile. If the projectile reaches this point without disturbance, one may safely

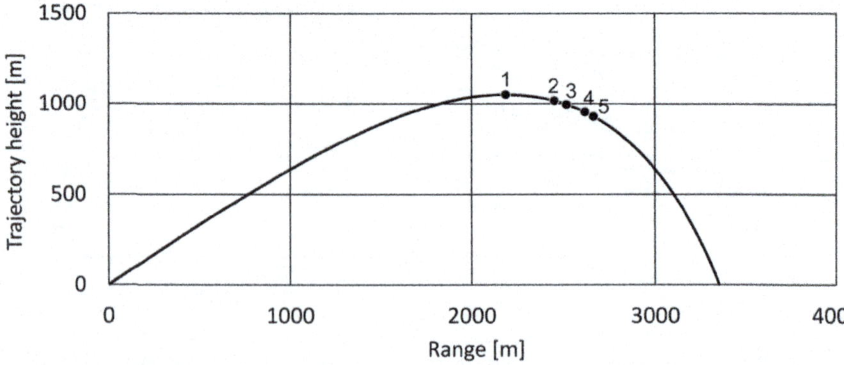


**Fig. 5.24.** Trajectory of a 5.56 × 45 bullet, showing the specific points. Elevation 600 mils. Trajectory and points shown to scale.

assume that it will maintain normal flight until the end of its trajectory. Please see 5.6, “Stability and tractability” for a more detailed discussion of tractability. The remaining three points have no direct relevance to ballistics.

## 5.4.6 Approximations

### *5.4.6.1 Approximations for flat trajectories*

A trajectory is classed as *flat* if the angle of shot is less than approximately 100 mils (approx. 5.6°). Calculation of such a trajectory is considerably simpler than in other cases, as acceleration due to gravity always acts approximately perpendicular to the direction of movement, the time of flight is short and atmospheric density and temperature are constant throughout the trajectory. The short time of flight implies a relatively small change in velocity, which in turn allows us to assume a constant drag coefficient $C_D$.

The starting point for calculating a flat trajectory is Eqn. 5.4:10/1, using Eqn 5.4:37 for deceleration.

$$\dot{v} = -C_D \cdot \frac{1}{2} \cdot \rho \cdot v^2 \cdot \frac{1}{q} , \qquad [\mathrm{m/s^2}]$$

with $\sin\theta \approx 0$. All fixed values can be grouped together in one constant, K.

$$(5.4{:}73) \qquad K = \frac{C_D \cdot \rho}{2 \cdot q} , \qquad [1/\mathrm{m}]$$

This results in the following integrable, time-dependent differential equation for velocity:

$$(5.4{:}74) \qquad \dot{v} = -K \cdot v^2 . \qquad [\mathrm{m/s^2}]$$

Separating variables and integrating with an initial condition $v(0) = v_0$ gives us the following solution:

$$(5.4{:}75) \qquad v(t) = \frac{v_0}{1 + K \cdot v_0 \cdot t} . \qquad [\mathrm{m/s}]$$

Applying the substitution of Eqn 5.4:39 to Eqn 5.4:74 yields the following time-dependent equation:

$$(5.4{:}76) \qquad v' = -K \cdot v . \qquad [1/\mathrm{s}]$$

This equation can be solved in the same manner and with the same initial condition.

$$(5.4{:}77) \qquad v(x) = v_0 \cdot e^{-K \cdot x} . \qquad [\mathrm{m/s}]$$

By solving Eqn 5.4:75 for time of flight, substituting Eqn 5.4:77 for velocity, we obtain an equation for time of flight as a function of distance.

(5.4:78) $$t(x) = \frac{1}{v_0 \cdot K} \cdot (e^{K \cdot x} - 1) \ . \qquad [s]$$

We can also derive an equation for the trajectory itself using the conditions for flat trajectories. Our starting point is Eqn 5.4:41/2:

$$\dot{v}_y = \ddot{y} = -\frac{F_D}{m} \cdot \frac{v_y}{v} - g = -K \cdot v \cdot v_y - g \ . \qquad [m/s^2]$$

We can rearrange this equation to use x instead of y for the second derivation over time

$$\ddot{y} = \left( y'' \cdot v_x + y' \cdot v_x' \right) \cdot v_x \ , \qquad [m/s^2]$$

and substitute the resulting expression into the above equation:

(5.4:79) $$\left( y'' \cdot v_x + y' \cdot v_x' \right) \cdot v_x = -K \cdot v \cdot v_y - g \ . \qquad [m/s^2]$$

For a flat trajectory $v = v_x$, so using Eqn 5.4:77 we obtain:

$$v_x = v_{0x} \cdot e^{-K \cdot x} \ , \qquad [m/s]$$

$$v_x' = -K \cdot v_{0x} \cdot e^{-K \cdot x} \ , \qquad [1/s]$$

and $$v_y = y' \cdot v_x \ . \qquad [m/s]$$

Substituting into Eqn 5.4:79 and simplifying, we obtain the differential equation for the trajectory

(5.4:80) $$y'' = -\frac{g}{v_{0x}^2} \cdot e^{2 \cdot K \cdot x} \ , \qquad [1/m]$$

which can be integrated as an elementary function. Using the initial condition $y'(0) = \tan \theta_0$, we first obtain an equation for the angle of trajectory

(5.4:81) $$y'(x) = \tan\{\theta(x)\} = \tan \theta_0 - \frac{g}{2 \cdot v_0^2 \cdot \cos^2 \theta_0 \cdot K} \cdot (e^{2 \cdot K \cdot x} - 1) \ . \qquad [-]$$

Integrating this with the initial condition $y(0) = 0$ yields the closed representation of the trajectory curve:

(5.4:82) $$y(x) = x \cdot \tan \theta_0 - \frac{g}{4 \cdot v_0^2 \cdot \cos^2 \theta_0 \cdot K^2} \cdot (e^{2 \cdot K \cdot x} - 2 \cdot K \cdot x - 1) \ . \qquad [m]$$

Eqns 5.4:77 and 78, plus 5.4:81 and 82 form a set of equations with which we can calculate the most important trajectory data directly. We can easily program this set of equations on a pocket graphing calculator. Table 5.5 shows a number of values for constant K. If the BC of a projectile (see 5.4.2.4) is known for a certain range of velocities, we can calculate K using

(5.4:83) $$K = 0.00068422 \cdot \frac{C_D(SP)}{BC} \ , \qquad [1/m]$$

**Table 5.5.** Values of K for approximate trajectory data calculations

| Calibre | Mass | Maximum range | Altitude [m above sea level] | | |
|---|---|---|---|---|---|
| | [g] | [m] | 0 | 200 | 400 |
| 5.56 mm NATO | 4.1 | 300 | 0.0011630 | 0.0011383 | 0.0011142 |
| 7.5 × 55 GP 11 | 11.3 | 300 | 0.0007333 | 0.0007182 | 0.0007035 |
| 7.62 × 39 Kalashnikov | 8.0 | 300 | 0.0017081 | 0.0016713 | 0.0016353 |
| 7.62 mm NATO | 9.5 | 300 | 0.0009654 | 0.0009451 | 0.0009253 |
| 338 Win. Mag. | 16.2 | 500 | 0.0008318 | 0.0008144 | 0.0007972 |
| 12.7 × 99 (50 Brown.) | 42.0 | 500 | 0.0005750 | 0.0005631 | 0.0005514 |

where $C_D(SP)$ is the mean drag coefficient for the G1 standard shell at the velocities concerned (SP = standard projectile). The coefficient incorporates conversion from US units to metric.

The simplifying assumptions do limit the scope of these equations. At velocities greater than the speed of sound, we can only assume a constant drag coefficient over the entire trajectory for velocity changes of up to 200 or 300 m/s. This corresponds to a maximum range of approximately 300 m, or 500 m in the case of larger calibres. At transsonic velocities, the change in $C_D$ is so large that these calculations can only be used for very short ranges (up to 50 m) and are therefore of little use.

### 5.4.6.2 Approximations for apex height

The maximum height of the projectile $y_{AP}$ (its apex) is also a significant exterior ballistics parameter. We cannot calculate this directly from the equations derived in the previous section. However, an approximation formula taken from in vacuo ballistics yields fairly accurate results. This is known as *HAUPT's equation.*

(5.4:84) $$y_{AP} = \frac{1}{8} \cdot g \cdot t_E^2 \ . \qquad [m]$$

Where g = acceleration due to gravity and $t_E$ = time of flight for the total distance $x_E$ for which we wish to know the apex height $y_{AP}$. For practical calculations, we can use the time of flight from Eqn 5.4:65. For the derivation of HAUPT's equation, please see 2.2.6.3 (Eqn 2.2:38).

Every trajectory is completely enclosed by two parabolas of the same distance, of which one has the same angle of departure as the trajectory and the other the same angle of descent (see Fig. 5.25). We derive the equations for the apices of the two parabolas from a combination of Eqns 2.2:35 and 37 (see 2.2.6.3):

$$y_{AP1} = \frac{x_E}{4} \cdot \tan\theta_0 \ , \qquad [m]$$

$$y_{AP2} = \frac{x_E}{4} \cdot \tan\omega_H \ , \qquad [m]$$

where $x_E$ = range, $\theta_0$ = elevation und $\omega_H$ = angle of descent relative to the horizontal.

The geometric mean of these two apices gives a fairly accurate value for the apex of the trajectory that they enclose:

$$(5.4{:}85) \qquad y_{AP} = \frac{x_E}{4} \cdot \sqrt{\tan\theta_0 \cdot \tan\omega_H} \,. \qquad [\mathrm{m}]$$

The closer the trajectory approaches a parabola in form (i.e. the higher the sectional density of the projectile and the lower its velocity) the more accurate the results of this approximation formula will be.

In the example of Fig. 5.25, a small-calibre projectile with a fairly asymmetric trajectory, the error is approximately 3.3%.

To use Eqn 5.4:85, we need not only the range but also the elevation and the angle of descent, which will probably not be known. The question therefore arises as to how useful such an equation would be. Existing diagrams of elevation (or range) and angle of descent based on measured angles of descent show the approximate range and (using Eqn 5.4:85) the apex height. For further details, please see 8.2.1, "Trajectory reconstruction".

### 5.4.6.3 Approximations for point of impact

The most import parameters as regards the point of impact are range, final velocity and angle of descent. It is virtually impossible to formulate simple approximation equations for range or final velocity. For range, it is worth mentioning the trajectory equation formulated by A.-H. PITON-BRESSANT (1852). This is derived from power series expansion of the trajectory equation, stopping after the third element (H. MOLITZ, R. STROBEL, 1963):

$$(5.4{:}86) \qquad y = x \cdot \tan\theta_0 - \frac{g \cdot x^2}{2 \cdot v_0^2 \cdot \cos^2\theta} \cdot (1 + Q \cdot x) \,. \qquad [\mathrm{m}]$$

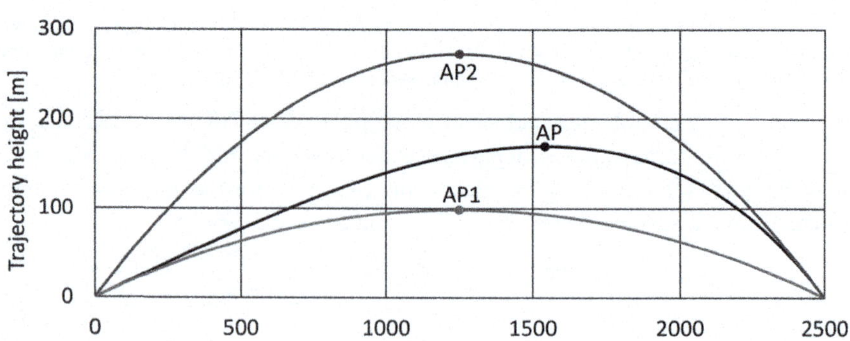


**Fig. 5.25.** Trajectory of a 5.56 × 45 rifle bullet, showing the two parabolas that enclose it and the apices of all three trajectories.

Q depends on the projectile, elevation and $v_0$, and must be determined empirically. If we know Q, we can calculate the trajectory fairly accurately using Eqn 5.4:86.

#### 5.4.6.4 Firing uphill and downhill

Weapons are generally zeroed with the barrel horizontal, or nearly so. If the weapon is then fired upwards or downwards at a steep angle, the trajectory will differ from that obtained during zeroing. The effective angle of departure (elevation) is the sum of the angle of shot and the angle of sight (see Fig. 5.15, p. 182).

When the weapon is discharged horizontally, acceleration due to gravity acts on the projectile perpendicular to the target line. If the direction of shot is rotated about an angle $\lambda$, only the $g \cdot \cos\lambda$ component of gravity acts on the projectile. The projectile experiences a smaller acceleration vertically to the target line, which means the point of impact will be higher than for a horizontal shot.

We can derive an approximation formula for relative vertical error from the in vacuo equations. This formula is as follows:

$$(5.4{:}87) \qquad \Delta\sigma = \frac{1000 \cdot \cos\lambda \cdot \sin\sigma}{\cos(\lambda+\sigma)} \cdot \left(1 - \frac{\cos\lambda}{1-\tan\lambda \cdot \tan\sigma}\right), \qquad \text{[mils]}$$

where $\Delta\sigma$ = change in angle of shot (and also the vertical error in per mille of range), $\sigma$ = angle of shot and $\lambda$ = angle of sight to the target.

Table 5.6 shows the expected vertical deviation for a number of angles of shot and angles of sight. Projectiles with higher initial velocities (and hence smaller angles of shot) deviate less than slower projectiles.

#### 5.4.6.5 Shots fired steeply into the air

There are always those who think firing into the air is a good way of celebrating New Year, birthdays, etc. A projectile fired steeply into the air accelerates during the descending branch of its trajectory, but without exceeding the terminal velocity of 5.4.5.2, Point 5.

**Table 5.6.** Effect of change in angle of shot [mils] when firing uphill or downhill, showing vertical error in per mille of range.

| Angle of sight λ | Angle of shot σ [mils] | | | | |
|---|---|---|---|---|---|
| [°] | 2 | 3 | 4 | 5 | 6 |
| 5 | 0.007 | 0.010 | 0.014 | 0.017 | 0.019 |
| 10 | 0.029 | 0.043 | 0.057 | 0.070 | 0.084 |
| 15 | 0.066 | 0.098 | 0.130 | 0.161 | 0.192 |
| 20 | 0.117 | 0.175 | 0.232 | 0.288 | 0.344 |
| 25 | 0.183 | 0.273 | 0.362 | 0.451 | 0.539 |
| 30 | 0.261 | 0.391 | 0.520 | 0.647 | 0.774 |

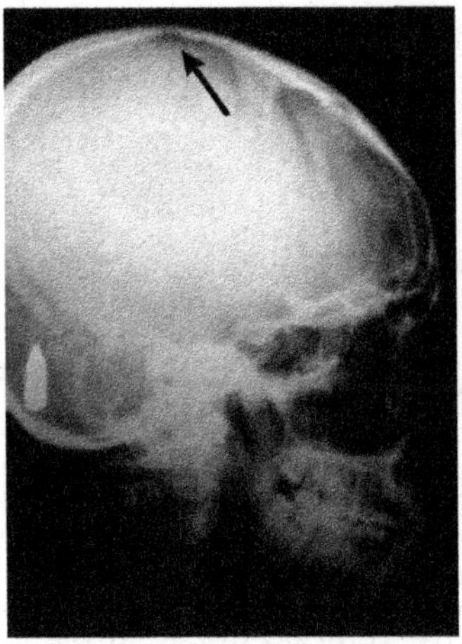

**Fig. 5.26.** Injury caused by a falling bullet. The 7.62 × 39 Kalashnikov bullet passed through the calvarium (see arrow) and ended up lodged in the base of the skull.

Projectiles with a high sectional density (artillery shells), travelling at high velocity, attain their relative maximum velocity at a point above muzzle height. In all other cases, the projectile will still be accelerating slightly when it reaches this point (Δv of approximately 1 m/s per 100 m drop or less), which corresponds to a virtually constant velocity of descent. Table 5.7 shows these values, along with corresponding values of energy and energy density, at an altitude of 400 m above sea level. The figures show that a falling projectile will exceed the energy and energy density levels needed to cause danger to persons (see B. P. KNEUBUEHL, 2022) in all cases, with the exception of an air rifle pellet or a shotgun pellet of no more than 5 mm diameter. In other words, virtually all falling projectiles are dangerous, and can kill.

Fig. 5.26 shows a typical wound caused by a falling bullet, a common type of injury in countries where guns are fired into the air during festivities.

A projectile fired into the air at an angle steeper than approximately 80° acquires too little tractability (see 5.6.4.6 and 5.6.5). As a result, it adopts a large angle of attack in the region of the apex and starts to tumble. As it accelerates during descent, it usually stabilizes again, but possibly with the tail pointing downwards.

**Table 5.7.** Ballistic data for falling projectiles

| Projectile | d [mm] | q [g/mm²] | $v_f$ [m/s] | $v_E$ [m/s] | $E_E$ [J] | ED [J/mm²] | $y_{AP}$ [m] |
|---|---|---|---|---|---|---|---|
| 4.5 mm Diabolo FN | 4.5 | 0.0332 | 30.3 | 26.1 | 0.13 | 0.008 | 149 |
| Shotgun pellet (lead) | 5.0 | 0.0377 | 35.4 | 34.5 | 0.44 | 0.022 | 271 |
| Shotgun pellet (lead). | 6.0 | 0.0452 | 38.8 | 37.8 | 0.91 | 0.032 | 314 |
| 22 L.R. LRN | 5.6 | 0.1035 | 80.6 | 77.8 | 7.73 | 0.314 | 882 |
| 9 mm Luger (FMJ RN) | 9.0 | 0.1258 | 94.4 | 91.3 | 33.4 | 0.524 | 1130 |
| 38 Spl. SWC | 9.0 | 0.1610 | 91.5 | 86.6 | 38.4 | 0.604 | 913 |
| 44 Rem. Mag. TMF | 11.2 | 0.1578 | 102.9 | 99.6 | 77.1 | 0.783 | 1335 |
| 5.56 mm NATO FMJ | 5.56 | 0.1647 | 123.4 | 120.6 | 29.1 | 1.198 | 2240 |
| 7.62 × 39 Kal. FMJ | 7.62 | 0.1754 | 115.2 | 112.6 | 50.7 | 1.112 | 1852 |
| 7.62 mm NATO FMJ | 7.62 | 0.2083 | 142.8 | 140.3 | 93.6 | 2.052 | 2632 |

d: calibre, q: sectional density, $v_f$: impact velocity (from Eqn 5.4:71), $v_E$: effective impact velocity, $E_E$: impact energy, ED: energy density, $y_{AP}$: apex height at a 1250 mil angle of departure (approx. 70°). Altitude: 400 m above sea level.

## 5.4.7 The boundary value problem

### *5.4.7.1 Outline of the problem*

Ballistic studies – be they theoretical or empirical – almost always use known starting values, because the aim is to find out what happens under precisely-defined conditions. For practical situations, however, one wishes to achieve a specific result and needs to identify the initial conditions that will lead to it. The only unknown starting parameters are line of departure (azimuth), elevation and muzzle velocity.

In most cases, we know muzzle velocity, either because the ammunition is in cartridge form and hence contains a fixed quantity of propellant, or because we are using a system of various prefabricated charges and will select the charge appropriate to the range.

Some mortars can be adjusted to one, two or three fixed elevations, with range set by varying muzzle velocity. This is achieved either by varying the charge or, in the case of pneumatic mortars and propellant mortars with adjustable load volume, by regulating pressure.

We also know the line of departure (to a first approximation), because we know the positions of the muzzle and the target, so we only need to determine the elevation.

One obvious solution is to use trajectory calculations based on known starting values to plot a *range-elevation curve* (see Fig. 5.27). However, a graph of this sort – or a table – can only provide a very approximate solution to the problem, as the calculations can only be carried out using a standard atmosphere, and the intervals between target altitudes will be fairly large.

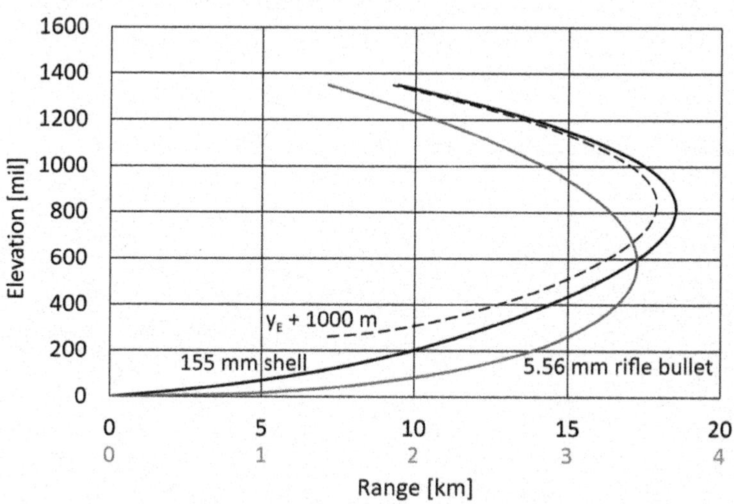


**Fig. 5.27.** Elevation curves for two projectiles with different sectional densities. For the rifle bullet (line and labelling in grey), maximum range is achieved at an elevation of approximately 600 mils, whereas the artillery shell (black line) travels furthest at an elevation of approximately 810 mils. The graphic also shows the curve for the artillery shell when aimed at a target at a height of 1000 m.

The boundary value problem of exterior ballistics can hence be formulated as follows: *For given target values $x_E$, $y_E$, $h_E$, measured muzzle velocity $v_0$, and known projectile ballistic data, calculate the elevation that will cause the projectile to hit the point of aim, taking account of the atmospheric conditions (atmospheric density and wind) obtaining at the time.* In the first instance, the line of departure will be determined separately.

This means that atmospheric pressure, temperature profile and wind speed and direction must be measured close to the trajectory shortly before firing. Artillery units therefore have their own weather services. In most cases, they also have a means of either measuring muzzle velocity or estimating this parameter from the temperature of the propellant and the internal dimensions of the barrel, which will change as it wears.

As solving the boundary value problem is essential for practical shooting, and as sufficiently fast computers did not exist until comparatively recently, a number of graphical and mechanical methods were devised over the years. A number of these methods are still in use (as backup). They are also sufficiently interesting from a ballistics point of view that it is worth examining them briefly.

### 5.4.7.2 Firing and range tables, trajectory maps and the Benoist Plan

Tables and sets of tables giving the relationship between range and elevation are known as *firing tables* or *range tables*. Firing tables are designed for indirect-fire weapons with a high trajectory (see Table 5.8). These not only link range and elevation but also contain all other information required for shooting under practical conditions. They are therefore a true solution of the boundary value problem. Range tables of this sort are generally fairly substantial documents, as they contain the data for all charges used with a particular gun.

**Table 5.8.** An extract from a firing table for an indirect-fire weapon (155 mm howitzer). Calculated for firing at sea level.

| Distance | Elev. | $\Delta x$ | Flight time | $\alpha_z$ | $\Delta\alpha_w$ | Range wind | | Atmospheric density | |
|---|---|---|---|---|---|---|---|---|---|
| | | per mil | | | ±10 m/s | +10 m/s | −10 m/s | +0.1 kg/m³ | −0.1 kg/m³ |
| [m] | [mil] | [m] | [s] | [mil] | [mil] | [mil] | [mil] | [mil] | [mil] |
| 13000 | 340.5 | 19.9 | 34.5 | 11 | 11 | −11.9 | 11.6 | 26.6 | −22.0 |
| 13100 | 345.5 | 19.7 | 34.9 | 11 | 11 | −12.2 | 11.9 | 27.0 | −22.5 |
| 13200 | 350.6 | 19.5 | 35.3 | 11 | 11 | −12.5 | 12.2 | 27.4 | −22.8 |
| 13300 | 355.8 | 19.3 | 35.7 | 12 | 11 | −12.9 | 12.5 | 28.0 | −23.2 |
| 13400 | 361.0 | 19.1 | 36.1 | 12 | 11 | −13.2 | 12.9 | 28.4 | −23.6 |
| 13500 | 366.2 | 18.9 | 36.6 | 12 | 11 | −13.6 | 13.2 | 28.9 | −24.0 |

The $\alpha_z$ (drift) and $\Delta\alpha_w$ (crosswind) columns are azimuth corrections. A second page contains columns for corrections to elevation for other muzzle velocities and for shooting uphill and downhill, together with apex height.

**Table 5.9**. An extract from a range table for a direct-fire weapon (338 Win. Mag.). Calculated for firing at sea level.

| Distance [m] | $\sigma$ [mil] | v [m/s] | E [J] | ED [J/mm²] | t [s] | $z_E$ [m] | $y_{AP}$ [m] | $x_{AP}$ [m] |
|---|---|---|---|---|---|---|---|---|
| 0 | 0.00 | 810 | 5315 | 96 | 0.00 | 0.00 | 0.00 | 0 |
| 100 | 0.80 | 748 | 4532 | 82 | 0.13 | 0.05 | 0.02 | 51 |
| 200 | 1.70 | 689 | 3844 | 69 | 0.27 | 0.21 | 0.09 | 103 |
| 300 | 2.70 | 633 | 3243 | 59 | 0.42 | 0.49 | 0.22 | 156 |
| 400 | 3.83 | 580 | 2721 | 49 | 0.58 | 0.91 | 0.42 | 211 |
| 500 | 5.10 | 529 | 2269 | 41 | 0.77 | 1.48 | 0.72 | 268 |

$\sigma$: angle of shot, E: energy, ED: energy density, $z_E$: crosswind deviation with a 10 m/s crosswind, $y_{AP}$: apex height, $x_{AP}$: apex width.

Producing such documents requires a very large number of trajectory calculations with different starting conditions. Before the advent of computers, range table corrections were calculated using ballistic disturbance calculations, which relied on approximation equations. For a comprehensive discussion, see H. MOLITZ, R. STROBEL (1963).

One point worthy of note from a ballistics point of view, and not immediately obvious, is the difference between the correction for a following wind and that for a headwind. The difference between the corrections for rising and falling atmospheric density is clear.

The second type, the range table, is used for flat-trajectory weapons (see Table 5.9). In addition to the range/elevation relationship, these contain ballistic data for the trajectory in question, such as velocity, energy, time of flight, crosswind deflection, apex height and apex width. Please see Fig. 5.15 on p. 182 regarding the difference between elevation and angle of shot. In contrast to the first type of table, those for flat-trajectory weapons generally take the form of simple datasheets. Angle of shot data are included primarily for the designing of sights.

*Graphic range tables* formed an intelligent and rapid means of solving the boundary value problem. These consisted of a transparent disc on which the range-elevation relationship was engraved in the form of lines, taking account of target altitude. See Fig. 5.28, p. 210. The user placed the origin over the location of the gun on a map (to the same scale as the instrument) and set the target altitude at the point of aim. Elevation could then be read off the instrument directly. Corrections for current conditions could be taken directly from the disc.

A *trajectory map* was a useful graphical tool in mountainous regions, where target altitude and muzzle altitude often differ by a significant amount. This consisted of a side view of the trajectory, in elevation steps of 20 mils, calculated under standard conditions. There was a page for each charge, and the pages were combined into a booklet. Separate booklets were often compiled for different atmospheric densities at the muzzle (corresponding to different muzzle altitudes), so that the appropriate card could be selected for the position of the gun. Please see Fig. 5.29 (on the next double page) for an example of such a trajectory card. This method also makes it possible to determine the elevation for a target uphill or downhill from the gun.

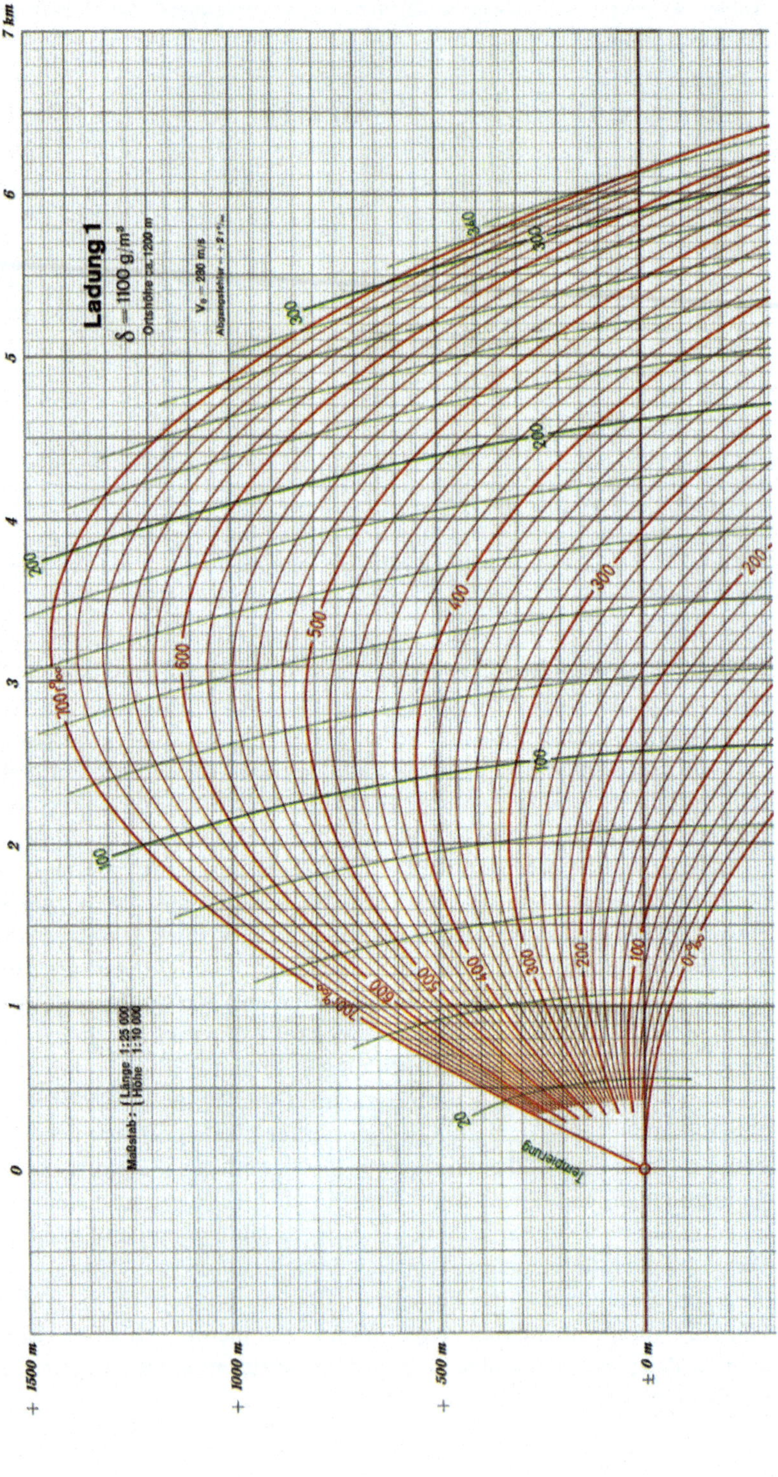
10,5 cm Kanone
Ladung 1
δ = 1100 g/m³
Ortshöhe ca. 1200 m
Maßstab: Länge 1:25 000
Höhe 1:10 000
Tempierung
7 km
+ 1500 m
+ 1000 m
+ 500 m
± 0 m

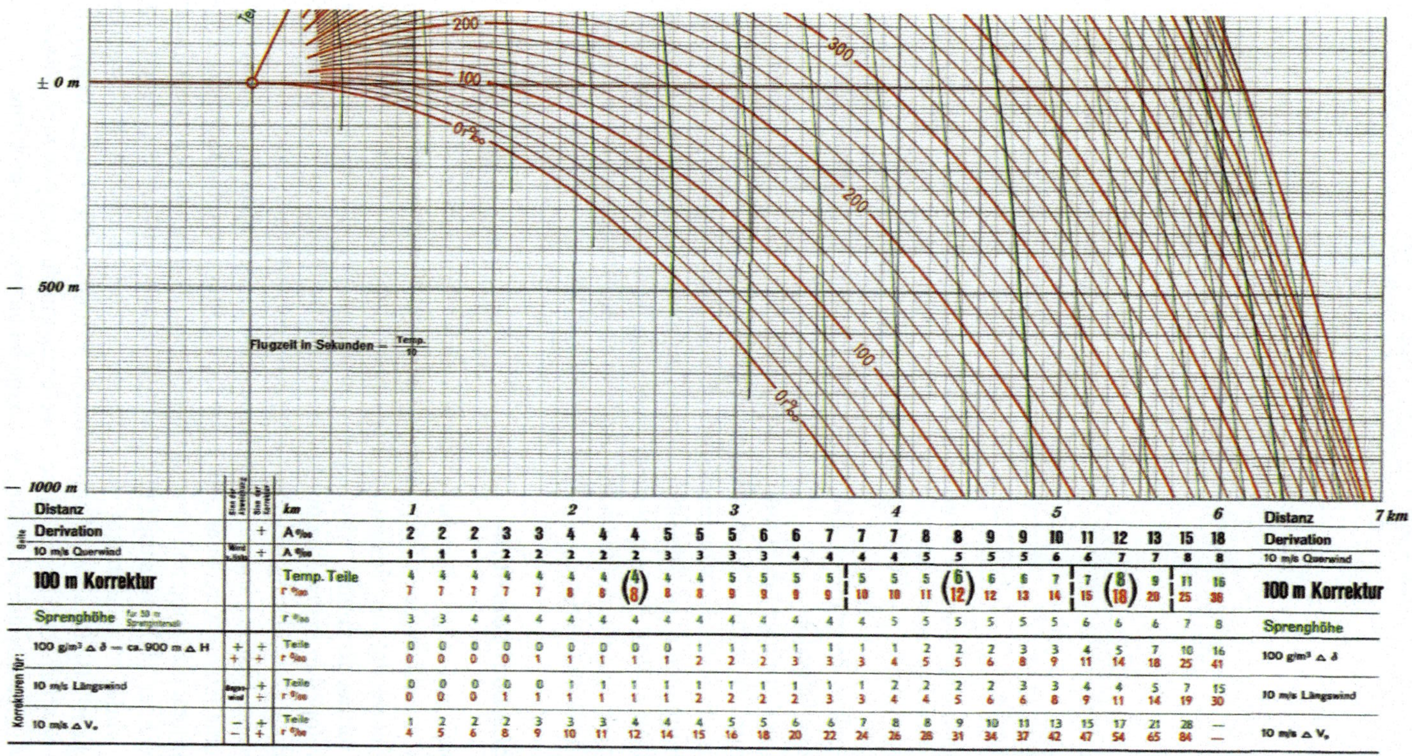


**Abb. 5.29**. Example of a trajectory map from a booklet for an artillery gun in 105 mm calibre. The original map is printed on one sheet and is spread over two pages here only to save space. The trajectories are drawn in red, the fuze setting curves (which correspond to the flight time in tenths of a second) in green. The correction values listed below are in the same colours for better differentiation. Elevations and corrections are given in an angular measure formerly used in artillery, which divided the full circle into 6300 parts (so-called radius per mil). The reason was the better approximation to $2\pi$ compared to mils. Translations of the terms in the map can be found in the appendix.

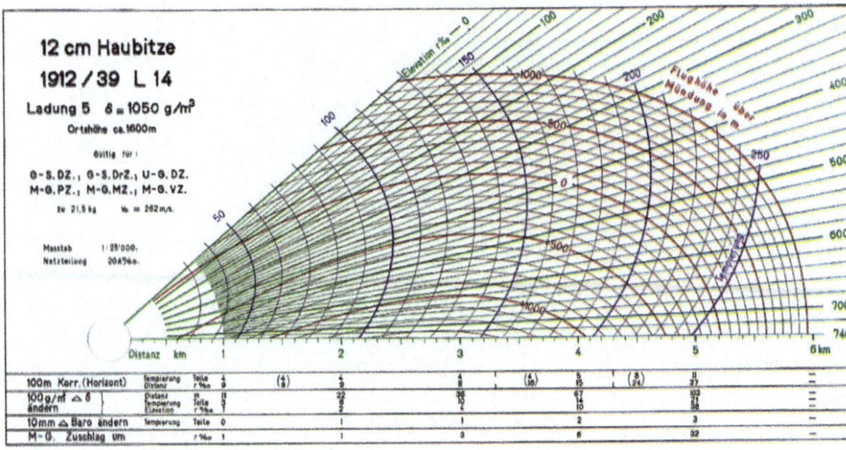


**Fig. 5.28.** Graphic range table. The origin is placed on the map at the location of the gun. The card is then rotated until the appropriate trajectory height line (red) passes through the target point. The green line that passes also through the target point gives the correct elevation.

Footnotes below the graphics give corrections for weather conditions, drift and different muzzle velocities.

However, neither range tables nor trajectory cards were capable of determining line of departure or range. To address this problem, Colonel A. BENOIST of the French artillery published a simple method of calculating these two parameters, even for an entire battery (A. BENOIST, 1922). This method, which has come to be called the "Benoist Plan", used two superimposed grids of coordinates, one Cartesian and one polar. It remained in service until fire control computers came into use, at the end of the 20th Century.

### 5.4.7.3 Mechanical solutions

Trajectory maps and range tables have a further major disadvantage, in addition to their inability to indicate line of departure or range; the correction values they provide are only correct at muzzle altitude. For targets higher or lower than this, they would need further footnotes, which would impair clarity. Useful devices were therefore developed in the middle third of the 20th Century that were also capable of more accurately calculating trajectories for different target altitudes.

The Swiss and Austrian armies, for instance, used a mechanical firing data computer (the "Schießelementerechner", SER, see Fig. 5.30) to perform accurate calculations of azimuth, elevation and time of flight. A mechanical correction computer ("Korrekturrechner", KR) was used to correct differences between target and point of impact, while a mechanical ballistic correction computer("Ballistischer Korrekturrechner", BKR) allowed corrections to be applied for changes in muzzle velocity and weather conditions. These three precision devices provided comprehensive and sufficiently accurate solutions to the boundary value problem.

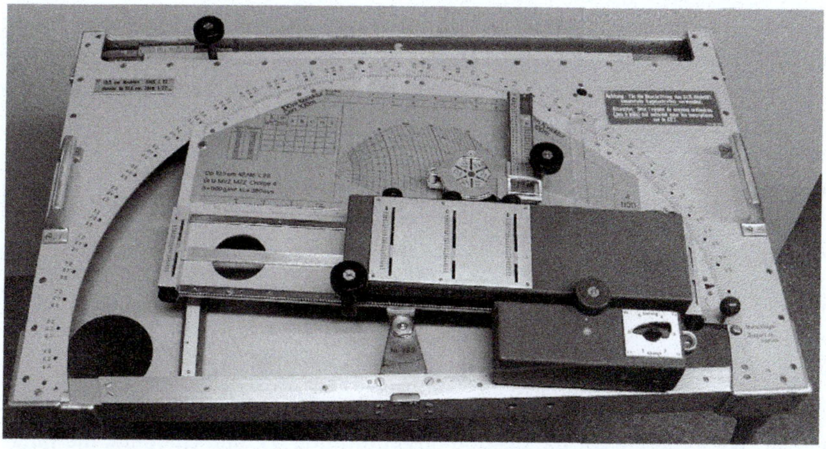

**Fig. 5.30.** Mechanical firing data computer ("Schießelementerechner", SER) with trajectory card in place. This device was capable of calculating the key parameters rapidly and accurately.

### 5.4.7.4 Calculated solutions

Finding solutions to a system of differential equations that have predefined values at the boundaries of the defined range is a well-known problem of numerical mathematics. As a result, solutions exist, including the "shooting method" (J. STOER, R. BULIRSCH, 1973). This involves using the true values of muzzle velocity and atmospheric conditions (including wind) to solve the initial value problem $f_A$ and plotting trajectory height y at range $x_E$ as a function of elevation θ.

$$f_A(\theta) = y\,, \quad \text{with} \quad x = x_E\,. \qquad [\mathrm{m}]$$

The difference with respect to target altitude yields a function $f_B$, for which the zero must be found.

$$(5.4{:}88) \qquad f_B(\theta) = f_A(\theta) - y_E = 0\,. \qquad [\mathrm{m}]$$

This can be achieved using the following iteration scheme, for instance:

$$(5.4{:}89) \qquad \theta_{i+1} = \theta_i - \frac{\theta_i - \theta_{i-1}}{f_A(\theta_i) - f_A(\theta_{i-1})} \cdot \left[ f_A(\theta_i) - y_E \right]\,, \qquad [\mathrm{m}]$$

The iteration is repeated until a predetermined error limit is exceeded (e.g. $10^{-6}$).

If it is a matter of finding the difference between two trajectories (in this instance, between the desired trajectory that passes through the target and an approximation), we can make use of in vacuo ballistics, because if we ignore drag, the two trajectories are subject to approximately the same error, which is cancelled out in the difference between them. Equations can be derived for ballistics in the absence of drag that can be transposed to real trajectories where drag is present.

By integrating the trajectory for the desired range $x_E$ with the true atmospheric conditions and calculating the vertical error $\Delta y_{E(i)}$, one obtains an iteration formula with which to calculate a new, more accurate elevation from the current elevation and the resulting vertical deviation.

$$(5.4{:}90) \qquad \theta_{i+1} = \theta_i - \arctan\frac{\Delta y_{E(i)}}{x_E\cdot\left(1-\tan^2\theta_i\right)} \,. \qquad [-]$$

Angles in radians.

We can derive a somewhat simpler equation from the swing of the trajectory.

$$(5.4{:}91) \qquad \theta_{i+1} = \theta_i + \left(\arctan\frac{y_E}{x_E} - \arctan\frac{y_{E(i)}}{x_E}\right) . \qquad [-]$$

If, however, we integrate for the desired target altitude $y_E$ and calculate the range error $\Delta x_{E(i)}$, we obtain the following iteration formula (again, derived from in vacuo ballistics):

$$(5.4{:}92) \qquad \theta_{i+1} = \frac{\pi}{2}-\frac{1}{2}\cdot\arcsin\frac{\sin(2\cdot\theta_i)}{1+\frac{\Delta x_{E(i)}}{x_E}} \,. \qquad [-]$$

The first two equations (5.4:90 and 91) give the lower of the two elevations that would achieve the required range, while the third (Eqn 5.4:92) gives the higher elevation for the same range.

It is possible to solve the boundary value problem in approximately three to four iterations using all three equations. The value of elevation initially chosen is of no importance, as long as the required range or target altitude is actually attained. At both high and low elevations, therefore, one starts with an elevation that yields close to maximum range. Because one is integrating with the true conditions, one obtains the solution of the boundary value problem directly.

If it is not possible for the projectile to reach the desired range, the elevations in Eqns 5.4:90 and 91 will not converge, but will increase in either an alternating or a uniform fashion. In the case of Eqn 5.4:92, this may become apparent as a result of a sine value of more than 1. When programming these functions, it is therefore necessary to set a suitable exit criterion. If it is possible for the projectile to reach the target, the methods generally converge quickly.

Another method of calculating elevation iteratively uses PITON-BRESSANT's trajectory formula (see Eqn 5.4:86), in which the parameter Q takes account of drag.

$$y = x\cdot\tan\theta_0 - \frac{g\cdot x^2}{2\cdot v_0^2\cdot\cos^2\theta}\cdot(1+Q\cdot x) \,. \qquad [m]$$

If point (x, y) = ($x_E, y_E$) is known, this formula takes the form of a quadratic equation for the tangent to the elevation, and can be solved directly.

$$\theta = \arctan\left(\frac{x_E}{2\cdot B} \pm \sqrt{\frac{x_E^2}{4\cdot B^2} - \left(1+\frac{y_E}{B}\right)}\right), \qquad (5.4:93) \quad [-]$$

with

$$B = \frac{g}{2\cdot v_0^2}\cdot x_E^2 \cdot (1+Q\cdot x_E)\,. \qquad (5.4:93a) \quad [m]$$

If the Piton-Bressant number Q is known for a given point, then the two solutions from Eqn 5.4:93 give the two possible elevations that will enable the projectile to reach the target at the velocity concerned, one solution for the upper elevation, the other solution for the lower.

If one calculates a trajectory under real conditions with any initial elevation, and obtains either the vertical error from the range or the range error from the target altitude, one can calculate the Piton-Bressant Q number for this approximated point (x, y) and the corresponding elevation.

$$Q(x,y) = \frac{(x\cdot\tan\theta - y)\cdot 2\cdot v_0^2\cdot\cos^2\theta}{g\cdot x^3} - \frac{1}{x}\,. \qquad (5.4:94) \quad [1/m]$$

Substituting into Eqn 5.4: 93a gives us B, which is then substituted into Eqn 5.4:93 to obtain the new elevation. We repeat this procedure until we have obtained a sufficiently accurate result.

If a point lies close to maximum range, the discriminant in Eqn 5.4:93 may become negative. In such cases, one sets it to zero and continues the calculation process using the resulting elevation. If this occurs a second time, then it is not possible to reach the target.

For a given initial elevation, this method requires approximately the same number of iterations to achieve a sufficiently accurate result as do Eqns 5.4:90–92. However, it does allow us to calculate the upper and lower elevations using the same system of equations, and it does take account of muzzle velocity. The Piton-Bressant approach also enables us to compile a set of approximate mean values of Q, from which we can derive a reasonably accurate initial elevation. This saves one or two iterations.

### 5.4.7.5 Further comments regarding the boundary value problem

When systems of differential equations for ballistics are integrated in order to solve boundary value problems, this usually occurs in a real-life situation. This means that current, measured meteorological data are available: atmospheric pressure and temperature at muzzle altitude, plus temperature and wind layer data (direction and velocity) to a level beyond that of the expected apex height. As mentioned in 5.4.3.4, one continues integration up to each layer boundary and then starts again using the new values.

The lowest wind layers have little effect on the flight of the projectile, as it is only in those layers for a short time, during the ascending and descending branches. The layers in the vicinity of the apex have a more significant effect. Situations may even occur in which no solution to the boundary value problem is possible. If a projectile falls just short of the target at the lower of the two possible elevations, and there is a tail wind at the apex, increasing elevation may mean that the apex occurs in the next layer, in which the projectile encounters a headwind, which would reduce its range despite the increased elevation. In such wind shear situations – which do occur – the only possible option would be to switch to the higher range of elevations, if the gun is capable of this.

### 5.4.8 The exterior ballistics of fragments

#### *5.4.8.1 Sectional density*

After being accelerated by the expanding gases of an explosive, a fragment will very rarely, if ever, maintain its original orientation as it moves along its trajectory (with the exception of spherical fragments). Rather, it is reasonable to suppose that when the detonation front acts upon the fragment, the line of action of the accelerating force does not pass through its centre of mass. The fragment therefore undergoes torque, causing it to rotate about an axis, and this axis will generally not be fixed. The cross-sectional area subjected to airflow will hence change constantly, which means that the sectional density will also change. We therefore have to use a mean value for trajectory calculations.

To calculate a representative cross-sectional area, we first replace the fragment by a convex body of the same mass and similar length, generally a cuboid. This is an acceptable approximation, as when air flows around a non-convex body, the concave areas are "smoothed out" by the small vortices that form within them.

We can make either of two assumptions as regards the rotation of the fragment:

- the rotational axis of the cuboid changes continuously and the fragment tumbles
- the fragment rotates about a fixed axis; this is generally the axis of the greatest moment of inertia.

In the first case, we can use CAUCHY's theorem, according to which the projected area of a convex body averaged over the entire solid angle is equal to one quarter of its surface S. In that case, the reference area that will determine drag is given by:

$$(5.4{:}95) \qquad A_0 = \frac{1}{4}\cdot S\,. \qquad [m^2]$$

If we assume that the fragment rotates about a fixed axis, then the mean shadow area (and hence the reference area) of a cuboid with sides a, b and c (with a parallel to the axis of rotation) is given by

$$A_0 = \frac{4 \cdot a}{2 \cdot \pi} \cdot \int_0^{\pi/2} (b \cdot \cos\alpha + c \cdot \sin\alpha) \cdot d\alpha \,, \qquad [m^2]$$

(5.4:96)
$$A_0 = \frac{2 \cdot a}{\pi} \cdot (b + c) \,. \qquad [m^2]$$

*Example*: For a cube calculated using Eqn 5.4:95, $A_m = {}^3/_2 \cdot a^2 = 1.5 \cdot a^2$. However, using Eqn 5.4:96 for a fragment where $a = b = c$, $A_m = {}^4/_\pi \cdot a^2 = 1.273 \cdot a^2$.

The value obtained using the second equation is therefore 15% smaller. The further the ratio of the sides departs from 1 (i.e. the further the form of the fragment deviates from that of a cube), the larger the difference.

In the first case (with a changing axis of rotation), we can assume that the projectile is subject to nutation, which will become less pronounced over a longer time of flight. When calculating the maximum distance that a fragment could travel (to determine safety distances) it is therefore appropriate to use Eqn 5.4:96, especially in view of the fact that the result will err on the safe side, as using a sectional density higher than the true value will mean that the calculated distance is greater than that which the fragment will actually travel. At shorter ranges (when conducting effect assessments to determine danger areas), Eqn 5.4:95 is more suitable.

### 5.4.8.2 Drag coefficient

The drag coefficients for spheres and for cubes are known to a good degree of accuracy. There is also a “fragment simulating projectile” (FSP), a cylinder of which the edges are bevelled at one end (see Fig. 5.31).

For a cube, one side is taken as the reference surface. This means that when we use these $C_D$ functions, we have to convert their values from the lateral face to one of the above reference areas.

Fig. 5.32 shows the $C_D$ curves for these three projectiles, obtained from various publications and from the author’s own measurements. The values for the sphere are from H. Molitz, R. Strobel (1963), R. Sutterlin (1965), A. May, W. R. Witt (1953) and from the author’s own measurements using 5.56 mm and 7.5 mm spheres. The values for the cube are from G. Hantsche, J. Rinehart (1952) und C. W. Whiteford, J. M. Regan (1961). Those for the FSP are from the author’s own measurements.

The $C_D$ curve for the sphere is valid only for a pre-fragmented, spherical fragment with a smooth surface. The curve for the cube can be used for virtually any irregularly-shaped fragment. This is possible because the

**Fig. 5.31.** A fragment simulating projectile (FSP), a standardized projectile used to simulate the effects of fragments. These are available in various calibres, and are used mainly to test protection against fragments. Each mark on the scale at the bottom of the photo is one millimetre.

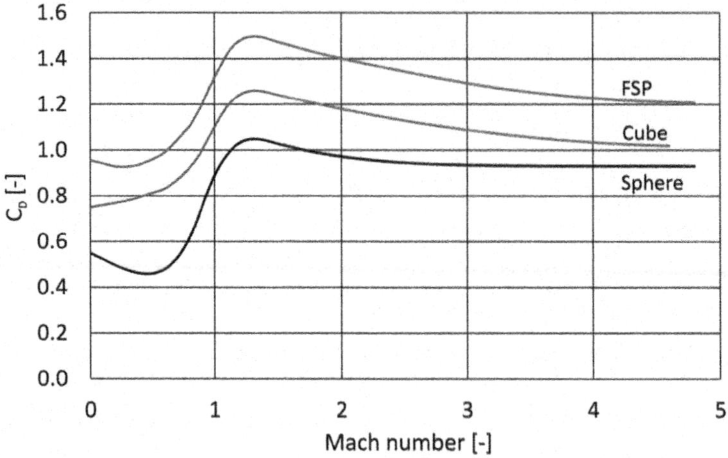


**Fig. 5.32.** $C_D$ curves for three fragment shapes that are suitable for calculating fragment trajectories.

215 depends on the form of the fragment, but not on its geometric dimensions. In the case of angular, blunt shapes, the flow will always separate, whether the fragment takes the form of a cube, a cuboid or an irregular object. At supersonic velocities, the shock wave will likewise be separated from the fragment, which means we can expect a certain similarity as regards flow. Because the flow separates at the edges of the object, much of the surface will be enveloped in turbulent boundary layers, so the coefficient of friction will not differ to a decisive extent.

### *5.4.8.3 Trajectory calculations*

We can calculate fragment trajectories using a two-dimensional system of equations, as it is virtually impossible to calculate lateral acceleration and the time of flight is too short for Coriolis acceleration to have any significant effect. In most cases, the trajectory will fall into the lower group of angles, so the system of Eqn 5.4:42 will be the most suitable.

In those unusual cases in which we need to perform calculations for fragments with large angles of departure (above approximately 600 mils), we will have to use the time-dependent system (Eqn 5.4:41), as the descending branch will be very steep, and we would have to use very small distance steps to prevent the calculation entering the unreachable zone.

We can calculate deviation due to crosswinds very accurately for all elevations using DIDION's formula.

For short distances (less than 10 m) and high initial velocities (over 1000 m/s) it is not usually necessary to carry out a complete trajectory calculation. In such cases, we can take $C_D$ as constant, with approximate values of 1 for spheres and cubes, and 1.3 for cylinders, and use the sectional density from Eqn 5.4:95. As atmospheric density will also be constant, we can use the approximation equations from 5.4.6.1 with K calculated according to Eqn 5.4:73:

$$K = \frac{C_D \cdot \rho}{2 \cdot q}, \qquad [1/m]$$

and differential equation Eqn 5.4:76. This gives us the following calculable equation for velocity as a function of distance:

$$(5.4{:}97) \qquad v(x) = v_0 \cdot e^{-K \cdot x}, \qquad [m/s]$$

and hence for the corresponding energy we can write:

$$(5.4{:}98) \qquad E(x) = E_0 \cdot e^{-2 \cdot K \cdot x}. \qquad [J]$$

The deceleration of a projectile in the first 10 m generally lies within the variation between the muzzle velocities for different firings and is therefore of no importance. Fragments, especially small fragments, lose a large amount of velocity and energy within this distance, with 0.1 g fragments losing up to 60%. This is often of critical importance when investigating incidents such as accidents or crimes involving such devices as hand grenades.

### 5.4.8.4 Maximum distances

Because of its low sectional density, a small fragment loses a lot of its velocity and energy along its trajectory. Fast-moving, light fragments may decelerate by as much as 100 m/s per metre of flight. Because small fragments decelerate so rapidly, they slow to subsonic speeds at a very early stage.

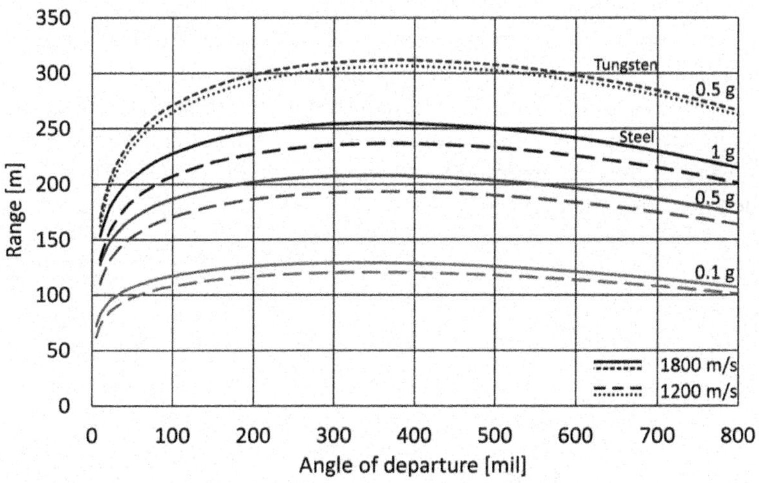


**Fig. 5.33.** Maximum ranges of small fragments as a function of angle of departure. Maximum range changes little at angles of departure from 200 to 600 mils. The influence of initial velocity is surprisingly small.

A 0.1 g fragment with an initial velocity of 1800 m/s will be travelling at less than the speed of sound after only 0.05 s. If it takes around 4 s to reach maximum range, the fragment will therefore be subsonic for almost 99% of its flight.

The initial velocity of a fragment therefore has a negligible effect on its maximum range, regardless of its shape and the material of which it is made. For instance, reducing the initial velocity of any small fragment by a third will reduce its maximum range by less than 10%. The main factors determining range are mean sectional density and shape (spherical or other). Fig. 5.33 shows the maximum ranges of a number of cuboid fragments as a function of angle of departure. If mass increases (which, for a given material, corresponds to an increase in sectional density) range increases significantly, but the difference in velocity for the same fragment is not particularly large. The effect of increased sectional density also becomes clear if a denser material such as tungsten is used, for the same mass.

The rapid deceleration of small fragments also results in a highly asymmetric trajectory. This in turn means that maximum range varies by less than 10% at angles of departure of between 200 mils and approximately 600 mils (see Fig. 5.33). Calculating the maximum range of smaller fragments is therefore fairly uncritical.

Large, heavy fragments (weighing a number of kilogrammes) have a relatively high sectional density, so at low velocities (100 to 300 m/s) they decelerate less and can travel further. This is particularly the case for plate-like fragments. If they are rotating about the axis of greatest moment of inertia, like a discus, they can travel for considerable distances.

155 mm artillery training shells break up into a relatively small number of large fragments, on account of the small amount of explosive they contain. Palm-sized fragments of this kind have been known to travel almost 2.5 km.

As it is difficult to determine the sectional density and drag coefficient resulting from the actual orientation of the fragment in flight, calculating the trajectories of such fragments is among the more difficult tasks a ballistician can face.

Disturbances such as rangewind and crosswind are proportional to the deceleration of the projectile. Small fragments can therefore be carried considerable distances by the wind, if they are exposed to it for long enough. This is of particular relevance when investigating damage.

# 5.5 Trajectory calculations with more than three degrees of freedom

## 5.5.1 Trajectory calculations with six-degrees of freedom

### *5.5.1.1 Reference systems*

We use three frames of reference to describe the motion of a three-dimensional, rigid projectile (see Fig. 5.34):

- An *earth-fixed ballistic coordinate system* (the right-handed system shown in Fig. 2.1), with unit vectors $(\vec{i}, \vec{j}, \vec{k})$, and with its origin generally at the muzzle.
- A *trajectory tangent system*, with unit vectors $(\vec{t}, \vec{s}, \vec{h})$, of which the first axis coincides with the tangent to the trajectory, the third $\vec{h}$ lies perpendicular in a horizontal plane and the second combines the other two to form a right-handed system. $\vec{s}$ and $\vec{h}$ therefore form a normal plane (A) to the tangent to the trajectory. The origin of the system is located in the projectile centre of mass.
- A *body-fixed system* of which the unit vectors $(\vec{c}, \vec{p}, \vec{q})$ form a right-handed system locked to the projectile and of which the origin is also located in the projectile centre of mass. Its primary axis $\vec{c}$ coincides with the projectile axis of symmetry, while the other two axes lie in a plane (B) perpendicular to the first.

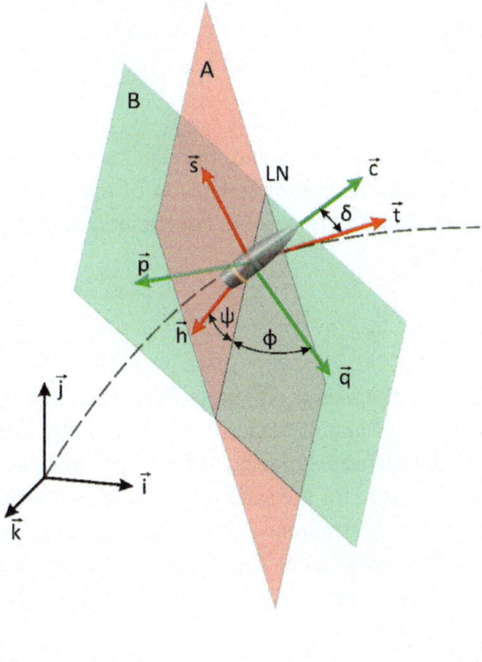


**Fig. 5.34.** Frames of reference for describing the complete motion of a projectile. An earth-fixed ballistic coordinate system $(\vec{i}, \vec{j}, \vec{k})$ describes the path of the centre of mass, the trajectory tangent system $(\vec{t}, \vec{s}, \vec{h})$ traces the direction of motion and the body-fixed system $(\vec{c}, \vec{p}, \vec{q})$ traces the orientation of the projectile in space. Vectors $\vec{s}$ and $\vec{h}$ form plane A, which passes through the centre of mass at right angles to the tangential vector. Vectors $\vec{p}$ and $\vec{q}$ form a normal plane B that passes through the projectile centre of mass at right angles to the projectile axis. The line of intersection between planes A and B is the line of nodes LN. The Eulerian angles are labelled $\delta$, $\phi$ and $\psi$. The resistance plane created by $\vec{t}$ and $\vec{c}$ is not shown.

Let θ be the angle of trajectory (the angle between the tangent to the trajectory and the horizontal), η the azimuth angle (the angle between the horizontal projection of the tangent to the trajectory and the i-axis) and the projectile centre of mass $\vec{r}$, then we can transfer any point $\vec{x}$ in the trajectory tangent system into the earth-fixed system using the following transformation:

$$\vec{x}_{(EF)} = \begin{pmatrix} \cos\theta\cdot\cos\eta & -\sin\theta\cdot\cos\eta & -\sin\eta \\ \sin\theta & \cos\theta & 0 \\ \cos\theta\cdot\sin\eta & -\sin\theta\cdot\sin\eta & \cos\eta \end{pmatrix} \cdot \vec{x}_{(BT)} + \vec{r}\ . \qquad \text{[m]} \tag{5.5:1}$$

The trajectory tangent system and the body-fixed system are linked via the *Euler angles* δ, ϕ and ψ (see Fig. 5.34). δ is the angle between the tangent to the trajectory and the projectile axis; this is also the angle between planes A and B. Angle ψ lies in plane A, and is the angle between the horizontal vector $\vec{h}$ and the line of nodes, while ϕ is the angle between the line of nodes and vector $\vec{q}$ in plane B. The body-fixed system can now be transferred into the trajectory tangent system by rotations about ϕ (axis of rotation $\vec{c}$), δ (angle of rotation vector $\vec{q}'$, which now lies on the line of nodes) and ψ (angle of rotation $\vec{c}''$, which coincides with $\vec{t}$). The product of the corresponding three rotation matrices gives us the transformation matrix T, which transfers the body-fixed system into the trajectory tangent system.

$$T = \begin{pmatrix} 1 & 0 & 0 \\ 0 & \cos\psi & \sin\psi \\ 0 & -\sin\psi & \cos\psi \end{pmatrix} \cdot \begin{pmatrix} \cos\delta & \sin\delta & 0 \\ -\sin\delta & \cos\delta & 0 \\ 0 & 0 & 1 \end{pmatrix} \cdot \begin{pmatrix} 1 & 0 & 0 \\ 0 & \cos\phi & \sin\phi \\ 0 & -\sin\phi & \cos\phi \end{pmatrix} . \tag{5.5:2}$$

Taking wind into account would require a third, wind-fixed system. This has been omitted in the interests of clarity, because we can proceed as described in 5.4.4.

### 5.5.1.2 Forces and moments

We have already described the magnitudes of the forces and moments acting on a spin-stabilized projectile in 5.3.3. To calculate the complete motion of the projectile, we need to take account of the relationship between the coefficients and the angle of attack, and to express both vectorially.

In representing drag using Eqn 5.3:20

$$F_D = C_D \cdot \frac{1}{2} \cdot \rho \cdot v^2 \cdot A_0\ , \qquad \text{[N]}$$

we implicitly assumed that flow is parallel to the axis of the projectile. If, however, the axis of the projectile assumes some angle of attack δ relative to the direction of movement (the tangent to the trajectory), the surface subjected to flow will increase. This will increase drag, as the stagnation pressure for the same velocity remains the same. The convention in ballistics is to retain the reference cross-sectional area $A_0$ and assign the increase in drag to the drag coefficient $C_D$, thereby making that coefficient a function of the angle of attack: $C_D = C_D(\delta)$. The same applies to the lift force coefficient $C_L$ and the overturning moment coefficient $C_M$.

These coefficient functions are generally expanded in power series within a certain interval around $\delta = 0$.

$$C_Z(\delta) = C_{Z0} + C'_Z \cdot \delta + C''_Z \cdot \frac{\delta^2}{2!} + C'''_Z \cdot \frac{\delta^3}{3!} + \dots , \quad (5.5:3) \quad [-]$$

where index Z stands for any index of Table 5.4, p. 167.

The function $C_D(\delta)$ is clearly an even function, as it increases at the same rate for a positive or negative angle of attack: $C_D(-\delta) = C_D(\delta)$. The other two functions, $C_L(\delta)$ and $C_M(\delta)$ are odd: $C_L(-\delta) = -C_L(\delta)$, in a manner similar to $C_M$. Fig. 5.35 shows the results of wind tunnel measurements on a model of a 7.5 × 55 bullet at Ma = 2.5.

The power series of Eqn 5.5:3 are usually halted after two terms. We therefore obtain:

$$C_D(\text{Ma},\delta) = C_{D0}(\text{Ma}) + C''_D(\text{Ma}) \cdot \frac{\delta^2}{2} , \quad (5.5:4) \quad [-]$$

$$C_L(\text{Ma},\delta) = C'_L(\text{Ma}) \cdot \delta + C'''_L(\text{Ma}) \cdot \frac{\delta^3}{6} , \quad (5.5:5) \quad [-]$$

$$C_M(\text{Ma},\delta) = C'_M(\text{Ma}) \cdot \delta + C'''_M(\text{Ma}) \cdot \frac{\delta^3}{6} . \quad (5.5:6) \quad [-]$$

For normal flight, we can assume angles of attack of just a few degrees, which means that in most cases we can omit the second terms in the three coefficient functions.

From Equations 5.3:20 and 5.5:4 we obtain the following expression for drag:

$$\vec{t} = \frac{\vec{v}}{v} , \quad (5.5:7) \quad [-]$$

$$\vec{F}_D = -\left( C_{D0} + C''_D \cdot \frac{\delta^2}{2} \right) \cdot \frac{1}{2} \cdot \rho \cdot A_0 \cdot v^2 \cdot \vec{t} . \quad (5.5:8) \quad [N]$$

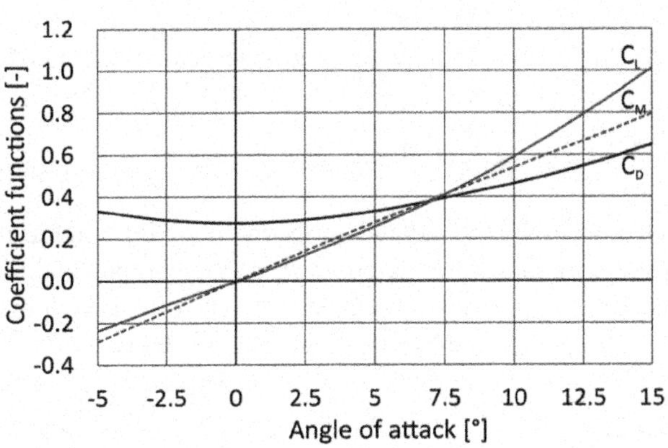


**Fig. 5.35.** Relationship between coefficient functions $C_D$, $C_L$ and $C_M$ and the angle of attack. Results of wind tunnel measurements on a projectile (Ma = 2.5).

Lift lies in the plane generated by $\vec{t}$ and $\vec{c}$, the *resistance plane*, and perpendicular to the drag vector, and hence coincides with $\vec{t}$. We can therefore write:

$$\vec{F}_L = |\vec{F}_L| \cdot [\vec{t} \times (\vec{c} \times \vec{t})] . \qquad [N]$$

Substituting Eqns 5.3:23 and 5.5:5, we obtain

$$(5.5{:}9) \qquad \vec{F}_L = \left( C'_L \cdot \delta + C'''_L \cdot \frac{\delta^3}{6} \right) \cdot \frac{\rho}{2} \cdot A_0 \cdot v^2 \cdot [\vec{t} \times (\vec{c} \times \vec{t})] . \qquad [N]$$

Overturning moment is perpendicular to the resistance plane, so

$$(5.5{:}10) \qquad \vec{M}_M = \left( C'_M \cdot \delta + C'''_M \cdot \frac{\delta^3}{6} \right) \cdot \frac{\rho}{2} \cdot v^2 \cdot A_0 \cdot d_0 \cdot (\vec{t} \times \vec{c}) . \qquad [N{\cdot}m]$$

The general equation linking a moment and the force that generates it is

$$(5.5{:}11) \qquad \vec{M}_M = \vec{e}_A \times \vec{F}_A , \qquad [N{\cdot}m]$$

where $\vec{e}_A$ is the vector between the centre of rotation (the centre of mass in this instance) and the (fictitious) point of application of the aerodynamic force $F_A$. In the case of a spin-stabilized projectile, this point lies on the axis, between the centre of mass and the tip of the projectile.

The pitch damping force is the result of rotation of the projectile about a transverse axis and has three components. The first is the rotation itself, with an angular velocity of $\omega_p$. The second is the result of the change in angle of attack over time. The third is the change in angular velocity $\omega_p$ itself. The last of these is usually ignored (R. L. McCoy, 1999) and the other two are sometimes combined (H. Molitz, R. Strobel, 1963). Earlier authors assume a force when deriving this equation (R. F. Lieske, R. L. McCoy, 1964) or simply omit this force altogether.

Expressed as a force, this factor is incorporated as indicated in Eqn 5.3:29, with the direction determined by the change over time of the unit vector of the projectile axis, which is known to be perpendicular to the vector that is changing:

$$(5.5{:}12) \qquad \vec{F}_E = -C_E \cdot \frac{\rho}{2} \cdot v^2 \cdot A_0 \cdot \omega_p \cdot \frac{d}{v} \cdot \dot{\vec{c}} . \qquad [N]$$

The direction of the corresponding pitch damping moment is perpendicular to both the direction of the force and the projectile axis:

$$(5.5{:}13) \qquad \vec{M}_H = -C_H \cdot \frac{\rho}{2} \cdot v^2 \cdot A_0 \cdot d_0 \cdot \omega_p \cdot \frac{d}{v} \cdot (\vec{c} \times \dot{\vec{c}}) . \qquad [N{\cdot}m]$$

The roll damping moment is generated by the radial component of the flow-induced frictional force on the surface. No lateral force is induced, as the components on opposite sides act in opposite directions and hence cancel out. The direction of the moment coincides with the projectile axis, and its magnitude is given by Eqn 5.3:31:

(5.5:14) $$\vec{M}_I = C_I \cdot \frac{\rho}{2} \cdot v^2 \cdot A_0 \cdot d_0 \cdot \frac{\omega \cdot d}{v} \cdot \vec{c} \ . \qquad [\text{N·m}]$$

Because the projectile is at an angle to the flow, that flow results not only in a lateral force but also in a transverse flow. As we saw in 2.3.4.5, rotation of the projectile about its longitudinal axis results in a Magnus force. When the angle of attack changes, the Magnus force also changes direction in a manner analogous to lift. As a result, the corresponding coefficient is also an odd function. Naturally, the same applies to the Magnus moment:

(5.5:15) $$C_K(Ma,\delta) = C_K'(Ma)\cdot\delta + C_K'''(Ma)\cdot\frac{\delta^3}{6} \ , \qquad [-]$$

(5.5:16) $$C_J(Ma,\delta) = C_J'(Ma)\cdot\delta + C_J'''(Ma)\cdot\frac{\delta^3}{6} \ . \qquad [-]$$

The Magnus force always acts perpendicular to the flow, which itself acts in the direction of the resistance plane. Using Eqn 5.3:32, we obtain:

(5.5:17) $$\vec{F}_K = \left(C_K' \cdot \delta + C_K''' \cdot \frac{\delta^3}{6}\right) \cdot \frac{\rho}{2} \cdot v^2 \cdot A_0 \cdot \frac{\omega \cdot d}{v} \cdot \left(\vec{t} \times \vec{c}\right) . \qquad [\text{N}]$$

The corresponding Magnus moment acts perpendicular to the corresponding force and to the projectile axis. From Equations 5.3:33 and 5.5:16 we obtain:

(5.5:18) $$\vec{M}_J = \left(C_J' \cdot \delta + C_J''' \cdot \frac{\delta^3}{6}\right) \cdot \frac{\rho}{2} \cdot v^2 \cdot A_0 \cdot d_0 \cdot \frac{\omega \cdot d}{v} \cdot \left[\vec{c} \times \left(\vec{t} \times \vec{c}\right)\right] . \qquad [\text{N·m}]$$

In a manner analogous to Eqn 5.5:11, we can also write an expression for the Magnus moment and Magnus force:

(5.5:19) $$\vec{M}_J = \vec{e}_K \times \vec{F}_K \ , \qquad [\text{N·m}]$$

where $\vec{e}_K$ points from the centre of mass to $P_K$, the (fictitious) point of application of the Magnus force. Unlike the point of application of the aerodynamic force, this point generally lies between the centre of mass and the base of the projectile.

The pitching movement about a transverse axis also produces a crossflow, which in conjunction with the rotation of the projectile about its longitudinal axis results in a Magnus effect. The corresponding force is known as the pitch Magnus force, the moment as the pitch Magnus moment. As it appears to have been impossible to demonstrate the influence of these two factors experimentally (R. L. McCoy, 1999), they are not used in calculations. They are mentioned in 5.3.3.4 for the sake of completeness.

#### 5.5.1.3 The equations of motion

The equations of motion for the 6-DOF model of projectile motion are based on the general laws of conservation. In accordance with the law of conservation of momentum, the vector differential equation is

$$\text{(5.5:20a)} \quad m\cdot\dot{\vec{v}} = \vec{F}_D + \vec{F}_L + \vec{F}_E + \vec{F}_K + m\cdot\vec{g} + m\cdot\vec{d}_z + m\cdot\vec{a}_c \,, \qquad \text{[N]}$$

which, in conjunction with

$$\text{(5.5:20b)} \quad \dot{\vec{r}} = \vec{v}\,, \qquad \text{[m/s]}$$

gives us the velocity and position of the centre of gravity. We first represent all forces in the trajectory tangent system and then transform them into the earth-fixed system.

The motion of the projectile about its centre of gravity is described using the law of conservation of angular momentum, in the body-fixed $(\vec{c}, \vec{p}, \vec{q})$ system.

$$\text{(5.5:20c)} \quad \dot{\vec{L}} + \vec{\Omega}\times\vec{L} = \vec{M}_I + \vec{M}_M + \vec{M}_H + \vec{M}_J \,, \qquad \text{[N·m]}$$

where the angular momentum vector $\vec{L}$ is given by

$$\vec{L} = \Theta\cdot\vec{\omega}\,, \qquad \text{[kg·m}^2\text{/s]}$$

where $\vec{\omega}$ is the vector of the angular velocity of the projectile and $\Theta$ is the tensor of inertia of the projectile, in which only the entries on the diagonals are non-zero, on account of the assumed axial symmetry

$$\Theta = \begin{pmatrix} J_a & 0 & 0 \\ 0 & J_r & 0 \\ 0 & 0 & J_r \end{pmatrix}, \qquad \text{[kg·m}^2\text{]}$$

where $J_a$ and $J_r$ are the axial and radial moments of inertia respectively. The vector $\vec{\Omega}$ is the angular velocity with which the body-fixed frame of reference $(\vec{c}, \vec{p}, \vec{q})$ rotates relative to the trajectory tangent system $(\vec{t}, \vec{s}, \vec{h})$. The actual orientations of the projectile axes in space are given by the fourth differential equation

$$\text{(5.5:20d)} \quad \dot{\vec{\varphi}} = T\cdot\vec{\Omega}\,, \qquad \text{[1/s]}$$

where T is the Eulerian transformation matrix of Eqn 5.5:2. Finally, both equations (5.5:20c and d) are transformed into the earth-fixed system using the transformation of Eqn 5.5:1.

Expanded into their components, the four vector equations (5.5:20a–d) form a system of 12 linear differential equations. If we have all the necessary data, we can solve these by a numerical method, with Runge-Kutta methods being the most commonly used. We shall not list the components here; they are presented in D. E. CARLUCCI, S. S. JACOBSON (2008), R. L. MCCOY (1999) (but be aware that this work contains typographical errors) and H. MOLITZ, R. STROBEL (1963).

Readers wishing to get seriously involved in programming this system will need to write out these equations and their transformations in detail.

#### 5.5.1.4 Further points

***Transformations.*** Geometrically speaking, transformation between the body-fixed and the tangent system using the Eulerian angles is very clear and can readily be programmed. However, it does have a significant disadvantage in practice. The sequence of rotations using the Eulerian angles can lead to “gimbal lock”.

A gimbal is the system used to mount such a device as a gyroscopic compass.

Gimbal lock can occur when performing spatial transformations using Eulerian angles, if the transformations are carried out in a particular sequence. The rotations are performed one after another, with each axis of rotation being partly determined by the previous representations. Under certain circumstances, it is possible for two axes to coincide, resulting in the loss of one piece of angle information. In a transformation this causes erroneous results, and in a physical gimbal system it causes the apparatus to lock up.

It is possible to avoid this problem by using Hamiltonian quaternions rather than Eulerian angles to transform the body-fixed into the trajectory-fixed system. However, this does render the calculation more complicated.

Quaternions were discovered by Irish mathematician W. R. HAMILTON in 1843, when he was attempting to extend the complex numbers $a + b \cdot i = (a, b)$ by a further dimension. Defining multiplication only became possible as a result of incorporating a further unit, and hence a fourth dimension.

A quaternion is a quadruple of real numbers ($a, b_1, b_2, b_3$); a is the real part and $b_i$ ($i = 1, 2, 3$) are the imaginary parts. However, the multiplication defined in this manner proved to be non-commutative, with the result that the field of quaternions forms a skew field. The usual representations of the quaternions are:

(5.5:21a) $$Q = a + b_1 \cdot i + b_2 \cdot j + b_3 \cdot k \,,$$ classical and

(5.5:21b) $$Q = \left(a \,, \vec{b}\right) ,$$ vectorial,

where $\vec{b} = (b_1 \cdot i, b_2 \cdot j, b_3 \cdot k)$ is a three-dimensional imaginary vector. They can also be written as four-dimensional complex vectors or as complex matrices. The calculation rules for the units i, j and k are:

(5.5:22a) $$i^2 = j^2 = k^2 = i \cdot j \cdot k = -1 \,,$$ and hence also

(5.5:22b) $$i \cdot j = k \,, j \cdot k = i \,,\; k \cdot i = j \quad \text{und} \quad j \cdot i = -k \,,\; k \cdot j = -i \,,\; i \cdot k = -j \,.$$

Adding two quaternions $Q_1 = (a_1, \vec{b}_1)$ and $Q_2 = (a_2, \vec{b}_2)$ gives us

(5.5:23) $$Q_1 + Q_2 = \left(a_1 + a_2 \,, \vec{b}_1 + \vec{b}_2\right) ,$$

from which the additive neutral element $(0, \vec{0})$ is also apparent. The formula for multiplication is

$$(5.5{:}24) \qquad Q_1 \cdot Q_2 = \left(a_1 \cdot a_2 - \vec{b}_1 \cdot \vec{b}_2 \,,\, \vec{b}_1 \times \vec{b}_2 + a_1 \cdot \vec{b}_2 + a_2 \cdot \vec{b}_1\right) .$$

When multiplying the vectors $\vec{b}_i$, we are referring to the scalar or cross product. The latter also explains why $Q_1 \cdot Q_2 \neq Q_2 \cdot Q_1$. In a manner analogous to the complex numbers, there is also a conjugated quaternion, defined by $\bar{Q} = (a, -\vec{b})$, and

$$(5.5{:}25) \qquad \text{Norm}(Q) = Q \cdot \bar{Q} = |Q|^2 = a^2 + \vec{b}^2 .$$

The root of the norm (i.e. $|Q|$) is described as the length of Q. The norm allows us to construct the inverse of a quaternion

$$(5.5{:}26) \qquad Q^{-1} = \frac{\bar{Q}}{|Q|^2} .$$

The unit quaternions play an important role in representing transformations. These are obtained from any quaternion by dividing by its length.

$$(5.5{:}27) \qquad Q^* = \frac{Q}{|Q|} = \left(\frac{a}{|Q|}, \frac{\vec{b}}{|Q|}\right) = \left(a^*, \vec{b}^*\right) .$$

A quaternion $P = (0, \vec{x})$ of which the real part is zero is termed pure imaginary (or a pure quaternion). Such a quaternion forms a three-dimensional vector. As for the complex numbers, there is also a polar representation for quaternions:

$$(5.5{:}28) \qquad Q = |Q| \cdot (\cos\varphi \,,\, \vec{\varepsilon} \cdot \sin\varphi)$$

where $\vec{\varepsilon}$ is a pure imaginary unit quaternion.

Following these preparations, we can now formulate the central equation for rotation in three-dimensional space by means of quaternions.

Let $\vec{x}$ be a vector in three-dimensional space and P the corresponding pure imaginary quaternion. The operation

$$(5.5{:}29) \qquad D = Q^* \cdot P \cdot \bar{Q}^*$$

with the unit quaternion $Q^*$,

$$Q^* = (\cos\varphi \,,\, \vec{\varepsilon} \cdot \sin\varphi) ,$$

rotates the vector $\vec{x}$ through an angle of $2 \cdot \varphi$ about the axis represented by the vector $\vec{\varepsilon}$. If the vectors in three-dimensional space are represented as pure quaternions, it will always be possible to find one unit quaternion with which we can perform the desired rotation.

Representing the rotations using quaternions has the advantage that gimbal lock cannot occur. However, this comes at the cost of greater programming effort, partly because we still need the matrices for the translations.

Spatial rotation is often calculated by means of quaternions when programming robots and computer games. One may assume that the necessary modules exist. We shall therefore refrain from presenting the corresponding transformation equations in detail. These are available in V. FLECK (2009), for instance. For further details regarding quaternions, please consult T. LEDOUX (2011) und M. LUST (2001), among others.

***Calculating the force and moment coefficients***. To numerically integrate the vector system of differential equations (5.5:20a–d), we need to know all force and moment coefficients as a function of Mach number and angle of attack. There are two ways of achieving this: by calculation on the basis of aerodynamic principles and by measurements using model or real-life experiments. We can use aerodynamic calculations to find the axial drag coefficient and, for small angles of attack, lift and overturning moment (see 5.7.3). Using special software, we can also calculate Magnus force, Magnus moment and pitch damping moment.

Drag coefficients can be ascertained using radar velocity-distance measurements. However, angle of attack is usually not measured during such experiments, which means we cannot differentiate between $C_{D0}$ and $C''_D$. We can measure lift and overturning moment in a wind tunnel, but it will not be possible to simulate rotation about the longitudinal axis, except with considerable effort. We can obtain certain coefficients on free-flight ballistic ranges, by measuring pitch movements (spatial angle of attack and the trajectory of the centre of mass) using spark photography or shadowgraphs. It is possible by this means to obtain the coefficients of drag, lift, overturning moment, Magnus moment and pitch damping moment. However, conducting such experiments and analysing the results is very labour intensive and is generally restricted to small- and medium-calibre projectiles. In the case of larger-calibre projectiles, it is possible to insert sensors that transmit the orientation of the axis and its rate of rotation by telemetry. We can then use these data to calculate the same coefficients. It is important to ensure that the physical characteristics of the measuring shell (shape, centre of mass, moment of inertia) are identical to those of the shell under investigation.

***Determining the initial values***. Before we can start to integrate the motion of the projectile, all four vector functions must have specific values for $t = 0$. We have to know the position and velocity vectors of the centre of mass, the direction of the projectile axis and its vector angular velocity at the moment the projectile leaves the barrel. In view of the barrel oscillation that occurs before that moment and the resulting lateral movement of the muzzle, and the effect of the propellant gas that passes over the projectile from behind after it leaves the muzzle, it is difficult or even impossible to obtain those values.

6-DOF models are used to circumvent the above problems, especially when conducting parameter studies (e.g. examining the influence of initial values and flight conditions on scatter) and resolving local motion problems (stability and tractability). In such cases one generally assumes a small angle of attack.

### *5.5.1.5 Equation for the complex angle of attack*

One way of finding an analytical solution for the motion of the projectile about its centre of mass is to use the *complex angle of attack*, where $\delta$ is the angle of attack (the angle between the tangent to the trajectory and the axis of the projectile) and $\psi$ is the angular displacement of the projectile axis about the tangent to the trajectory (see Figures 5.36 and 5.34):

(5.5:30) $$\zeta = \delta\cdot(\cos\psi + i\cdot\sin\psi) = \delta\cdot e^{i\cdot\psi} .$$ [-]

We can define the complex angle of trajectory in similar fashion:

(5.5:31) $$\varepsilon = \theta + i\cdot\eta .$$ [-]

If we derive the two equations for time twice, substitute the resulting values from Eqn 5.5:20a–d and perform a series of rearrangements and mathematical tricks, which include making trajectory length as a function of calibre the independent variable instead of time (see E. CARTOUX, 1974, R. L. MCCOY, 1999)

(5.5:32) $$\lambda = \int_0^t \frac{v}{d}\cdot dt ,$$ [-]

we obtain a second-order differential equation for the complex angle of attack $\zeta$:

(5.5:33) $$\zeta'' + (\alpha_1 - i\cdot\beta_1)\cdot\zeta' + (\alpha_0 - i\cdot\beta_0)\cdot\zeta = \alpha_3 - i\cdot\beta_3 .$$ [-]

The above assumes a rotationally-symmetric projectile, a flat trajectory, a small angle of attack $\delta$, low angular velocities $\dot\theta$ and $\dot\eta$ and coefficients that are only related to the angle of attack in a linear fashion.

In addition to the fixed physical parameters of the projectile, the real and imaginary parts of the coefficients comprise velocity, angular velocity, angle of trajectory and the coefficients of the forces and moments. These are dimensionless and are as follows:

(5.5:34a) $$\alpha_0 = -C'_M\cdot\frac{\pi\cdot\rho}{8}\cdot\frac{d^5}{J_r}$$ [-]

(5.5:34b) $$\beta_0 = \frac{\pi\cdot\rho}{8}\cdot\frac{d^3}{m}\cdot\frac{J_a}{J_r}\cdot\frac{\omega\cdot d}{v}\cdot\left(C'_L - C'_J\cdot\frac{d^2\cdot m}{J_a}\right)$$ [-]

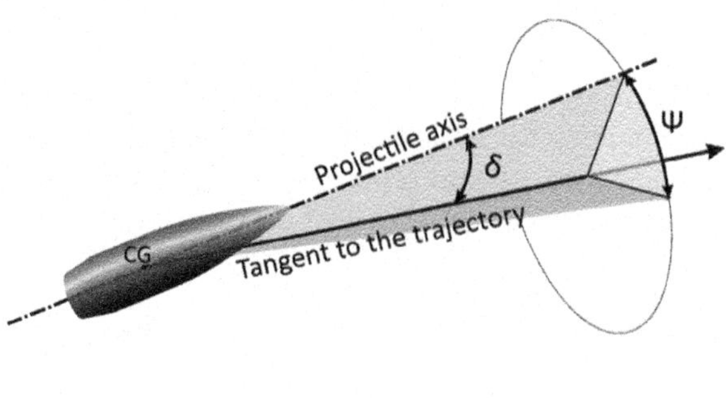


**Fig. 5.36.** Definition of the complex angle of attack.

(5.5:34c) $$\alpha_1 = \frac{\pi \cdot \rho}{8} \cdot \frac{d^3}{m} \cdot \left( C'_L - C_{D0} + C_H \cdot \frac{d^2 \cdot m}{J_r} \right)$$ [-]

(5.5:34d) $$\beta_1 = \frac{J_a}{J_r} \cdot \frac{\omega \cdot d}{v}$$ [-]

(5.5:34e) $$\alpha_3 = \left( \frac{g \cdot d \cdot \cos\theta}{v^2} \right)' + \frac{g \cdot d \cdot \cos\theta}{v^2} \cdot \frac{\pi \cdot \rho \cdot d^5}{8} \cdot \left( \frac{C_H}{J_r} - \frac{C_I}{J_a} \right)$$ [-]

(5.5:34f) $$\beta_3 = g \cdot \frac{J_a}{J_r} \cdot \frac{\omega \cdot d^2}{v^3} \cdot \cos\theta$$ [-]

Eqn 5.5:33 is a simplified equation for the complete motion of the projectile about its centre of gravity for small angles of attack and flat trajectories. It is also called the *linearized equation of motion*. Under certain conditions, it can be solved analytically using complex exponential functions (see the literature cited above) and it also forms the starting point for examining stability and for making simplifications that eliminate certain disadvantages of the point-mass model without the effort required to produce a complete 6-DOF model.

## 5.5.2 The modified point-mass models

### *5.5.2.1 Introduction*

At the start of the computer era, calculating the complete motion of the projectile was impossible not only because coefficients were not available but above all because solving the associated system of differential equations was extremely slow. Efforts were therefore made to find compromise solutions that would describe the motion of the projectile more precisely than did the point-mass model while requiring less computing capacity than a full 6-DOF model. The equation for the complex angle of attack was particularly suitable for the purpose, having already undergone a number of simplifications. The resulting *modified point-mass models* are still of interest today, as they require fewer projectile data, which means that one is spared the effort of obtaining those data.

One obvious error in a trajectory calculated using a point-mass model is the apex height calculated for higher elevations. The trajectory is calculated using a drag coefficient based on a mean angle of attack for the trajectory as a whole, because when $C_D$ is determined empirically, the component dependent on the angle of attack, $C''_D(\delta)$ is generally not included. In reality, the angle of attack is smaller in the ascending and descending branches than it is when the trajectory is changing direction at its apex. The consequence of this is that the drag used to calculate the ascending part of the trajectory is higher than the true value. This results in the apex height of the calculated trajectory being lower than the true value. In the case of an artillery shell, discrepancies of several tens of metres can occur. The primary aim of a modified point-mass model is to take account of the effect of the angle of attack – and hence of lift and Magnus force – along the trajectory.

### *5.5.2.2 The decrease in angular velocity*

The angular velocity of a spin-stabilized projectile has a significant effect on its trajectory, as the stability of the gyroscopic motion and the associated angle of precession exert a major influence on the angle of attack. An extension of the point-mass model that took account of the decrease in speed of rotation would make it possible to estimate the stability of the projectile along its trajectory. Furthermore, angular velocity contributes to some of the coefficients in the equation for the complex angle of attack. To use those coefficients, it is therefore necessary to know how angular velocity changes along the trajectory.

The only factor responsible for slowing the spin of a spin-stabilized projectile (without fins) is the moment of friction drag (the roll damping moment). This is calculated on the basis of the frictional force per unit area of projectile surface, bearing in mind that friction drag increases with boundary layer thickness and hence with distance from the tip of the projectile. Assuming

$$F_F \propto x^{0.8} ,$$

for turbulent flow (H. SCHLICHTING, 1965):

$$(5.5{:}35) \qquad F_F(x) = 1.8 \cdot C_F \cdot \frac{\rho}{2} \cdot \left(v^2 + r^2(x) \cdot \omega^2\right) \cdot \frac{d^2 \cdot \pi}{4 \cdot A_S \cdot \ell_p^{0.8}} \cdot x^{0.8} , \qquad [\mathrm{N/m^2}]$$

where x = distance from the tip of the projectile, $\ell_p$ = its length, r(x) is the meridian and $A_S$ = projectile surface area. So for the moment about the projectile axis we can write:

$$(5.5{:}36) \qquad dM_I = F_F(x) \cdot \frac{r(x) \cdot \omega}{\sqrt{v^2 + r^2(x) \cdot \omega^2}} \cdot r(x) \cdot dA_S . \qquad [\mathrm{N \cdot m}]$$

Substituting Eqn 5.5:35 into Eqn 5.5:36, summarizing and simplifying yields:

$$dM_I = C_F \cdot \frac{9 \cdot \rho \cdot d^2 \cdot \pi}{40 \cdot A_S \cdot \ell_p^{0.8}} \cdot r^2(x) \cdot \omega \cdot v \cdot \sqrt{1 + \frac{r^2(x) \cdot \omega^2}{v^2}} \cdot x^{0.8} \cdot dA_S . \qquad [\mathrm{N \cdot m}]$$

As $r \cdot \omega \ll v$, we can linearize the root:

$$(5.5{:}37) \qquad dM_I = C_F \cdot \frac{9 \cdot \rho \cdot d^2 \cdot \pi}{40 \cdot A_S \cdot \ell_p^{0.8}} \cdot \omega \cdot v \cdot r^2(x) \cdot \left(1 + \frac{r^2(x) \cdot \omega^2}{2 \cdot v^2}\right) \cdot x^{0.8} \cdot dA_S . \qquad [\mathrm{N \cdot m}]$$

Integrating over the surface of the projectile gives

$$(5.5{:}38) \qquad M_I = C_F \cdot \frac{\rho}{2} \cdot \omega \cdot v \cdot \left(C_1 + \frac{\omega^2}{v^2} \cdot C_2\right) , \quad \text{with} \qquad [\mathrm{N \cdot m}]$$

$$C_1 = \frac{0.45 \cdot d^2 \cdot \pi}{A_S \cdot \ell_p^{0.8}} \cdot \oint r^2(x) \cdot x^{0.8} \cdot dA_S \,, \tag{5.5:39}$$ $[m^4]$

$$C_2 = \frac{0.225 \cdot d^2 \cdot \pi}{A_S \cdot \ell_p^{0.8}} \cdot \oint r^4(x) \cdot x^{0.8} \cdot dA_S \,. \tag{5.5:40}$$ $[m^6]$

Integration of the friction torque is therefore purely geometric. Please see 5.7.3.5 regarding calculation of the coefficient of friction $C_F$.

Angular velocity is then calculated using the equation of motion for the rotating body, where $J_a$ = axial moment of inertia.

$$J_a \cdot \dot{\omega} = -M_I \,. \tag{5.5:41}$$ $[N \cdot m]$

This equation can be integrated numerically along with the equations of motion for the point mass, yielding the instantaneous angular velocity at each point along the trajectory:

$$\left| \begin{array}{lcll} m \cdot \dot{\vec{v}} & = & \vec{F}_D + m \cdot \vec{g} + m \cdot \vec{d}_z + m \cdot \vec{a}_c \,, & [N] \\ \dot{\omega} & = & -\dfrac{M_I}{J_a} \,, & [1/s^2] \\ \dot{\vec{r}} & = & \vec{v} \,. & [m/s] \end{array} \right. \tag{5.5:42}$$

This is often referred to as a 4-DOF exterior ballistics model, on account of the additional differential equation.

### 5.5.2.3 The BRL approach

In the 1960s, the problem of long computing times for the 6-DOF model prompted the Computing Laboratory of the US Army Ballistic Research Laboratories (BRL) to look for a method that would at least provide a better way of solving the above-mentioned problem of the angle of attack. Report 1314 (R. F. LIESKE, M. L. REITER, 1966) includes an equation for the angle of attack that uses the simplifications described in 5.5.1.5.

One can achieve the same result by ignoring the angular accelerations and velocities $\zeta''$ and $\zeta'$ in the linearized equation of motion (Eqn 5.5:33). The complex angle of attack $\zeta$ and hence also angles $\delta$ and $\psi$ can be calculated directly:

$$\zeta = \frac{\alpha_3 - i \cdot \beta_3}{\alpha_0 - i \cdot \beta_0} = \frac{(\alpha_3 - i \cdot \beta_3) \cdot (\alpha_0 + i \cdot \beta_0)}{\alpha_0^2 + \beta_0^2} \,. \tag{5.5:43}$$ $[-]$

This incorporates expressions 5.5:34a and b or 5.5:34e and f, which also include angular velocity. The differential equation (Eqn 5.5:41) must therefore also be integrated. If the angles are known, it is possible to calculate lift force and Magnus force. In combination with Eqn 5.5:43, we obtain the following system:

$$(5.5{:}44)\quad \begin{aligned} m\cdot\dot{\vec{v}} &= \vec{F}_D+\vec{F}_L+\vec{F}_K+m\cdot\vec{g}+m\cdot\vec{d}_z+m\cdot\vec{a}_c\,, && [\mathrm{N}] \\ \dot{\omega} &= -\frac{M_I}{J_a}\,, && [1/\mathrm{s}^2] \\ \dot{\vec{r}} &= \vec{v}\,. && [\mathrm{m/s}] \end{aligned}$$

In this model, the projectile has a fixed angle of attack throughout the trajectory, an angle determined by velocity and angle of trajectory.

A detailed description of this model (also known as the *modified static point-mass model*) is contained in R. L. McCoy (1999) and in Report 1314.

#### 5.5.2.4 The Fleck and Molitz approach

In a paper they presented to the 3rd International Symposium on Ballistics, V. Fleck and H. Molitz (1977) suggested ignoring only the second derivation of the complex angle of attack in the linearized equation of motion. This results in a first-order linear differential equation for $\zeta$

$$(5.5{:}45)\quad \zeta' + \frac{\alpha_0 - i\cdot\beta_0}{\alpha_1 - i\cdot\beta_1}\cdot\zeta = \frac{\alpha_3 - i\cdot\beta_3}{\alpha_1 - i\cdot\beta_1}\,, \qquad [-]$$

which can be numerically integrated together with the system of Eqn 5.5:44. This is known as the *modified dynamic point-mass model.* Here again, the expressions of Eqn 5.5:34a–f are used. Assuming that $C_D$ is constant within each step, we can hence estimate the complex angle of attack (i.e. $\delta$ and $\psi$) for each integration step. The angle of attack calculated using this model is much closer to the 6-DOF value than the angle obtained using the BRL method, as we can see from Figs 5.37 and 5.38, which are taken from the paper mentioned above.

V. Fleck (2009) presents comparison calculations showing the differences between the point-mass models and between those models and the 6-DOF model. Calculations were performed for a 155 mm artillery shell with a muzzle velocity of 800 m/s, fired at an elevation of 1120 mils. Compared with the 6-DOF model, the

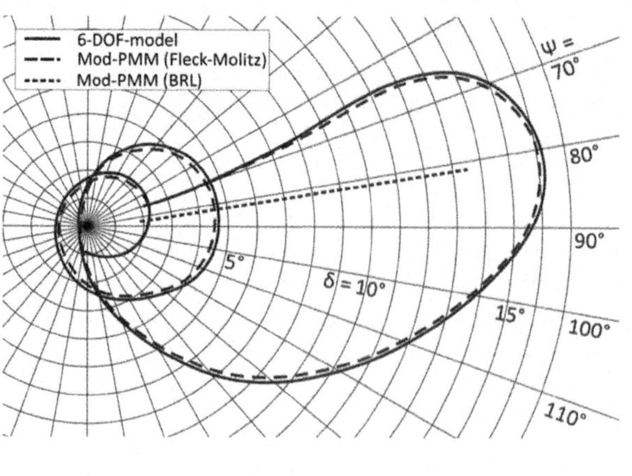


**Fig. 5.37.** The complex angle of attack with components $\delta$ and $\psi$, calculated using the two modified point-mass models (mod-PMM) and compared with the 6-DOF model. The dynamic model conforms closely to the 6-DOF model, whereas in the static model the angular displacement of the projectile about the tangent to the trajectory remains constant, with only the angle of attack changing.

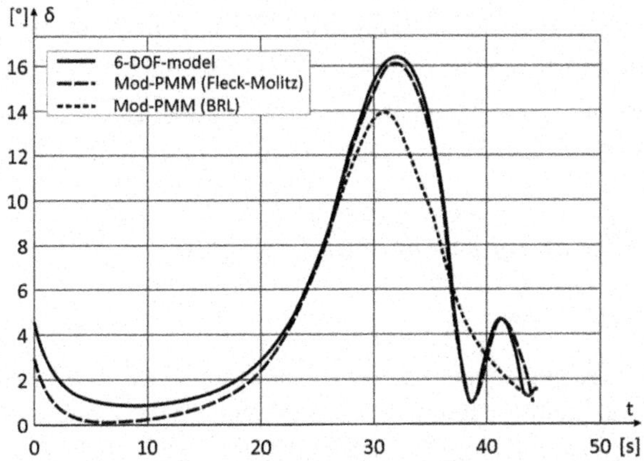


**Fig. 5.38.** Angle of attack as a function of time of flight. Same comparison between models as in Fig. 5.37.

static model had a range error of +5.0‰, while that of the dynamic model was only +0.2‰. The lateral errors were –8.6‰ and –12.0‰, and time of flight errors were +1.0‰ and 0.4‰. The modified models are clearly accurate enough for practical purposes.

## 5.6 Stability and tractability

### 5.6.1 What is stability?

Generally speaking, every physical system has a state of rest. This state of rest corresponds to a particular equilibrium of forces and moments, and the system will remain in this state unless that equilibrium is disturbed. If a disturbance (a force or a torque) acts briefly on the system, the system can react in three different ways:

- If the induced motion becomes less pronounced with time, the system is said to be *stable*. In such a case, there will be a counterforce (generally induced by the disturbing force).
- If the movement becomes more pronounced with time, the system is said to be *unstable*. In such cases forces are often present that are amplified by the disturbance.
- If the movement continues, becoming neither more nor less pronounced with time, the system is said to be *neutral*. When the disturbing force is removed,

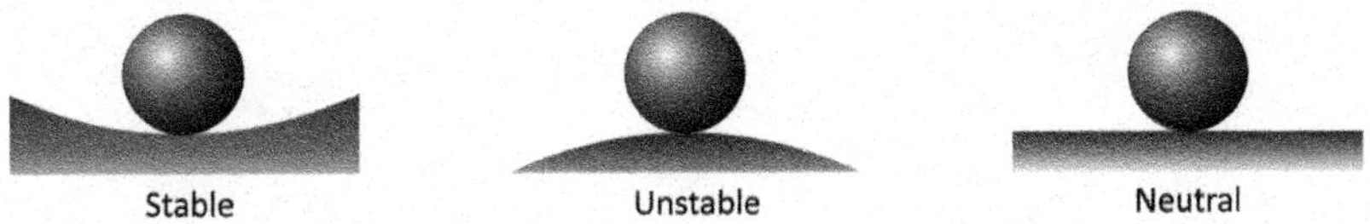


**Fig. 5.39.** The possible stability states. We can readily imagine the motion of these spheres when subjected to a brief lateral force.

the system enters a new state of rest. This may differ from the original state, but under certain conditions the system will return to that original state.

These three cases can be illustrated by a sphere lying on the concave side of a hemisphere, on the convex side of a hemisphere and on a plane surface (see Fig. 5.39).

## 5.6.2 Principles of stability as they relate to a projectile

### *5.6.2.1 Spin-stabilized projectiles*

The stabilizing characteristic of a gyroscope, whereby it does not yield to a force but evades it perpendicularly, is the most frequently used method of ensuring that a projectile remains stable along its trajectory. This does require very high rotation speeds, but these are fairly simple to achieve by means of a rifled barrel. The principle is explained in 2.2.8.4 and 5.3.3.1 and in Fig. 5.10.

### *5.6.2.2 Stabilizing non-spinning projectiles*

If a projectile is fired with little or no spin, the stabilizing moment must be provided by aerodynamic force. This requires that the point of application of the aerodynamic force lie behind the projectile centre of mass, in order to create a windvane effect. That can be achieved by one of the following methods:

- So distributing the mass as to move the centre of gravity to a position forward of the point of application of the aerodynamic force (arrow stabilization, see Fig. 5.40 a).
- Displacing the point of application of aerodynamic force rearwards by applying additional aerodynamic forces at the tail. There are two ways of achieving this:
  - Increase the lift force at the tail by adding fins (fin stabilization, see Fig. 5.40 b).
  - Increase drag at the tail by adding a conical extension or a disc of larger diameter (rear-drag stabilization, Fig. 5.40 c).

Rear-drag stabilization has the disadvantage that the projectile loses a considerable amount of energy over its trajectory because of the high level of drag. Arrow stabilization is used in such applications as firework rockets, arrows, crossbow bolts

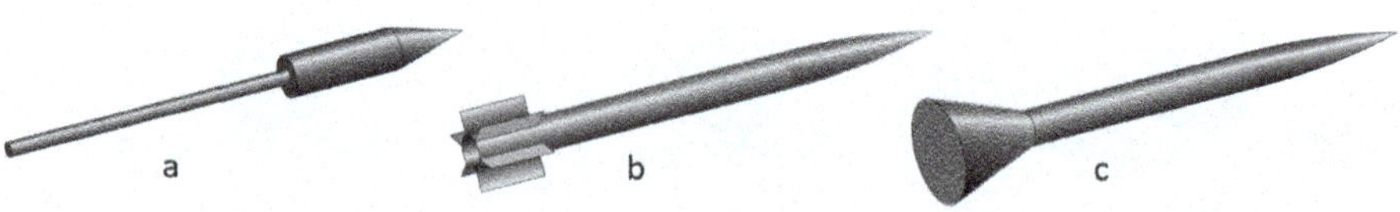


**Fig. 5.40.** Methods of achieving stabilization without spin: **a.** Arrow stabilization. **b.** Fin stabilization. **c.** Drag stabilization.

(sometimes in conjunction with rear-drag stabilization) and certain shotgun slugs.

The two fins on certain arrows and crossbow bolts are not intended primarily as a means of stabilization, as the lift they provide changes too much with the rotational orientation of the projectile. Their main role is to induce slow rotation, to compensate for the effects of asymmetry in mass distribution and forces.

The most common form of non-spin stabilization is the use of fins, as arrow-stabilized and drag-stabilized projectiles are subjected to a higher level of drag.

#### *5.6.2.3 Shoulder stabilization*

Another type of stabilization comes into play in the case of projectiles with a larger frontal surface area, regardless of whether the projectile is spinning or not. At small angles of attack, the asymmetric distribution of pressure on the face of the projectile due to stagnation pressure generates a stabilizing moment (see Fig. 5.41 a). Normally, the flow separates on the edge of the face. As a result, there are no compressive forces acting on the rest of the projectile. At larger angles of attack, this moment has a destabilizing effect (see Fig. 5.41 b). The angle at which the transition from stabilization to destabilization occurs depends on the diameter of the face and its distance from the centre of gravity. Fig. 5.42 shows two examples of shoulder stabilized projectiles for smooth-bore hunting weapons (shotguns).

### 5.6.3 Gyroscopic stability

We can assess the stability of a spin-stabilized projectile using gyroscope theory. The motion of a gyroscope is described by means of Eulerian gyroscope equations. For a body-fixed system of major axes, these are as follows:

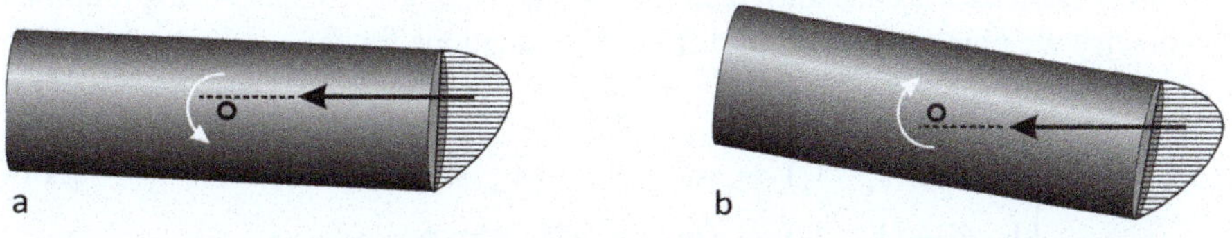


**Fig. 5.41**. Shoulder stabilization: At small angles of attack, the asymmetry in the distribution of pressure produces a stabilizing moment, while at large angles of attack it produces a destabilizing moment.

**Fig. 5.42**. Two examples of shoulder-stabilized projectiles for shotguns.
**a**. A Balle Blondeau slug with shoulder stabilization only.
**b**. A Brennecke slug with both shoulder and arrow stabilization.

(5.6:1) $$\frac{d\vec{L}}{dt} = \frac{\partial \vec{L}}{\partial t} + \vec{\omega} \times \vec{L} = \vec{T} \, .$$ [N·m]

Where $\vec{L}$ = angular momentum, $\vec{\omega}$ = angular velocity and $\vec{T}$ = total torque.

Because of the system of major axes used, the tensor of inertia is limited to the diagonals, and in component form Eqn 5.6:1 becomes:

(5.6:2) $$\left| \begin{array}{lcl} J_c \cdot \dot{\omega}_c + (J_p - J_q) \cdot \omega_p \cdot \omega_q & = & T_c \\ J_p \cdot \dot{\omega}_p + (J_q - J_c) \cdot \omega_q \cdot \omega_c & = & T_p \\ J_q \cdot \dot{\omega}_q + (J_c - J_p) \cdot \omega_c \cdot \omega_p & = & T_q \end{array} \right| .$$ [N·m]

We can generally assume rotational symmetry in the case of a projectile, so $J_p = J_q = J_r$ and $J_c = J_a$.

In the case of a spin-stabilized projectile, torque $\vec{T}$ stems mainly from aerodynamic force (ignoring other moments such as the Magnus moment and pitch damping moment). This torque acts perpendicular to the projectile (gyroscope) axis $\vec{c}$ and the tangent to the trajectory (the precession axis) $\vec{t}$, and its direction determines the line of nodes K (see Fig. 5.34, p. 219). In the body-fixed system, therefore

(5.6:3) $$\vec{T} = M_M \cdot (0, -\sin\phi, \cos\phi) \, .$$ [N·m]

By inserting the Eulerian angles $\phi$, $\psi$ and $\delta$, we obtain the following equation for angular velocity.

(5.6:4) $$\vec{\omega} = \begin{pmatrix} \omega_c \\ \omega_p \\ \omega_q \end{pmatrix} = \begin{pmatrix} \dot{\phi} + \dot{\psi} \cdot \cos\delta \\ -\dot{\delta} \cdot \sin\phi + \dot{\psi} \cdot \sin\delta \cdot \cos\phi \\ \dot{\delta} \cdot \cos\phi + \dot{\psi} \cdot \sin\delta \cdot \sin\phi \end{pmatrix} .$$ [1/s]

If precession is assumed to be regular (i.e. the angle $\delta$ between the figure axis and the precession axis is constant), then from Eqns 5.6:2 and 5.6:3 and using rotational symmetry we obtain a system of differential equations for $\vec{\omega}$:

(5.6:5) $$\left| \begin{array}{lcl} J_a \cdot \dot{\omega}_c & = & 0 \\ J_r \cdot \dot{\omega}_p + (J_r - J_a) \cdot \omega_q \cdot \omega_c & = & -M_M \cdot \sin\phi \\ J_r \cdot \dot{\omega}_q + (J_a - J_r) \cdot \omega_c \cdot \omega_p & = & M_M \cdot \cos\phi \end{array} \right| .$$ [N·m]

By multiplying the first equation by $\omega_c$, the second by $\omega_p$ and the third by $\omega_q$ and then adding the three equations, we obtain:

$$J_a \cdot \dot{\omega}_c \cdot \omega_c + J_r \cdot (\dot{\omega}_q \cdot \omega_q + \dot{\omega}_p \cdot \omega_p) = M_M \cdot (-\sin\phi \cdot \omega_p + \cos\phi \cdot \omega_q) \, .$$

Rearranging the left-hand side, substituting 5.6:4 and integrating yields the law of conservation of energy

(5.6:6) $$J_r \cdot (\omega_q^2 + \omega_p^2) + J_a \cdot \omega_c^2 - 2 \cdot M_M \cdot \delta = \text{const} \, .$$ [J]

Replacing angular velocities $\omega_p$ and $\omega_q$ using Eqn 5.6:4 yields

$$(5.6:7)\qquad J_r \cdot \dot{\psi}^2 \cdot \sin^2\delta + J_a \cdot \omega_c^2 - 2 \cdot M_M \cdot \delta \;=\; \text{const} \;. \qquad [\text{N·m}]$$

Because of the first equation in 5.6:5, $\omega_c$ is constant (this was assumed for $\delta$). $\dot{\psi}$ is also constant on account of Eqn 5.6:7, and according to the last equation in 5.6:4 this also applies to $\dot{\phi}$. Under these conditions, the derivation of the second equation in 5.6:4 is:

$$\dot{\omega}_p \;=\; -\,\dot{\psi} \cdot \dot{\phi} \cdot \sin\delta \cdot \sin\phi \;. \qquad [1/\text{s}^2]$$

Substituting into the second equation in 5.6:5 and rearranging gives us

$$(5.6:8)\qquad J_r \cdot \cos\delta \cdot \dot{\psi}^2 - J_a \cdot \omega_c \cdot \dot{\psi} + \frac{M_M}{\sin\delta} \;=\; 0 \;. \qquad [\text{N·m}]$$

This is a quadratic equation for angular velocity $\dot{\psi}$, which must have real solutions. The discriminant must therefore be positive

$$J_a^2 \cdot \omega_c^2 - 4 \cdot J_r \cdot \cos\delta \cdot \frac{M_M}{\sin\delta} \;>\; 0 \;, \qquad [\text{kg}^2\text{·m}^4/\text{s}^2]$$

or

$$\frac{J_a^2 \cdot \omega_c^2}{4 \cdot J_r \cdot \dfrac{M_M}{\tan\delta}} \;>\; 1 \;. \qquad [-]$$

The expression to the left of the inequality sign is known as the *gyroscopic stability number* s.

$$(5.6:9)\qquad s \;:=\; \frac{J_a^2 \cdot \omega_c^2}{4 \cdot J_r \cdot \dfrac{M_M}{\tan\delta}} \;>\; 1 \;. \qquad [-]$$

The magnitude of moment $M_M$ is calculated using Eqn 5.5:10. The linearized representation of the overturning moment coefficient $C_M$ is sufficiently accurate for this purpose.

$$M_M \;=\; C'_M \cdot \delta \cdot \frac{\rho}{2} \cdot v^2 \cdot A \cdot d \;. \qquad [\text{N·m}]$$

For small values of $\delta$ (less than approximately 10°) we can assume $\tan\delta \approx \delta$. This being so, we obtain the following expression from Eqn 5.6:9:

$$(5.6:10)\qquad s \;=\; \frac{2}{\pi} \cdot \frac{J_a^2 \cdot \omega^2}{J_r \cdot C'_M \cdot \rho \cdot v^2 \cdot d^3} \;>\; 1 \;. \qquad [-]$$

It should be noted that in deriving the gyroscopic stability number s, we assumed that the angle of attack was constant. Precession therefore need not decay. Inequality 5.6:10 is hence a necessary but not a sufficient condition for stable flight.

This was already pointed out in H. MOLITZ (1957) and in H. MOLITZ, R. STROBEL (1963). In most cases, however, we need only calculate s (see B. P. KNEUBUEHL, 1983[1]).

### 5.6.4 The Molitz stability criterion

#### 5.6.4.1 Introduction

MOLITZ' stability theory has been the basis for assessing the flight of a spin-stabilized projectile since the middle of the 20th Century. MOLITZ started by using the classical gyroscopic stability number of a top for his theory. As we have seen, this is a necessary but not a sufficient condition for the stability of a projectile. He suggested a further stability criterion, which would constitute the missing condition.

However, the original literature only hints at the physical and mathematical background to this theory, especially as regards the second part. Examination of those principles reveals aspects of the validity and limitations of stability theory that could be useful as means to understanding it.

#### 5.6.4.2 Mathematical definition of stability

Consider a movement described by the following system of differential equations:

(5.6:11) $$\dot{\vec{x}} = \vec{f}(\vec{x})$$

This movement has a state of equilibrium $\vec{b}$ (e.g. the rest position in the frame of reference used).

Such a state of equilibrium $\vec{b}$ is termed stable if an initial condition (close to $\vec{b}$ $\left|\vec{\xi}-\vec{b}\right|<\tau$) always yields a solution that remains close to the state of equilibrium at all times. The formal definition is as follows (L. S. PONTRJAGIN, 1965):

*Definition*: Consider a system of differential equations in accordance with 5.6:11, of which the partial derivations

$$\frac{\partial f_i}{\partial x_k} = f_{i,k}$$

exist and are continuous. Let $\vec{y}(t,\vec{\xi})$ be a solution to the system at 5.6:11 that satisfies the initial conditions $t=0$ and $\vec{x}=\vec{\xi}$. A state of equilibrium $\vec{b}$ as per 5.6:11 is termed stable according to LJAPUNOV if the following conditions are fulfilled:

1 A positive number $\tau$ exists that is sufficiently small that the solution $\vec{y}(t,\vec{\xi})$ is defined for all cases where $t>0$, as long as .

2 For every positive number $\varepsilon$ there exists a positive number $\delta<\tau$ such that $\left|\vec{y}(t,\vec{\xi})-\vec{b}\right|<\varepsilon$ where $\left|\vec{\xi}-\vec{b}\right|<\delta$ and $t>0$.

A state of equilibrium that is stable in accordance with the above definition is described as asymptotically stable if a positive number $\sigma < \tau$ exists that is sufficiently small that for $\left|\vec{\xi} - \vec{b}\right| < \sigma$

$$\lim_{t\to\infty} \left| \vec{y}(t, \vec{\xi}) - \vec{b} \right| = 0 .$$

### 5.6.4.3 Stability criteria

Let $\vec{b}$ be a state of equilibrium for the system at 5.6:11 and let $\Delta\vec{x}$ be a disturbance. We can then write

(5.6:12) $$\vec{x} = \vec{b} + \Delta\vec{x} .$$

Substituting into 5.6:11 and splitting off the linear term of the Taylor series gives

(5.6:13) $$\Delta\dot{x}_i = f_i(\vec{b}) + \sum_{j=1}^{n} \frac{\partial f_i(\vec{b})}{\partial x_j} \cdot \Delta x_j + R_i ,$$

or with $f_{i,j} = A = (A_{ij})$ in vector form:

(5.6:14) $$\Delta\dot{\vec{x}} = \vec{f}(\vec{b}) + A \cdot \Delta\vec{x} + \vec{R} ,$$

where vector $\vec{R}$ contains only terms of order greater than 1. LJAPUNOV's theorems can then be expressed as follows:

*Theorem 1* If all characteristic values of Matrix A have negative real parts, then the state of equilibrium $\vec{b}$ of the system

$$\dot{\vec{x}} = \vec{f}(\vec{x})$$

is asymptotically stable.

*Note.* The characteristic values of Matrix A are equal to the roots of the characteristic equation of the linearized system $\Delta\dot{\vec{x}} = \Delta\vec{x}$ .

However, LJAPUNOV's theorem can be made somewhat more rigorous. To do so, we shall assume that in the equilibrium state $\vec{b} = 0$. This can always be achieved by means of a transformation. We then have:

(5.6:15) $$\Delta\dot{\vec{x}}_i = A \cdot \Delta\vec{x} + \vec{R}$$

*Theorem 2* If all characteristic values of Matrix A have negative real parts, and if the functions $R_i$ satisfy the condition

(5.6:16) $$\left|R_i(t, \vec{x})\right| \le M \cdot \left\{ \sum_{j=1}^{n} x_j^2 \right\}^{\frac{1}{2}+\alpha} ,$$

then the trivial solution of 5.6:15 is stable.

*Note.* For oscillations, the formulation of Theorem 1 and the first condition of Theorem 2 are clear: The real part of the solution corresponds to the exponent of maximum amplitude $e^{\beta t}$. Amplitude decreases only if $\beta < 0$.

The stability problem therefore becomes a matter of drawing up conditions under which an algebraic equation only yields solutions with negative real parts.

In order for all solutions of the algebraic equation

$$(5.6{:}17)\qquad \sum_{j=0}^{n} c_j \cdot x^j = 0$$

to have a negative real part, then for real coefficients $c_j$ the well-known Routh-Hurwitz stability criterion must be satisfied (A. HURWITZ, 1895). For complex coefficients, one can consult H. BILHARZ (1944), which extended the Hurwitz theorem to include complex coefficients.

*Theorem 3* (BILHARZ) An nth-degree algebraic equation as per 5.6:17, with any number of complex coefficients, will exclusively have solutions with negative real parts if and only if all the principal subdeterminants of Matrix C are positive. C = $(\gamma_{jk})$ is given by:

$$\left.\begin{aligned} \gamma_{j,2k-1} &= (-i)^{j-k} \cdot c_{j-k} \\ \gamma_{j,2k} &= -i^{j-k+1} \cdot \overline{c}_{j-k} \end{aligned}\right\rangle \quad k = 1,\ldots,j;\ j = 1,\ldots,2n\ ,$$

$$\gamma_{j,m} = 0 \qquad 2j \le m \le 2n\,;\ j = 1,\ldots,2n\ .$$

For a quadratic equation we therefore have the following two principal subdeterminants, which must be positive:

$$(5.6{:}18)\qquad \begin{vmatrix} c_0 & -i \cdot \overline{c}_0 \\ -i \cdot c_1 & \overline{c}_1 \end{vmatrix} > 0\ ,$$

$$(5.6{:}19)\qquad \begin{vmatrix} c_0 & -i \cdot \overline{c}_0 & 0 & 0 \\ -i \cdot c_1 & \overline{c}_1 & c_0 & -i \cdot \overline{c}_0 \\ -c_2 & i \cdot \overline{c}_2 & -i \cdot c_1 & \overline{c}_1 \\ 0 & 0 & -c_2 & i \cdot \overline{c}_2 \end{vmatrix} > 0\ .$$

### 5.6.4.4 Ballistic stability

To describe the pitching motion of a projectile (precession and nutation) it is helpful to use the complex angle of attack $\zeta$ introduced at 5.5.1.5.

$$(5.6{:}20)\qquad \zeta = \delta \cdot (\cos\psi + i \cdot \sin\psi)$$

where $\delta$ is the angle between the tangent to the trajectory and the projectile axis and $\psi$ lies in the plane normal to the trajectory and describes the rotation of the projectile about the tangent to the trajectory.

In the rest position, the projectile axis and the tangent to the trajectory coincide ($\zeta = 0$).

If we make the assumptions described in 5.5.1.5, we can obtain the complex angle of attack (5.6:20) from the following equation:

(5.6:21) $$\zeta'' + (\alpha_1 - i \cdot \beta_1) \cdot \zeta' + (\alpha_0 - i \cdot \beta_0) \cdot \zeta = \alpha_3 - i \cdot \beta_3 \, . \qquad [-]$$

Coefficients $\alpha_0 - i \cdot \beta_0$ and $\alpha_1 - i \cdot \beta_1$ are assumed to be locally constant. $\alpha_3 - i \cdot \beta_3$ is a function of velocity and angle of attack. The independent variable is trajectory length, which is measured in calibres and is hence dimensionless.

Because the coefficients are constant, $\vec{R} = 0$ in Eqn 5.6:15, and Eqn 5.6:21 fulfils the conditions (5.6:16) of Theorem 2. The normal position of the complex angle of attack (5.5:20) $\zeta = 0$ and hence the rest position of the projectile axis are stable if all characteristic values of the corresponding matrix have negative real parts. According to Theorem 3, this is the case if all principal subdeterminants of the matrix defined therein are positive, i.e. if in the case of Eqn 5.6:21 determinants 5.6:18 and 5.6:19 are both positive. With

$$c_0 = \alpha_0 - i \cdot \beta_0 \quad \text{and} \quad c_1 = \alpha_1 - i \cdot \beta_1 \, ,$$

and after some calculation, we obtain the following inequalities:

(5.6:22) $$\alpha_0 \cdot \alpha_1 + \beta_0 \cdot \beta_1 > 0 \, ,$$

(5.6:23) $$\alpha_1 \cdot (\alpha_0 \cdot \alpha_1 + \beta_0 \cdot \beta_1) - \beta_0^2 > 0 \, ,$$

and from these we can conclude that

(5.6:24) $$\alpha_1 > 0 \, .$$

Coefficients $\alpha_I$ and $\beta_i$ are listed at 5.5.1.5 (Eqns 5.5:34a–d):

(5.6:25a) $$\alpha_0 = -C'_M \cdot \frac{\pi \cdot \rho}{8} \cdot \frac{d^5}{J_r} \, , \qquad [-]$$

(5.6:25b) $$\beta_0 = \beta_1 \cdot \frac{\pi \cdot \rho}{8} \cdot \frac{d^3}{m} \cdot \left( C'_L - C'_J \cdot \frac{d^2 \cdot m}{J_a} \right) , \qquad [-]$$

(5.6:25c) $$\alpha_1 = \frac{\pi \cdot \rho}{8} \cdot \frac{d^3}{m} \cdot \left( C'_L - C_{D0} + C_H \cdot \frac{d^2 \cdot m}{J_r} \right) , \qquad [-]$$

(5.6:25d) $$\beta_1 = \frac{J_a}{J_r} \cdot \frac{\omega \cdot d}{v} \, . \qquad [-]$$

If we use Eqns 5.6:25a and 5.6:25d to form the expression:

$$-\beta_1^2 / (4 \cdot \alpha_0) \, , \qquad [-]$$

we obtain the gyroscopic stability coefficient s (5.6:10):

(5.6:26) $$s = -\frac{\beta_1^2}{4 \cdot \alpha_0} = \frac{2}{\pi} \cdot \frac{J_a^2 \cdot \omega^2}{J_r \cdot C'_M \cdot \rho \cdot v^2 \cdot d^3} \, . \qquad [-]$$

MOLITZ generated a further stability coefficient in H. MOLITZ, R. STROBEL (1963):

$$(5.6{:}27)\qquad \hat{s} = \frac{\beta_0}{\alpha_1 \cdot \beta_1}\ . \qquad [-]$$

Substituting the expressions of Eqn 5.6:25b–d gives us the *Molitz stability factor*, also known as the *dynamic stability factor*:

$$(5.6{:}28)\qquad \hat{s} = \frac{C'_L - C'_J \cdot \dfrac{d^2 \cdot m}{J_a}}{C'_L - C_{D0} + C_H \cdot \dfrac{d^2 \cdot m}{J_r}}\ . \qquad [-]$$

Inequality 5.6:22 gives us a first stability criterion for a spin-stabilized projectile. If we divide this by $\alpha_1$ and extend the second addend by including $\beta_1$, then in conjunction with Eqn 5.6:27 we obtain the inequality

$$(5.6{:}29)\qquad \alpha_0 + \beta_1^2 \cdot \hat{s} > 0\ . \qquad [-]$$

Substituting Eqn 5.6:25a and d and performing the necessary calculations yields the inequality linking the gyroscopic and dynamic stability coefficients:

$$(5.6{:}30)\qquad s > \frac{1}{4 \cdot \hat{s}}\ .$$

From the inequality of Eqn 5.6:23, we can derive a second, more rigorous condition linking the two stability coefficients:

$$(5.6{:}31)\qquad s > \frac{1}{4 \cdot \hat{s} \cdot (1 - \hat{s})}\ . \qquad [-]$$

As mentioned above, the inequality for the gyroscopic stability coefficient (5.6:10) forms a necessary but insufficient condition for stable flight. The Molitz inequality (Eqn 5.6:31) provides a condition that is both necessary and sufficient.

### 5.6.4.5 The Molitz stability triangle

MOLITZ used a clear, easily-memorized graphical representation of the relationship between the two stability coefficients in his 1963 publication that we have mentioned on a number of occasions. He first inserted a variable transformation

$$(5.6{:}32)\qquad \sigma = \sqrt{1 - \frac{1}{s}}\ , \qquad [-]$$

resulting in the following form for inequality 5.6:31:

$$(5.6{:}33)\qquad \sigma > |2 \cdot \hat{s} - 1|\ . \qquad [-]$$

The graphical representation of this absolute value inequality consists of a triangle that has its apex at $(\hat{s}, \sigma) = (0.5, 0)$ and widens as the value of $\sigma$ increases. This is

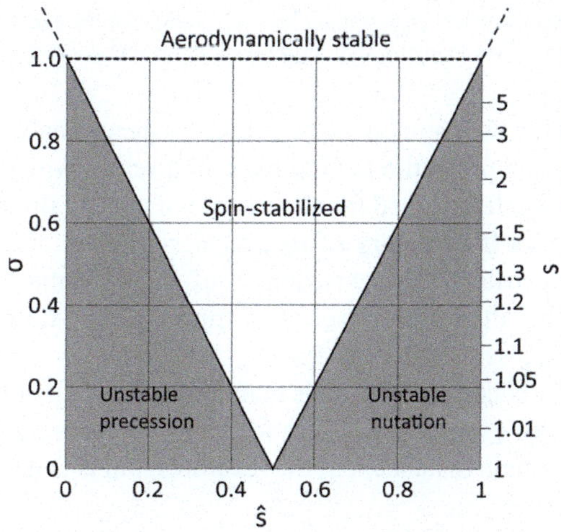


**Fig. 5.43.** Molitz stability triangle. If one knows the $(\hat{s},\sigma)$ values of a projectile, one can see whether it flies stable or not. Because the stability coefficient changes along the trajectory, the path of the "point of stability" $(\hat{s},\sigma)$ should be entirely enclosed by the white triangle.

known as the *Molitz stability triangle* (see Fig. 5.43). If a projectile is stable, all of its $(\hat{s},\sigma)$ values will lie within this triangle.

### 5.6.4.6 Stability problems. Unstable projectiles.

We must distinguish between three quite different cases when dealing with what are often referred to as "unstable" elongated projectiles.

In the first case, the projectile is not rotating about its longitudinal axis and therefore does not experience a stabilizing gyro effect. This occurs when an elongated projectile is fired from a smooth barrel. The barrel oscillation described in 4.3.1 and the gases that follow the projectile as it leaves the muzzle will induce a torque that will cause the projectile to start rotating about a transverse axis. In accordance with the principle of conservation of angular momentum (Fig. 5.44), the projectile will maintain this rotation throughout its trajectory. This leads to a phenomenon known as *keyholing*. At intervals along its trajectory, the projectile will create a round hole almost indistinguishable from that of a stable projectile (see Fig. 5.44, 6 m).

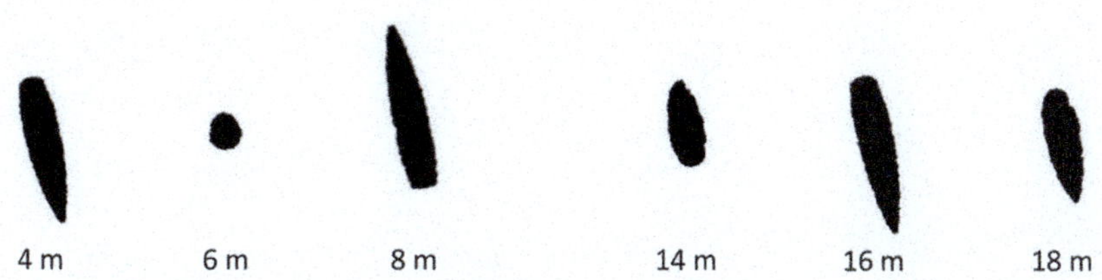


**Fig. 5.44.** Holes punched through paper targets by a rifle bullet without spin. The angular momentum generated by barrel oscillation and the propellant gas behind the projectile is maintained throughout the trajectory. The axis of rotation is the axis of greatest moment of inertia. See Fig. 7.21 (p. 383), which shows a stable but disturbed projectile. Distance is measured from the muzzle.

During one experiment consisting of 10 shots, the orientation of the plane of rotation (of the vector of angular momentum) only varied by ± 10.5°. This indicates that barrel oscillation was consistent between shots.

In the second case, the bullet receives too little spin in the barrel, and is hence unable to attain stable flight. This occurs when a projectile is fired from a barrel for which it was not designed, e.g. if an artillery shell designed for an 8° angle of twist is fired from a barrel with a 6° angle of twist, or if a military rifle bullet designed for a 7″ (178 mm) twist length is fired from a rifle with a twist length of 12″ (305 mm). Under such conditions, the projectile will exhibit gyroscopic motion, but with a large and often increasing angle of precession (see Fig. 5.45).

Projectiles fired from worn barrels behave in a similar manner, as do shells of which the driving bands slip because of excessive acceleration. In most cases, the shell does engage in the rifling for a certain distance, but the resulting spin is insufficient.

The angular velocity $\omega_c$ with which the projectile rotates about its own axis can be estimated from the projection of the vector of angular momentum onto the projectile axis and the corresponding moment of inertia:

(5.6:34) $$\omega_c = \frac{|\vec{L}|}{J_a} \cdot \cos\gamma \,, \qquad [1/s]$$

where $J_a$ = axial moment of inertia of the projectile and $\gamma$ = angle between angular momentum vector and projectile axis.

This means that as the projectile adopts a position increasingly perpendicular to the direction of travel, its speed of rotation about its own axis decreases. If the projectile axis is perpendicular to the axis of spin (i.e. approximately perpendicular to the direction of travel), then $\omega_c$ drops to zero, i.e. the projectile ceases to rotate about its longitudinal axis and instead rotates about a transverse axis. If $\alpha$ becomes greater than 90°, the projectile will begin to rotate in the opposite direction, and after rotating about its transverse axis by 180° will begin to rotate about its longitudinal axis at approximately the angular velocity that it would experience in normal flight, but in the opposite direction (see the "tippe top" in F. KUYPERS and Ch. UCKE, 1994).

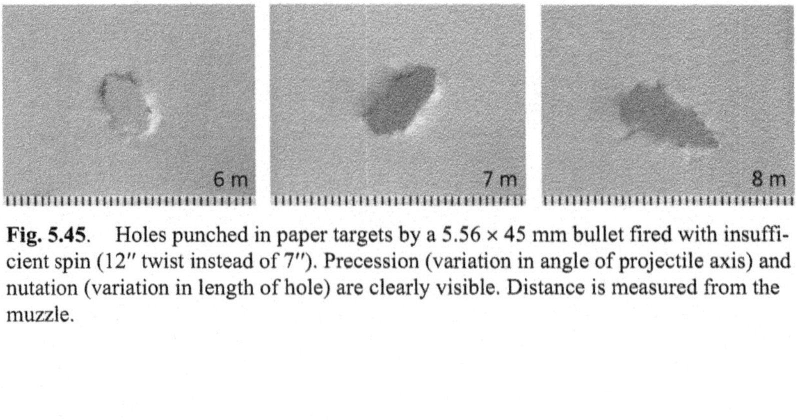


**Fig. 5.45.** Holes punched in paper targets by a 5.56 × 45 mm bullet fired with insufficient spin (12″ twist instead of 7″). Precession (variation in angle of projectile axis) and nutation (variation in length of hole) are clearly visible. Distance is measured from the muzzle.

This means that when the projectile tumbles, we can expect the direction of rotation in the body-fixed system to be reversed.

An artillery live-firing exercise provided empirical evidence of this. A shell acquired insufficient spin in the barrel (for whatever reason) and hit the ground after only a third of its expected range, owing to instability. Surprisingly, the fuze was missing; it had unscrewed itself and was found lying a few hundred metres from the rest of the shell. The thread on a fuze is so designed that normal spin will tighten it. The only way the fuze could have come unscrewed was for the shell to have changed its direction of spin in flight, which meant it must have rotated about a transverse axis at least once.

The third case is that of a projectile with normal spin and sufficient stability at the muzzle, which becomes unstable in flight. Because gyroscopic stability increases with distance, this can only occur under particular circumstances.

The regular increase in stability along the trajectory is immediately apparent from Eqn 5.6:10. Angular velocity $\omega$ decreases considerably more slowly than velocity. As a result, $\omega^2/v^2$ increases rapidly. On curved trajectories, this phenomenon is accentuated by the decrease in atmospheric density with altitude.

The only variable affecting the stability coefficient that can become critical is the moment coefficient derivative $C''_M$. At around $M = 1$, this coefficient function can increase (see Fig. 5.46), which will lead to a noticeable decrease in stability. If the projectile is a tracer round, and the tracer is just burning out at that point, the increase in drag could cause the projectile to tumble. Fig. 5.47 (next page) shows such an example.

That figure plots velocity against elapsed time of flight for a 7.5 × 55 tracer bullet fired at a sleeve target during air defence training. Approximately one third of the rounds fired were hitting the ground before they were expected to, for no apparent reason. Velocity measurements using a tracking radar system showed that the projectiles were decelerating very sharply after reaching the speed of sound (the visible kink on the normal time-velocity curve). This was causing those rounds to reach the apex of their trajectory AP′ prematurely, after approximately 8 s. The projectiles then fell to the ground with a mean constant velocity of approximately 40 m/s. This descent velocity would indicate that the projectile was unstable during its fall. For a normal shot, the range to the base of the trajectory was approximately 3.85 km, whereas the disturbed projectiles were only reaching approximately 1.3 km.

An otherwise stable projectile may also become unstable at the apex of a very steep trajectory (with an elevation of more than approximately 70° or 1250 mils). Because

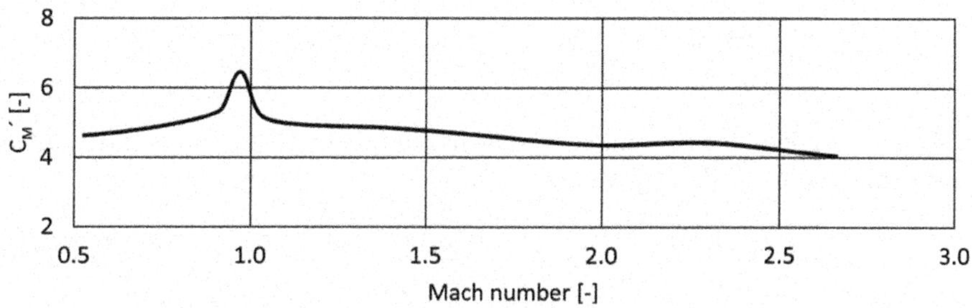


**Fig. 5.46.** Moment coefficient derivative vs Mach number for a 7.5 × 55 bullet, measured in transsonic and supersonic wind tunnels.

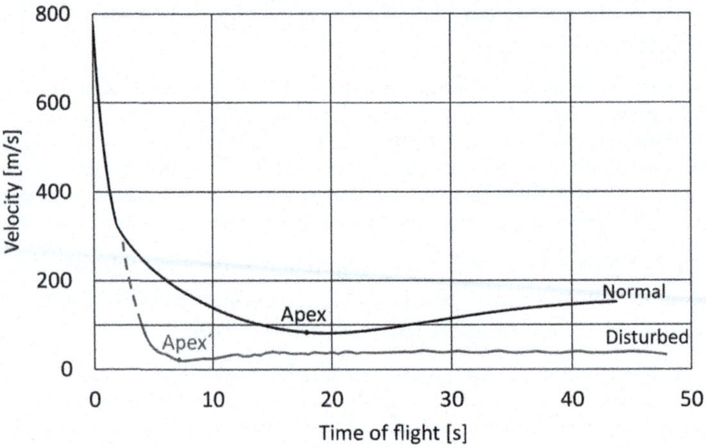


**Fig. 5.47.** Velocity-time graph of two tracer bullets, one flying stable, the other unstable. Elevation 1000 mils. See text for details.

the projectile axis follows the tangent to the trajectory, its angle of attack increases in the vicinity of the apex. The resulting lift is capable of causing tumbling, at least in small-calibre projectiles. As the projectile continues its flight, it will become stable again, although it may then fly tail-first (see 5.4.6.5 and Fig. 5.26).

### 5.6.5 Tractability

#### *5.6.5.1 Definition of tractability*

In stability theory, the ability of a bullet to keep its axis aligned with the tangent to the trajectory is known as *tractability*. The more tractable a bullet, the smaller the angle between the tangent to the trajectory and the axis of the bullet at all points along the trajectory. This is the case if the change in the angle of trajectory $\theta$ over time is at all times smaller than the change in the precession angle $\psi$. If we take the quotients of these two changes in angle we obtain a dimensionless number, known as *tractability* f. This must fulfil the following condition:

$$(5.6{:}35)\qquad f = \frac{|d\psi/dt|}{|d\theta/dt|} = \frac{|\dot{\psi}|}{|\dot{\theta}|} > 1 . \qquad [-]$$

For the expression in the numerator, we can use the quadratic equation of Eqn 5.6:8 for $\dot{\psi}$. The solution to that equation,

$$\dot{\psi} = \frac{J_a \cdot \omega_c}{2 \cdot J_r \cdot \cos\delta} \cdot \left(1 \pm \sqrt{1 + \frac{4 \cdot M_M \cdot J_r \cdot \cos\delta}{J_a^2 \cdot \omega_c^2 \cdot \sin\delta}}\right), \qquad [1/s]$$

can be linearized thus:

$$\dot{\psi} = -\frac{M_M}{J_a \cdot \omega_c \cdot \sin\delta}, \qquad [1/s]$$

where $M_M$ = overturning moment, $J_a$ = the axial moment of inertia of the projectile and $\delta$ = its angle of attack.

The expression in the denominator is taken from the system of differential equations for trajectory geometry, Eqn 5.4:10:

$$\dot{\theta} = -\frac{g}{v} \cdot \cos\theta \qquad [1/s]$$

Substituting into Eqn 5.6:35 we obtain:

$$(5.6:36) \qquad f = \left| \frac{M_M \cdot v}{J_a \cdot \omega \cdot g \cdot \cos\theta \cdot \sin\delta} \right| . \qquad [-]$$

Low angular velocities and small axial moments of inertia have a positive effect on tractability.

If we expand the right-hand side of Eqn 5.6:36 to include $J_a \cdot \omega$, we can insert the gyroscopic stability number:

$$(5.6:37) \qquad f = \frac{J_a \cdot v \cdot \omega}{4 \cdot J_q \cdot g \cdot \cos\theta} \cdot \frac{1}{s} > 1 . \qquad [-]$$

Tractability is hence inversely proportional to stability.

### *5.6.5.2 When are tractability problems likely to arise?*

If we estimate the order of magnitude of the tractability number near the muzzle using rough values for the variables of Eqn 5.6:37, we obtain values of between $10^4$ and $10^5$ for small-calibre projectiles, and between $10^2$ and $10^3$ for artillery shells. This will not change significantly at small values of trajectory angle $\theta$ ($\cos\theta \approx 1$), which means that a flat-trajectory projectile will generally be tractable. It is often claimed that overstabilization can occur (i.e. an increasing angle of attack on the descending branch of the trajectory), but that is impossible. This is apparent from the stability and tractability numbers for the 7.62 mm NATO bullet, shown in Table 5.10 (next page). In order to increase angular velocity (Eqn 5.6:36) to a point at which tractability problems could occur, one would have to impart more spin to the bullet than the barrel is capable of, because of the rifling groove forces that would be generated in the process.

From Eqn. 5.6:37 it is apparent that tractability problems can only occur if trajectory velocity is low relative to the surrounding air and the stability coefficient is high. Critical values occur only in the vicinity of the apex of very steep trajectories (with elevations of approximately 70° or more), at the point of lowest tractability (see 5.4.5.4 and Fig. 5.24).

**Table 5.10.** Trajectory data, showing stability and tractability numbers. Bullet: 7.62 mm NATO, $v_0$: 830 m/s, mass: 9.5 g

| x [m] | v [m/s] | ω [1/s] | s [-] | f [-] |
|---|---|---|---|---|
| 0 | 830 | 2729 | 1.37 | 34021 |
| 50 | 791 | 2711 | 1.49 | 29649 |
| 100 | 754 | 2692 | 1.62 | 25837 |
| 150 | 718 | 2673 | 1.76 | 22485 |
| 200 | 683 | 2654 | 1.92 | 19519 |
| 250 | 649 | 2634 | 2.09 | 16883 |
| 300 | 616 | 2614 | 2.29 | 14538 |

Examples include the bullets discussed in 5.6.4.6, which were fired steeply into the air and generally, after tumbling, fell to earth in stable flight, either tip or tail forward. There is also the example of a technical firing test with 155 mm artillery shells fired at high elevation. The apex of the trajectory lay in a jet stream. Under normal conditions, this steep trajectory would have been sufficiently tractable (see Table 5.11). The jet stream was flowing in approximately the same direction as that in which the projectile was moving, leading to a very low relative velocity at the peak, which in turn reduced the tractability number almost to 1 (see Table 5.12). The resulting instability at the apex meant that the shell had a lateral deviation of more than 2000 m when it reached muzzle height, rather than the usual 1370 m.

**Table 5.11.** Trajectory data, showing stability and tractability numbers, normal trajectory. 155 mm artillery shell, v: 684 m/s, mass: 43 kg, elevation: 1280 mil.

| t [s] | y [m] | v [m/s] | s [-] | f [-] |
|---|---|---|---|---|
| 10 | 4503 | 400 | 12.3 | 320.0 |
| 20 | 7193 | 267 | 32.1 | 62.2 |
| 30 | 8865 | 188 | 86.2 | 11.6 |
| 40 | 9664 | 136 | 176.3 | 3.1 |
| 44.5 | 9747 | 130 | 195.8 | 2.45 |
| 50 | 9619 | 138 | 167.8 | 3.2 |
| 60 | 8745 | 187 | 79.6 | 12.7 |

**Table 5.12.** Trajectory data, showing stability and tractability numbers. Measured wind profile. 155 mm artillery shell, $v_0$: 684 m/s, mass: 43 kg, elevation: 1280 mils.

| t [s] | y [m] | v [m/s] | s [-] | f [-] |
|---|---|---|---|---|
| 10 | 4463 | 394 | 12.2 | 309.0 |
| 20 | 7108 | 267 | 37.4 | 48.1 |
| 30 | 8756 | 189 | 107.5 | 7.8 |
| 40 | 9534 | 141 | 286.0 | 1.4 |
| 44.2 | 9608 | 136 | 333.3 | 1.06 |
| 50 | 9468 | 145 | 251.3 | 1.7 |
| 60 | 8574 | 194 | 90.8 | 9.8 |

## 5.7 Determining force and moment coefficients

### 5.7.1 Preliminary remarks

Thanks to the concept of dimensionless parameters, calculating trajectories requires us only to obtain the force and moment coefficients and their derivatives with respect to the angle of attack. Two basic approaches are available to us: we can calculate these values using the aerodynamic flow around the projectile or we can obtain them empirically. If we choose the empirical option, we can conduct measurements in free flight or in a wind tunnel.

What interests us most in practical terms is the drag coefficient. If we know that, we can use a point-mass model as per Eqn 5.4:36 or 5.4:40 to calculate the trajectory values with sufficient accuracy. If we require further information regarding stability and the angle of attack, we will also need the coefficient of friction and the slope of the overturning moment coefficient.

The next few sections will describe a number of methods of obtaining these key coefficients. To calculate other coefficients, one generally requires suitable computer software.

### 5.7.2 Experimental and empirical methods

#### *5.7.2.1 Drag coefficient*

***Time-distance measurements***. Before the invention of Doppler radar, this was one of the main methods of measuring a trajectory. Ballistic cameras were used to record the path of a tracer projectile from two locations. The motion of the projectile over time was recorded by synchronized, periodic opening and closing of the apertures. Radar was used to measure distance, lateral angle and vertical angle, from which it was possible to calculate the position of the projectile at any point in time. Today, it is also possible to make time-distance recordings using a built-in GPS device.

Once we have plotted the *distance-time graph* in fixed time steps $\Delta t$, we can use subtraction to calculate first speed and then acceleration in accordance with the following formulae:

$$[x_i(t_i), y_i(t_i)], \quad i = 1, \ldots, n \,, \qquad [\mathrm{m}]$$

$$(5.7{:}1) \qquad v_{i+1}\left(t_i + \tfrac{1}{2}\cdot\Delta t\right) = \frac{\sqrt{(x_{i+1} - x_i)^2 + (y_{i+1} - y_i)^2}}{\Delta t} \,, \qquad [\mathrm{m/s}]$$

$$(5.7{:}2) \qquad a_{i+1}\left(t_{i+1}\right) = \frac{v_{i+2} - v_{i+1}}{\Delta t} \,. \qquad [\mathrm{m/s^2}]$$

As there are always measuring errors, subtracting twice produces very large variations in the deceleration values.

To demonstrate this, a real measuring error with a standard deviation of only 10 cm was added to the calculated trajectory of a 7.62 mm bullet for a range of 1100 m. The first subtraction operation (see Eqn 5.7:1) produced a speed curve that followed the true curve very closely (see Fig. 5.48). However, the second subtraction, to calculate deceleration (see Eqn 5.7:2), produced a wildly varying curve, despite the small measuring error (see Fig. 5.49).

However, it is still possible to obtain usable results if one applies mathematical smoothing techniques (see C. Reinsch, 1967 and H. Rutishauser, 1976).

When the results of the first subtraction operation are smoothed, they differ from the calculated values by no more than 0.17%. The smoothed acceleration values differ from the calculated values by less than 5%. We can therefore use time-distance measurements to calculate the $C_D$ function of a projectile.

***Distance-velocity measurements.*** We can obtain much better results if we measure velocity directly. This eliminates the second subtraction operation, which suffers significantly from error propagation.

From the equation for drag (Eqn 5.3:29) we obtain the following expression for deceleration:

$$\frac{dv}{dt} = \frac{F_D}{m} = -C_D \cdot \frac{\rho}{2} \cdot v^2 \cdot \frac{1}{q}$$

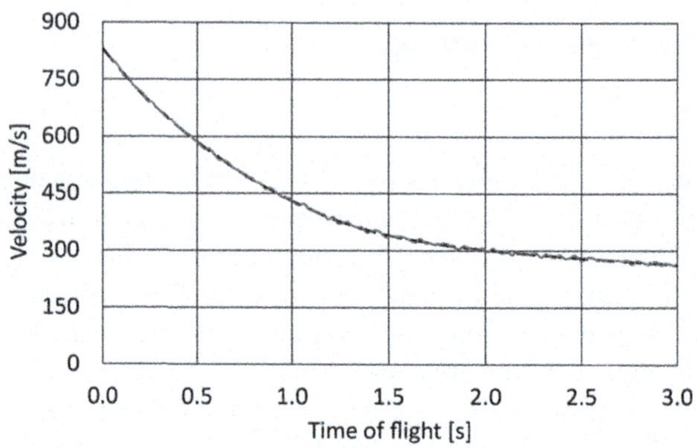


**Fig. 5.48.** Speed calculated by subtracting successive distances obtained from a time-distance recording. *Grey line*: original values. *Dashed line*: smoothed values. The difference between these and the calculated values is very small.

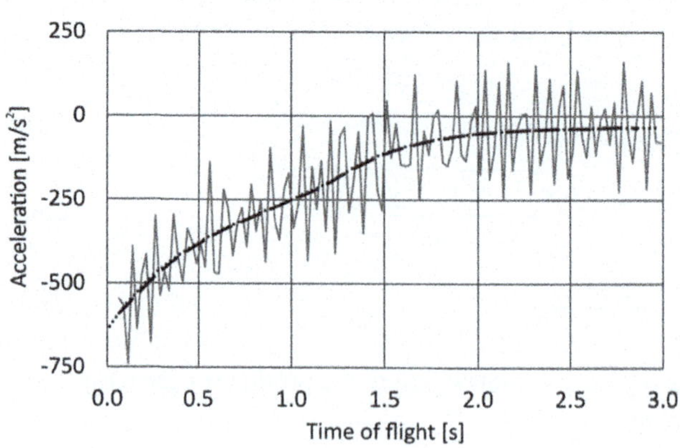


**Fig. 5.49.** Deceleration plotted by double subtraction using data points from a time-distance recording. *Grey line*: original values. *Dashed line*: values obtained by spline smoothing. *Dotted line*: calculated deceleration (coincides almost entirely with the dashed line).

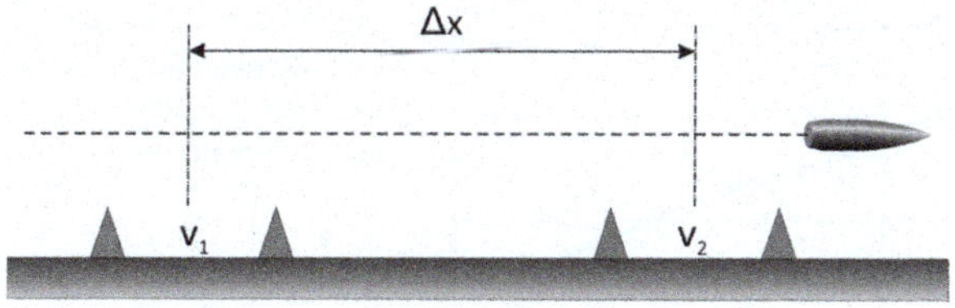


**Fig. 5.50.** Determining $C_D$ from distance-velocity measurements.

Using Eqn 5.3:20 on the left-hand side and then discretizing the equation yields:

$$(5.7:3) \qquad \frac{\Delta v}{\Delta x} = C_D \cdot \frac{\rho \cdot v_m}{2 \cdot q}, \qquad [1/s]$$

and one obtains the following formula for the drag coefficient:

$$(5.7:4) \qquad C_D = \frac{2 \cdot \Delta v \cdot q}{\Delta x \cdot \rho \cdot v_m}, \qquad [-]$$

where $\Delta v$ = deceleration over a distance $\Delta x$, $\rho$ = atmospheric density, $v_m$ = mean velocity over $\Delta x$ and q = sectional density of the projectile.

To determine the drag coefficient using Eqn 5.7:4, we only need two pairs of photoelectric beams. We can calculate $\Delta v$, $\Delta x$ and $v_m = \frac{1}{2} \cdot (v_1 + v_2)$ using the arrangement shown in Fig. 5.50. We already know the sectional density of the projectile and we can calculate atmospheric density from the current values of pressure and temperature using Eqn 5.2:2.

Eqn 5.7:4 gives us $C_D$ for velocity $v_m$. By firing from different points along the range or using different charges, we can obtain the $C_D$ curve for a larger range of velocities. The accuracy of the results, especially in the difficult transsonic range, is apparent from the $C_D$ curve for a 9 mm Luger bullet plotted using this method. (Fig. 5.51, taken from B. P. KNEUBUEHL, 1983[2]).

***Doppler radar measurements***. Doppler radar uses the difference in frequency between the signal transmitted to the projectile and the signal reflected by it to calculate its speed. This is known as the Doppler effect. That method is therefore ideal

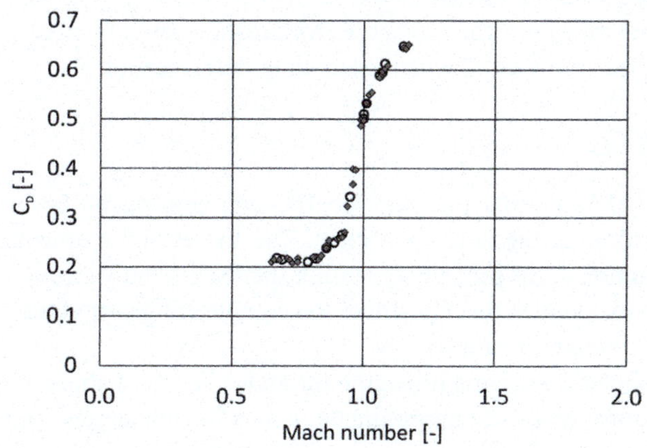


**Fig. 5.51.** $C_D$ values for a 9 mm Luger bullet, calculated from distance-velocity measurements. The circles are average values calculated from 10 individual measurements. The dots are individual measurements.

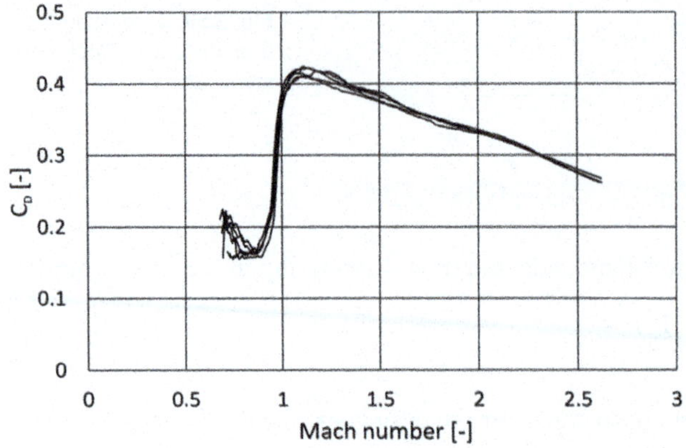


**Fig. 5.52.** $C_D$ curves for five identical 6 × 47 bullets, derived from radar measurements.

for measuring $C_D$, as a single subtraction operation is sufficient to calculate deceleration. Modern equipment is capable of tracking even small-calibre projectiles over several kilometres. Radar measurements have therefore become the usual means of determining drag coefficients. Fig. 5.52 shows $C_D$ curves for a 6 mm bullet obtained using radar measurements. Discrepancies between individual bullets are visible between Mach 1.0 and Mach 1.3, and below Mach 0.9, but these are of the order of a few percent.

***Wind tunnel measurements***. If continuous-flow supersonic and transsonic wind tunnels are available, we can plot $C_D$ curves for practically the entire trajectory of a projectile. The models must be fabricated to a specific scale, which will depend on the cross-section of the wind tunnel. In particular, it will be necessary to adapt surface roughness appropriately. For small-calibre projectiles, it is usually possible to use a model larger than the original, whereas models of larger-calibre projectiles will often have to be smaller than the original.

Increasing the size of the model has a positive effect on the measurements, as this results in flow parameters such as Reynolds and Nusselt numbers that correspond more closely to free flight than they do in the case of reduced-scale models.

Because force transducers are inserted into the model from the rear, using a "sting", flow will be disturbed around the tail of the projectile. As a result, it will be necessary to apply appropriate correction factors when calculating drag coefficients.

### 5.7.2.2 Further coefficients

***Free flight measurements***. Determining the other coefficients empirically is far more labour-intensive than finding the drag coefficient. The forces and moments responsible for the proper motion of the projectile manifest themselves only through the change over time of the two angles that constitute the complex angle of attack. We therefore need to obtain these two values.

Earliest results were obtained by firing a projectile through a series of sheets of paper. However, this gave only relatively approximate values for the angles, and the projectile was disturbed each time it passed through a sheet. Modern ranges use up to 25 measuring points, at which shadowgraph images of the projectile are created in plan and side view. Once we have a record of the two angles, we can use a non-linear corrector formula to determine the coefficients of Eqn 5.5:33, with which we can then calculate coefficients $C'_L$, $C'_M$, $C'_J$ and $C_H$ using Eqn 5.5:34.

For a comprehensive treatment of this method, please see R. L. McCoy (2009).

***Wind tunnel measurements***. Using a wind tunnel with continuous flow, the same model can be adjusted to provide various angles of attack, allowing lift and overturning moment to be measured, together with their dependency on angle, from which we obtain the coefficients $C'_L$ and $C'_M$.

There are also facilities in which the envelope of the model projectile can be rotated about the model's longitudinal axis. This makes it possible to determine the Magnus moment and the coefficient of friction, but high rotation speeds are required for such measurements.

## 5.7.3 Calculating the drag coefficient

### *5.7.3.1 The components of the drag coefficient*

If a projectile (or indeed any object) is subjected to airflow, changes in velocity and pressure will occur within that flow. Air particles are first accelerated as they meet the tip of the projectile and then decelerated. The layers closer to the projectile undergo a greater change in velocity than those further away.

This can be explained clearly using the laws of conservation. For frictionless flow, the laws of conservation of energy and momentum require that the velocity of a given layer of air be the same after passing over the body as before. The air particles close to the object have to cover a greater distance than those further away, and must therefore accelerate temporarily.

The work required for that acceleration is taken from the pressure energy. Velocity increases until the point of maximum cross-section. At the same time, pressure falls (in accordance with Bernoulli's Law). In frictionless flow, the kinetic energy generated is converted back into pressure energy as the flow converges again behind the body, and the velocity of the flow decreases accordingly.

The kinetic and pressure energy of the flow are conserved throughout this process – in the case of frictionless flow. In such a situation, the body would not lose any velocity or energy (d'Alembert's paradox).

However, friction does occur, between the air particles (or molecules), as a result of the different velocities within the layers of air flowing round the object, and some of the kinetic energy is converted into heat. The remaining energy is not sufficient to re-create the original pressure and velocity conditions behind the projectile. A vortex is created at the rear of the projectile, in which the pressure is lower than at the front. This difference in pressure generates a force that opposes the forward

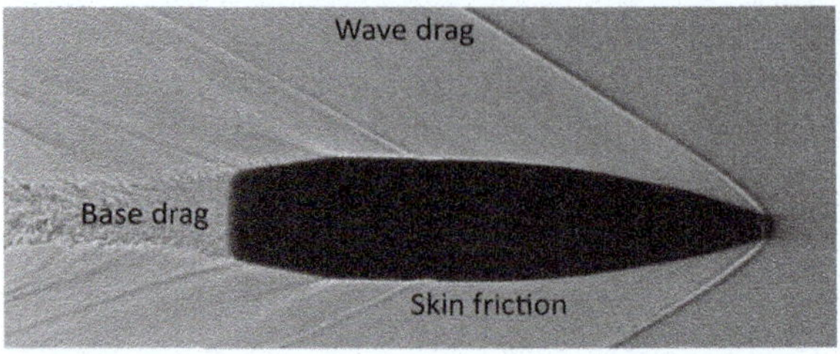


**Fig. 5.53.** Shadowgraph of a projectile in flight. The vortex at the base, the boundary layer at the surface and the shock waves are all clearly visible.

motion of the projectile. That force is generally known as *base drag*. Base drag accounts for the majority of total drag (85 to 90% at subsonic velocities and 50 to 60% at supersonic velocities).

Any gas (and any fluid) will adhere to the surface of the body over which it is flowing. Effective fluid velocity is attained in a very thin layer (the *boundary layer*). Differences in velocity within the boundary layer are very large. This results in friction and hence in the conversion of kinetic energy into heat, causing the projectile to slow down. The force responsible for this deceleration is known as *skin friction drag*. This accounts for between 5 and 15% of total drag, depending on the nature and area of the surface.

If the projectile is travelling faster than the speed of sound, a shock wave forms at the tip (see 2.3.3.4 and Fig. 5.53). Fluid velocity suddenly becomes subsonic at the shock front. At the same time, pressure and density both increase. This compression shock is what causes sonic boom. It possesses a certain amount of energy (in the form of sound and heat), which it takes from the kinetic energy of the projectile. However, extracting energy causes the projectile to decelerate. The force responsible for this phenomenon is known as *wave drag*. It only occurs at supersonic speeds, but at those speeds it accounts for some 40 to 50% of total drag.

### 5.7.3.2 Wave drag. Method of axial singularities.

To calculate wave drag, we use a cylindrical coordinate system (x, r) with its origin in the theoretical tip of the projectile and its x-axis identical to the axis of the projectile. The projectile is assumed to be slender. By this we mean that for every point along the meridian

$$(5.7{:}5) \qquad \frac{x}{r} < \sqrt{Ma^2 - 1} = \cot\alpha \,, \qquad [-]$$

where $\alpha$ = the Mach angle of the flow. Typical ogival projectiles generally fulfil this condition.

If we assume that the three-dimensional flow is stationary, isentropic and irrotational, then flow velocity can be represented by a potential function. For small disturbances to laminar flow, the potential is

$$(5.7{:}6) \qquad \Phi(x,r) = u_\infty \cdot x + \phi(x,r) + v_\infty \cdot r + \overline{\phi}(x,r) \,, \qquad [m^2/s]$$

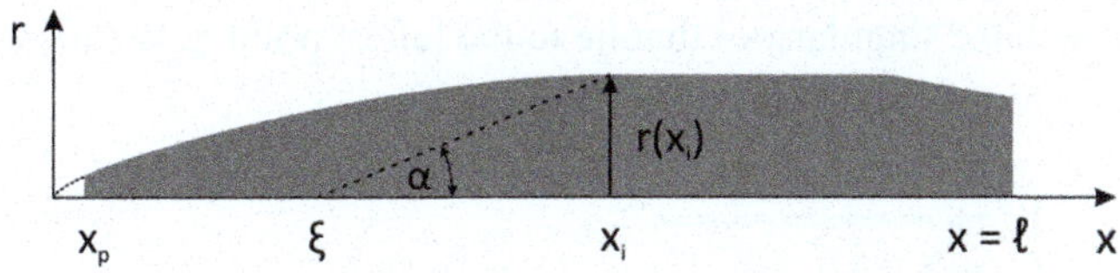


**Fig. 5.54.** Coordinate system for pressure calculations.

where $\phi$ is the interference potential of the axial flow and $\overline{\phi}$ is that of the angular flow.

In the first instance, we are primarily interested in the fundamental value of the drag coefficient for axial flow, with the axis of the projectile parallel with the axis of flow. This eliminates the two terms $v_\infty$ and $\overline{\phi}$ from Eqn 5.7:6. Under this condition, we can represent fluid velocity using a potential function. We are only dealing with small disturbances to parallel flow, as we are assuming a slender body, so the corresponding differential equation can be linearized for the potential function, the simplest option being to consider only the velocity interference potential. This gives us (in cylinder coordinates):

$$(5.7:7) \qquad \left(1-\mathrm{Ma}_\infty^2\right)\cdot\phi_{xx} + \phi_{rr} + \frac{1}{r}\cdot\phi_r \;=\; 0\,. \qquad [1/s]$$

The coordinate system is arranged such that its origin lies at the theoretical tip of the projectile and its x-axis is identical to the axis of the projectile (Fig. 5.54).

The velocity potential $\phi$ is calculated using the singularity method, bearing in mind that the streamlines should be tangential to the outline of the body. Singularities (sources and sinks) are so distributed over the axis of the projectile that precisely this flow is created over its surface.

Let the meridian of the projectile be described by $r = r(x)$. The boundary condition is then

$$(5.7:8) \qquad r' \;=\; \frac{\phi_r}{u_\infty + \phi_x}\,. \qquad [-]$$

Where $u_\infty$ = the velocity of undisturbed flow and $\phi_x$ and $\phi_r$ are the corresponding interference velocities (as the derivation of their potential function).

From potential theory we know that Eqn 5.7:7 can be solved by

$$(5.7:9) \qquad \phi(x,\,r) \;=\; \int_0^{\ell_p} \frac{-\,k(\xi)\cdot d\xi}{\sqrt{(x-\xi)^2 - a^2\cdot r^2}}\,, \qquad [m^2/s]$$

where $k(\xi)$ is the distribution of singularities along the x axis and

$$a \;=\; \sqrt{\mathrm{Ma}_\infty^2 - 1} \;=\; \cot\alpha\,, \qquad [-]$$

where $\alpha$ = the Mach angle. In the case of subsonic flow, all singularities will influence all points along the meridian, which is why we integrate from 0 to $\ell_p$. In the case of supersonic flow, however, the point with coordinate $x_i$ (see Fig. 5.54) will only be influenced by those singularities that lie to the left of point $\xi$. We must therefore sum only from 0 to $x - r \cdot \cot\alpha$, i.e. $x - a \cdot r$:

$$\text{(5.7:9a)} \qquad \phi(x, r) = \int_0^{x-a\cdot r} \frac{-k(\xi)\cdot d\xi}{\sqrt{(x-\xi)^2 - a^2 \cdot r^2}} \, . \qquad [\mathrm{m^2/s}]$$

With k(0) = 0, we obtain the following two equations by making the substitution

$$x - \xi = a \cdot r \cdot \cosh\lambda$$

$$\text{(5.7:10)} \qquad \phi_x = -\int_0^{x-a\cdot r} \frac{-k'(\xi)\cdot d\xi}{\sqrt{(x-\xi)^2 - a^2 \cdot r^2}} \, , \qquad [\mathrm{m/s}]$$

$$\text{(5.7:11)} \qquad \phi_r = \frac{1}{r} \cdot \int_0^{x-a\cdot r} \frac{-k'(\xi)\cdot (x-\xi)\cdot d\xi}{\sqrt{(x-\xi)^2 - a^2 \cdot r^2}} \, . \qquad [\mathrm{m/s}]$$

As the disturbances are assumed to be small, we can treat $\phi_x$ as being negligible by comparison with $u_\infty$ in Eqn 5.7:8. Substituting Eqn 5.7:11, we obtain

$$\text{(5.7:12)} \qquad r \cdot r' = r \cdot \frac{\phi_r}{u_\infty} = \frac{1}{u_\infty} \cdot \int_0^{x-a\cdot r} \frac{-k'(\xi)\cdot (x-\xi)\cdot d\xi}{\sqrt{(x-\xi)^2 - a^2 \cdot r^2}} \, , \qquad [\mathrm{m}]$$

where $\xi = x - a \cdot r$. For a given meridian r(x), this is a Volterra integral equation for the derivative of the singularity distribution k'($\xi$). In order to solve it numerically, that equation is written down as a differential equation, with k'($\xi$) constant over the interval under consideration:

$$\text{(5.7:13)} \qquad k'(\xi) = A_j \cdot u_\infty \, , \qquad \text{for} \qquad \xi_{j-1} < \xi < \xi_j \, . \qquad [\mathrm{m/s}]$$

We can therefore write

(5.7:14)

$$r_k \cdot \frac{r_k - r_{k-1}}{x_k - x_{k-1}} = A_k \int_{\xi_{k-1}}^{x_k - a\cdot r_k} \frac{(x_k - \xi)\cdot d\xi}{\sqrt{(x_k-\xi)^2 - a^2 \cdot r_k^2}} + \sum_{j=1}^{k-1} A_j \cdot \int_{\xi_{j-1}}^{\xi_j} \frac{(x_k - \xi)\cdot d\xi}{\sqrt{(x_k-\xi)^2 - a^2 \cdot r_k^2}} \, . \qquad [\mathrm{m}]$$

The two integrals in Eqn 5.7:1 are solvable by elementary methods. They give the following results:

$$\int_{\xi_{k-1}}^{x_k - a\cdot r_k} \frac{(x_k - \xi)\cdot d\xi}{\sqrt{(x_k-\xi)^2 - a^2 \cdot r_k^2}} = a \cdot r_k \cdot \sqrt{\left(\frac{x_k - \xi_{k-1}}{a \cdot r_k}\right)^2 - 1} \, , \qquad [\mathrm{m}]$$

$$\int_{\xi_{j-1}}^{\xi_j} \frac{(x_k - \xi)\cdot d\xi}{\sqrt{(x_k-\xi)^2 - a^2 \cdot r_k^2}} =$$

$$= a \cdot r_k \cdot \left[ \sqrt{\left( \frac{x_k - \xi_{j-1}}{a \cdot r_k} \right)^2 - 1} - \sqrt{\left( \frac{x_k - \xi_j}{a \cdot r_k} \right)^2 - 1} \right] . \qquad [m]$$

Substituting into Eqn 5.7:14 and solving for $A_k$, we obtain a recursion formula for these coefficients:

(5.7:15)

$$A_k = \frac{\left\{ \frac{r_k - r_{k-1}}{a \cdot (x_k - x_{k-1})} - \sum_{j=1}^{k-1} A_j \cdot \left[ \sqrt{\left( \frac{x_k - \xi_{j-1}}{a \cdot r_k} \right)^2 - 1} - \sqrt{\left( \frac{x_k - \xi_j}{a \cdot r_k} \right)^2 - 1} \right] \right\}}{\sqrt{\left( \frac{x_k - \xi_{k-1}}{a \cdot r_k} \right)^2 - 1}} , \qquad [-]$$

which is initiated using the following value

(5.7:16) $$A_1 = \frac{r_1}{a \cdot x_1} \cdot \frac{1}{\sqrt{\left( \frac{x_1}{a \cdot r_1} \right)^2 - 1}} . \qquad [-]$$

We now need the relationship between the potential and the pressure coefficient. From the one-dimensional Eulerian equation

(5.7:17) $$\rho \cdot \left( \frac{\partial u}{\partial t} + u \cdot \frac{du}{dx} \right) + \frac{dp}{dx} = 0 , \qquad [N/m^3]$$

in view of the assumption that the flow is stationary, i.e. $\partial u / \partial t = 0$, we can write

$$dp = -\rho \cdot u \cdot du , \qquad [N/m^2]$$

and after integration we obtain

$$p - p_0 = -\rho \cdot \frac{u^2 - u_0^2}{2} = -\rho \cdot \frac{(u + u_0) \cdot (u - u_0)}{2} . \qquad [N/m^2]$$

Using the approximation $u + u_0 \approx 2 \cdot u_0$,

$$p - p_0 = -\rho \cdot u_0 \cdot (u - u_0) = -\rho \cdot u_0 \cdot \Delta u , \qquad [N/m^2]$$

and rearranging gives us

(5.7:18) $$C_{PR} = \frac{p - p_0}{\frac{1}{2} \cdot \rho \cdot v_0^2} = -\frac{2}{u_\infty} \cdot \phi_x . \qquad [-]$$

Substituting Eqn 5.7:10, we obtain

(5.7:19) $$C_{PR}(x_k) = \frac{2}{u_\infty} \cdot \int_{\xi=0}^{x_k - a \cdot r} \frac{k'(\xi) \cdot d\xi}{\sqrt{(x_k - \xi)^2 - a^2 \cdot r_k^2}} , \qquad [-]$$

and with the assumption of Eqn 5.7:13:

(5.7:20)

$$C_{PR}(x_k) = 2 \cdot A_k \cdot \int_{\xi_{k-1}}^{x_k - a \cdot r_k} \frac{d\xi}{\sqrt{(x_k - \xi)^2 - a^2 \cdot r_k^2}} + 2 \cdot \sum_{j=1}^{k-1} A_j \cdot \int_{\xi_{j-1}}^{\xi_j} \frac{d\xi}{\sqrt{(x_k - \xi)^2 - a^2 \cdot r_k^2}} \ . \quad [-]$$

These integrals can also be solved by elementary methods:

$$\int_{\xi_{k-1}}^{x_k - a \cdot r_k} \frac{d\xi}{\sqrt{(x_k - \xi)^2 - a^2 \cdot r_k^2}} = \ln \frac{x_k - \xi_{k-1} + \sqrt{(x_k - \xi_{k-1})^2 - a^2 \cdot r_k^2}}{a \cdot r_k} \ , \quad [-]$$

$$\int_{\xi_{j-1}}^{\xi_j} \frac{d\xi}{\sqrt{(x_k - \xi)^2 - a^2 \cdot r_k^2}} = \ln \frac{x_k - \xi_{j-1} + \sqrt{(x_k - \xi_{j-1})^2 - a^2 \cdot r_k^2}}{x_k - \xi_j + \sqrt{(x_k - \xi_j)^2 - a^2 \cdot r_k^2}} \ , \quad [-]$$

and substituted into Eqn 5.7:20:

(5.7:21)

$$C_{PR}(x_k) = 2 \cdot A_k \cdot \ln \frac{x_k - \xi_{k-1} + \sqrt{(x_k - \xi_{k-1})^2 - a^2 \cdot r_k^2}}{a \cdot r_k} + \quad [-]$$

$$+ \ 2 \cdot \sum_{j=1}^{k-1} A_j \cdot \ln \frac{x_k - \xi_{j-1} + \sqrt{(x_k - \xi_{j-1})^2 - a^2 \cdot r_k^2}}{x_k - \xi_j + \sqrt{(x_k - \xi_j)^2 - a^2 \cdot r_k^2}} \ . \quad [-]$$

Taken in conjunction with Eqn 5.7:15, this gives us the pressure distribution along the body of the projectile. To calculate the wave drag coefficient, we require one more integral:

(5.7:22)

$$C_{D,W} = \frac{4}{d^2} \cdot \int_{x_p}^{\ell} C_{PR} \cdot 2 \cdot r \cdot r' \cdot dx \ . \quad [-]$$

This can be achieved most simply, and sufficiently accurately, using the trapezoidal rule.

For practical calculations, one should take account of the following points:

- Calculation of the velocity and pressure distribution starts from a fictitious projectile tip at ($x = 0$, $r = 0$), which is also the origin of the coordinate system. However, integration of drag (Eqn 5.7:22) only starts at the true projectile tip $x_p$ (see Fig. 5.54).
- The effects of driving bands and similar projections can vary considerably. If the cylindrical part of the projectile extends to a point at least 1 calibre in front of the driving band, and the nose merges with the cylindrical part at an angle of at least 2° (which is generally the case for such projectiles) we can ignore driving bands when calculating pressure drag (see 5.8.2.1). They only contribute to total drag via the suction effect on the rearward-facing sides.

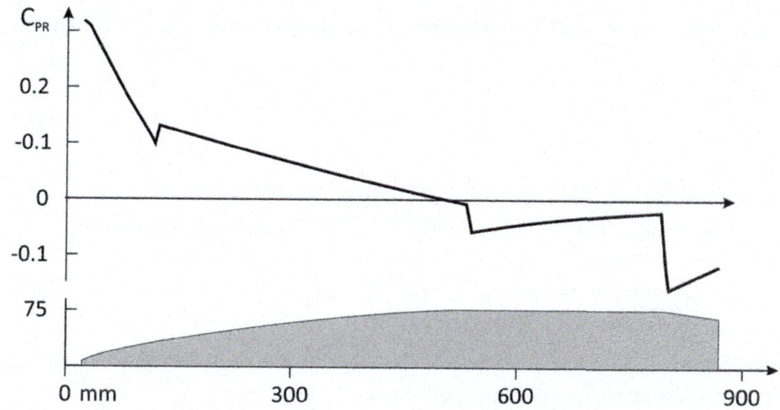


**Fig. 5.55.** Relative pressure distribution along an artillery shell. The sudden changes in pressure occur at points of discontinuity in the gradient of the surface of the projectile.

Fig. 5.55 presents an example of calculated pressure distribution along an artillery shell. The sudden changes in pressure occur at the points of discontinuity in the gradient of the projectile surface, i.e. at the transitions between fuze and body, nose and cylindrical section, and cylindrical section and boattail.

### *5.7.3.3 Wave drag. Newtonian theory.*

In classical Newtonian theory, a flowing medium consists of particles travelling in a straight line. On striking a body, they retain only the tangential component of their momentum and transfer the normal component to the body. In this model, the side of the body away from the flow experiences no force.

Fluid velocity normal to the body is given by:

$$u_{\infty,n} = u_\infty \cdot \sin\tau \,, \qquad (5.7{:}23) \qquad [\mathrm{m/s}]$$

where $\tau$ is the angle of inclination of the body tangent at the point under consideration (see Fig. 5.56). From Eqn 5.6:11, we obtain the pressure on the surface:

$$p - p_0 = \rho \cdot (u_\infty \cdot \sin\tau)^2 \,, \qquad [\mathrm{bar}]$$

which in turn gives the following expression for the pressure coefficient:

$$C_{PR} = \frac{p - p_0}{\frac{1}{2} \cdot \rho \cdot u_\infty^2} = 2 \cdot \sin^2\tau \,. \qquad (5.7{:}24) \qquad [-]$$

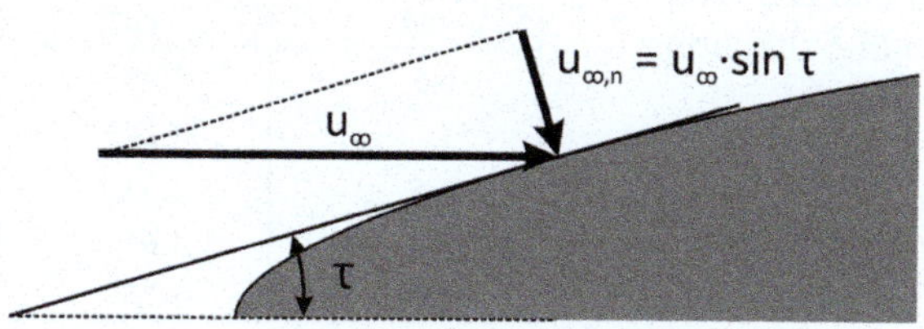


**Fig. 5.56.** Derivation of NEWTON's theory.

Here again, the body is assumed to be slender. This allows us to make the following approximation:

$$\sin^2\tau = \frac{r'^2}{1+r'^2} \approx r'^2 \qquad (5.7{:}25) \quad [-]$$

If we substitute Eqn 5.7:25 into Eqn 5.7:24 and then substitute that into Eqn 5.7:22, we obtain the following expression for body drag (h = length of the nose):

$$F_{DS} = \frac{F_D}{\frac{1}{2}\cdot\rho_0\cdot u_\infty^2} = 4\cdot\pi\cdot\int_0^h r\cdot r'^3\cdot dx\ . \qquad (5.7{:}26) \quad [m^2]$$

Given the concepts of Newtonian flow, one can imagine that the validity of the resulting equations increases with fluid velocity.

### 5.7.3.4 The drag coefficient of the nose flat

For practical reasons, almost all projectiles have a flat at the tip. So far, we have ignored the effect of this on total drag.

According to HAACK, the drag due to this flat – which can be responsible for a significant proportion of total drag, especially for a small-calibre projectile – can be estimated as follows.

PRANDTL gives the coefficients shown in Table 5.13 for the stagnation pressure at point S. These can be approximated by means of a polynomial

$$C_{PR,NF}(Ma) = 0.7340+0.6931\cdot Ma-0.1179\cdot Ma^2\ . \qquad (5.7{:}27) \quad [-]$$

Pressure is assumed to be distributed parabolically across the flat, equal to the stagnation pressure in the centre and falling to zero at the edge.

$$C_{D,NF}(Ma) = \frac{1}{2}\cdot C_{PR,NF}(Ma)\cdot\left(\frac{d_p}{d}\right)^2, \qquad (5.7{:}28) \quad [-]$$

where d = nominal calibre and $d_p$ = diameter of the flat.

### 5.7.3.5 Coefficient of friction

To calculate the coefficient of friction, we shall use the well-known theory of laminar friction. The coefficient for the total surface, with a turbulent boundary layer that is initially laminar, is given by the Prandtl-Schlichting drag law:

$$C_F = \frac{0.455}{(\log Re)^{2.58}} - \frac{B}{Re}\ . \qquad (5.7{:}29) \quad [-]$$

**Table 5.13.** PRANDTL's coefficient for stagnation pressure

| Ma | $c_{ps}$ |
|---|---|
| 0.5 | 1.065 |
| 1 | 1.275 |
| 1.5 | 1.53 |
| 2 | 1.65 |
| 3 | 1.75 |

Where Re is the Reynolds number for undisturbed flow as a function of projectile length.

The constant B depends on the position of the boundary layer separation point. This is obtained by replacing the turbulent drag between tip and separation point by the laminar drag, calculated according to Blasius' Law:

$$B = 0.074\cdot\left(Re_{crit}\right)^{0.8} - 1.328\cdot\left(Re_{crit}\right)^{0.5}\ . \qquad (5.7:30)\quad [-]$$

To facilitate calculation of the turbulent component of B, one uses the 1/7 power law of velocity distribution (H. SCHLICHTING, 1965).

However, the friction drag calculated using Eqn 5.7:29 is valid only for incompressible flow. At flow velocities of over Mach 1 we must take compressibility into account. This is achieved by calculating the following quotient for turbulent boundary layers:

$$\beta = \frac{C_{F,k}}{C_{F,ink}}\ . \qquad (5.7:31)\quad [-]$$

The numerator uses VAN DRIEST's equation for the friction coefficient of a turbulent boundary layer in a compressible fluid. That equation (rearranged to facilitate numerical treatment) is

$$C_{F,k} = \frac{0.242^2\cdot\left(1-\lambda^2\right)\cdot\left(\arcsin\lambda\right)^2}{\lambda^2\cdot\left[\log\left\{Re\cdot C_{F,k}\cdot\left(1-\lambda^2\right)^{1.1}\right\}\right]^2}\ , \qquad (5.7:32)\quad [-]$$

with

$$1-\lambda^2 = \left(1+0.2\cdot Ma^2\right)^{-1}\ . \qquad [-]$$

In this form, we can solve the equation iteratively for $C_{F,k}$.

The first summand from Eqn 5.7:29 is used for the coefficient of friction for incompressible flow in the denominator of Eqn 5.7:31. Fig. 5.57 shows how β is related to Ma and Re.

As the cross-sectional area of the nominal calibre is almost always used as the cross-sectional area for ballistics calculations, we need to convert the coefficient of friction accordingly:

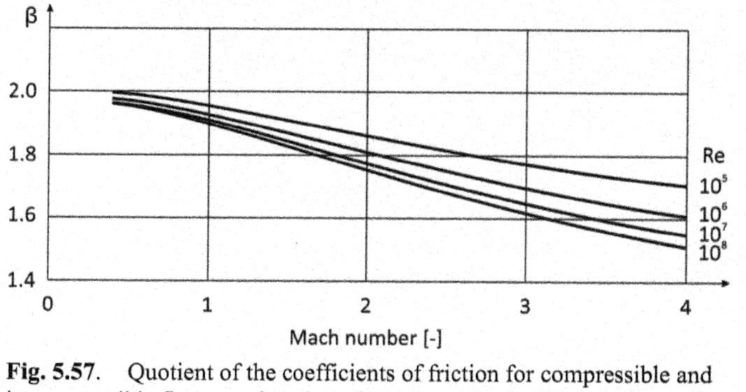


**Fig. 5.57.** Quotient of the coefficients of friction for compressible and incompressible flow as a function of Mach number and Reynolds number.

$$C_{D,F} = \beta \cdot C_F \cdot \frac{4 \cdot A_S}{d^2 \cdot \pi}, \qquad \text{[-]} \tag{5.7:33}$$

taking $C_F$ from Eqn 5.7:29. $A_S$ is the total surface subjected to flow.

For practical calculations, we also need to determine the critical Reynolds number at the separation point (at a distance $\ell$). From shadowgraphs, we can see that in the case of small-calibre projectiles the boundary layer remains laminar until the start of the boattail (see Fig. 5.58). In some cases, however, turbulence already starts at the crimp groove. Schlieren/shadowgraph images also reveal that in the case of large-calibre projectiles, we can assume that the boundary layer will become turbulent 5–15 cm behind the tip (see Fig. 5.59). Roughness and grooves have a marked influence on the point at which turbulence begins.

The critical Reynolds number $Re_{crit}$ required for Eqn 5.7:30

$$Re_{crit} = \frac{\rho \cdot u \cdot \ell}{\eta}, \qquad \text{[-]} \tag{5.7:34}$$

is obtained from the following local elements:

$$u = u_\infty \cdot \sqrt{1 - c_p}, \qquad \text{[m/s]} \tag{5.7:35}$$

$$\rho = \rho_0 \cdot \left(\frac{p}{p_0}\right)^{\frac{1}{\gamma}}, \qquad \text{[kg/m}^3\text{]} \tag{5.7:36}$$

$$\eta = \eta_0 \cdot \left(\frac{T}{T_0}\right)^{0.6}, \quad \text{with} \quad T = \frac{p}{R \cdot \rho} \qquad \text{[N·s/m]} \tag{5.7:37}$$

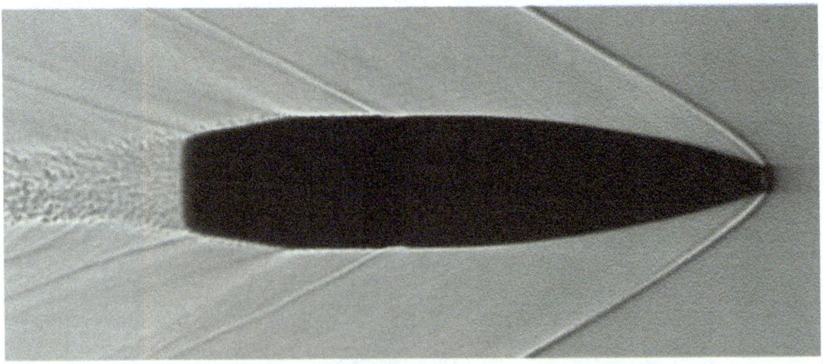

**Fig. 5.58.** Shadowgraph of a 7.5 × 55 rifle bullet. The boundary layer remains laminar up to the beginning of the boattail. A slight disturbance occurs at the crimp groove.

**Fig. 5.59.** Shadowgraph of the nose of a 155 mm artillery shell. The changes in the boundary layer and the start of the transition from laminar to turbulent flow are visible.

### *5.7.3.6 Negative pressure effects at the base*

Calculating flow at the base of a projectile is rather difficult. However, one can use experimental data. For a comprehensive collection and summary of empirical results, please see W. R. SCHMIDT and E. HERPFER (1973). That work contains a large quantity of measurement data grouped by boattail angle and ratio of base diameter $d_B$ to the diameter of the cylindrical part of the projectile. The graphics go up to Mach 6 and cover boattail angles from 1° to 12°. This publication divides base drag into two components: tail drag (base and boattail), which includes the negative pressure at the boattail, and base drag (only base), which for the authors' purposes encompasses only the drag at the base of the projectile. New projects can readily be inserted into these graphics and interpolated.

When calculating wave drag, one should use the graphs of base drag coefficient if the negative pressure drag of the boattail has already been taken into account, as will be the case if calculations are performed using the singularities (see Fig. 5.55). If that is not the case, e.g. if the Newtonian method is used to calculate drag, one should use the tail drag coefficient.

As the graphs were plotted from the aerodynamic point of view, where the reference dimension is the maximum diameter of the projectile, it is necessary for ballistics purposes to convert the coefficients to take account of nominal calibre.

The effect of a driving band was ignored, but a low pressure effect will also occur at its rear face. We assume that the base pressure that acts on a cylindrical base acts on the circular ring area that projects beyond the body of the projectile.

The total drag coefficient for negative pressure effects at the base is given by:

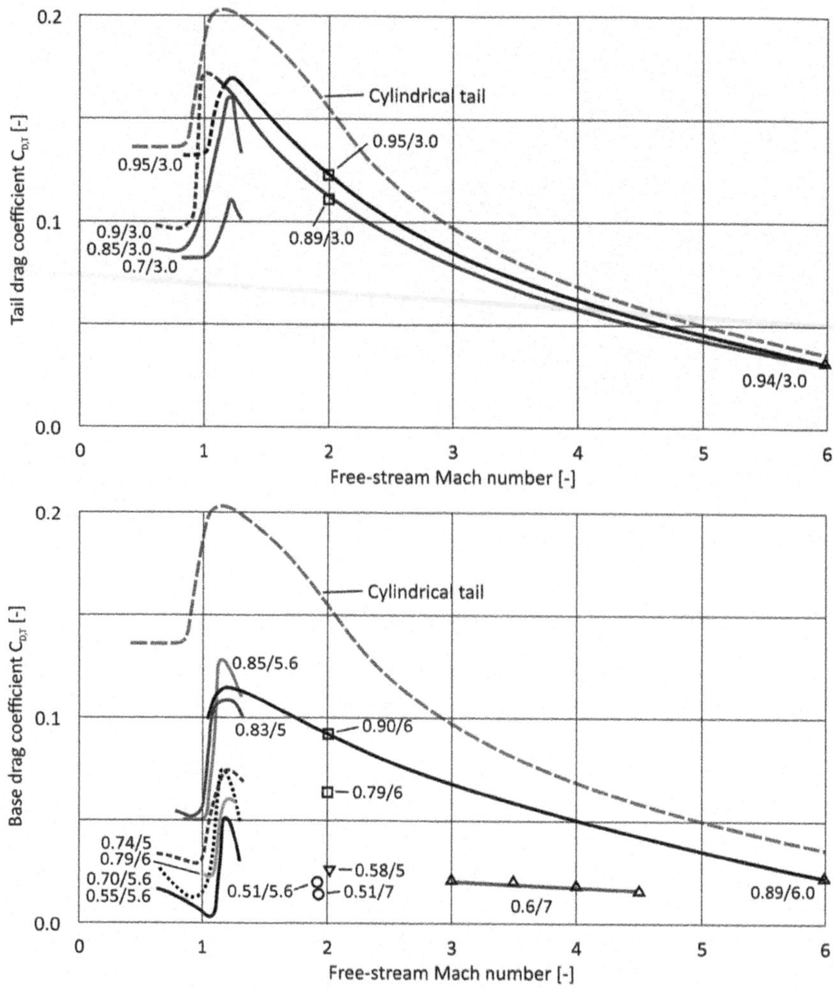


**Fig. 5.60.** Two examples from the collection of graphs of tail drag and base drag coefficients (from W. R. SCHMIDT, E. HERPFER, 1973, partly redrawn). **a** *Top:* Tail drag with a 3° boattail angle. **b**. *Bottom:* Base drag with a 4° to 7° boattail angle. The first number is the base diameter measured in projectile diameters, the second is the boattail angle.

(5.7:38) $$C_{D,B} = C^*_{D,B} \cdot \frac{d^2_{max}}{d^2} + C_{D,B,cyl} \cdot \frac{d^2_F - d^2_{max}}{2 \cdot d^2} \, . \qquad [-]$$

where $C^*_{D,B}$ = base drag coefficient according to SCHMIDT, HERPFER, $d_{max}$ = maximum projectile diameter, $d_F$ = diameter of the driving band and d = reference calibre.

The factor of 2 in the denominator of the second addend takes account of the fact that the rear face area of the driving band is approximately halved by the lands of the barrel cutting into it.

In interpolating within these graphs, the parameter $d_B/d_{max}$ was taken into account first, the boattail angle second. This makes sense in view of the measurement results represented in these graphs, as clearly the ratio of the two diameters has a greater influence on base drag than does the boattail angle.

### *5.7.3.7 Examples*

The drag coefficients of various projectiles were determined using the methods described in 5.7.3.2 and 5.7.3.4–6 and compared with the results of firing tests and wind tunnel measurements. Small-, medium- and large-calibre projectiles of various shapes were used.

Although no account was taken of the angle of attack, the point-mass model calculations produced usable $C_D$ curves that were more suitable than those generated using standard projectiles and ballistic coefficients (BC). Fig. 5.61 shows the results for two rifle bullets and an artillery shell, along with the corresponding measurements.

The effect of ignoring the angle of attack is particularly noticeable at high Mach numbers (in the part of the trajectory closest to the muzzle). The measured values for that region are higher than the calculated values (slightly in some cases, significantly in others) because of the initial pitch.

It is surprising that in the case of the artillery shell the calculated values are higher than those recorded in free-flight tests, even though the calculations ignored angle of attack. This could be due to the calculated drag values for the fuze plate or the driving band being too high.

For a newly-developed artillery shell, for which the $C_D$ curve was calculated using the method described, the range error at first firing was approximately 300 m, at a range of approximately 20 km. Given the simplicity of the model used to calculate the $C_D$ curve (the graphs of Fig. 5.60, plus a simple BASIC program), the results were surprisingly good.

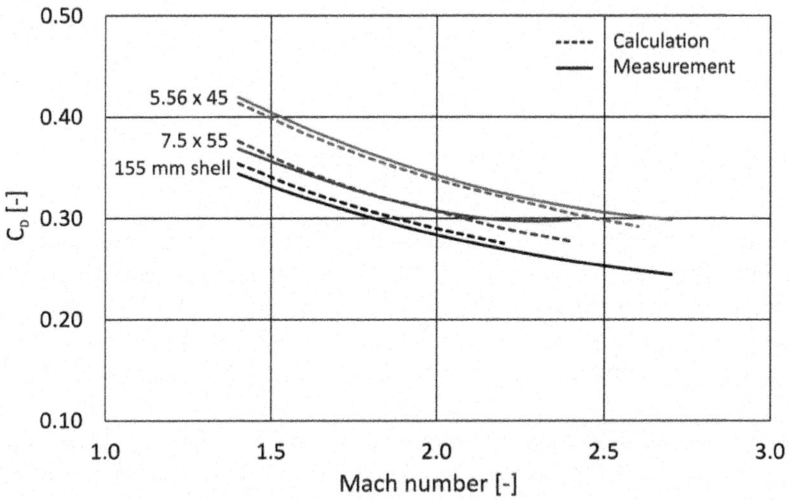


**Fig. 5.61.** Comparison between measured drag coefficients and coefficients calculated using the method described. The increase in drag coefficient resulting from initial pitch is clearly visible in the case of the small-calibre projectiles.

#### 5.7.3.8 Subsonic flow

We can use the potential equation (Eqn 5.7.9) to calculate the pressure distribution in subsonic flow. This time, however, we shall integrate over the entire length of the projectile, as all singularities will influence the flow around it.

$$\text{(5.7:39)} \qquad \phi(x, r) = \int_0^{\ell_p} \frac{-k(\xi)\cdot d\xi}{\sqrt{(x-\xi)^2 + b^2\cdot r^2}}, \qquad [m^2/s]$$

where $b = \sqrt{1 - Ma_\infty^2}$ . [-]

Furthermore, the boundary condition of Eqn 5.7:8, whereby the flow must follow the contours of the body, is also applicable here

$$\text{(5.7:40)} \qquad r' = \frac{\phi_r}{u_\infty}. \qquad [-]$$

That expression already incorporates the approximation $u_\infty + \phi_x \approx u_\infty$ for slender projectiles. Including r in the equation and substituting and evaluating the derivative of Eqn 5.7:39 with respect to r yields the relationship between the singularity distribution and the meridian curve

$$\text{(5.7:41)} \qquad k(x) = \frac{u_\infty}{2}\cdot r\cdot r' = u_\infty \cdot \frac{q'(x)}{4\cdot\pi}, \qquad [m^2/s]$$

where q(x) is the cross-sectional area of the rotating body at point x. Substituting into the potential equation and taking the derivative gives us the formulae for the two velocity components at the surface

$$\text{(5.7:42)} \qquad \phi_x = \frac{u_\infty}{4\cdot\pi}\cdot \int_0^{\ell_p} \frac{q'(\xi)\cdot(x-\xi)\cdot d\xi}{\sqrt{\{(x-\xi)^2 + b^2\cdot r^2\}^3}}, \qquad [m/s]$$

$$\text{(5.7:43)} \qquad \phi_r = \frac{u_\infty\cdot b^2\cdot r}{4\cdot\pi}\cdot \int_0^{\ell_p} \frac{q'(\xi)\cdot d\xi}{\sqrt{\{(x-\xi)^2 + b^2\cdot r^2\}^3}}. \qquad [m/s]$$

The two equations can be solved by discretization, in a manner analogous to the method used to calculate wave drag in 5.7.3.2, although this does result in complicated integrals. The other approach is to use a numerical process. If we assume that the changes in cross-sectional area q(x) are small, and that $q(0) = q(\ell_p) = 0$ (i.e. both ends of the body are pointed), we can also derive closed approximation formulae (J. ZIEREP, 1991).

At subsonic speeds, drag consists only of friction and base suction effects. For practical purposes, it may suffice if one collects the calculations and data described in 5.7.3.5 and 5.7.3.6, avoiding the need for the labour-intensive numerical integration described above.

### 5.7.4 Calculating other coefficients

#### *5.7.4.1 Lift force and overturning moment coefficients*

If the axis of the projectile adopts a certain (small) angle to the direction of flow, the potential equation (Eqn 5.7:6)

$$\Phi(x,r) = u_\infty \cdot x + \phi(x,r) + v_\infty \cdot r + \bar{\phi}(x,r) \qquad [\mathrm{m^2/s}]$$

has to take account of the interference potential of the non-parallel flow. This requires us to replace Eqn 5.7:7 by the following differential equation:

$$(5.7{:}44) \qquad \left(1 - Ma_\infty^2\right) \cdot \bar{\phi}_{xx} + \bar{\phi}_{rr} + \frac{1}{r} \cdot \bar{\phi}_r - \frac{1}{r^2} \cdot \bar{\phi} = 0 \,. \qquad [1/\mathrm{s}]$$

As for axial flow, observing the corresponding boundary conditions gives us the following expression for supersonic flow

$$(5.7{:}45) \qquad \bar{\phi}(x,r) = u_\infty \cdot \sin\delta \cdot \frac{1}{\pi \cdot r} \cdot \int_0^{x - a\cdot r} \frac{q'(\xi) \cdot (x-\xi) \cdot d\xi}{\sqrt{(x-\xi)^2 - a^2 \cdot r^2}} \,, \qquad [\mathrm{m^2/s}]$$

and for subsonic flow

$$(5.7{:}46) \qquad \bar{\phi}(x,r) = u_\infty \cdot \sin\delta \cdot \frac{b^2 \cdot r}{2 \cdot \pi} \cdot \int_0^{\ell_p} \frac{q(\xi) \cdot d\xi}{\sqrt{\left\{(x-\xi)^2 + b^2 \cdot r^2\right\}^3}} \,. \qquad [\mathrm{m^2/s}]$$

By deriving the potential equations with respect to x or r, we obtain the two disturbance velocities for the situation in which the projectile is at an angle to the flow, which must then be superimposed on those that occur when that is not the case. We can calculate pressure distribution – and hence the forces – from the velocities at the surface of the projectile.

For a comprehensive treatment of the singularity method, please see J. ZIEREP (1991).

If we only consider the axial velocity interference at the surface of the projectile, we obtain a simple equation for overturning moment, based exclusively on geometric variables, at both subsonic and supersonic speeds. This equation was originally developed for airships by M. MUNK (1924), and is known as *MUNK's airship equation*.

$$(5.7{:}47) \qquad M_M = \frac{\rho}{2} \cdot v^2 \cdot \sin(2\cdot\delta) \cdot \left[V - (\ell_p - x_{CM}) \cdot A_B\right] . \qquad [\mathrm{N \cdot m}]$$

Where $\rho$ = density of the air, v = projectile velocity, $\delta$ = angle of attack, V = projectile volume, $\ell_p$ = projectile length, $x_{CM}$ = distance between centre of mass and tip of projectile and $A_B$ = surface area of the projectile base.

If we set the right-hand sides of Eqns 5.3:26 and 5.7:47 equal to each other and use the linear approximation of Eqn 5.5:6, we obtain the following formula for $C'_M$ at small angles of attack ($\sin\lambda \approx \lambda$):

(5.7:48) $$C'_M = \frac{8 \cdot \left[ V - (\ell_p - x_{CM}) \cdot A_B \right]}{\pi \cdot d^3} \quad . \qquad [-]$$

The results obtained using this formula can be verified by means of wind tunnel measurements on known projectile shapes. Those measurements reveal that MUNK's formula over-estimates $C'_M$ at supersonic velocities. The formula under-estimates projectile stability, but that is preferable to over-estimating this factor.

Please see W. HAACK (1959) and G. BRUHN (1961) for further information on the calculation of lift coefficients and overturning moment coefficients.

#### 5.7.4.2 Other coefficients

The remaining coefficients of interest are primarily those that we must know in order to study the stability of a projectile inside the Molitz stability triangle. These are the moment coefficient derivative for the gyroscopic stability number, the derivatives of the lift coefficient and of the Magnus moment coefficient, and the pitch damping coefficient.

To calculate these coefficients, we require software capable of calculating the forces acting on the projectile from the equations for flow. If one does not have such software, one can make do with the Molitz approximations

(5.7:49) $$C_H = k \cdot \ell_p^2 \, , \qquad k = 0.4 \ldots 0.5 \, , \qquad [-]$$

(5.7:50) $$C'_J = -0.02 \cdot \left( \ell_p - \frac{h}{2} \right)^2 \, , \qquad [-]$$

where $\ell_p$ = projectile length and h = nose length, both in calibres.

## 5.8 Ballistically optimal projectiles

### 5.8.1 Aerodynamically optimal projectile designs

#### 5.8.1.1 HAACK's projectile theory

To avoid the relatively labour-intensive numerical integration of the integral equation (Eqn 5.7:12), HAACK (1941) suggested the use of the Kàrmàn approximation. KÀRMÀN assumed that in this equation, a·r was negligible by comparison with x und x–ξ.

$$r \cdot r' = \frac{1}{u_\infty} \cdot \int_0^{x - a \cdot r} \frac{-k'(\xi) \cdot (x - \xi) \cdot d\xi}{\sqrt{(x-\xi)^2 - a^2 \cdot r^2}} \quad . \qquad [m]$$

This assumption is justified by the slenderness of the body, which is already assumed in deriving that equation. Furthermore, we may assume that in the case of a projectile

$$a = \sqrt{Ma_\infty^2 - 1} = \cot\alpha \qquad [-]$$

is sufficiently small. Using this approximation, and setting $r^2 = q$, then from the above equation we have

$$(5.8{:}1) \qquad r \cdot r' = \tfrac{1}{2} \cdot q'(x) = \frac{1}{u_\infty} \cdot k(x) \,. \qquad [m]$$

We can use Eqn 5.7:18 to calculate the pressure coefficient

$$(5.8{:}2) \qquad C_{PR} = \frac{p - p_0}{\frac{1}{2} \cdot \rho_0 \cdot u_\infty^2} = -\frac{2}{u_\infty} \cdot \phi_x \,, \qquad [-]$$

which in turn gives the following expression for wave drag:

$$(5.8{:}3) \qquad F_{DS} = \frac{F_D}{\frac{1}{2} \cdot \rho_0 \cdot u_\infty^2} = 2 \cdot \pi \cdot \int_0^h C_{PR} \cdot r \cdot r' \cdot dx \,. \qquad [m^2]$$

Substituting Eqns 5.8:1, 5.8:2 and 5.7:10, and applying KÀRMÀN's approximation to the last of these gives

$$(5.8{:}4) \qquad F_{DS} = \pi \cdot \int_0^h q'(x) \cdot \left\{ \int_0^x \frac{q''(\xi) \cdot d\xi}{x - \xi} \right\} \cdot dx \,. \qquad [m^2]$$

This establishes a direct relationship between wave drag and the shape of the projectile (because $q = r^2$). In turn, that relationship forms the basis for designing a Haack nose shape.

Drag $F_{DS}$ as a function of stagnation pressure can be minimized if the singularity distribution ahead of and behind the projectile disappears. This results in the following constraints:

$$k(0) = 0 \;;\; k(h) = 0 \,, \qquad [m^2/s]$$

and hence from Eqn 5.8:1:

$$(5.8{:}5) \qquad q'(0) = 0 \;;\; q'(h) = 0 \,. \qquad [m]$$

In addition, the nose of the projectile should originate on the projectile axis and end with a cross-section determined by the projectile diameter $d_P$. This results in:

$$(5.8{:}6) \qquad q(0) = 0 \;;\; q(h) = \left(\frac{d_P}{2}\right)^2 = \frac{S}{\pi} \,, \qquad [m^2]$$

where S = the cross-sectional area of the cylindrical part of the projectile. We also need to impose the constraint of a fixed volume V:

$$(5.8{:}7) \qquad V = \int_0^h q(x) \cdot dx \,. \qquad [m^3]$$

The result is a constrained variational problem for the function q(x) of the squared projectile meridian. Following extended calculations (shown in detail in I. SZABÓ, 1972) we obtain the following parametric representation:

(5.8:8a) $$x(\beta) = \frac{h}{2}\cdot(1-\cos\beta) \; ; \; 0 \le \beta \le \pi ,$$ [m]

(5.8:8b) $$q(\beta) = \frac{2}{\pi^2}\cdot\left[\frac{S}{2}\cdot\left(\beta-\frac{1}{2}\cdot\sin 2\cdot\beta\right)+\frac{4}{3}\cdot\left(2\cdot\frac{V}{h}-S\right)\cdot\sin^3\beta\right] ,$$ [m²]

and the drag for this nose shape is given by

(5.8:9) $$F_{DS} = \frac{4}{\pi\cdot h^2}\cdot\left(9\cdot S^2+32\cdot\frac{V^2}{h^2}-32\cdot\frac{S\cdot V}{h}\right) .$$ [m²]

Eqns 5.8:8 and 5.8:9 constitute a set of formulae with which one can obtain the profile with the lowest possible wave drag using the variables projectile diameter (calibre), nose length and volume.

If only two of these three parameters are fixed, one can adjust the third for minimum drag. This is achieved by zeroing the derivative of Eqn 5.8:9 with respect to the non-fixed variable. There are three possibilities:

I. Calibre and length are known. Minimum wave drag will occur when

$$\frac{d F_{DS}}{dV} = 0 ,$$ [1/m]

which gives the following relationship between V, S, and h:

(5.8:10) $$V = \tfrac{1}{2}\cdot S\cdot h .$$ [m³]

II. Volume and length are fixed. The condition for minimum wave drag is

$$\frac{d F_{DS}}{dS} = 0 ,$$ [-]

and the relationship between S, V and h is given by:

(5.8:11) $$S = \frac{\pi}{4}\cdot d^2 = \frac{16\cdot V}{9\cdot h} .$$ [m²]

III. Calibre and volume are fixed. Minimum drag will be achieved when the following condition is satisfied:

$$\frac{d F_{DS}}{dh} = 0 ,$$ [m]

which gives us the following expression for the relationship between V, S, and h:

(5.8:12) $$h = \frac{8\cdot V}{3\cdot S} .$$ [m]

As all three equations (5.8:10, 5.8:11 and 5.8:12) can be solved for any of the three variables, fixing two variables will always result in three nose shapes, one of which will have the lowest wave drag.

This means that for a given projectile diameter (calibre) and length, we obtain one of the nose shapes shown in Fig. 5.62.

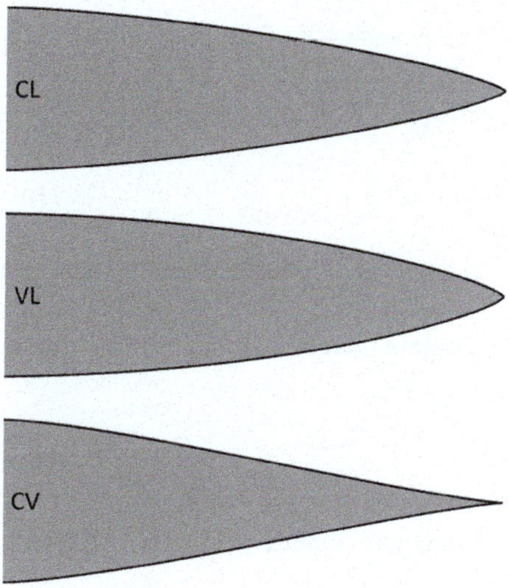


**Fig. 5.62** Projectile noses optimized using Haack's theory.
Calibre-length: Calibre and nose length fixed.
Volume-length: Volume and nose length fixed.
Calibre-volume: Calibre and volume fixed.
All three nose shapes have a horizontal tangent where they widen to nominal projectile diameter.

### *5.8.1.2 Applying Newtonian theory*

A Newtonian projectile nose shape is based on Eqn 5.7:26:

(5.8:13) $$F_{DS} = 4 \cdot \pi \cdot \int_0^h r \cdot r'^3 \cdot dx \, .$$ [m²]

The search for a meridian curve r(x) that will result in minimum drag leads us to the general extremal problem

(5.8:14) $$F_{DS} = \int_a^b f(x, r, r') \cdot dx = \min \, .$$ [m²]

The following Eulerian differential equation is an essential criterion for achieving a minimum:

(5.8:15) $$\frac{d}{dx} f_{r'} - f_r = 0 \, .$$ [-]

If we replace f by the function $r{\cdot}r'^3$ from Eqn 5.8:13, then after some brief calculations we obtain:

(5.8:16) $$r \cdot r'^3 = \text{const.}$$ [m]

To calculate the constants, we now need to include the boundary conditions. First, we assume that the projectile is pointed, i.e.

(5.8:17) $$r(0) = 0 \, .$$ [m]

If calibre and length are fixed, we can also write

$$(5.8{:}18)\qquad r(h) = \frac{d}{2}\,. \qquad [m]$$

Integrating Eqn 5.8:16 with the conditions of Eqns 5.8:17 and 18 yields:

$$(5.8{:}19)\qquad r(x) = \frac{d}{2}\cdot\left(\frac{x}{h}\right)^{3/4}, \qquad 0 \le x \le h\,. \qquad [m]$$

The nose shape based on Newtonian theory is hence also referred to in the literature as a 3/4 power-law nose.

If volume and length are fixed, we can calculate the calibre using the following formula:

$$(5.8.20)\qquad \frac{d}{2} = \sqrt{\frac{5\cdot V}{2\cdot\pi\cdot h}}\,. \qquad [m]$$

If calibre and volume are fixed, we can calculate length using the following formula:

$$(5.8{:}21)\qquad h = \frac{10\cdot V}{\pi\cdot d^2}\,. \qquad [m]$$

In contrast to the Haack nose shapes, however, there is no significant difference between the three variants. Fig. 5.63 shows the shape of a projectile nose designed using Newtonian principles.

### *5.8.1.3 Scope of application of the optimizations*

If we examine Eqns 5.8:9 and 5.8:13 for calculating the drag of the projectile nose, we note that it is independent of velocity. This stems from the Kàrmàn approximation in the case of the Haack nose, and from the Newtonian approximation in the case of the Newtonian nose.

HAACK (W. HAACK, 1941), studied the variational problem for the nose profile in a form that did take account of velocity. The conclusion was that the optimum profile was virtually independent of velocity within the limits of validity of the linear theory. This means that if all conditions are observed, the Haack nose shapes can be used between approximately Mach 1.3 and Mach 2.7. The optimal range for the use of Newtonian profiles begins approximately where that for Haack nose profiles ends (see 5.8.3.2).

One additional consequence of the simplifications adopted is that the Haack calibre-length and volume-length shapes, and the Newtonian shape, are all blunt, i.e.

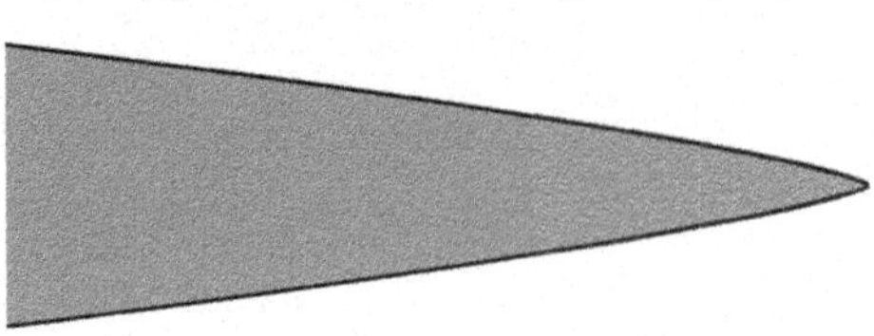

**Fig. 5.63** Nose optimized according to the Newtonian flow assumption.

they have a vertical tangent at x = 0. That contradicts the assumption that half the cone angle of the tip should be smaller than the Mach angle. This makes no difference in practice, as projectile tips are always flattened on account of other design features such as the fuze plate.

### *5.8.1.4 Optimizing the tail of the projectile*

We mentioned in 5.7.3.1 that base suction effects account for the majority of total drag. This means that we must also consider the shape of the tail if we wish to design a drag-optimized projectile.

When we are designing the tail of a projectile, we can manipulate two main parameters: the angle of the boattail and its length. We also know that the off-flow angle for a cylindrical base at typical Mach numbers is approximately 10°–12° (see Fig. 5.64). A steeper cone angle would therefore be pointless, as the flow would separate at the corner and the entire surface of the base would be subjected to negative pressure (see Fig. 5.65). We can therefore expect that for a given forebody, a boattail angle of somewhere between 0° and 12° will result in minimum drag.

The author was able to demonstrate this minimum empirically. Tail cones were added to twelve 6.45 × 48 bullets with identical nose shapes. Cone lengths of ½, ¾ and 1 calibre were used, with angles of 5°, 7°, 9° and 11°. The values of $C_D$ were then measured in free flight between Mach 1.8 and Mach 2.7. These experiments yielded the following results:

- Drag decreased as cone length increased.
- Minimum drag occurred at a boattail angle of between 6° and 8°, regardless of Mach number (see Figs 5.66 and 5.67 next page).
- Between these two angles, drag decreased linearly with increasing cone length.

R. L. McCoy (1999) obtained the same results, also reporting a linear drag curve with a minimum at an angle of between 5° and 9°. He noted that at angles greater than 10° one must expect the boundary layer to separate, which would negate the advantage of the boattail.

**Fig. 5.64** Flow behind a cylindrical tail. The free off-flow angle is approximately 10.5°.

**Fig. 5.65** Off-flow angle for a projectile with a "turbo" tail. Here again, the angle is approximately 10.5°.

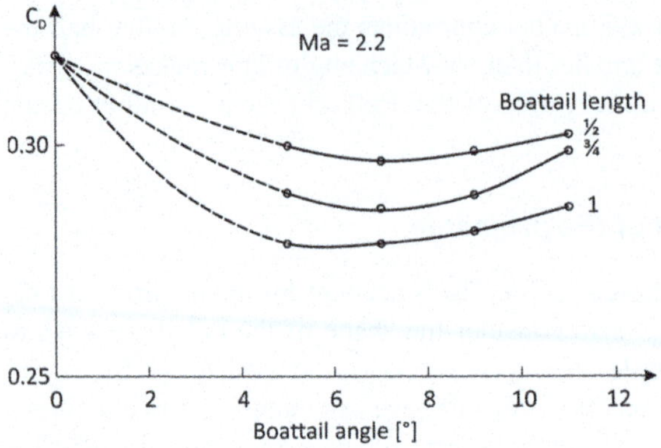


Fig. 5.66 Drag coefficient as a function of boattail angle at Ma = 2.2.

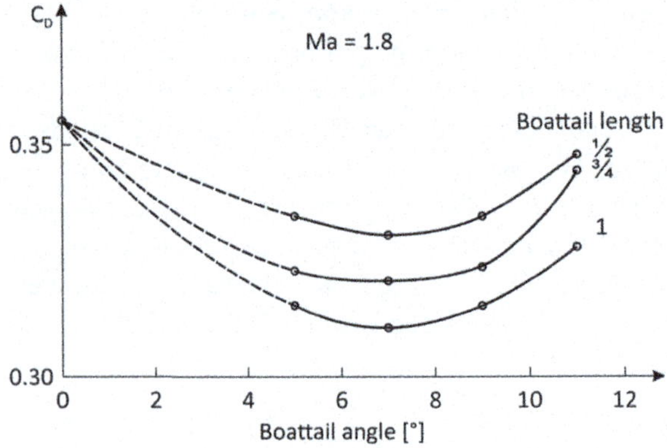


Fig. 5.67 Drag coefficient as a function of boattail angle at Ma = 1.8.

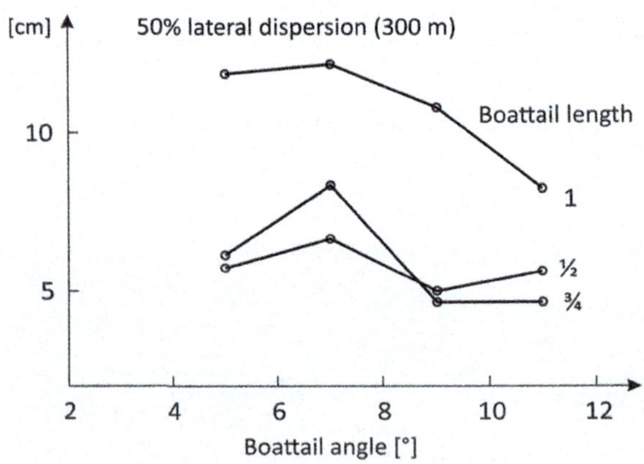


Fig. 5.68 Relationship between dispersion on the one hand, and boattail angle and length on the other.

The experiments also showed that a long boattail has a negative effect on stability. This is apparent if we look at vertical and horizontal dispersion, which are both significantly greater for the bullet with the 1-calibre boattail.

## 5.8.2 Practical design of an optimum projectile shape

### 5.8.2.1 Nose shape

***Haack profiles.*** When one is designing a projectile nose shape in practice, the calibre – and hence the diameter of the projectile – is fixed, which limits the options to calibre-length (CL) and calibre-volume (CV) projectiles. However, the CV shape is problematic in terms of construction and manufacture, which means that only CL projectiles are used in practice.

The parametric representation of the meridian of a CL nose is

$$x = \frac{h}{2}\cdot(1 - \cos\beta)\,, \qquad (5.8{:}22) \quad [\text{m}]$$

$$r^2(\beta) = \frac{S}{\pi^2}\cdot\left[\beta - \tfrac{1}{2}\cdot\sin(2\cdot\beta)\right]. \qquad (5.8{:}23) \quad [\text{m}^2]$$

If the projectile has a fuze, this will bring further constraints.

These are generally the radius of the base of the fuze $r_f$ and its height $h_f$, which require the meridian to pass through a specific point ($h_f$, $r_f$).

We calculate V from $h_f$ and $r_f$, using Eqns 5.8:8a and 5.8:8b, and that value is then used to calculate the meridian.

If the projectile has a driving band, the transition from the nose to the cylindrical section will not be smooth, but will include a tangent jump of approximately 2°.

One can estimate that angle fairly accurately from the thickness of the driving band and the length of the cylindrical section between the driving band and the transition from nose to cylindrical section.

The turbulent boundary layer will be thickened to the point at which the driving band lies entirely inside it. This prevents the boundary layer separating at the leading edge of the driving band, which would cause increased tail drag (see Fig. 5.69). The nose length and diameter required to calculate the Haack nose shape will need to be modified accordingly.

**Fig. 5.69** Boundary layer around a 20 mm projectile with driving band. The small angle between the nose and the cylindrical section thickens the boundary layer, preventing separation at the driving band.

***Newtonian nose shape***. We can use the formula of Eqn 5.8:19 to calculate this type of nose shape. If the volume is fixed, one starts by calculating the length or calibre from Eqn 5.8:20 or 5.8:21 respectively.

One factor that makes the use of such a nose shape more difficult is that the tangent of the meridian is not horizontal at x = h; even a very slender nose connects to the cylindrical section at an angle of 5° or more. It will be necessary to round off the transition between the nose and the cylindrical section, resulting in a smooth transition for projectiles that do not have a driving band and a tangent jump of about 2° for those that do. To facilitate calculation of this curve, one can make it a quadratic parabola or an arc of a circle, for instance.

Another complicating factor is that no constraints can be accepted for the incorporation of a fuze, as the potential function (5.8:19) has no further parameters.

#### 5.8.2.2 Comments regarding the cylindrical section and the tail

The cylindrical part of a projectile has to be so designed that the projectile moves along the barrel correctly. For the projectile to be fired correctly, the cylindrical part of the projectile that ensures centering in the barrel must be at least 1.2 calibres long.

During development of new rifle ammunition, it was noted that some bullets had grooves cut into them by the rifling even at the tip. The reason was that those bullets had already become skewed in the barrel. The cylindrical section of the bullet available to guide it through the barrel was approximately 1.1 calibres long.

As the cylindrical section of a projectile has fulfilled its role once the projectile leaves the barrel, it may appear logical to eliminate it altogether and use a centering ring to provide guidance in the barrel, discarding it after the projectile passes through the muzzle. This makes it possible to design the entire length of the projectile as a tip, achieving a high measure of slenderness. That in turn results in very low drag at supersonic speeds.

W. HAACK indeed recommended such projectiles. However, the discarded fragments of the centering ring create a hazard for anyone standing in front of the gun, which means that it is generally not possible to use such projectiles

It is useful to design the projectile with a boattail 0.5–0.75 calibres long and a boattail angle of approximately 6°–8° (see 5.8.1.4 regarding optimal tail design).

#### 5.8.2.3 Stability and tractability

The higher the slenderness ratio of the nose (h/d, where h = nose length and d = calibre), the lower the wave drag on the projectile. However, maximum slenderness ratio is limited by stability considerations. A shell for an artillery gun with a typical twist of 6°–7° can be no more than 6 calibres long (see 5.8.4.3), of which no more than 4–4.5 calibres can be used for the nose.

One can expect stable flight if the ratio of the radial moment of inertia $J_r$ to the axial moment of inertia $J_a$ satisfies the following condition:

$$\frac{J_r}{J_a} \leq 10\ldots11 \,. \qquad (5.8:24) \qquad [-]$$

At the muzzle, the gyroscopic stability number s should be at least 1.3 and ideally about 1.4. However, it is not easy to give an upper limit. Experience shows that a projectile with s = 2.0 is still sufficiently tractable to be fired at high elevation. The value of s can be even higher for projectiles fired from flat-trajectory weapons, as there is no risk of overstabilization and associated problems – despite what is sometimes said.

## 5.8.3 Examples of optimized projectiles

### *5.8.3.1 Verifying designs experimentally; general comments*

It is only possible to confirm that a projectile shape is indeed optimal by using projectile models of which the cylinder and tail are of the same shape as on the original and creating the same conditions of flight (in terms of pitch) as far as possible, by distributing mass appropriately within the model. Such experiments can facilitate selection of the optimum profile.

If a new projectile is to be developed or an existing projectile is to be enhanced ballistically, one will want not only to redesign the nose but also to take the opportunity to extend range by other means, such as increasing muzzle velocity, increasing sectional density and redesigning the cylinder and boattail. One can only establish whether one has actually achieved any improvement by comparing the drag coefficients.

### *5.8.3.2 Studies using model projectiles*

The following factors make model projectiles in small arms calibres a good choice for exterior ballistics studies:

- They are cheap. This means one can conduct tests with a sufficiently large number of firings.
- It is possible to measure the drag coefficient very accurately over a wide range of Mach numbers on a free-flight ballistic range.
- Tests can be carried out on small ranges with comparatively few personnel and limited equipment.
- Wind tunnel measurements can be carried out on large-scale models, which is advantageous in terms of the aerodynamic similarity law.

To verify the theoretical conclusions, model projectiles were therefore fabricated in a calibre of 6.35 mm (0.25"), with the following nose shapes:

1 Ogive, 9.45 mm radius.

2 Haack nose shape, nose length 2.5 calibres.

3 Newtonian nose shape, nose length 2.5 calibres.

The following experiments were carried out using these three nose shapes:

***Wind tunnel measurements***. For each nose shape, a model was fabricated for the supersonic wind tunnel of which the cannelure, grooves and surface roughness corresponded as closely as possible to the original.

The scale of the model was so chosen that one could expect Mach and Reynolds numbers to be close to those of the original.

Drag, lift and overturning moment coefficients and the corresponding slopes were measured at between Mach 1.5 and Mach 3.5.

***Free-flight tests***. Free-flight tests were conducted using brass projectiles with each of the above three profiles. Despite low stability (because of unfavourable mass distribution) and corresponding angles of attack during the first part of the trajectory, these tests confirmed the conclusions from the wind tunnel tests.

Fig. 5.70 shows the three $C_D$ curves from the wind tunnel measurements. Between Mach 1.5 and Mach 2.8, the Haack nose shape has the lowest drag. The expectation of 5.7.3.3 that the Newtonian nose shape would be especially suited for high velocities was confirmed. The ogive shape performed significantly worse than the other two, despite its higher slenderness ratio.

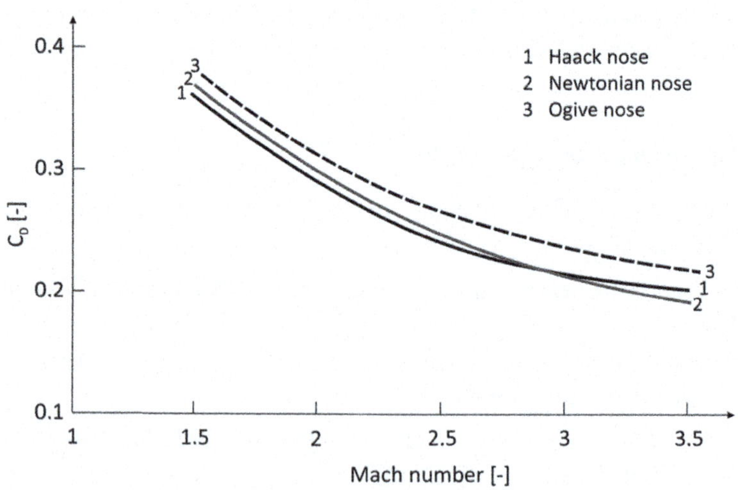


**Fig. 5.70** Wind tunnel measurements on three projectile models with different nose shapes. At up to approximately Mach 2.9, the Haack nose shape has the lowest drag, while at higher Mach numbers the Newtonian nose shape performs better in this respect. Both of these are visibly better than the circular arc ogive.

#### *5.8.3.3 Artillery shells*

***105 mm shell.*** A 105 mm Haack-profile artillery shell was developed and entered service with the Swiss Army in the mid-20th Century. The new shell was intended to provide increased range and less scatter.

The 105 mm "Spitzgranate" (SpG, pointed shell) was a calibre-length projectile with an overall length of 6 calibres and a nose length of 4.2 calibres, giving it a very slender nose. Because of its slender shape, the new shell was lighter than the standard shell and its muzzle velocity was somewhat higher.

The range increase achieved by the shape of the projectile was calculated using a point-mass model in which the old and the new shell were each assigned the same mass and $v_0$. As the stability coefficients at the muzzle were sufficiently high for both types of shell, one could expect similar flight characteristics. The calculated range increase was slightly over 16%, and the actual increase slightly greater.

Firing test results for both shells were available, along with drag, lift and overturning moment coefficients and their slopes, measured at transonic and ultrasonic speeds in the wind tunnel. At subsonic speeds, the wind tunnel results corresponded closely to those of free-flight tests. At supersonic speeds, the wind tunnel values were initially lower than those for firing tests, on account of the zero angle of attack.

Both wind tunnel measurements and firing tests indicated a 20–25% reduction in drag coefficient as a result of the modified shell profile. As expected, the Haack nose shape conferred no advantage at subsonic speeds.

***155 mm shell.*** In the early 1980s, the Swiss Army once again required an artillery shell with greater range. Once again, aerodynamic optimization was the obvious approach. The barrel of the gun for which the new shell was intended (the 155 mm Pz Hb M109) has a very high twist angle. It was therefore possible to design a shell 6 calibres long (0.93 m), with a nose length of 3.8 calibres, a 1.2 calibre cylindrical section and a base-bleed tail (the 155 mm SpG, see Fig. 5.71).

The new projectile shape extended the range of the gun by up to 15%. Fig. 5.72 (next page) shows the $C_D$ curves for the M107 shell and the SpG. The new profile, the addition of a base-bleed gas generator and increased muzzle velocity extended the range of the gun from 18 km to over 30 km, which was a considerable range for the time (around 1980).

**Fig. 5.71**
155 mm SpG.

#### *5.8.3.4 Small-calibre projectiles*

One might well ask whether drag optimization is worthwhile in the case of small-calibre projectiles, given their shorter range and times of flight. However, reducing deceleration due

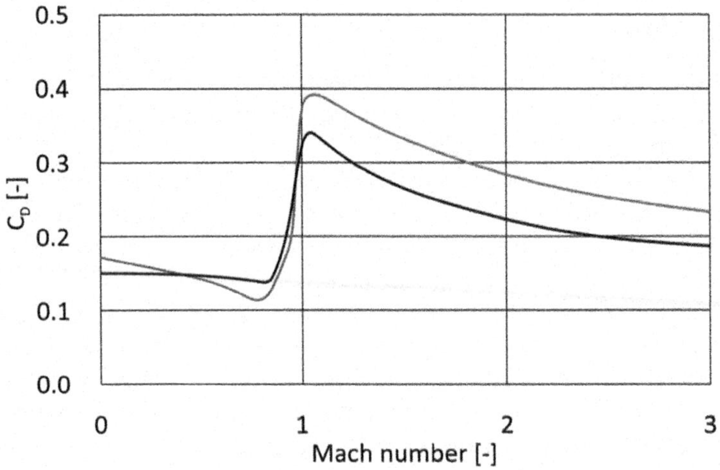


**Fig. 5.72** Drag coefficients for two 155 mm artillery shells: conventional M107 shell (grey line) and the optimized shell shown in Fig. 71 (black line).

to drag conserves energy and reduces the effects of disturbances. Table 5.14 shows potential improvements that could be achieved by using a Haack-profile bullet instead of either an ogival pointed bullet or a conventional hunting bullet, modifying only the shape. It is only slightly more difficult to produce a bullet with a Haack nose than with an ogival pointed nose.

Given the increasing use of bullets without a lead core, which often have a lower sectional density and hence higher deceleration, this could conserve energy along the trajectory.

## 5.8.4 Optimizing stability

### *5.8.4.1 Optimum projectiles and gyroscopic stability*

We can also reduce deceleration due to drag by making sectional density as high as possible, as we can see from Eqn 5.4:37:

$$\frac{F_D}{m} = a = C_D \cdot \frac{1}{2} \cdot \rho \cdot v^2 \cdot \frac{1}{q} \,. \qquad [\mathrm{m/s^2}]$$

**Table 5.14.** Ballistic improvements achieved by optimizing bullet shape.

| Ballistic parameter | Range | Improvement by comparison with: | |
|---|---|---|---|
| | [m] | Pointed bullet | Hunting bullet |
| Energy | 300 | +8% | +13% |
| | 500 | +17% | +23% |
| Sensitivity to crosswind | 300 | -14% | -22% |
| | 500 | -16% | -24% |

For a given calibre, this implies a heavy and therefore long projectile. Long, heavy projectiles will inevitably be subjected to greater radial moments of inertia, which has a negative effect on stability.

Designing an optimum projectile in terms of exterior ballistics therefore involves finding the shape that will ensure greatest stability. This could make it possible to design a longer and hence heavier projectile. In considering that question, we must first recall the equation for gyroscopic stability (Eqn 5.6:10):

$$(5.8:25) \qquad s = \frac{2}{\pi} \cdot \frac{J_a^2 \cdot \omega^2}{J_r \cdot C'_M \cdot \rho \cdot v^2 \cdot d^3} \; . \qquad [-]$$

In this equation, we must replace velocity v, angular velocity $\omega$ and the slope of the overturning moment coefficient $C'_M$ by parameters that will enable us to produce a geometrical projectile design.

With a few exceptions, the stability of a projectile increases along its trajectory ($\omega$ decreases significantly more slowly than v), so the stability coefficient is lowest at the muzzle. An optimized projectile design will therefore have to be based on this value. At the muzzle, angular velocity can be expressed in terms of muzzle velocity and twist length $\Lambda$:

$$(5.8:26) \qquad \omega = \frac{2 \cdot \pi \cdot v}{\Lambda} \; . \qquad [1/s]$$

Substituting into Eqn 5.8:21, we obtain an expression for the stability coefficient in which the only unknown is the slope of the overturning moment coefficient $C'_M$:

$$(5.8:27) \qquad s = \frac{8 \cdot \pi}{\rho} \cdot \frac{J_a^2}{J_r \cdot C'_M \cdot \Lambda^2 \cdot d^3} \; . \qquad [-]$$

It is interesting to note that s does not depend on velocity directly.

If calculating $C'_M$ in accordance with 5.7.4.1 appears too labour-intensive, we can use the equation based on MUNK's airship formula (Eqn 5.7:48), which uses only geometric projectile data. Substituting into Eqn 5.8:27 we obtain:

$$(5.8:28) \qquad s = \frac{\pi^2}{\rho} \cdot \frac{J_a^2}{J_r \cdot \Lambda^2 \cdot \left[V - (\ell_p - x_{CM}) \cdot A_B\right]} \, , \qquad [-]$$

where $\rho$ = atmospheric density, $\Lambda$ = twist length, V = projectile volume, $\ell_p$ = projectile length, $x_{CM}$ = distance between centre of mass and tip of projectile, $A_B$ = area of projectile base and $J_a$ and $J_r$ are the axial and radial moments of inertia respectively.

The remaining free parameter describing the geometry of the projectile is nose length h, expressed as the ratio $h/\ell_p$, i.e. as a proportion of total projectile length. To establish the influence of this parameter on gyroscopic stability, wind tunnel measurements were conducted at various Mach numbers and $C'_M$ calculated using four model projectiles with $h/\ell$ values of 0, 0.5, 0.67 and 1 (see Fig. 5.73). Fig. 5.74

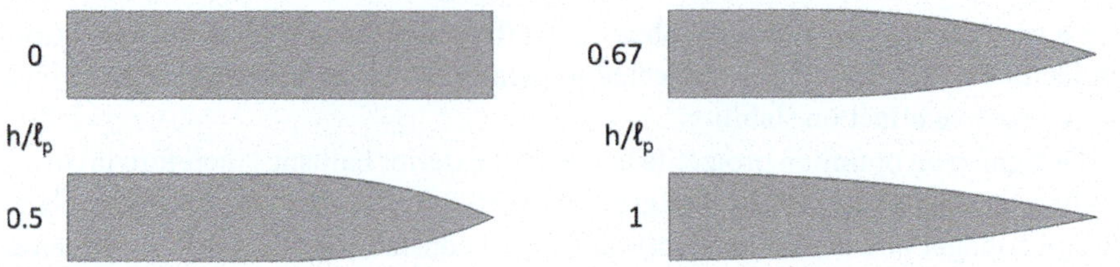


**Fig. 5.73** Model projectiles for wind tunnel measurements to investigate the influence of nose length $h/\ell_p$.

presents the results and Fig. 5.75 shows the corresponding stability coefficients. The values for MUNK's airship formula are shown on the two graphs as dotted lines.

The first thing to note from the graphs is that MUNK's approximation produces too steep a slope for the overturning moment coefficient, except at velocities close to the speed of sound, where $C'_M$ can apparently increase significantly.

This matches the measurements of moment coefficient derivative for a small-calibre projectile shown in Fig. 5.46, which also showed a sharp peak in the vicinity of Mach 1.

As one would expect, the excessively high values obtained using the Munk approximation result in stability coefficients that are too low, as mentioned in 5.7.4.1. These model tests also revealed the existence of a maximum gyroscopic stability number, which depended on nose length as a proportion of total projectile length (see Fig. 5.75).

### 5.8.4.2 Projectiles with maximum stability

This result was verified by carrying out calculations for real boattail projectile shapes with conical, ogive and Haack noses. In the case of the ogive projectile,

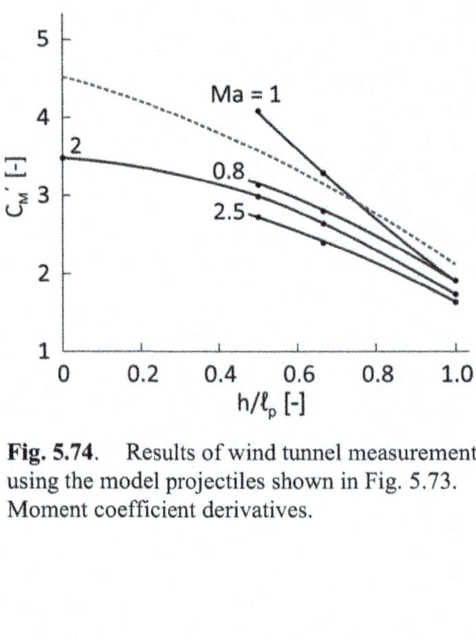


**Fig. 5.74.** Results of wind tunnel measurements using the model projectiles shown in Fig. 5.73. Moment coefficient derivatives.

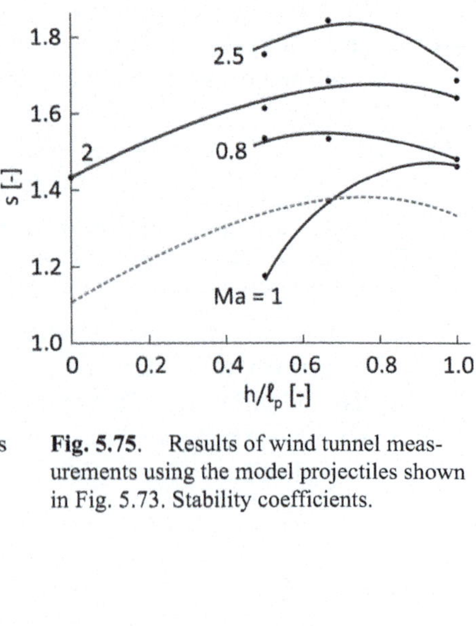


**Fig. 5.75.** Results of wind tunnel measurements using the model projectiles shown in Fig. 5.73. Stability coefficients.

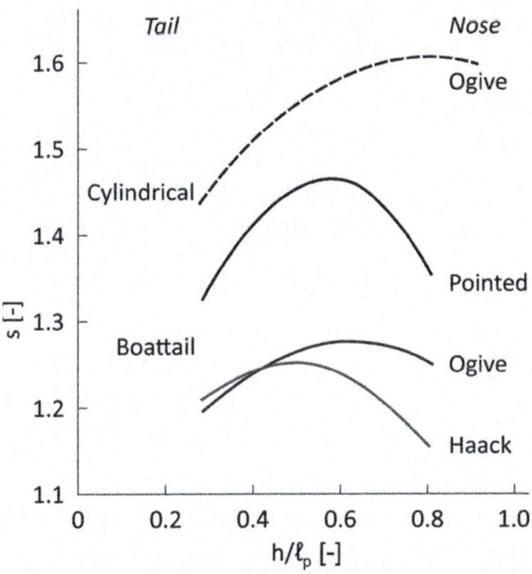


**Fig. 5.76.** Gyroscopic stability as a function of nose length $h/\ell_p$. See text for details.

calculations were also performed for a cylindrical base. All projectiles were assumed to be 4 calibres long.

If one varies the ratio of nose length to total projectile length, each nose shape has a very clear maximum stability value (see Fig. 5.76). This maximum occurs at a ratio of approximately 0.5 in the case of the Haack nose and about 0.6 in the case of the conical and ogive noses. The ogive projectile with cylindrical base would only reach maximum stability if the ratio were over 0.8, which is not achievable in practice.

Initial results of this study show that adding a boattail reduces stability. For a given ratio of nose length to total length, a more pointed projectile will have a higher stability coefficient than will a blunter projectile. The conical nose results in the highest stability coefficient, while the rounded noses are noticeably less stable.

At first sight this is surprising, as the centre of pressure of a pointed projectile is further to the rear, and one would therefore expect a higher aerodynamic force moment. Two factors counteract this tendency, however. Firstly, while the centre of mass is indeed further forward in the case of a blunter projectile, so is the point of application of the aerodynamic force, which means a larger moment can still be created. Secondly, the radial moment of inertia has a significant influence on the gyroscopic stability number, and that moment is higher for a rounded projectile, as the nose is heavier.

Blunt projectiles are sometimes essential, e.g. in the case of hunting bullets, and of course it is possible to increase the stability of such projectiles by appropriate design measures. These include reducing nose mass by incorporating a void or using a lighter material.

#### *5.8.4.3 What is the maximum length of a spin-stabilized projectile?*

So far, we have always assumed that the projectile is spinning fast enough, and is therefore stable enough. In practice, one will either be designing a projectile for an

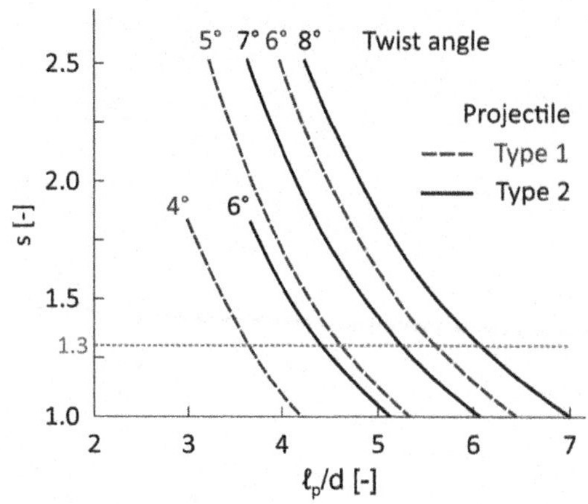


**Fig. 5.77.** Maximum length of a spin-stabilized projectile.
Type 1: Small-calibre projectile.
Type 2: Artillery shell.
*Example*: For a stability coefficient of 1.3 and a twist angle of 5°, the maximum length of a small-calibre projectile is approximately 4.8 calibres.

existing barrel, with a certain twist length or else (if weapon and projectile are being designed in parallel) one will need to tell the weapon designer the twist length required for the new projectile. To state the maximum possible length of a spin-stabilized projectile in general terms, one uses the (dimensionless) angle of twist Γ rather than the calibre-dependent twist length Λ. The relationship between these two values is given by:

(5.8:29) $$\Gamma = \arctan\frac{d\cdot\pi}{\Lambda}\,, \qquad [°], [-]$$

where d = calibre.

If we assume a realistic projectile, optimized in accordance with the principles set out above, Eqns 5.8:27 and 5.8:29 together give us the relationship between angle of twist Γ and the gyroscopic stability coefficient, with the total length of the projectile (in calibres) as an additional parameter. This enables us to calculate either the maximum projectile length for a given angle of twist and target stability coefficient, or the required angle of twist for an existing projectile and a given stability coefficient. In the latter case, we must bear in mind that the characteristics of the materials used impose an upper limit on the angle of twist. A compromise will be required between the weapon and projectile developers.

The stability coefficient should not be lower than 1.3. In the case of projectiles for flat-trajectory weapons, values of 1.4 to 1.5 are preferable. The angle of twist should not exceed 6° for a small-calibre barrel and 6°–9° for an artillery barrel.

Fig. 5.77 shows the relationships between the various factors in graphical form. The graph shows results for two types of the commonly-encountered ogive nose. Type 1 is a small-calibre projectile with a dense lead core and a thin steel or copper jacket, while Type 2 is an artillery shell with a less dense core (containing explosive) and thicker (steel) jacket.

**Fig. 5.78.** Firing of a base-bleed shell. The photo shows the flame from the gas generator (top right), the smoke trail and the ignition trail.

## 5.8.5 Reducing base drag

### *5.8.5.1 The "base bleed" principle*

There are limits as to how far one can go in optimizing the shape of a projectile tail in order to reduce base drag; the boattail angle must not exceed 10°–12° and the boattail must not be longer than about ¾ calibre because of the increased scatter that will result. This is explained in detail in 5.8.1.4. However, it is possible to further reduce drag related to the base of the projectile by reducing the negative pressure at the base that results from airflow.

To achieve this, a gas generator (or base bleed unit) is incorporated into the tail of the projectile. During an extended part of the trajectory, this ejects gas through openings in the projectile base.

Fig. 5.78 shows the firing phase of a base bleed shell. The flame from the gas generator (top right) and the associated smoke trail are clearly visible. One can also see the darker ignition smoke cloud near the muzzle.

Typical designs have either a single central opening (see Fig. 5.79) or several smaller holes arranged in a circle. The gas is generally supplied by composite ammonium perchlorate propellants. See 3.2.4.2 and E. M. YOUSSEF et al., 2015).

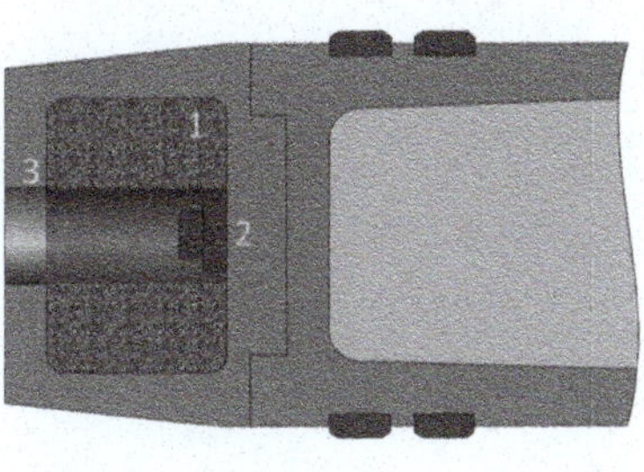


**Fig. 5.79.** Schematic drawing of a projectile base with base-bleed unit.
1: Composite propellant.
2: Igniter.
3: Gas exit.

The increase in pressure at the base of the projectile resulting from the gas jet reduces base drag by about 30–50% at supersonic speeds, and because base suction accounts for approximately 50% of drag at those speeds, total drag falls by approximately 15–25%. The reduction in drag may be even greater at subsonic speeds, where base suction accounts for 80–90% of total drag. The increase in range depends to a large extent on the time for which the gas generator operates.

The first base bleed designs extended the range of the artillery shell described in 5.8.3.3 by 20%. In the meantime, design enhancements and longer burn times have increased the ranges by up to 30%.

### *5.8.5.2 Tracer projectiles*

Tracer rounds are designed to enable one to "trace" the path of the projectile by means of a pyrotechnic charge in the base, rather than to give extended range, but we shall mention them for the sake of completeness. Like the base bleed gas generator, the pyrotechnic charge of a tracer projectile reduces drag. As a result, the $C_D$ values for a tracer projectile are often lower than those for a normal round of the same shape (see Fig. 5.80).

We can assess the combustion of the pyrotechnic charge in qualitative terms by measuring $C_D$. The ignition phase of the 7.5 × 55 test bullet in Fig. 5.80 was too slow, and the bullet started to topple slightly when the charge burned out, possibly because of the overturning moment coefficient curve at Mach 1 (Fig. 5.46). The irregular burning of the pyrotechnic charge is directly visible in the $C_D$ curve.

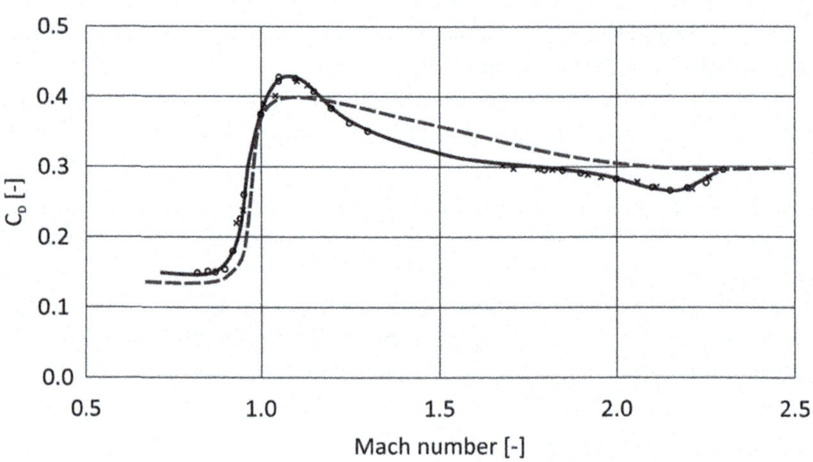


**Fig. 5.80.** $C_D$ for a tracer projectile (solid line) and for an FMJ projectile of the same shape without tracer (dashed line).

## 5.9 Elementary exterior ballistics of unguided rockets

### 5.9.1 Preliminary remarks

Rockets carry with them all or part of the energy they require for propulsion. This gives them the advantage that they only require light launching platforms, which can be carried by a single person even in the case of larger calibres. As a result, the projectile is subject to significantly greater forces in flight, and these lead to increased scatter. For this reason, rockets are often steered in flight by applying forces, either aerodynamically by means of remote-controlled fins or by using additional control jets. Because guided rockets (i.e. missiles) are influenced by steering inputs they do not belong to the field of classical (physical) exterior ballistics. This section will therefore only discuss unguided rockets.

We shall impose another limitation on this section. Rockets are often long, fin-stabilized projectiles, which have a tendency to pitch, especially at the start of their trajectory. This modifies the direction of propulsion, requiring complex calculations that can only be performed using a complete 6-DOF model. A number of sources cover that topic, such as H. MOLITZ, W. STROBEL, 1963 and W. WOLFF, 1976. We shall simply use the point-mass model, which produces fairly good results for stable rockets fired with a certain muzzle velocity, such as portable anti-tank weapons – rocket-assisted projectiles.

### 5.9.2 Rocket trajectories

#### *5.9.2.1 General*

The trajectory of an unguided rocket always consists of two stages: In the first stage, the rocket continues to experience the thrust responsible for its acceleration, in addition to the usual aerodynamic forces. This acceleration or boost phase is part of interior ballistics, and was already covered in 3.6. The second stage begins when propellant combustion ends. From this point on, the projectile is subject only to drag, and we can calculate its trajectory using the method described in 5.4. As a result, calculating the trajectory of a rocket is also a two-stage process.

#### *5.9.2.2 The boost phase*

In 3.6.3, we saw that the boost acting on a rocket has two components: the moment per unit time generated by its motor, and the pressure difference between the external streaming pressure exerted on the body of the projectile and the pressure at the nozzle exit:

(5.9:1) $$F_T = \frac{\Delta m}{\Delta t} \cdot v_a + (p_e - p_0) \cdot A_e \,, \qquad [\mathrm{N}]$$

where $\Delta m/\Delta t$ = ejected mass per unit time, $v_a$ = exhaust velocity, $p_e$ = nozzle pressure, $p_0$ = outside pressure and $A_e$ = nozzle area.

It is not possible to include this pressure difference in a point-mass trajectory model, so the two components are combined to make up the total thrust. The corresponding time-thrust curve can only be determined empirically. As mentioned in 3.6.3.3, there are two ways of achieving this.

The first is to measure static thrust on a rocket-engine test stand. This has the disadvantage that there is no airflow around the rocket, which means that the pressure conditions at the nozzle exit differ from those that occur in flight, and the propellant burns differently. The pressure difference between the nose and tail of the projectile is also absent. One must therefore determine the drag coefficient to use in conjunction with static thrust from the effective acceleration measured in free flight. The second method involves measuring velocity along the trajectory using Doppler radar. Acceleration is then calculated from the velocity plot, using the drag coefficient from the midcourse phase. In both cases, the missing pressure difference is added to drag.

The following elements are therefore available for calculating the trajectory during the boost phase: a curve of thrust versus time $F_T(t)$ that includes the reduction in mass m(t), and a corresponding $C_D$ curve. Because thrust varies over time, it is preferable to take time as the independent variable. This results in the following system of differential equations:

$$
(5.9{:}2)\quad \left|\begin{array}{lll}
\dot{v}_x = \dfrac{F_T(t)-F_D}{m(t)}\cdot\dfrac{v_x}{v} + g_x + a_{cx}\,, & & [\mathrm{m/s^2}] \\
\dot{v}_y = \dfrac{F_T(t)-F_D}{m(t)}\cdot\dfrac{v_y}{v} + g_y + a_{cy}\,, & & [\mathrm{m/s^2}] \\
\dot{v}_z = \dfrac{F_T(t)-F_D}{m(t)}\cdot\dfrac{v_z}{v} + g_z + d_z + a_{cz}\,, & & [\mathrm{m/s^2}] \\
\dot{x} = v_x\,, & & [\mathrm{m/s}] \\
\dot{y} = v_y\,, & & [\mathrm{m/s}] \\
\dot{z} = v_z\,, & & [\mathrm{m/s}]
\end{array}\right.
$$

with aerodynamic force

$$
(5.9{:}3)\qquad F_D = C_D\cdot\frac{1}{2}\cdot\rho\cdot v^2\cdot A\,. \qquad [\mathrm{N}]
$$

Integration is halted at the point where the thrust diagram ends. These final values form the starting values for calculating the coasting segment of the trajectory.

#### 5.9.2.3 The midcourse phase

We calculate the trajectory of a rocket in the midcourse phase using a point-mass model in the same way as for any other projectile. The only difference is that the starting values (including the trajectory coordinates) are the final values from the boost phase.

If the point-mass model is programmed appropriately, one need only input the coefficient function for the drag at all-burnt and continue the calculation.

#### 5.9.2.4 General trajectory characteristics

In 5.4.5, we discussed the trajectory of a free-flying projectile and the corresponding equations. In the case of a rocket, the sign in these equations is reversed. We therefore have to assume that the corresponding characteristics are also reversed, if the projectile is propelled throughout its trajectory. In particular, its velocity will be higher at a given height in the descending branch than at the same height in the ascending branch, and the apex will occur before the projectile has travelled half the distance to its target. Furthermore, the special points along the trajectory (see 5.4.5.4) will occur in the opposite sequence.

Having said which, propulsion will rarely continue through to the end of the trajectory. As we can see from Fig. 5.81, this reversed asymmetry will only occur if the rocket undergoes acceleration for an extended period.

The two trajectories in Fig. 5.81 were modelled for the same initial velocity (100 m/s) and elevation (100 mils). The rocket motor for Trajectory 1 burns intensively but briefly, whereas that for Trajectory 2 burns for longer. The motor for Trajectory 1 burns out after 0.7 s at 243 m/s. For Trajectory 2, the corresponding figures are 2.1 s and 794 m/s. Both projectiles travel the same distance. As one can readily imagine, the angle of trajectory at the end of burning also plays an important role.

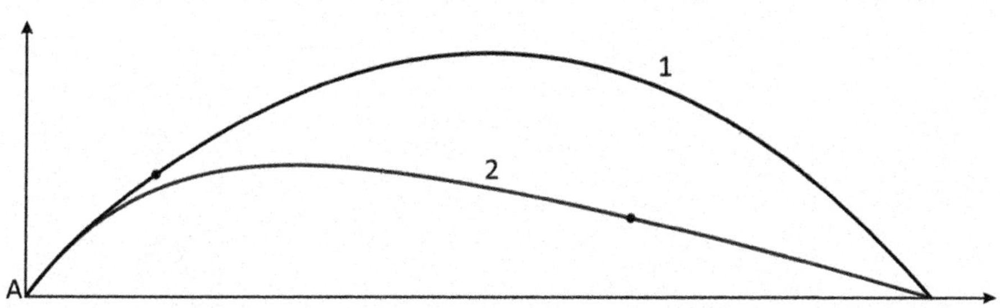


**Fig. 5.81.** Trajectories of two rockets with the same initial velocities and elevations. See text for a description of the trajectories. The dots indicate "all-burnt". If burning time is increased, the trajectory becomes more asymmetric. The trajectories are shown in highly exaggerated form.

### 5.9.3 The behaviour of rockets subjected to crosswinds

We discussed the lateral deviation of a projectile under the influence of a crosswind in 5.4.4. Transforming the initial values into a wind-axis reference system and transforming the end point back again gave us DIDION's crosswind formula (Eqn 5.4:64).

$$z_E = v_{wx} \cdot \left( t_E - \frac{x_E}{v_{0x}} \right), \qquad [\mathrm{m}]$$

where $v_{wx}$ = crosswind velocity, $v_{0x}$ = horizontal component of muzzle velocity and the values with subscript E refer to the end point.

Because no assumptions regarding the point-mass model equations were made in deriving this equation, it is valid for both decelerating and accelerating projectiles. It can be discussed simply in qualitative terms if time of flight is ignored in the above equation.

$$(5.9{:}4) \qquad z_E = v_{wz} \cdot t_E \cdot \left( 1 - \frac{\overline{v}_x}{v_{0x}} \right), \qquad [\mathrm{m}]$$

where $\overline{v}_x = x_E / t_E$ is the mean (horizontal) trajectory velocity. As usual, the point-mass model assumes that the projectile axis coincides with the lines of action of both thrust and drag. Qualitative behaviour under the influence of a crosswind can now be discussed using the expression in brackets.

*Case 1*. If the sum of drag and thrust is always negative (which is the case for all passive projectiles) the resultant will include one component $F_{WV}$ that impedes the forward motion of the projectile and another component $F_{WQ}$ that acts in the direction of the wind. As a result, the projectile will drift in the direction of the wind. The horizontal component of trajectory velocity will decrease over the trajectory

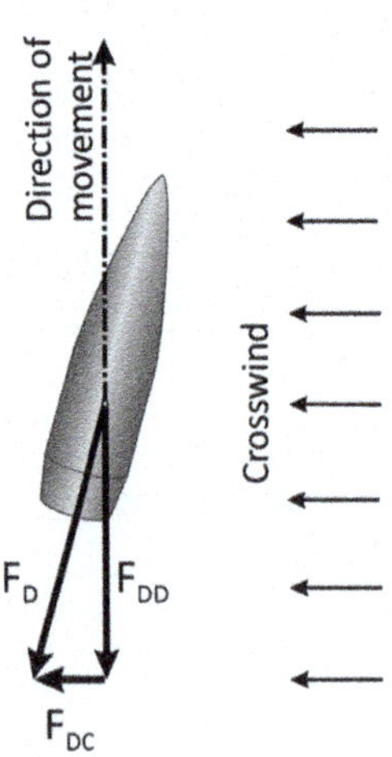


**Fig. 5.82.** *(Left)*. If drag is greater than thrust, one force $F_{DD}$ will act against the forward motion of the projectile and another force $F_{DC}$ will act in the same direction as the crosswind. Result: the projectile drifts laterally in the direction of the wind.

**Fig. 5.83.** *(Right)*. If thrust is greater than drag, one force $F_{TA}$ will act in the direction of travel, accelerating the projectile, and another force $F_{TC}$ will act in the opposite direction to the wind. Result: the projectile drifts laterally against the wind.

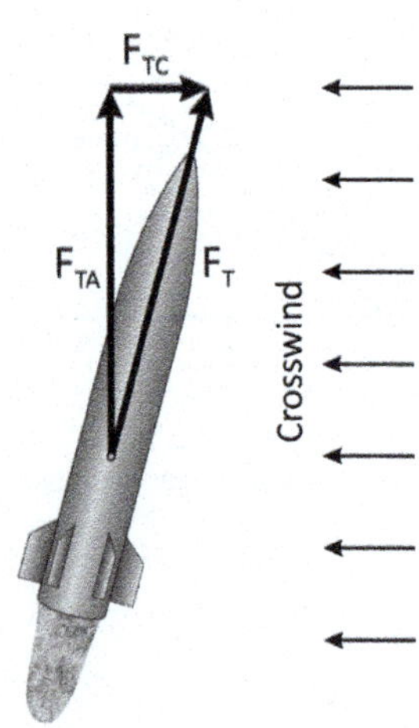

(see 5.4.5.2). The subtrahend in thc expression in brackets will always be less than 1, and the expression will be positive. As a result, wind direction and projectile deviation will also have the same sign in the formula.

*Case 2*. If the sum of thrust and drag is a positive value, the parallelogram of forces will include an accelerating force $F_{SB}$ acting in the direction of travel and another force $F_{SQ}$ acting in the opposite direction to the wind (see Fig. 5.83). As a result, the projectile will drift against the wind. On account of the acceleration, the mean trajectory velocity in the subtrahend of Eqn 5.9:4 will now be greater than the horizontal component of muzzle velocity and the expression in brackets will become negative; the projectile will now move laterally against the wind.

*Case 3*. It is possible to design a rocket in such a way that its thrust exactly compensates for drag. This is known as *vacuum thrust*. Airspeed will then remain constant and the expression in brackets in Eqn 5.9:4 will evaluate to zero. The rocket will not drift, even under the influence of a high crosswind.

This kind of motor design has been used in portable anti-tank rockets and it is surprising – even for a specialist – to see how such projectiles still fly straight in a storm.

A rocket motor does not usually provide acceleration throughout the trajectory, which is therefore divided into a powered and a free-flight phase. It is worth bearing in mind that by the time its motor burns out, the rocket has attained a significant velocity directed against the wind, and this must fall to zero before it can start to drift in the direction of the wind. As a result, the wind drift that occurs during the boost phase will virtually never be cancelled out by drift in the opposite direction during the midcourse phase.

Section 5.4.4.5 describes how crosswind drift can be determined empirically. The method described can also be used for rockets. Experiments will result in drift against the wind as described under “Case 2” above.

# 6 An outline of terminal ballistics

## 6.1 Introduction

### 6.1.1 Preliminary remarks

From a physical point of view, an impact between a projectile and a solid object is a highly complex process. It takes place over a very short timespan and involves large forces and pressures (see Table 2.8, p. 49). In this environment, materials often behave differently from what one is used to in relatively slow processes. Those are based on known material properties determined under static or quasi-static conditions, whereas very different values of strain, strength, etc. may occur during highly dynamic processes.

This is largely due to the rapidity of these events, which does not allow the materials the time to react "normally". The process of a rifle bullet striking a steel plate lasts only about 50 μs, for instance.

The consequence of this is that most terminal ballistics processes largely elude our imagination, and that often leads to mistaken attempts at explanation. One should therefore exercise caution when interpreting interactions between projectiles and materials.

This chapter is not a detailed discussion of terminal ballistics; it merely provides an insight into typical qualitative and empirical laws. Detailed treatments of terminal ballistics processes can be found in Z. ROSENBERG, E. DEKEL (2012) and J. A. ZUKAS et al. (1992) for hard materials and in A. BHATNAGAR et al. (2016) and R. C. LAIBLE et al. (1980) for ballistic fabrics.

### 6.1.2 Basic laws

Penetration and perforation processes and the potential for destruction of material are determined by two sets of laws related only to the energy of the projectile and its sectional density. The energy density is one important parameter:

$$(6.1{:}1) \qquad ED = \frac{\frac{1}{2}\cdot m\cdot v^2}{A} = \tfrac{1}{2}\cdot q\cdot v^2 \qquad [J/mm^2]$$



B. P. Kneubuehl, *Ballistics*,
https://doi.org/10.1007/978-3-662-69237-0_6

In the above equation, A denotes the contact area between the projectile and the material. If the longitudinal axis of the projectile coincides with its path, A is the cross-sectional area of the projectile or the calibre cross-sectional area; if the projectile is moving with a large angle of attack, A is the shadow area in the direction of motion. The sectional density q is of course calculated using the same area A.

The following two basic rules apply:

– The *penetration depth* $s_e$ of a projectile in a material is approximately proportional to its (instantaneous) energy density and hence directly proportional to its sectional density:

$$(6.1{:}2) \qquad s_e \propto ED = \tfrac{1}{2} \cdot q \cdot v^2 \qquad [m], [cm]$$

– The *local damage potential* of a projectile corresponds to the energy that it transfers locally to the material, which is approximately proportional to its instantaneous energy and indirectly proportional to its sectional density:

$$(6.1{:}3) \qquad W(x) = \frac{\Delta E_x}{\Delta x} \propto \frac{E(x)}{q} \qquad [J/cm]$$

Where $W(x)$ = local damage potential at location x, $\Delta E_x$ = energy delivered at x over distance $\Delta x$ and $E(x)$ = remaining energy of projectile. $W(x)$ therefore corresponds to the gradient of the graph of energy transfer as a function of penetration depth.

Penetration depth and damage potential are therefore inversely proportional to each other; a projectile that penetrates deeply will destroy relatively little along its path, while a projectile of similar impact energy that causes a larger amount of damage after penetration will penetrate less deeply.

### 6.1.3 Types of material

Solid materials can be divided into two groups as regards their reaction to projectiles:

– Glass and vitreous (brittle) materials such as ceramics
– All other materials (including transparent plastic)

This classification, which may at first seem somewhat unusual, is based on the way in which a material responds to a local load (energy input). In glass, cracks form and propagate at high speed (4000–6000 m/s). They therefore precede the projectile, which hence encounters only fragmented glass as it passes through. In addition, radial cracks radiate from the shot hole and annular cracks form around the point of impact. Destruction occurs in a large area around the shot channel (Fig. 6.1 a). In all other building materials, the damage caused by the projectile remains confined to a small radius about its path (Fig. 6.1 b), along which the projectile continuously encounters undestroyed material.

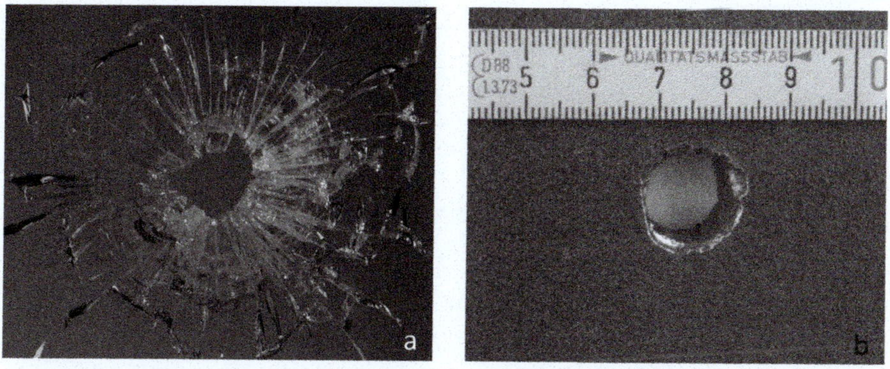


**Fig. 6.1.** **a**. Bullet hole in glass. Radial cracks form, usually accompanied by an annular crack at a certain distance from the centre. The glass inside the annular crack is badly damaged.
**b**. Bullet hole in steel. The area damaged is only slightly larger than the diameter of the bullet.

Perforation of these two classes of material is governed by different physical properties of the projectile. While *impact energy* plays a central role in glass, penetration depth in other materials depends primarily on *energy density* at the point of impact (see Eqn 6.1:2).

### 6.1.4 Terms and definitions

In addition to impact energy, energy density and the material properties of the target and projectile, the angle at which the projectile strikes the target plays an important role. The obvious approach is to specify the *angle of incidence* $\alpha$, i.e. the angle between the *line of arrival* and the tangential plane at the point of impact (labelled *Target plane* in Fig. 6.2). It is a simple matter to specify this angle if the target is either vertical or inclined by some angle about an axis of rotation perpendicular to the line of arrival.

Determining the angle of incidence becomes somewhat more complex if the target plane is arbitrarily positioned in space. If it is inclined from the vertical position by an angle $\kappa$ and rotated by $\delta$, the result is

$$(6.1{:}4) \qquad \alpha = \arctan\left\{\tan(90° - \kappa)\cdot\cos\delta\right\}, \qquad \kappa > 0 \qquad [-]$$

In practice, however, the complementary angle to the perpendicular to the target plane is usually used, rather than the angle of incidence. This angle $\beta$, which we shall refer to as the *NATO angle*, can be easily measured at the target if the line of arrival is known, without the need to know the tilt and rotation angles (see Fig. 6.2, next page). A perpendicular impact therefore means a NATO angle of 0°.

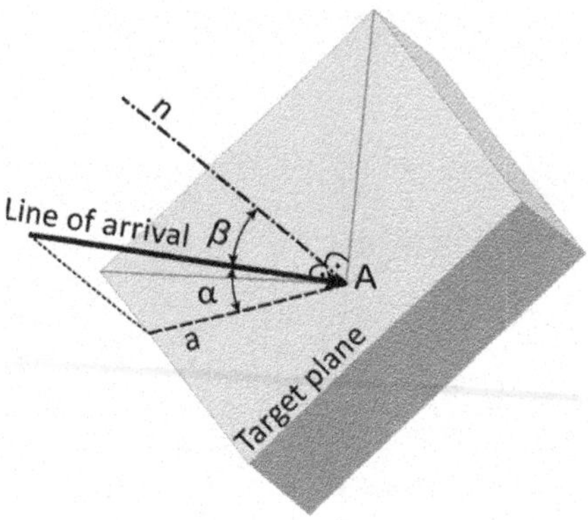


**Fig. 6.2.** Angle of incidence and NATO angle for an arbitrarily inclined target. n is the perpendicular, a the projection of the line of arrival onto the target plane. The angle of incidence is denoted by α and the NATO angle by β. The line of arrival lies with n and a on a plane normal to the target plane.

### 6.1.5 The Backman-Goldsmith diagram

Terminal ballistics is the ballistic subdiscipline that deals with the physical processes of interaction between a projectile and another (solid or liquid) body – the target. The projectile can do one of three things:

- Perforate the target, i.e. pass through it and emerge on the rear side, either intact or fragmented
- Remain embedded in the target
- Rebound from the target (i.e. ricochet), either intact or fragmented.

Which of these things happens depends primarily on the angle of incidence and the impact energy of the projectile. The structure and design of the projectile and the material and extent of the target also play a significant role. We must always remember that the projectile and the target influence each other.

The phase diagram produced by M. E. Backman and W. Goldsmith (1978) provides an excellent overview of the interaction between projectile and target (see Fig. 6.3). This diagram shows whether a projectile will perforate, ricochet or remain embedded, depending on its NATO angle and impact velocity rather than its energy.

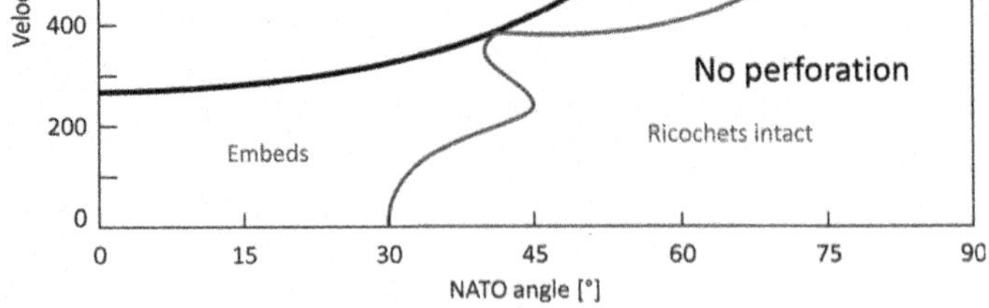


**Fig. 6.3.** Backman-Goldsmith diagram for a 6.35 mm pointed bullet and a 6.35 mm aluminium plate. The curves were partly calculated and partly determined experimentally (from M. E. Backman, W. Goldsmith, 1978).

Given the decisive role played by target material and projectile design, each projectile-target combination requires its own phase diagram.

Producing a Backman-Goldsmith diagram for a particular combination requires extensive and elaborate firing tests, as one must gather data not only for the entire range of angles but also for a wide range of velocities. In their paper, the above-mentioned authors presented such a diagram for a 6.35 mm (.25″) calibre pointed bullet and a 6.35 mm aluminium plate (Fig. 6.3). This would appear to be the only such diagram created.

## 6.2 Perforation models

### 6.2.1 The plugging model

Thin plates of hard material, of which the thickness is less than the length of the projectile, will deform little when perforated. At impact velocities of less than approximately 500 m/s, the projectile removes a disc or plug from the plate. The diameter of the resulting hole is usually slightly greater than that of the projectile (see Fig. 6.4, next page).

In a first, rough approximation, the shear fracture stress and shear area can be used to determine the shear work, which corresponds to the energy expended by the projectile to achieve perforation. We therefore obtain the following equation:

(6.2:1) $$E_S = C_S \cdot (\pi \cdot d \cdot D) \cdot D = C_S \cdot \pi \cdot d \cdot D^2 \qquad [J]$$

where $E_S$ = shear energy, d = calibre, D = plate thickness, and $C_S$ = the material-dependent shear fracture stress typical under the conditions obtaining.

Rearranging and extending Eqn 6.2:1, we obtain an equation for the maximum thickness perforated as a function of energy density ED:

(6.2:2) $$D = \frac{1}{2 \cdot \sqrt{C_S}} \cdot \sqrt{ED \cdot d} = K_S \cdot \sqrt{ED \cdot d} \qquad [m]$$

with a constant value of $K_S$. The maximum thickness perforated is therefore proportional to the square root of energy density times calibre. For a given energy density, a larger-calibre bullet has better perforation characteristics.

### 6.2.2 The displacement model

A projectile creates a permanent channel when fired through a plastically deformable object, by displacing the material laterally. The volume created in the process is proportional to the energy that the projectile transfers to the object; this is true not only overall but also locally.

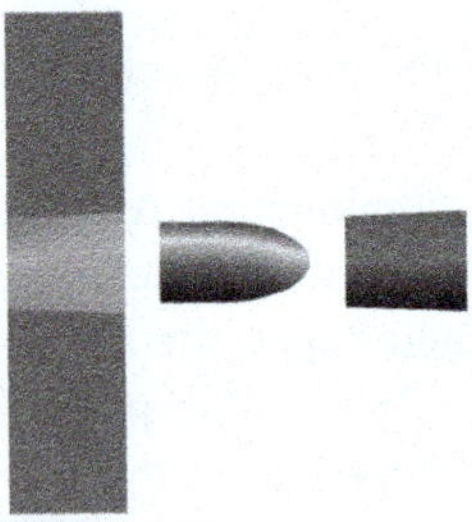

**Fig. 6.4.** Plugging. The energy required for plugging results in a decrease in projectile velocity.

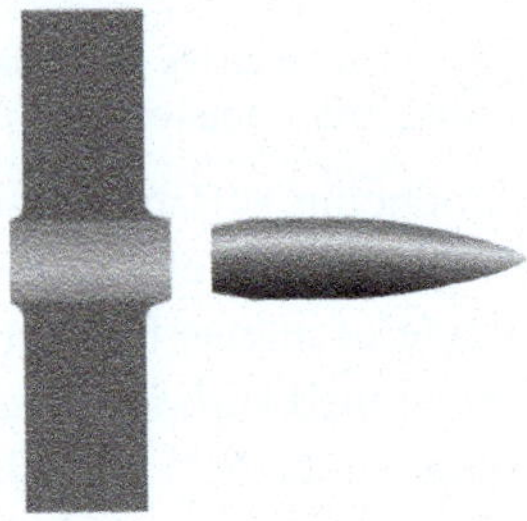

**Fig. 6.5.** Displacement. The energy expended is proportional to the volume created in the plate and results in a decrease in projectile velocity.

This law (MARTEL's Law) is attributed to French ballistician E. MARTEL, who was active at the turn of the 19th and 20th Centuries. Unfortunately, it has not been possible to find any specific document on this. When discussing the plugging model, M. E. BACKMAN, W. GOLDSMITH (1978) simply mention MARTEL as having published in 1897.

However, it was possible to clearly verify MARTEL's law by conducting test firings into ballistic soap using a wide variety of projectiles (see B. P. KNEUBUEHL et al., 2008).

Above a certain velocity, a projectile passing through a thin metallic plate clearly displaces material laterally rather than punching out a plug (Fig. 6.5). Using MARTEL's Law, we obtain

$$(6.2{:}3) \qquad E_V = C_V \cdot \frac{\pi}{4} \cdot d^2 \cdot D \qquad [\mathrm{J}]$$

where, d = calibre, D = plate thickness and $C_V$ is the material-dependent proportionality factor.

If we divide Eqn 6.2:3 by the calibre cross-sectional area, it becomes apparent that maximum thickness perforated is proportional to energy density:

$$(6.2{:}4) \qquad D = \frac{1}{C_V} \cdot ED \qquad [\mathrm{m}]$$

It is interesting to note that the maximum thickness perforated does not depend on the calibre of the projectile, in contrast to the plugging model.

The displacement model can be applied to most hard, deformable materials at higher impact velocities (approx. 600 to 1000 m/s for metals). In practice, we see that the direct relationship between energy density and perforation capacity holds good for projectiles that behave in a similar manner in the target material as regards deformation or tendency to fragment.

### 6.2.3 Perforation of thin sheets of material

When a projectile passes through thin plates or layers of metals or other materials (including fabrics), we often observe the following behaviour:

A sheet of a material just stops a bullet with impact energy $E_{thr}$ (the threshold energy). At a slightly higher impact energy $E_{thr+}$, the bullet passes through the sheet. The residual energy behind the sheet is $\alpha \cdot E_{thr+}$ ($\alpha < 1$) (see Fig. 6.6). The energy $E_P$ required to perforate the sheet is therefore:

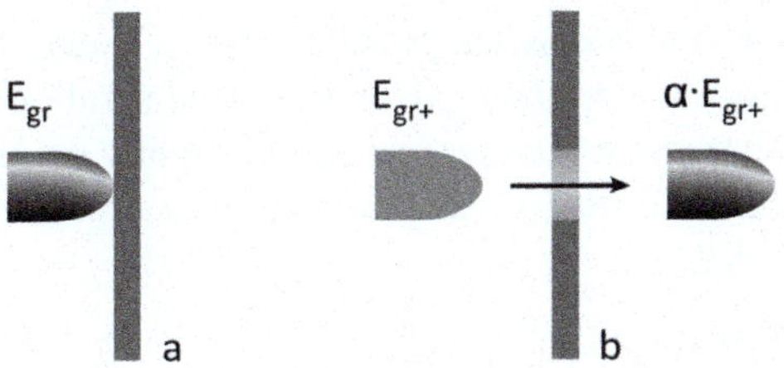


**Fig. 6.6.** Energy loss as a projectile passes through thin sheets of material.
**a.** A sheet of a material just stops a bullet with impact energy $E_{thr}$.
**b.** At $E_{thr+}$ the bullet just perforates the sheet. Residual energy $\alpha$ is likely to be approximately 0.7 to 0.9.

(6.2:5) $$E_P = E_{thr+} - \alpha \cdot E_{thr+} = (1-\alpha) \cdot E_{thr+} \quad [J]$$

$\alpha$ generally lies between 0.7 and 0.9. Because $E_{thr+}$ is only slightly larger than $E_{thr}$, the following is not necessarily true, contrary to expectations:

(6.2:6) $$E_P \ll E_{thr} \quad [J]$$

In other words, the target absorbs less of the projectile's energy if it passes all the way through than if it remains embedded. This is because as the projectile starts to pass through the sheet, a rupture line forms around the channel it is creating, preventing further energy being transferred from it to the sheet. As a result, it retains more residual energy.

If the energy of the projectile is greater, i.e. if its velocity is higher, the rupture line will form more quickly. This means that the proportion of energy transferred to the target sheet is even lower. Indeed, as velocity increases $\alpha$ also increases.

These relationships can explain some initially surprising phenomena. For instance, one will never find a bullet immediately under the victim's skin on the entry side. Either the skin will stop the bullet or else it will still have enough energy to penetrate at least some 2 cm into the underlying tissue. Likewise, a bullet will either be stopped by a protective vest or, if it perforates it, it will still be capable of causing noticeable injury.

## 6.2.4 The erosion model

Author: Dipl.-Ing. ETH Wilhelm ODERMATT

### *6.2.4.1 Description of model*

In addition to the plugging and displacement models, there is another form of projectile-target interaction, which occurs at high velocities and with projectiles made of heavy metals (sintered tungsten alloys or depleted uranium). Projectile and target material are deformed plastically at impact velocities of approx. 1000 m/s and higher and thus behave approximately like liquids.

In contrast, solid tungsten carbide projectiles displace the target material (armour steel) even at very high impact velocities and produce a channel of slightly less than their own diameter (in accordance with the displacement model, 6.2.2). They remain intact or break up, but do not deform.

The projectile material flows outwards in a mushroom shape and the target material is ejected via the entry hole. This is known as hydrodynamic or erosive penetration. The tip of the (shortening) projectile penetrates the target material at slightly more than half the impact velocity, while the (still intact) rear part continues to move almost at impact velocity (see Figs. 6.7–6.9).

However, deceleration does occur, in steps, on account of the shock waves that pass through the penetrator and are reflected at the tail.

The crater diameter is about twice that of the projectile. This results in a much larger crater volume for a given penetration depth than with the displacement model. Consequently, about four times the kinetic energy is required per unit penetration depth. To achieve the required penetration depth, long rod penetrators are used, which must be fin-stabilized for stable flight (see 5.6.2.2). These are known as long-rod projectiles, and the heavy metal part is called a *penetrator*. This type of sub-calibre projectile must be enclosed in a sabot for firing, as described in 3.7.6.3. If the impact energy is just sufficient for complete perforation of the target (the perforation limit),

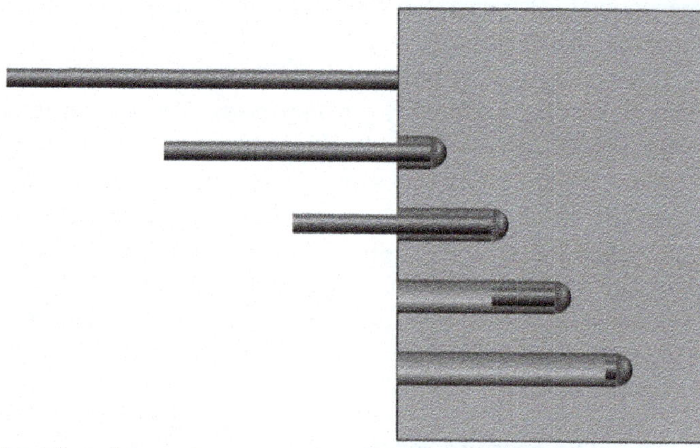

**Fig. 6.7.** Schematic representation of a long rod penetrator penetrating a semi-infinite target.

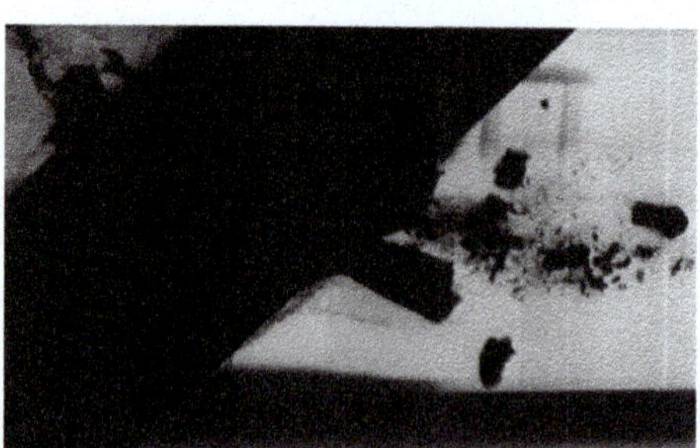

**Fig. 6.8.** X-ray flash image of perforation by a log rod penetrator.

**Fig. 6.9.** Stub remaining after the penetrator has perforated an inclined plate at the perforation limit.

a residual stub with a length of one to one and a half times the penetrator diameter emerges from the rear of the target plate, producing a plug and a large number of splinters.

#### *6.2.4.2 Hydrodynamic approach*

If we ignore dynamic strength and assume uniform penetration, we can describe the equilibrium of the pressure at the stagnation point using Bernoulli's equation as follows:

$$(6.2{:}7) \qquad \frac{1}{2}\cdot\rho_P\cdot(v-u)^2 = \frac{1}{2}\cdot\rho_T\cdot u^2 \qquad [\mathrm{J/m^3}]$$

where $\rho_P$ = penetrator density, $\rho_T$ = density of target material, v = impact velocity and u = penetration velocity.

This gives us the penetration velocity:

$$(6.2{:}8) \qquad u = \frac{v}{1+\sqrt{\dfrac{\rho_T}{\rho_P}}} \qquad [\mathrm{m/s}]$$

We can now equate the time that elapses until the penetrator is consumed with the time for the entire penetration process

$$(6.2{:}9) \qquad \frac{L}{v-u} = \frac{P}{u} \qquad [\mathrm{s}]$$

where L = penetrator length and P = penetration depth, both in metres.

If we substitute Eqn 6.2:8 into this simultaneity of time, a brief calculation yields the formula known as the *square-root-density law*:

$$(6.2{:}10) \qquad P = L\cdot\sqrt{\frac{\rho_P}{\rho_T}} \qquad [\mathrm{m}], [\mathrm{mm}]$$

The thickness penetrated is therefore proportional to the square root of the ratio of the density of the penetrator material to the density of the target material.

V. P. ALEKSEEVSKI (1966) and A. TATE (1967) independently modified Bernoulli's equation by including the dynamic strength limits:

$$(6.2{:}11) \qquad \frac{1}{2}\cdot\rho_P\cdot(v-u)^2+Y_P = \frac{1}{2}\cdot\rho_T\cdot u^2+R_T \qquad [\mathrm{N/m^2}]$$

where $Y_P$ = strength of penetrator and $R_T$ = strength of target in the case of erosive penetration, u = the velocity of the penetrator tip and v = velocity at the rear of the penetrator. Using this relationship, TATE discussed three possible cases of hydrodynamic penetration.

If the strength of the target is greater than that of the penetrator ($R_T > Y_P$), the penetrator behaves like a liquid until the end of the penetration process, which is complete when the rear of the penetrator has decelerated to the following velocity:

$$v = \sqrt{\frac{2 \cdot |R_T - Y_P|}{\rho_P}} \qquad [m/s] \tag{6.2:12}$$

However, if the strength of the penetrator exceeds that of the target ($R_T < Y_P$), the rear of the penetrator attains the velocity of the tip (given by Eqn 6.2:12) at the same time. From this point until the end of the penetration process, the penetrator behaves like a solid body.

If the strengths of the two materials are equal, hydrodynamic penetration continues until the tip and the rear of the penetrator have come to rest.

A. TATE (1967) includes a detailed account of this approach, albeit without taking into account the aspect ratio of the penetrator or the obliquity of the target.

#### 6.2.4.3 Reference targets

The semi-infinite target and the oblique armour plate can be used as reference targets for long-rod projectiles. The semi-infinite target (a single steel plate of sufficient width and depth that no interference from free surfaces can occur) is widely used, because one test generally gives a valid result (see Fig. 6.10). However, the plate must be cut open to measure the penetration depth. The crater is usually clogged with projectile and target plate debris.

For large-calibre long-rod projectiles, the penetration depth is often so great that it is necessary to arrange several plates in a stack (Fig. 6.11). In this type of target, only compressive stresses are transmitted across the interfaces between the plates, while tensile and shear stresses are not. As a result, perforation performance is often

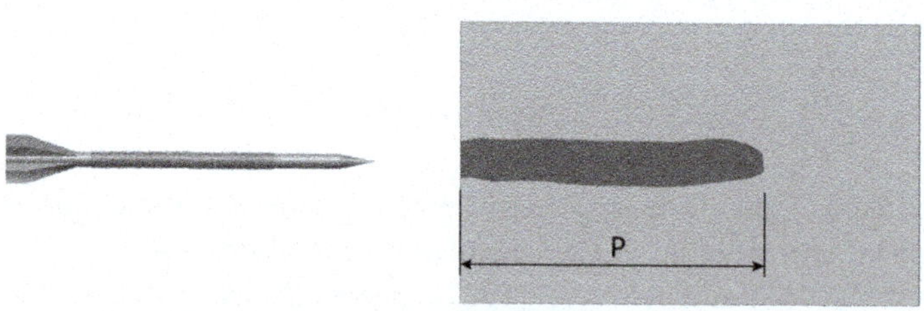


**Fig. 6.10.** Semi-infinite target (schematic representation).

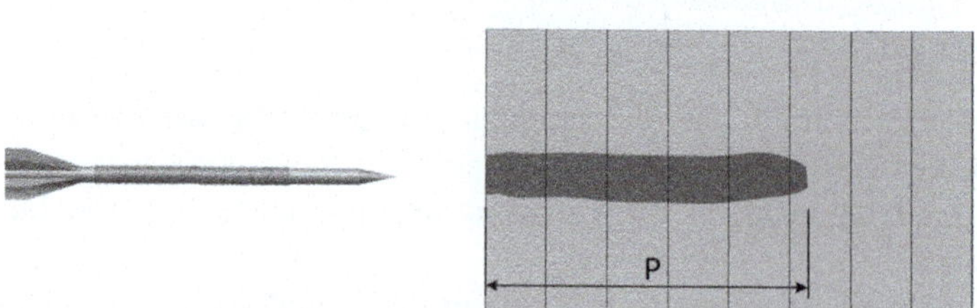


**Fig. 6.11.** Target in the form of a stack of plates (schematic representation).

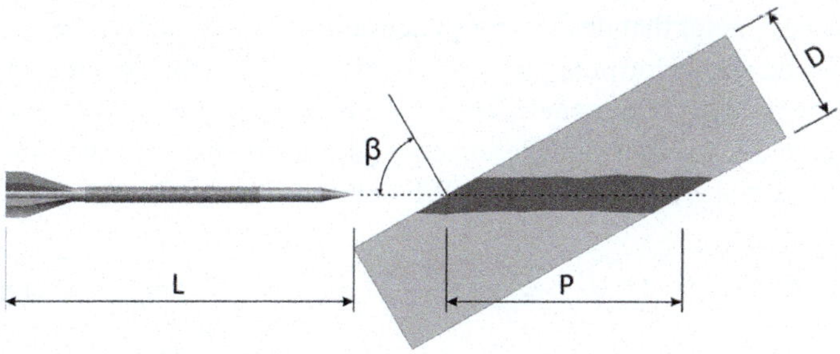


**Fig. 6.12.** The oblique single target (schematic representation); β is the NATO angle (see Fig. 6.2). P = D/cos β.

massively overestimated. Results will be reproducible only with the same ballistic conditions, target arrangement and material properties (same number of plates of the same thickness and the same strength). A stack of plates cannot be compared with a semi-infinite target.

It is relatively easy to determine the perforation limit when using an *oblique single target* (Fig. 6.12). The long rod should penetrate the armour plate so narrowly that the residual penetrator has virtually no energy left. There are two ways to achieve this: adjust the obliquity of the plate until the long rod just perforates it, while maintaining impact velocity constant, or keep the angle of attack constant and adjust the velocity of the long rod until it just perforates the target. The results are reproducible and, as will be shown below, are even transferable to other situations; they can be transposed to specified conditions if required.

P = penetration depth. For an oblique single target, the perforation length is defined as the plate thickness divided by the cosine of the NATO angle. This is also known as line of sight perforation.

### 6.2.4.4 Most significant parameters, dimensional analysis

For a long rod, the parameters that determine penetration depth P for a semi-infinite target, or the perforation limit for a single plate, are diameter d, length L, density $\rho_P$ and impact velocity v. For the target, the parameters are thickness D, NATO angle β, density $\rho_T$ and tensile strength $\sigma_T$. The semi-infinite target is generally placed vertically for this type of test, so the NATO angle is 0°.

One method for deriving formulae from these parameters is to use the *π Theorem* (E. BUCKINGHAM, 1914). Dimensionless terms can be determined from the initial variables, even if the exact relationship in the form of an equation is not yet known.

One can describe the dimensionless, relative perforation limit (the dependent variable) as follows:

$$(6.2{:}13) \qquad \frac{P}{L} = f\left(\frac{L}{d}, \beta, \frac{\rho_P}{\rho_T}, \frac{\sigma_P}{\sigma_T}, \frac{\sigma_T}{\rho_P \cdot v^2}\right) \qquad [-]$$

Impact velocity is the parameter that determines penetration. Consequently, the last dimensionless term in the above formula plays an important role in the penetration process. Analysis of many tests with penetrators of identical aspect ratio (L/d), on oblique targets of identical obliquity but different strength, shows that the following exponential approach gives the perforation limit P/L very accurately:

(6.2:14) $$\frac{P}{L} = a \cdot e^{\frac{-c \cdot \sigma_T}{\rho_P \cdot v^2}}$$ [-]

This approach takes account of the S-shaped curve of penetration as a function of impact velocity. We can assume the parameter c to be constant for common armour plates (with tensile strengths of 700–1000 N/mm$^2$ or Brinell hardnesses of 200–300). If we also keep the penetrator density and the plate tensile strength constant, we obtain a 2-parameter fit:

(6.2:15) $$\frac{P}{L} = a \cdot e^{-\left(\frac{b}{v}\right)^2}$$ [-]

It is possible to determine parameters a and b reliably from a small number of test results, from which we can estimate penetration depth as a function of impact velocity. Fig. 6.13 shows the typical curve of perforation limit vs impact velocity for a selected penetrator type.

The influence of the densities of penetrator and target material can be taken into account as described in 6.2.4.2 (Eqn 6.2:10).

(6.2:16) $$\frac{P}{L} \sim \sqrt{\frac{\rho_P}{\rho_T}}$$ [-]

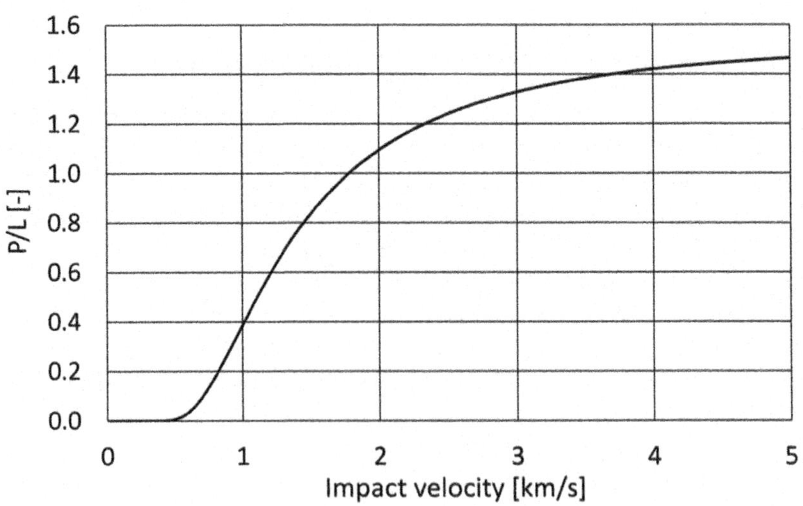


**Fig. 6.13.** Relative perforation limit for a generic long-rod penetrator.

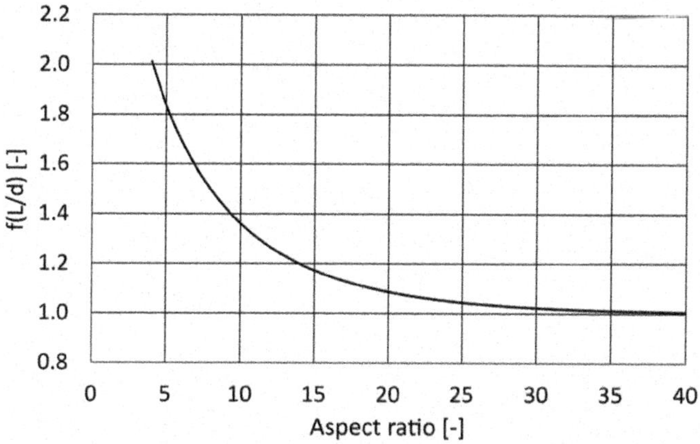


**Fig. 6.14.** Influence of the aspect ratio from Eqn 6.2:17. Coefficients $b_0$ and $b_1$ are taken from Table 6.1, p. 317.

The ratio of the strength of the heavy metal penetrator to that of the steel target material is of virtually no significance, because of the large difference in density ($\rho_P/\rho_T \sim 2.3$).

The factor c in the exponent of Eqn 6.2:14 ceases to be a constant in the upper range of the tensile strength of the target, but can be represented very well by a linear strength function (see 6.2.4.5).

The influence of the aspect ratio L/d can be described using the following approach, which has been adapted to the results of over 100 tests using tungsten, depleted uranium and steel penetrators (see Fig. 6.14):

$$(6.2{:}17) \qquad f\left(\frac{L}{d}\right) = \frac{1}{\tanh\left(b_0 + b_1 \cdot \frac{L}{d}\right)} \qquad [-]$$

This is valid for values of L/d from 4 to ∞ (asymptotic approximation to 1). If the aspect ratio L/d is lower, perforation length will be greater than for purely hydrodynamic penetration. For modern long-rod projectiles with an L/d greater than 30, this influence is negligible (2%).

The influence of the NATO angle can be described using a modified cosine law:

$$(6.2{:}18) \qquad f(\beta) = \cos^m \beta \qquad \text{m according to Table 6.1.} \qquad [-]$$

The apparently greater perforation depth at larger NATO angles (see Fig. 6.15) is related to the definition of $P = D/\cos\beta$. However, penetration is linear only up to

**Table 6.1.** Parameters for Eqn 6.2:19

| Penetrator material | $b_0$ | $b_1$ | m | a | $c_0$ | $c_1$ |
|---|---|---|---|---|---|---|
| Tungsten | 0.283 | 0.0656 | -0.224 | 0.994 | 134.5 | -0.148 |
| DU | | | | 0.825 | 90.0 | -0.0849 |

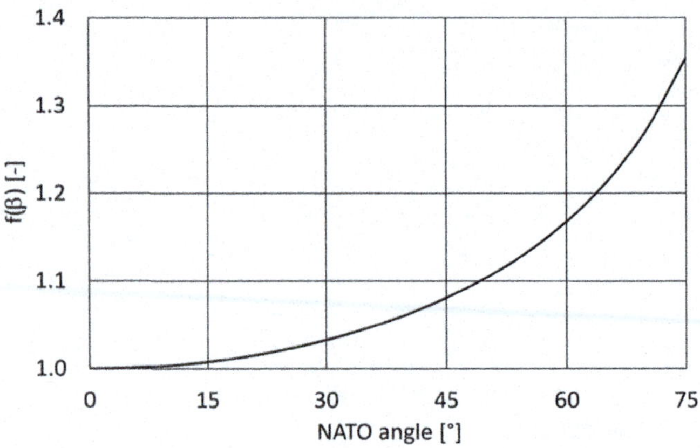


**Fig. 6.15.** Influence of the NATO angle from Eqn 6.2:18

the point at which the free surface of the rear of the target begins to exert an influence. As the perforation channel nears the rear face of the plate, the angle between perforation channel and plate surface approaches 90°. This effect on perforation occurs earlier at more pronounced angles of inclination than with plates that are vertical or only slightly inclined.

### 6.2.4.5 Complete approximation formula for the perforation threshold

The parameters discussed in the previous section can now be combined into a single equation that is valid for both tungsten and DU penetrators. For system analysis studies, it is advantageous that the individual parameters act multiplicatively and do not influence each other. When fitting the parameters, it proved convenient to use the Brinell hardness of the target material instead of the tensile strength (the "Odermatt equation"):

(6.2:19) $$\frac{P}{L} = a \cdot \frac{1}{\tanh\left(b_0 + b_1 \cdot \frac{L}{d}\right)} \cdot \cos^m \beta \cdot \sqrt{\frac{\rho_P}{\rho_T}} \cdot e^{-\frac{s^2}{v^2}} \qquad [-]$$

with $$s^2 = \frac{(c_0 + c_1 \cdot HBW_T) \cdot HBW_T}{\rho_P}, \qquad [m^2/s^2]$$

lengths in mm, v in km/s and densities in kg/m$^3$. Table 6.1 lists constants a–c for two penetrator materials.

For cylindrical penetrators, L is the total rod length. However, the tip of the penetrator is usually conical. In such cases, one replaces the cone with a cylinder of the same mass as the cone and the same diameter as the penetrator.

The lower limit of validity for Eqn 6.2:19 is determined by the velocity, which can be calculated from:

(6.2:20) $$\frac{s^2}{v^2} = \frac{\left(c_0 + c_1 \cdot HBW_T\right) \cdot HBW_T}{\rho_P \cdot v^2} = 1.5 \qquad [-]$$

from which we can calculate the velocity limit:

(6.2:21) $$v_{limit} = \sqrt{\frac{2 \cdot \left(c_0 + c_1 \cdot HBW_T\right) \cdot HBW_T}{3 \cdot \rho_P}} \qquad [m/s]$$

It is noteworthy that this limit depends entirely on material parameters. There is no real upper limit, but calculated values may differ slightly from results if the impact velocity is greater than 2.5 km/s. The maximum velocities of the test results used to fit the parameters lay just below 2 km/s.

Lower bound examples:

| | | | |
|---|---|---|---|
| Tungsten | Density 17,500 kg/m³ | $HBW_T$ 250 | v > 965 m/s, |
| Depleted uranium | 18,600 kg/m³ | 250 | v > 785 m/s. |

## 6.3 The behaviour of various materials

### 6.3.1 Basic considerations

This section presents the results of terminal ballistics tests involving a number of target materials. Testing was conducted using small arms projectiles, i.e. bullets with a maximum energy of approximately 15 kJ.

One should regard the resulting rules purely as non-binding guidelines or indications. Projectile materials and designs often have a significant influence on terminal ballistics behaviour. If no reliable results are available (such as a test certificate from an official test centre or a proof house), testing is usually essential.

One should also bear in mind that a test certificate only confirms that the specimen submitted to the inspection body has passed the test. In most cases, test guidelines define the criteria for these specimens in precise terms. Whether the tested material will behave in the same way in a real installation as in the test is often uncertain. Bullet-resistant glass, for example, is tested using a 50 × 50 cm sample. The author is not aware of any study proving that a 2 × 3 m glass wall offers the same, better or worse protection when subjected to the same shot.

Terminal ballistic processes are always *interactions* between projectile and target. This extremely important point sometimes makes for astonishing and initially inexplicable events. For instance, it may happen that bullets start to pass through a target at a certain velocity, as one would expect, but that if one increases velocity still further, they are once again stopped by the same target. This is because the interaction has reached an energy level at which the projectile is so deformed that its sectional density, and thus its energy density, have fallen below the perforation threshold. What also plays a decisive role is the duration of the process, which decreases as speed increases. As a result, the projectile deforms faster than it penetrates. Bullets striking the target at some still higher velocity will once again perforate it, and

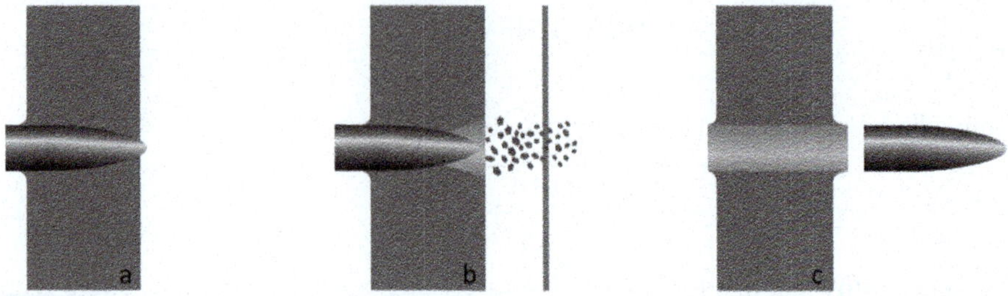


**Fig. 6.16.** Possible definitions of perforation. **a**. The bullet is visible on the rear of the target. **b**. Parts of the projectile or target perforate a witness foil. **c**. The bullet completely exits the rear side of the target.

perforation will continue to occur however much velocity increases.

This behaviour may be observed with textile protective vests or steel plates, for example, and also depends on the projectile.

The question also arises as to when a shot is considered to have perforated a target. This definition includes not only the situation in which the entire projectile emerges from the rear of the target (Case c in Fig. 6.16) but also those situations where the projectile presents a potential hazard (Case b), or where certain statistical criteria are fulfilled (Case a). (If the bullet is visible at the rear of the plate, complete perforation is quite possible within the usual dispersion of impact velocity and other parameters). In addition to these three definitions, a number of others appear in test guidelines, in particular for the case where the bullet does not remain in the shot channel but has rebounded or fractured on the firing side.

### 6.3.2 Steel

The maximum thickness of steel plate that a bullet can perforate is always less than its own length. This means that the path of the bullet within the target material is too short for it to tumble (e.g. as a result of the angle of attack at impact) and thus change its sectional density. As a result, perforation is little affected by impact parameters other than impact energy (angle of attack, for instance).

Under the displacement model approach (6.2.2), penetration depth depends primarily on energy density and is proportional to it in accordance with Eqn 6.2:4. Results of shooting tests against steel plates collected over many years even show a linear relationship between energy density and perforation threshold, with correlation coefficients in the range of 0.90–0.98. See Fig. 6.17. The solid lines are from tests with jacketed lead-core bullets, the dashed line is for steel and hard-core bullets ("AP" stands for "armour piercing").

This clearly shows the important role played by the material properties of the target plates in the perforation process, with the tensile strength (or hardness) of the steel being the primary factor. The harder the steel, the more gradual the increase

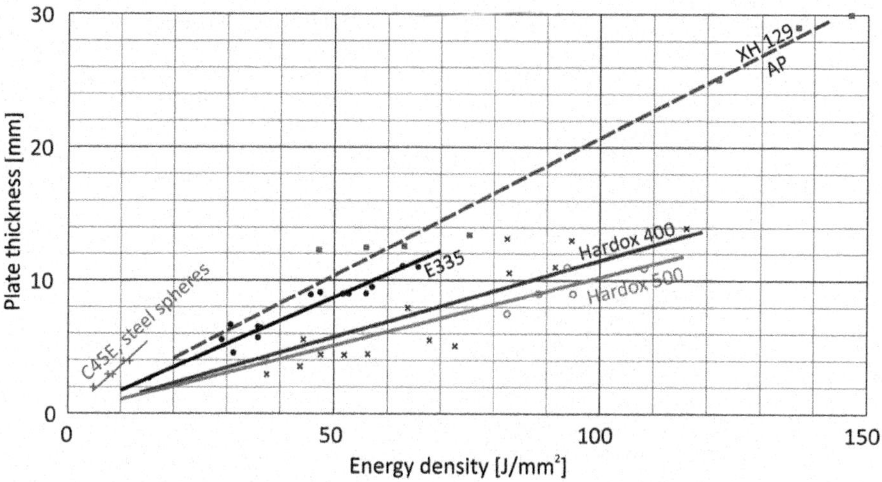


**Fig. 6.17.** Relationship between maximum thickness of steel plate perforated and the energy density of the impacting projectile. The relationship is almost linear. The dashed line is for steel-core bullets, while the other lines are for lead-core bullets. The influence of steel quality is clear.

in average perforation threshold (for a given bullet type) with increasing energy density.

For example, a 7.62 mm NATO FMJ lead-core bullet with an energy density of 63 J/mm$^2$ will perforate 11 mm of steel with a strength of 650 N/mm$^2$ but will only just perforate 7 mm of high-grade steel (1250 N/mm$^2$). Table 6.2 lists the strengths of typical steels and the associated Brinell hardness.

When using the graph in Fig. 6.17, one should bear in mind that the lines represent the mean perforation threshold. The design and construction of a projectile can significantly improve or degrade its perforation capability. This means that one must always use plate that is thicker than that indicated in the graph by one or two standard thicknesses.

In the above example with the NATO soft-core bullet, one must use 15 mm of low-strength (E335) or 10 mm of high-quality (Hardox 400) steel, even though the graph indicates perforation thresholds of 11 mm and 7 mm respectively for these steel grades.

**Table 6.2.** Steel grades

| Designation | Tensile strength [N/mm$^2$] | Brinell hardness [-] |
|---|---|---|
| S235 (St 37-2) | 340–470 | 102–140 |
| E335 (St 60-2) | 570–710 | 181–211 |
| C45E (Ck 45) | 700–850 | 208–252 |
| Hardox 400 | 1220–1500 | 360–440 |
| Hardox 500 | 1550–1920 | 450–550 |
| XH 129 | 1660–1840 | 480–530 |

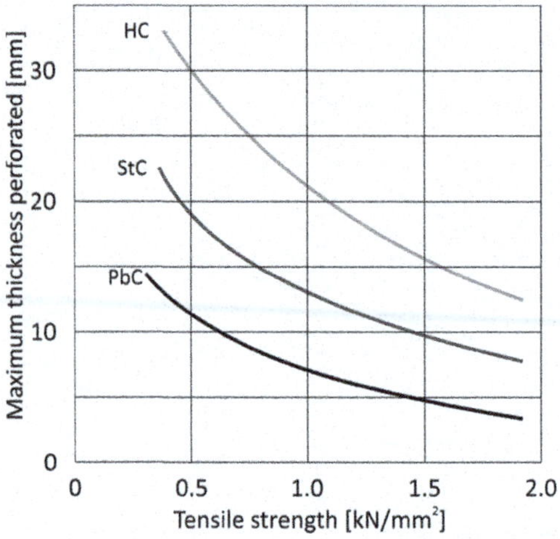


**Fig. 6.18.** Relationship between maximum thickness perforated and the tensile strength of the steel plate. Experimental results for soft core (PbC), steel core (StC) and tungsten carbide core (HC) bullets in 7.5 × 55.
Other calibres and bullets yield different curves.

If no reliable results for the projectile-target combination in question are available (e.g. a test certificate from a test centre), experimental testing is highly recommended and may indeed be essential.

Fig. 6.18 shows the influence of steel quality even more clearly, using experimental results for 7.5 × 55 bullets. As strength decreases, the plate thickness required for protection increases significantly. The harder the bullet material, the more rapidly the required thickness increases. The curves show how high-strength steels can reduce the weight of bullet-resistant structures by up to half.

It happens from time to time that someone unfamiliar with terminal ballistics replaces the high-strength steel required for protection with structural steel without saying anything, because it is cheaper and easier to work with. They believe that as long as the same thickness is used, the plate will provide the same protection. A test shot – which is always recommended – will quickly reveal that the wrong steel has been substituted.

### 6.3.3 Aluminium

In addition to steel, aluminium is also used for ballistic protection. Here again, one can initially assume that maximum thickness perforated is proportional to energy density. However, tests reveal not a linear relationship but a curve which, in contrast to that for steel, becomes approximately horizontal with increasing energy density (see Fig. 6.19). This makes aluminium useful at higher impact energies. Table 6.3 lists typical grades of aluminium.

The graph shows that as for steel plates, aluminium plates intended to provide protection against steel projectiles (Curves 1 and 2) must be thicker than those designed to protect against standard soft-core projectiles (Curve 3). However, the advantage of aluminium being almost three times less dense than steel almost disappears at low energy densities (below about 60 J/mm$^2$) due to the plate thicknesses

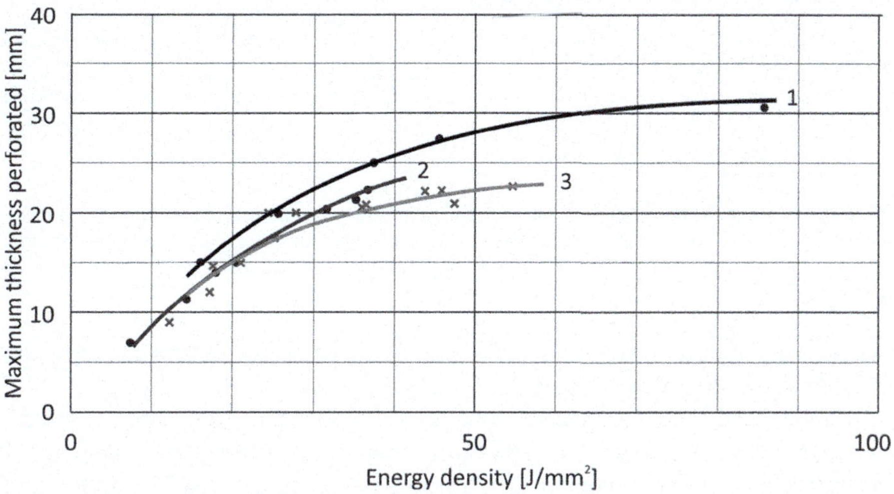


**Fig. 6.19.** Relationship between maximum thickness of aluminium plate perforated and the energy density of the impacting projectile. In contrast to steel, the curve becomes almost horizontal at higher energy densities. *Curve 1*: 12.5 mm steel spheres against Avional. *Curve 2*: 5.56 mm and 7.5 mm steel spheres against Anticorodal. *Curve 3*: Rifle bullets against Peraluman. Here again, material strength plays a role.

required. This is probably one reason why aluminium is rarely used for protection against bullets.

### 6.3.4 Concrete and masonry

Sand-lime bricks, clay bricks and non-reinforced concrete are only capable of absorbing low tensile and shear stresses. When a bullet hits such a material, the high pressure around the point of impact causes stresses that lead to conical spalling on the impact side. If one or two further shots strike approximately the same spot, cracking may occur on the rear side (depending on the wall thickness), allowing the next shot to cause material to break away on the rear side. The following shot will generally perforate the masonry completely (Fig. 6.20, next page).

Table 6.4 shows the results of a number of such experiments. Shots were fired at specially fabricated 1.2 × 1.5 m wall sections and concrete supplied by a concrete product manufacturer. The distance between weapon and target was 10 m for hand-

**Table 6.3.** Aluminium grades

| Designation | Tensile strength [N/mm²] | Brinell hardness [-] |
|---|---|---|
| Avional 152 (150) EN AW-2024 | min. 440 | min. 100 |
| Anticorodal 100 EN AW-6082 | 295–300 | 95 |
| Peraluman EN AW-5083 | 275 | 69 |

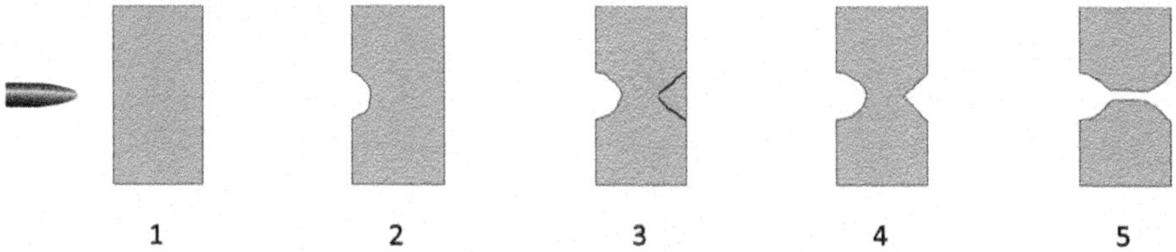


**Fig. 6.20.** Stages of perforation in masonry and non-reinforced concrete. The first shot produces conical spalling on the impact side (2). A few shots later, a similar crack pattern appears at the rear (3), and a piece breaks out following a further shot (4). The following shot usually results in perforation (5).

gun ammunition and 100 m for rifle ammunition. The energy figures quoted are the average values for these distances.

Neither sand-lime masonry nor concrete slabs offer much resistance to bullets. It is possibly to significantly improve protection by preventing the brick (or slab) from breaking out on the rear side. This is achieved by covering the back with a thin steel sheet (e.g. 3 mm). See Fig. 6.21. The bullets no longer reach the rear side, as they consume all their energy in crushing the material. This can double or even

**Table 6.4.** Results of tests on masonry and concrete (see text for explanations).

| Target material | Bullet | E [J] | Results |
|---|---|---|---|
| Sand-lime brick, 9 cm, no hole | 9 mm Luger FMJ | 580 | Spalling from rear after 6 shots. No perforation. |
|  | 44 Rem. Mag. semi-jacketed | 1340 | Spalling from rear after 2 shots. Perforation on 3rd shot. |
| Sand-lime brick, 12 cm, hole | 44 Rem. Mag. semi-jacketed | 1340 | Spalling from rear after 8 shots. No perforation after 10 shots. |
|  | 7.5 × 55, FMJ | 2575 | Spalling from rear after 1 shot. Penetration depth 8 cm. Perforation on 2nd shot. |
| Sand-lime brick, 15 cm, no hole | 7.5 × 55, FMJ | 2575 | Spalling from rear after 3 shots. No perforation. Penetration depth 7 cm. |
| | 7.62 mm NATO, steel core | 2505 | First shot: penetration depth 8 cm. Second and third shots: perforation. |
| Concrete slab, 6 cm, HPC 350 (0-18) | 5.56 × 45, FMJ | 1380 | Spalling from rear after 3 shots. Perforation on 4th shot. |
| | 7.5 × 55, FMJ | 2575 | Perforation on 1st shot. |
| Concrete slab, 7 cm | 5.56 × 45, FMJ | 1380 | Rear side still intact after 2 shots. Perforation on 3rd shot. |
| Concrete slab, 10 cm | 7.5 × 55, FMJ | 2575 | Crack on rear side after 2 shots. Perforation on 3rd shot. |
| | 7.62 mm NATO, steel core | 2505 | First shot: penetration depth 6 cm. Second shot: penetration depth 8 cm. Perforation on 3rd shot. |

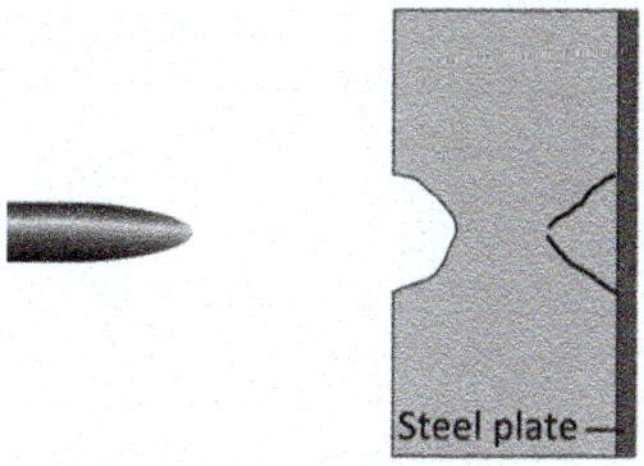


**Fig. 6.21.** Preventing material from breaking out at the rear of sand-lime brickwork and concrete slabs, e.g. by using a thin steel plate, increases protection against bullet perforation considerably.

triple the number of shots required to achieve perforation. Such protection is suitable for locations in which only a small number of hits are to be expected, e.g. at a certain distance from a target line on a shooting range. It is of course essential to conduct firing tests using bullets and building materials corresponding to the real-life situation for which protection is required. Special attention should be paid to lead-free (monobloc) bullets.

Brick masonry with a thickness of 18 cm is perforated by a single FMJ rifle bullet (distance 50 m, 3200 J). Such masonry only provides a certain degree of protection at projectile energy levels of up to 1700 J.

9 mm Luger bullets penetrate YTONG® blocks of aerated concrete (25 cm thick) to a depth of 16–18 cm. FMJ rifle bullets with over 3000 J of energy perforate them completely.

### 6.3.5 Wood

If the bullet can penetrate a material to a depth corresponding to its own length, penetration depth is determined not only by energy density (as per 6.2.2) but also by the stability of the bullet and its angle of attack at the point of impact. When passing through wood, long bullets usually rotate about a transverse axis, often more than once (i.e. they tumble). The more stable the bullet at the moment of impact and the smaller its angle of attack, the longer it retains a high energy density and the greater the depth to which it penetrates before tumbling begins. If a bullet rotates so that it is oriented perpendicular to its direction of travel, it presents a larger cross-sectional area, and both its energy density and its penetration depth decrease.

A bullet's ability to perforate wood therefore depends to a large extent on how soon it begins to rotate following penetration. The angle of attack at the point of impact and the stability of the bullet therefore influence penetration depth. Energy density alone does not allow us to even estimate penetration depth, in contrast to the case of a bullet striking a steel plate.

Fig. 6.22 (next page) shows experimentally determined values for dry, knot-free spruce and six military bullets: three 5.56 × 45 (223 Rem.) and three with calibres of 7.5 mm–7.62 mm. It is both surprising and noteworthy that two of the smaller calibres penetrate deeper into wood than the larger bullets at a range of 300 m,

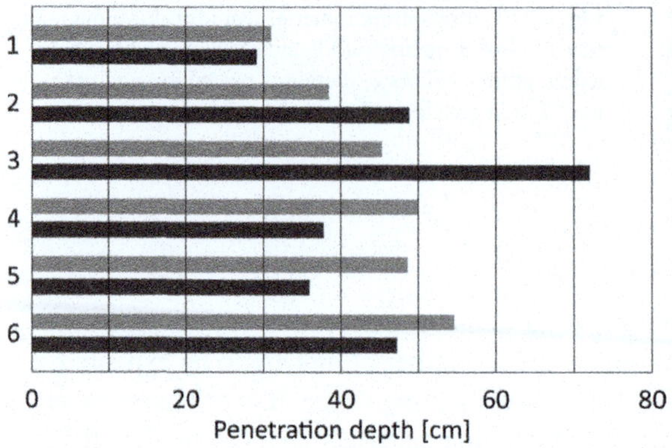


**Fig. 6.22.** Average penetration depths in knot-free spruce.
Light grey: 100 m.
Dark grey: 300 m.
1 223 Rem. SS92
2 5.56 mm NATO, SS109
3 5.6 mm GP 90
4 7.62 × 39 FMJ
5 7.62 mm NATO soft core
6 7.5 × 55 GP 11

despite their energy densities being approximately 25% lower (with 60% less energy).

This is at least partly because angle of attack is greatest shortly after the bullet leaves the muzzle, on account of the disturbances resulting from firing (see 4.2 and 4.3). Angle of attack decreases with distance. This can result in certain bullets penetrating deeper into wood at longer ranges than at shorter ranges, despite their lower energy and energy density. At still greater ranges, penetration depth will decrease due to the lower level of energy. There is therefore a zone in which the perforation capacity of a projectile attains a maximum.

For bullets 4–6 in Fig. 6.22, this zone occurs at approximately 100 m; for bullets 2 and 3, it is more likely to occur at approximately 300 m.

Full metal jacket bullets undergo surprisingly little deformation in soft wood (such as spruce), despite often tumbling several times. In most cases they do not deform at all. Deformation may occur in hard wood or if a bullet hits a knot.

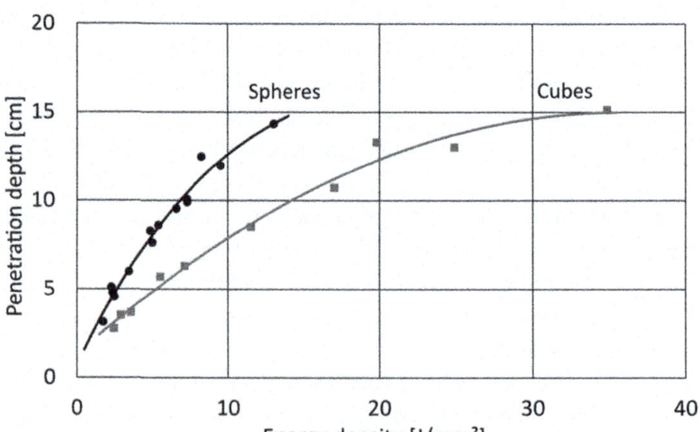


**Fig. 6.23.** Penetration depths of steel spheres and cubes in spruce (with a density of 0.46 kg/dm$^3$) as a function of energy density.

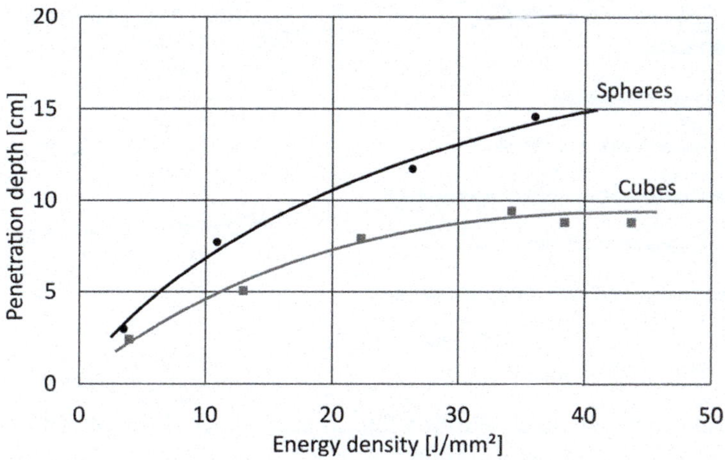


**Fig. 6.24.** Penetration depths of steel spheres and cubes in beech with a density of 0.66 kg/dm$^3$ as a function of energy density.

This being so, it is to be expected that the penetration behaviour of projectiles with a constant sectional density would depend mainly on energy density. Experiments with steel spheres of two diameters and cubes of two edge lengths show very close correlation between energy density and penetration depth in spruce (see Fig. 6.23). The spheres were 5 mm and 8 mm in diameter and the cubes had edge lengths of 5.65mm and 8-mm. Projectile velocities ranged from approx. 330 to 1420 m/s. As energy density increased, the gradient of the curve decreased, in a manner similar to that observed for aluminium. The curves for beech were even flatter (see Fig. 6.24).

The density and moisture content of the wood also have a significant effect on penetration depth. The ballistic relationships shown here are of course valid under other conditions, but at different scales.

### 6.3.6 Sand and soil

The influence of stability and angle of attack is even more evident in sand and soil than in wood. When penetrating these materials, high-velocity jacketed bullets (v greater than approx. 500–600 m/s) quickly rotate until they are perpendicular to their direction of travel, and the resulting bending stress rips open the jacket. This hugely increases the effective cross-sectional area and reduces penetration depth accordingly. At lower velocities, bullets remain compact and attain greater penetration depths despite their lower energy. The test results in Fig. 6.25 show that for all six military bullets tested, penetration in sand is greater at 300 m than at 100 m.

The difference between steel-jacketed bullets 3 and 6 is particularly striking.

The effect of distance was very similar in relatively dry, stone-free soil. However,

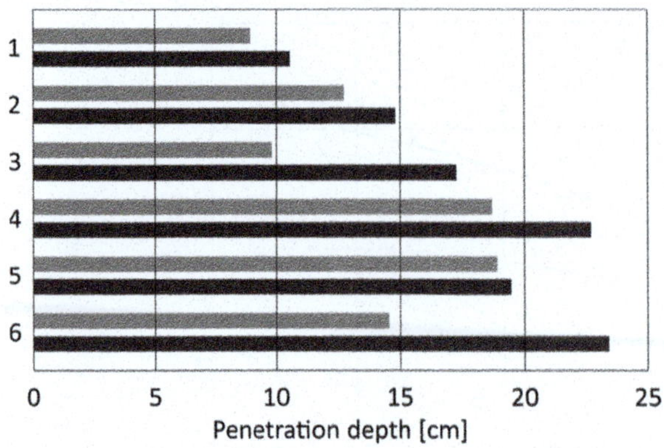


**Fig. 6.25.** Average penetration depths in dry sand.
Light grey: 100 m.
Dark grey: 300 m.
1 223 Rem. SS92
2 5.56 mm NATO, SS109
3 5.6 mm GP 90
4 7.62 × 39 FMJ
5 7.62 mm NATO soft core
6 7.5 × 55 GP 11

including the results here would be of little value, as the characteristics of soil can vary widely, producing very different results.

FMJ pistol and revolver bullets do not usually disintegrate in sand or soil. They therefore penetrate deeper (at short range) than rifle bullets. Solid lead bullets also remain intact at low velocities, although their surface is eroded. As velocity increases, they begin to deform.

Penetration depth in these materials is very much dependent on their composition and moisture content. It is therefore almost impossible to derive general rules. Penetration depth does not correlate with impact energy, energy density or momentum. Correlation with momentum might perhaps be conceivable because of the possibility of momentum being transferred to the grain structure. Comparative tests are best carried out using dry sand and sieved soil to ensure that results are indeed comparable.

## 6.4 Protective systems

### 6.4.1 Inclined plates

One of the main problems of ballistic protection is its weight, which is a consequence of the mass that must be used to block the projectile. It is therefore logical to incline protective plates, thereby extending the path of the projectile within the material.

If a projectile strikes a plate of thickness D that is inclined at an angle β, the geometric path of the projectile within the plate $D_g$ increases according to the following law:

$$(6.4{:}1) \qquad D_g = \frac{D}{\cos\beta} \qquad [\text{mm}]$$

This would allow the conclusion that an inclined plate could be thinner by the cosine of the NATO angle, meaning that it would be possible to halve the thickness

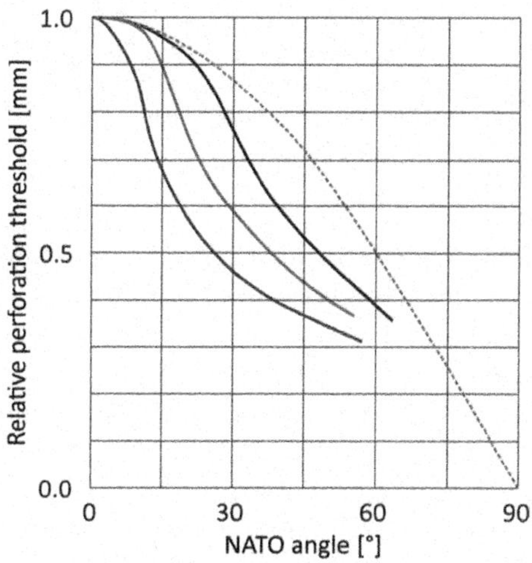


**Fig. 6.26.** Critical angle diagrams for three different projectiles.
The theoretical cosine curve is shown dotted. The other three curves are taken from experimental results and are calculated relative to the perforation threshold for a perpendicular impact (NATO angle = 0°). Other projectiles and target plates result in other curves.

of a plate if the NATO angle β were 60° (i.e. angle of incidence α = 30°). This theoretical law would result in the dotted line of Fig. 6.26.

In reality, the curves of maximum thickness perforated as a function of NATO angle (the *critical angle diagram*), almost always lie below the cosine function curve. Fig. 6.26 shows curves for three different projectiles relative to the perforation threshold for a perpendicular impact. For one projectile, plate thickness could be halved at a NATO angle of less than 30°. The curves end at the angle at which the projectile starts to ricochet. This in turn depends on the projectile and the plate (see also Fig. 6.3).

The experimentally determined critical angle diagrams already take account of an additional effect that comes into play when a projectile strikes a plate at an angle other than 90°; in many cases, rather than penetrate the protection in a straight line, the projectile is "refracted" in a manner similar to a beam of light (see Fig. 6.27).

This could be explained by the small torque that occurs when the first point of contact with the target is slightly eccentric to the direction of flight.

Many critical angle curves are close to the cosine curve at NATO angles of up to approximately 10°–15°. Because the cosine of these angles is close to 1 (deviation at 15° is just under 3.5%), a significant improvement in protection (or reduction in weight for the same protection) only results at a NATO angle of more than 15°–20°.

Clearly, no generally applicable rule can be established for perforation of inclined plates. Perforation depends on the projectile and its design, and on the quality

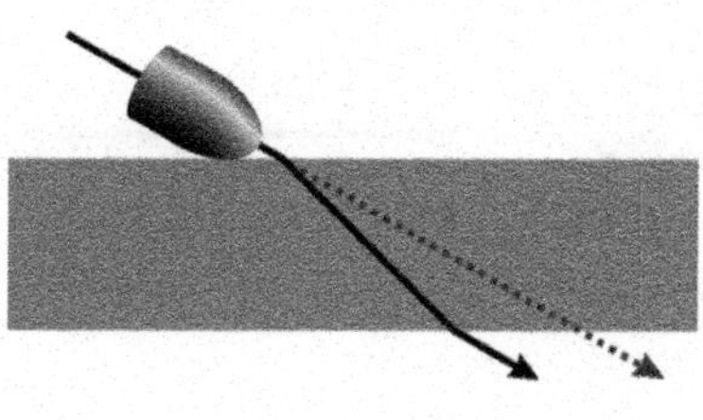

**Fig. 6.27.** A projectile does not usually perforate an inclined plate in a straight line, but is "refracted" like a beam of light.

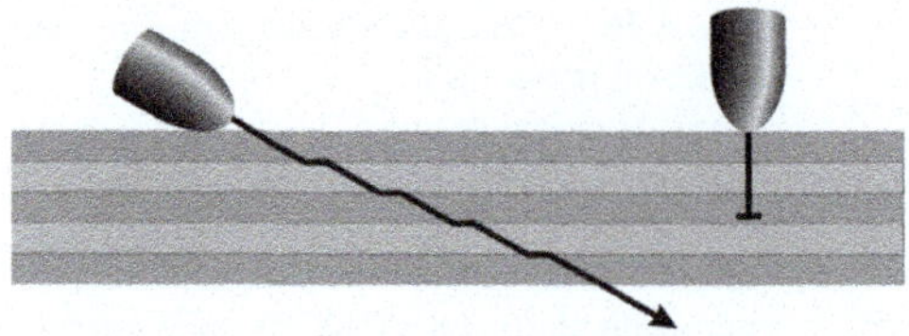

**Fig. 6.28.** Under certain conditions, a shot may penetrate a layered, inclined target even though the same target stops the same shot at a NATO angle of 0° (see text for details).

of the plate; both can vary widely. This means that one has to either rely on existing results or establish maximum perforation thickness by means of test firings.

A surprising phenomenon can occur with layered protection (e.g. textile protective vests or layered, moulded plastic panels). Under certain conditions, a bullet striking at a relatively large NATO angle (60°–70°) may perforate the plate or the protection package (see Fig. 6.28), even though a bullet striking at 90° is stopped.

In the case of a perpendicular hit, all layers obstruct the bullet simultaneously, whereas if the bullet strikes at an angle, it passes through the layers one by one, generally experiences virtually no deformation, loses much less energy per layer (see 6.2.3 and 6.4.2) and thus achieves perforation.

Layered protection must therefore be tested not only perpendicular to its surface, but also at a relatively large NATO angle.

### 6.4.2 *Spaced plates*

It may also be possible to improve protection by dividing a solid, thick protective plate into two or more thinner plates with space between them. These are known as spaced plates.

However, this design does not always result in better protection. For an explanation of the exceptions, see 6.2.3 on perforating thin plates, where it was stated that a plate absorbs much less energy if it is perforated than if it stops the bullet.

The principles behind the behaviour of projectiles as they penetrate spaced plates are most easily explained by examples taken from real experiments.

- A 3 mm sheet steel plate of sufficient strength will stop a short pistol bullet (9 mm Luger) (Fig. 6.29 a, upper picture). The same bullet will completely

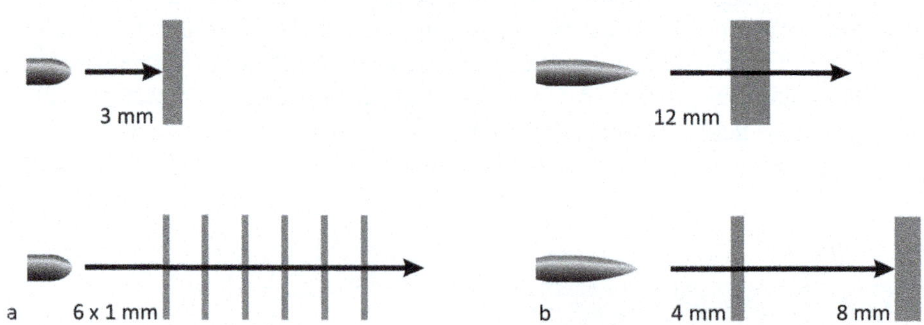


**Fig. 6.29.** Differences between short and long projectiles on impact with spaced plates. See text for details.

perforate a set of up to six 1 mm plates of the same quality (Fig. 6.29 a, lower picture).

See also Fig. 6.28 and the corresponding explanation.

- A steel-core 7.62 mm NATO rifle bullet that penetrates a 12 mm steel plate of a given quality (Fig. 6.29 b, upper picture) can be stopped by two plates of the same quality and overall thickness, placed 15–20 cm apart (Fig. 6.29 b, lower picture).

The first case is a direct consequence of the perforation of thin plates described in 6.2.3. Relatively little energy is transferred to these plates. Because the bullet is relatively short, its effective cross-sectional area hardly changes even if it continues its path with a larger angle of attack after passing through the plate. It strikes the second plate with slightly less energy but similar sectional density, and perforates it. This process continues until the bullet no longer has sufficient energy to perforate a 1 mm plate.

In the second case, the longer bullet, which easily perforates the 12 mm plate, acquires a large angle of attack after perforating the first plate and will immediately adopt a side-on position on striking the second plate. The effective cross-sectional area increases significantly and energy density decreases accordingly, so that the bullet is unable to perforate the second plate.

It would probably be possible to make the second plate thinner; this could easily be determined experimentally.

In this case, the first plate merely disrupts the flight of the bullet and can be relatively thin. The distance between the plates should be large enough to allow the bullet to acquire a sufficiently large angle of attack in the space between them. Splitting a protective plate into two or more thinner plates is particularly useful if as a result it so disturbs the bullet as to increase its cross-sectional area, thereby reducing its sectional density.

This design principle can also be applied to other materials.

Retrofitting a certain office building with high-grade bullet-resistant windows would have required reinforcement of the façade because of the additional load imposed by the glass. By taking account of the capacity of the existing triple glazing to disturb the bullet, and placing the bullet-resistant glass 20 cm inside the existing windows, it was possible to halve the thickness of the protective glass, obviating the need for façade reinforcement.

A 7.5 × 55 rifle bullet completely perforates a package of 90 layers of Aramid, decelerating by about 20% in the process. But if one suspends the 90 layers over a distance of 1.5 m, the bullet stops at approximately the 70th layer and falls out at the bottom.

### *6.4.3 Asymmetric arrangements*

The perforation behaviour of materials with homogeneous or symmetrically layered structures is the same in both directions. With an asymmetrical design, however, a plate can be made to stop a bullet in one direction and be perforated by the same

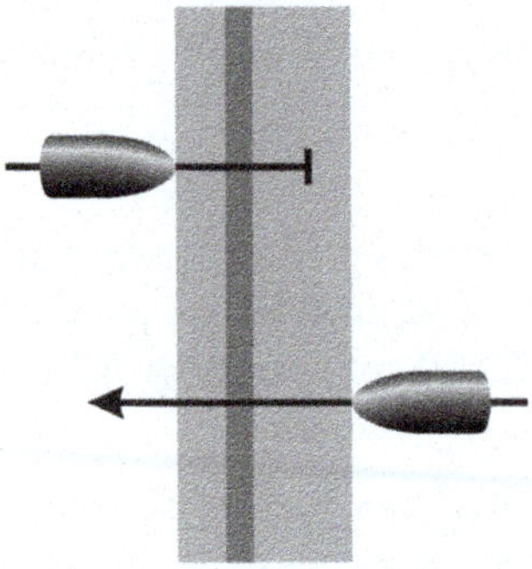

**Fig. 6.30.** Directionally-dependent perforation properties in asymmetric plates.

bullet in the other (see Fig. 6.30). This can result from the bullet being deformed differently in each direction.

Textile protective vests (e.g. those with stab protection) may exhibit this behaviour. In such cases, the attack side must be clearly designated.

## 6.5 Problems related to ballistic protection

### 6.5.1 General

#### *6.5.1.1 Personal protection and protection of material assets*

Perforation-inhibiting materials are primarily used for personal protection, as the means of attack – firearms and stabbing weapons – are primarily directed against the person. Targeted material damage by firearms is less common. Here it is mainly important and sensitive equipment (radar systems, computers, power systems, control systems, etc.) that require protection.

Personal protection can be divided into two categories:

- Protection worn on the body (protective vests, helmets)
- Protection of space occupied by persons

Protection of the body is usually only partial, as mobility must not be impaired excessively and there are limits on how much weight is acceptable.

Protected spaces are created in buildings and parts of buildings (such as counter areas and computer rooms) and in vehicles. As a rule, the materials used must not emit fragments into the protected area. Fragment-free protection of material assets is only required if fragments pose a threat to those assets.

#### *6.5.1.2 The threat potential*

The possible threats to persons from firearms are so varied that protective equipment is always designed for a specific threat profile. Probably the most significant threat is that from small arms bullets, which derive their potential effect from their kinetic energy.

"Small arms" are usually defined as weapons that can be operated with one or two hands and have a calibre less than 13 mm. This category of weapon also includes shotguns, which may have a barrel diameter of up to about 20 mm in the case of a 10-bore.

Although there are many types of rifles, pistols, revolvers and related ammunition (there are some 300 different long weapon calibres and about 100 handgun calibres, with dozens of different bullets in certain calibres), this delineation circumscribes the attack potential quite clearly. To keep weapons manageable, limits are imposed on both muzzle energy and muzzle impulse; this in turn imposes limits on bullet mass and muzzle velocity. These limits are generally:

| | | | |
|---|---|---|---|
| Muzzle energy | < | 12 kJ | (18 kJ) |
| Muzzle impulse | < | 25 Ns | (40 Ns) |
| Bullet mass | < | 35 g | (50 g) |
| Muzzle velocity | < | 1200 m/s | |

However, repeaters and semi-automatics in .50 Browning (12.7 × 99) (USA) and 12.7 × 108 (CIS) go well beyond these limits, as indicated by the values in parentheses. Wearable protection against these types of ammunition would be so heavy and cumbersome as to be impractical.

#### *6.5.1.3 Threat potential parameters*

Muzzle energy and muzzle impulse can be used to classify the threat in terms of the means of attack and its ease of use. Please see 6.1.3 regarding the assessment of protection. Solid materials can be divided into two general groups: vitreous materials (glasses and ceramics), where energy is primarily responsible for perforation, and all other materials, for which energy density is the main factor.

Energy and energy density are therefore of central importance in the evaluation of ballistic protection. Testing of protective equipment that uses both material groups will be especially complex.

### 6.5.2 Threat and protection classes

#### *6.5.2.1 General*

Protection against the entire range of small-arms threats mentioned in 6.5.1.2 is only possible using fixed equipment. For everyday wearable body protection, one must limit the requirements. This means, however, that before procuring protective equipment one must precisely specify the attacks against which it is to protect. The general threat potential must therefore be divided into classes that take account of real attack scenarios. This will make it possible to define concrete requirements.

First, we can divide firearms into long weapons and handguns. Handguns can be further subdivided into common utility weapons and heavy handguns. In the case of long weapons, the most obvious distinction is between military and hunting weapons. Shotguns should be considered separately, as they combine high energy

**Table 6.5.** Common weapon categories: typical values of muzzle energy $E_0$

| Weapon category | Calibre range [mm] | Mean $E_0$ [J] | Maximum $E_0$ [J] |
|---|---|---|---|
| Common utility handguns | 5.5–12 | 150–500 | 750 |
| Heavy handguns | 9–12 | 1000 | 2500 |
| Long weapons (military) | 5.5–6.5 | 1600 | 1800 |
| | 6.5–9 | 3000–3500 | 4000 |
| 12.7 × 99 / 12.7 × 108 | 12.7 | 18000 | |
| Long weapons (hunting rifles) | 5.5–6.5 | 2000 | 3000 |
| | 6.5–12 | 4000–5500 | 12000 |
| Long weapons (shotguns) | 12 (18.2 mm) | 2500 | 3500 |

with unusually low energy density. For military weapons, the 12.7 mm calibres mentioned in 6.5.1.2 must also be taken into account. Table 6.5 shows such a breakdown, indicating the calibre range and the mean and maximum muzzle energies. For practical application, these categories need to be specified in more detail and further subdivided (see also B. P. KNEUBUEHL, 2003).

### 6.5.2.2 Handguns

If we consider the nominal muzzle energy of 26 of the most common handgun ammunition calibres, we see clear steps at around 250 J, 500 J and 750 J. Very few calibres exceed this level of muzzle energy, with energy increasing relatively sharply from calibre to calibre among those that do. If we order these calibres by increasing energy density, we observe steps at 5, 8, and 11 J/mm$^2$. The steps are less pronounced than those for energy.

This allows us to define four handgun threat classes by energy and four by energy density (see Table 6.6). One can assign certain calibres to different classes, depending on which of the two measured values is decisive. It will therefore be necessary to test certain protective equipment with more than one calibre.

**Table 6.6.** Threat classes and associated handgun calibres

| Class | Energy | Calibre | Energy density | Calibre |
|---|---|---|---|---|
| I | Up to 250 J | 22 L.R.<br>7.65 Browning | Up to 5 J/mm$^2$ | 7.65 Browning |
| II. | Up to 500 J | 9 mm Luger, 38 Spl.,<br>22 Win. Mag. | Up to 8 J/mm$^2$ | 9 mm Luger<br>22 L.R. |
| III. | Up to 750 J | 10 mm auto | Up to 11 J/mm$^2$ | 10 mm auto |
| IV | Over 750 J | 357 Mag.<br>44 Rem. Mag. | Over 11 J/mm$^2$ | 357 Mag., 44 Rem. Mag.<br>22 Win. Mag. |

**Table 6.7.** Energies and energy densities of the most important military cartridges

| Calibre | Energy [J] | Energy density [J/mm²] |
|---|---|---|
| 5.45 × 39 Kalashnikov | 1400 | 60.0 |
| 223 Remington (5.56 × 45) | 1660 | 68.4 |
| 5.56 mm NATO | 1700 | 70.0 |
| 7.62 × 39 Kalashnikov | 2070 | 45.4 |
| 7.62 mm NATO (308 Win.) | 3270 | 71.7 |
| 30-06 (7.62 × 63) | 3400 | 74.6 |
| 7.62 × 54 R | 3840 | 84.2 |
| 300 Winchester Magnum | 5000 | 109.6 |
| 338 Lapua Magnum | 6500 | 117.3 |

### 6.5.2.3 Long weapons

As shown in Table 6.5, long weapons are first divided into two categories: military and hunting. Military rifles are generally designed to fire larger numbers of rounds and are usually capable of automatic fire. To ensure that these weapons achieve a reasonable service life, military cartridges are limited in power.

Over the last hundred years, the required operational range of military rifles has shrunk from over 1000 metres to a few hundred. This has resulted in a reduction in calibre, which means that today we can divide munitions into two calibre ranges when looking at threat potential, one around 5.5 mm, the other around 7.5–8.5 mm (see Table 6.7). It is noteworthy that the smaller calibres have similar energy densities despite developing only half as much energy near the muzzle. This must be taken into account when testing protective equipment.

Hunting rifles are either single-shot weapons or repeaters. Round counts are low, so the cartridges used can provide more energy. This is clear from Table 6.8. They

**Table 6.8.** Energies and energy densities of some typical hunting weapons

| Calibre | Energy [J] | Energy density [J/mm²] |
|---|---|---|
| 5.6 × 57 | 2600 | 105.6 |
| 6 mm Remington | 3030 | 107.2 |
| 6.5 × 68 | 4100 | 123.6 |
| 7 × 64 | 4050 | 105.2 |
| 7 mm Remington Magnum | 4800 | 124.7 |
| 8 × 68 S | 5700 | 113.4 |
| 9.3 × 62 | 4800 | 70.7 |
| 460 Weatherby Magnum | 9600 | 93.6 |
| 12/70 slug | 2800 | 10.7 |

clearly surpass military weapons in terms of energy density, and therefore have considerable threat potential.

Shotguns firing slugs constitute a special case. A shotgun slug has relatively high energy but low energy density. They present as much of a threat to vitreous materials as a rifle but only as much as a handgun when fired at other hard materials.

### 6.5.3 Threat probability

#### *6.5.3.1 Factors making up threat probability*

The division of threat potential into different classes means that one of these classes must be selected when specifying protection. However, this means that one may also have to take other threats into account. As a result, threat potential will always involve probabilities, which will depend on several factors. The determining factors will be:

- the probability that weapons of a given class are available, with frequency being a contributing factor
- the probability that weapons of a given class will be used (hunting weapons, for example, rarely pose a threat)
- the probability that the weapon used displays the ballistic characteristics of the class in question.

#### *6.5.3.2 Availability and frequency*

The availability and frequency of a weapon in a particular threat class play a decisive role in determining the probability that a given threat will occur.

Availability and frequency are not necessarily correlated; military weapons are very common in many countries, but it is difficult to acquire them.

The geographic distribution of a weapon type and regional differences must be taken into account.

Common utility handguns as listed in Table 6.5 are much more common in Europe than heavy handguns. As a result, the threat probability from pistols in 9 mm Luger is much higher than that from those in 44 Rem. Mag., for example.

In Germany, the typical Swiss Army 7.5 × 55 (GP 11) will have a very low threat potential as it is uncommon in that country, whereas its threat potential will be higher in Switzerland, where it is one of the most widely used types of ammunition.

The threat potential of a particular weapon can also change rapidly. The 7.62 × 39 Kalashnikov is a good example. Until 1989, it posed practically no threat in Western Europe, as it was almost absent; it did not appear in the testing guidelines for ballistic protection. After 1989, however, it rapidly became more widespread, and so did the potential threat. A decade later, it was assigned its own class of shot.

### 6.5.3.3 Threat probability for handguns

The discussion above is based on the most common muzzle energy in each calibre. This is usually a standard value for common barrel lengths and commercially available cartridges. When designing and testing protection in a certain protection class, one must bear in mind that higher energy levels are often achieved using longer barrels (e.g. in the case of submachine guns) or special cartridges. It is therefore necessary to find out what muzzle energies (or muzzle velocities) actually occur in connection with the test calibre in a given protection class. This is especially important for handguns, where there is a wide variety of barrel lengths and ammunition types.

One must not rely on manufacturers' documentation, as their data are usually obtained using test barrels of standard lengths.

The velocity distribution shown in Fig. 6.31 is based on over 60 measurement series of at least 6 shots each, using 8 g bullets in 9 mm Luger, 18 makes of cartridge and barrel lengths of between 90 and 225 mm. This distribution corresponds very closely to a normal distribution (mean 363.4 m/s, standard deviation 24.2 m/s).

A velocity of 360 m/s for 9 mm Luger bullets would cover no more than 50% of the muzzle energies actually produced by weapons of this calibre. A velocity of 395 m/s covers 90% of the energy levels achievable with this calibre and 420 m/s covers 99%.

Level 3 ballistic protection (VPAM, 2006) against 9 mm Luger rounds is tested at a velocity of 415 ± 10 m/s. The lower limit covers about 96% of the possible muzzle velocities. However, since the frequency of weapons with higher velocities is low, the probability that such a threat will occur is also low.

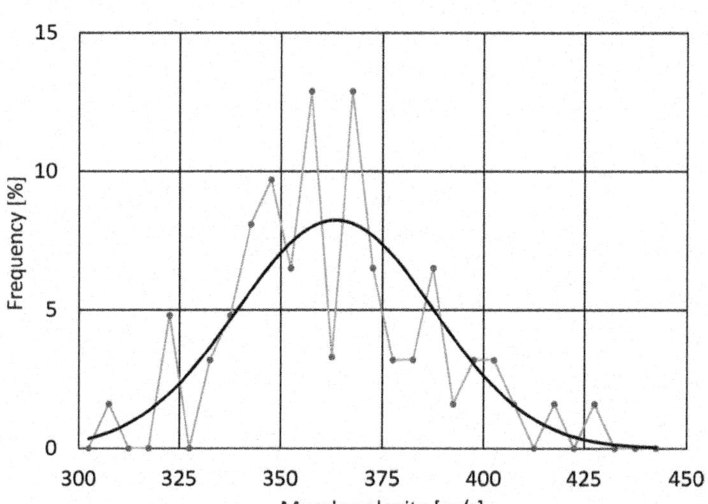


**Fig. 6.31.** Distribution of muzzle velocities of 8 g 9 mm Luger bullets, for different barrel lengths and makes of cartridge.

Similar distributions of muzzle velocity occur with other handgun calibres. Less variability in barrel length and ammunition type produces narrower curves. In summary, we can conclude as follows:

- Standard handgun velocities are suitable for threat classification, but not for determining effective attack potential.
- When determining threat probability, one must take account of attack potential and frequency of occurrence.

#### *6.5.3.4 Threat probability for rifles*

The barrel lengths of military rifles vary only slightly (< ± 5%), to ensure standardization. The ammunition used with these weapons also varies little. This means that it is virtually impossible to calculate a velocity distribution. As a general rule, the nominal muzzle velocities of the various calibres are taken to constitute the most likely threat potential and testing guidelines use those velocities.

Hunting weapons appear not to be considered a threat; the threat probability is so low that ballistic protection guidelines make no reference to these calibres.

### 6.5.4 Protection probability

#### *6.5.4.1 Perforation probability of ballistic protection*

Probabilities must be calculated not only for the threat but also for protection. It is commonly assumed that protection either will or will not protect against a given attack. In practice, however, the picture is different.

If protective equipment is subjected to shots of increasing velocity (and hence increasing energy), it will initially stop the projectiles, but at a certain velocity it will be perforated for the first time. However, at higher impact velocities it may

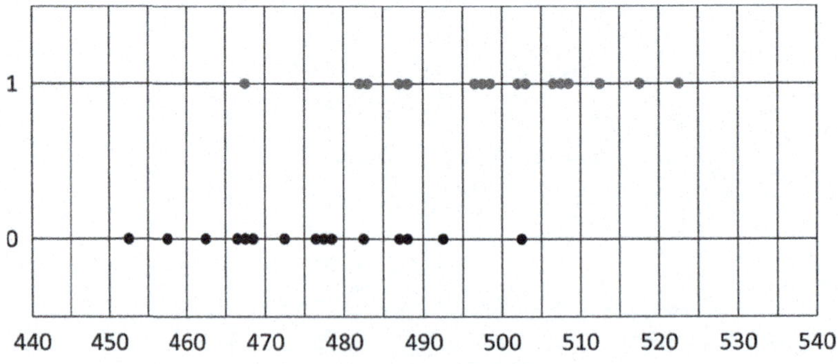


**Fig. 6.32.** Perforation diagram for a real protective vest, divided into velocity classes of 5 m/s. 0 = stopped. 1 = perforated. Perforations and stopped rounds overlap.

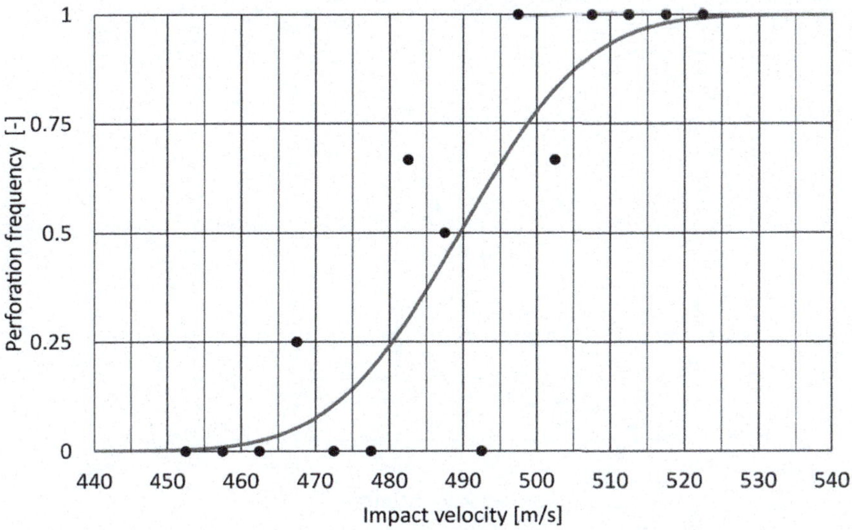


**Fig. 6.33**. Perforation frequencies derived from the perforation diagram of Fig. 6.32. The resulting normal distribution is also shown.

well stop the projectile again. There is always a velocity range in which stopped rounds ("non-perforations") and perforations overlap. Fig. 6.32 shows the results of 31 shots fired at a textile protective vest. Below 465 m/s all rounds were stopped; above 505 m/s all perforated. If the velocities in between are divided into (statistical) classes of, for example, 5 m/s, a shot in a given class may either be stopped or produce a perforation.

The relative perforation frequency for each class can be calculated and recorded. This results in a graph of the type shown in Fig. 6.33. As the number of shots increases, it becomes increasingly clear that the normal distribution function is a good model for predicting perforation. Five specimens of the protective vest for which results are shown in Figs. 6.32 and 6.33 were subjected to a total of 151 shots. The resulting perforation frequencies are shown in Fig. 6.34. There is a clear tendency for the results to follow a normal distribution curve.

Statistical tests that examine whether a dataset is normally distributed or not confirm that perforation frequencies are closely approximated by a normal distribution.

Because a normal distribution can be completely described by two parameters alone – mean and standard deviation – this also applies to the perforation probability of ballistic protection. The mean corresponds to the velocity at which 50% perforation is to be expected. This is generally labelled $v_{50}$. The standard deviation (s) is a measure of the steepness of the slope of the perforation probability curve; the higher the standard deviation, the flatter the slope between stopped rounds and perforation.

Any ballistic protection – be it bullet-resistant glass, a protective vest or a steel plate – has a certain perforation probability. The steepness of the curve will depend on the type and nature of the material. Homogeneous and well-defined materials

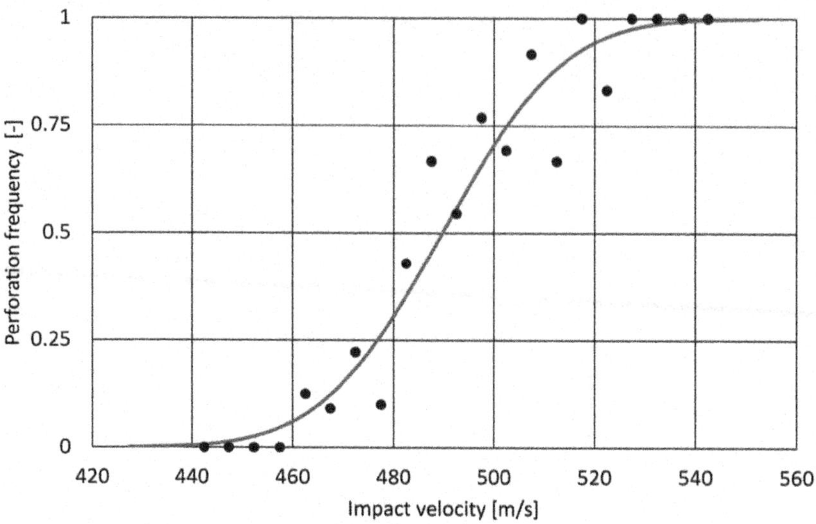


**Fig. 6.34.** Perforation diagram of the protective vest from Fig. 6.33, after testing 5 specimens with a total of 151 shots. The normal distribution is very clear.

(e.g. steel) tend to have steep gradients, while those of highly structured materials such as fabrics tend to be flatter.

It can sometimes be difficult or impossible to determine perforation probability parameters for certain materials (ceramics, glass, etc.).

Having determined the values of $v_{50}$ and the standard deviation of the distribution of perforations for a given ballistic protection, (see procedure in 6.5.5.3), one can calculate the probability of perforation for any velocity.

In the real example shown in Fig. 6.34, the mean perforation velocity is 489.7 m/s and the standard deviation is 19.1 m/s. The probability of perforation at VPAM Level 3 (415 m/s) mentioned in 6.5.3.3 is therefore $4.7 \cdot 10^{-5}$, i.e. 4.7 perforations per 100,000 shots.

The risk of perforation occurring at the test velocity specified by a guideline or standard is an important factor, and indeed is decisive as far as the user is concerned. This risk can be specified as part of the procurement procedure, for instance, and the product tested accordingly.

### 6.5.4.2 Protection probability

A shot either perforates ballistic protection or is stopped by it. The probability $P_S$ of the protection stopping a shot is thus the complementary probability of the shot achieving perforation $P_P$:

$$(6.5{:}1) \qquad P_S = 1 - P_P \qquad [-]$$

In 6.5.4.1, the probability of perforation did not take into account how often an attack at a certain velocity may occur. According to 6.5.3.3, this attack probability

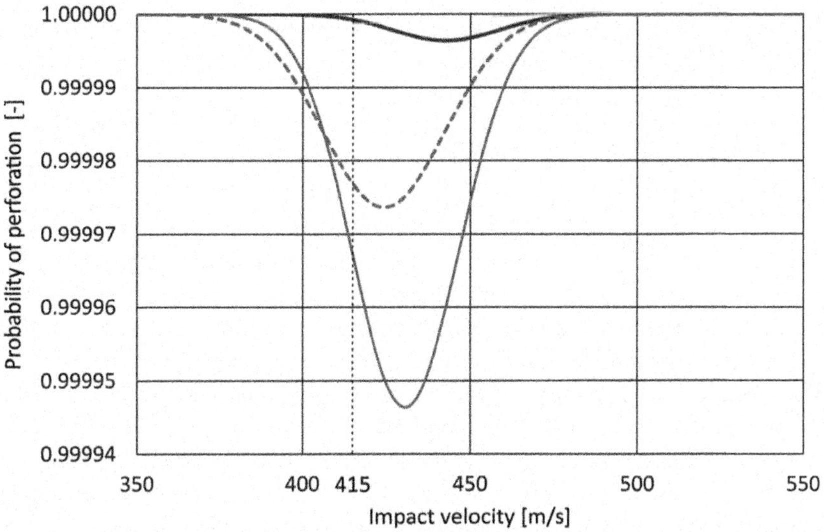


**Fig. 6.35.** Protection probability vs impact velocity, taking the frequency of the threat potential into account. The black line corresponds to the example in Fig. 6.34, the light grey line corresponds to a 20 m/s decrease in $v_{50}$ and the dashed line to an increase of 6 m/s in standard deviation. The velocity corresponding to VPAM Level 3 is also indicated.

$P_{AT}$ is approximately normally distributed. It must be taken into account when specifying protection probability. From Eqn 6.5:1:

(6.5:2) $$P_S = 1 - P_P \cdot (1 - P_{AT})$$ [-]

Fig. 6.35 shows this protection probability for the example in Fig. 6.34. At a $v_{50}$ of 490 m/s and a standard deviation of 19.1 m/s, taking account of threat potential probability indicates 99.9999424% protection at Level 3 (415 m/s).

Fig. 6.35 also shows that there is a minimum protection probability (the heavy solid line). This minimum (of 99.99964%) occurs at a velocity of approx. 442.6 m/s.

By varying the parameters, it is also possible to investigate the sensitivity of protection probability to changes in $v_{50}$ and s. This is exemplified in Fig. 6.35 by the light grey line (20 m/s decrease in $v_{50}$ with the same s) and the dashed line (same value of $v_{50}$, 6 m/s increase in s). The risk of perforation at Level 3 increases by a factor of 43 in the first case and by a factor of 30 in the second. The minimum also decreases significantly in both cases.

In summary, we can conclude as follows:

- The mean perforation velocity of ballistic protection must be significantly higher than the expected impact velocity in order to ensure a high probability of protection.
- The dispersion of perforation velocity is a quality characteristic of the protection. It should be as small as possible.

- Similar studies can of course be carried out for all calibres, as long as it is possible to determine the parameters.

## 6.5.5 Testing ballistic protection

### 6.5.5.1 General

At first glance, testing ballistic protection appears to be a simple matter: one just fires a few shots into it from the weapon against which one wishes to provide protection. If none of them perforate the protection, one is "good to go" (see Fig. 6.36). However, such an approach would overlook two important points.

***Basic conditions***. Tests must be reproducible; they must always produce the same result and the result must not depend on where and by whom they are performed. These two basic conditions – reproducibility and independence – mean that all details of a test must be precisely specified. This applies not only to the attack (the projectile and its velocity), but also to the way the specimen is presented and secured.

For example, if a sheet of bullet-resistant glass is held in a frame rigidly during one test and flexibly during another, it may stop a bullet in one case and be perforated by the same bullet in the other.

All textile protective vests will be perforated if they are mounted on spruce.

***Test standards and guidelines***. For the above reasons, firing tests always take place under idealized conditions. The projectile type, velocity and firing distance are precisely specified, the specimen (of which the dimensions are standardized) is held in a specified target structure and environmental conditions such as temperature and humidity must be within certain limits. All these requirements are set out in test standards and guidelines (EN, 1999, VPAM, 2009, NIJ, 2008 and others) and must be observed by the tester.

Tests are therefore not primarily designed to replicate "reality", but to compare different products with a given standard (the test standard) under identical conditions. It goes without saying that this standard must be based on reality (i.e. on the effective attack potential), as explained in 6.5.3. However, direct trans-

**Fig. 6.36.** Testing a protective vest in 1931. Original caption: "Bullet-proof armour is a German invention and is worn by the Berlin Criminal Investigation Department in clashes with criminals." (See illustration credits for source).

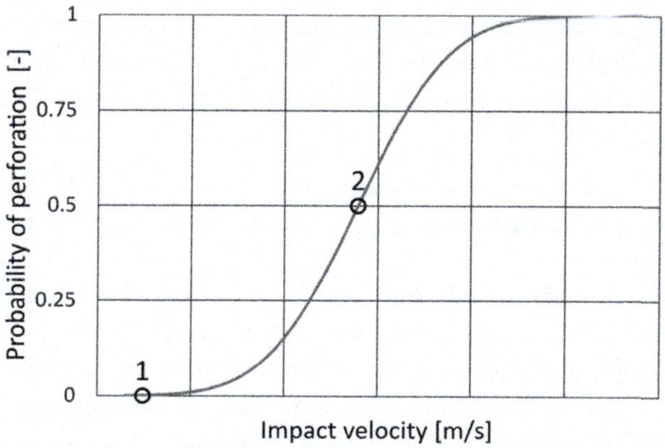


**Fig. 6.37.** Starting points for two different test methods. See text for details.

position of the test results into "reality" is often possible only to a limited degree, as the dimensions and design of real protective equipment rarely correspond to those of the test specimens (see also 6.3.1).

***The measuring equipment problem***. The testing of ballistic protection is a special subcategory of materials testing. The main difference is that the measuring equipment (the projectile) interacts with the specimen and can itself be irregularly deformed. This occasionally leads to strange results that require individual assessment.

When a steel plate is tested against a bullet with a hardened steel core, the bullet is initially stopped. When velocity is increased, the bullet perforates the plate, probably with its core relatively intact. If the impact velocity is increased further the bullet will once again be stopped, possibly because its core breaks up, leading to a decrease in energy density. If velocity increases further, the bullet will probably perforate the plate again.

Similar behaviour can be observed with textile protective vests, where at very high velocities the bullet is deformed to such an extent that it no longer perforates.

As a rule, such results are observed only at velocities that do not occur in reality (or only under very special conditions). Conventional test procedures do not normally need to take such velocities into account. Nevertheless, when one identifies a perforation limit it is important to remember that there may be a second limit above or below it. This is especially true with steel targets and solid projectiles.

***Test methods***. Requirements for ballistic protection must be based on the possible testing procedures, which are also specified in testing guidelines. Given the perforation probability curve mentioned above (see 6.5.4.1 and Fig. 6.37) the following three options are available:

1 Specify a (high) impact velocity at which the protection must not be perforated by any of a finite number of test shots (non-perforation test, Point 1 in Fig. 6.37).

2 Specify a minimum mean perforation velocity $v_{50}$ measured using the appropriate test method (Point 2 in Fig. 6.37).

3 As for 1, but specifying a maximum perforation probability at the impact velocity chosen. This method requires one to determine both the mean perforation velocity and the associated standard deviation, so that the probability of perforation can be calculated.

Options 2 and 3 require both perforations and stopped projectiles. They are particularly suitable for testing materials where both situations occur within a certain velocity range.

Option 1 is primarily chosen where dispersion in perforation velocity is small. This is the case when there is only a small difference between velocities at which there is a high probability that no perforations will occur and those at which the protection is almost always perforated. It is rare for both perforations and non-perforations to occur at a given projectile velocity with such materials.

### 6.5.5.2 Testing for non-perforation

Testing ballistic protection for non-perforation involves some rather thorny issues, which require special attention.

***Reliability of testing***. This test method is based on a certain number of shots per specimen. If one round perforates, the specimen is rejected; if no round perforates, it is accepted. There are always two ways in which such a "good-bad" test can produce a false result, and there will be a certain probability of this occurring.

1 The specimen is accepted after n stopped rounds (where n is the number of rounds specified by the testing guideline), although it actually does not meet the real requirements, e.g. because round n+1 would have been a perforation.

2 The specimen is rejected although it would meet the real requirements. This is the case when perforation occurs within the n test shots, but all subsequent rounds would have been stopped, even if one had fired 100,000 shots.

*Case 1.* For a given number of test shots n and a given effective probability of perforating the specimen $P_P$, the probability $P_A$ that the specimen will be accepted is given by a binomial distribution:

(6.5:3) $$P_A = \binom{n}{k} \cdot P_P^k \cdot (1-P_P)^{n-k} \qquad [-]$$

where k is the number of perforations. Because the specimen will only be accepted if no perforation has occurred, k is always zero.

Fig. 6.38 shows the probability of acceptance derived from Eqn 6.5:3 as a function of the number of test shots, for four probabilities of perforation.

The graph shows that a specimen with a perforation probability of 1% (1 perforation per 100 shots) will be accepted in 90% of cases in a 10-shot acceptance procedure, and a specimen with one perforation in 20 shots or fewer will be accepted in about 73% of 6-shot tests.

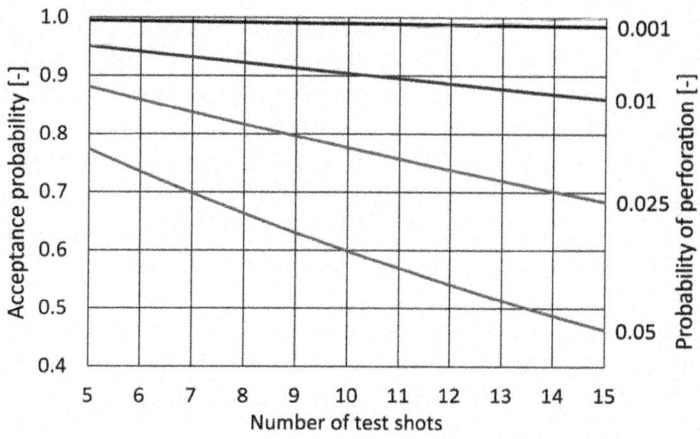


**Fig. 6.38.** Ballistic protection acceptance probability for a given perforation probability as a function of number of test shots stopped.

Increasing the number of shots leads to a lower probability of mistaken acceptance for a given perforation probability. However, for a realistic number of test shots the probability of acceptance remains high, and we learn almost nothing regarding the true degree of protection provided by the specimen.

*Case 2.* The second possible error involves the probability $P_R$ of rejecting a specimen following a test with n shots, even though the object would satisfy the requirement not to exceed a certain probability of perforation $P_P$. This probability is given by the expression

$$(6.5:4) \qquad P_R = 1 - [\,1 - P_P\,]^n \qquad [-]$$

Evaluation of this equation shows that such an error in a test for non-perforation can only occur very rarely and becomes even rarer as the quality of the ballistic protection increases.

***How many shots are required?*** A key issue in non-perforation testing is how many shots must be stopped for protection to be considered bullet-resistant against the threat in question. The steel plate shown in Fig. 6.39 is an example of this. Should the examiner have classed the plate as providing the required level of protection after it had stopped 25 shots?

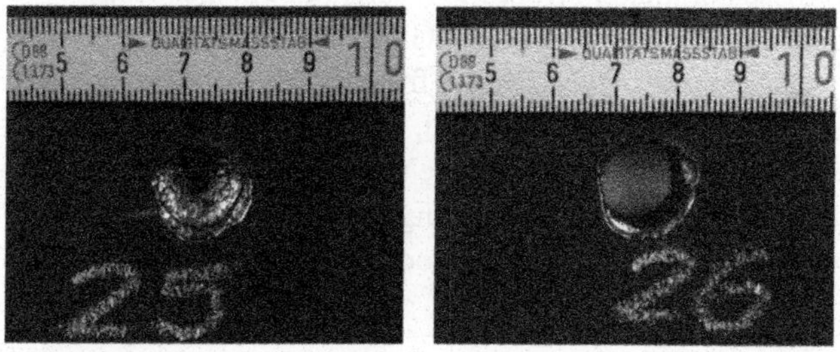


**Fig. 6.39.** Real test on a steel plate. The plate stopped 25 shots but was perforated by the 26th.

For many years, protective vests were tested with a total of 10 shots. The European standard for testing windows, doors, shutters and blinds (EN, 1998) requires three shots under the same conditions at all critical points. For technical reasons, it is only possible to carry out a single test shot on certain points, such as catches. However, the test procedure does allow the laboratory to repeat the test with a new specimen "if the results do not permit an unambiguous interpretation". If one thinks back to the paragraph "Reliability of testing" at the beginning of this section, one cannot help but conclude that there must be other qualitative, difficult-to-describe assessment options when testing for non-perforation.

***Additional assessment options***. In many cases, there are indications of how close a stopped round was to perforating. With hard materials, for example, one can inspect the rear face, from which one can often see whether or not perforation was imminent.

If Fig. 6.39 had also shown the rear face of the steel plate, one would have seen that Shot 25 had caused a clearly visible bulge. This must have been the case right from the first shots and would indicate that the plate is too thin.

For textile protection panels, it is possible to determine between which layers the bullet has been stopped and use this as an indication as to whether protection is borderline.

However, care must be taken here because a bullet is never found in approximately the last quarter of the layers. Either it is stopped beforehand, or it perforates all layers. The reason for this can be derived from the explanations in 6.4.

Such indications are mostly empirical values that cannot easily be transposed into precisely measurable rules (e.g. size of bulge, number of textile layers), because they are not generally valid but depend on the test situation. In such cases, one will have to rely on the experience of the examiner. Tests for non-perforation cannot therefore be performed purely according to a fixed, predetermined set of rules.

### 6.5.5.3 Determining the distribution parameters

According to 6.5.4.1, the perforation probability of ballistic protection can be approximated by a normal distribution function. We can therefore calculate the probability of perforation at any given impact velocity if we know the two distribution parameters – mean perforation velocity ($v_{50}$) and standard deviation (s). There are various ways of deriving these parameters from the results of ballistics tests.

***The STANAG method***. NATO standard STANAG 2920 describes a simple, approximate method for determining $v_{50}$ when testing fragmentation protective body armour. An even number of shots is fired (at least 6), half of which must be perforations. The whole dispersion must lie within a range determined by the total number of shots. The value of $v_{50}$ is obtained by calculating the mean value for the shots fired. It is not possible to calculate s using this method.

***Determining $v_{50}$ and s directly.*** We can estimate both parameters of the distribution function for perforation probability using normal distribution:

$$(6.5{:}5) \qquad v_m = \int_{-\infty}^{\infty} v \cdot p(v) \cdot dv \qquad [\text{m/s}]$$

$$(6.5{:}6) \qquad \sigma^2 = \int_{-\infty}^{\infty} (v - v_m)^2 \cdot p(v) \cdot dv \qquad [\text{m}^2/\text{s}^2]$$

where $v_m$ is the mean, $\sigma^2$ is the variance, and p(v) is the density function of perforation probability.

Firing tests are always performed with a finite number of shots. Equations 6.5:5 and 6.5:6 are therefore discretized and the probability function p(v) estimated.

This is done by dividing the velocity into classes, in which the probability function can be estimated from the relative frequency $f_{cl}$. Let the class midpoint be $v_{cl}$. From 6.5:5 and 6.5:6 we obtain:

$$(6.5{:}7) \qquad v_{50} \approx \sum v_{cl} \cdot f_{cl} \qquad [\text{m/s}]$$

$$(6.5{:}8) \qquad s \approx \sqrt{\sum (v_{cl} - v_{50})^2 \cdot f_{cl}} \qquad [\text{m/s}]$$

Simulations show that estimating s using Eqn 6.5:8 yields values that are too small if fewer than 100 shots are fired. These values can be corrected using

$$(6.5{:}9) \qquad s_{corr} = s \cdot [1.71 - 0.151 \cdot \ln(n)] \qquad [\text{m/s}]$$

Please see B. P. KNEUBUEHL (2003) and VPAM (2014) for detailed guidance on the execution and evaluation of such firing tests.

***Other procedures***. Instead of the method described above, other similarly based methods can be used, such as probit analysis (described in D. J. FINNEY, F. TATTERSFIELD, 1952), LANGLIE's method (H. J. LANGLIE, 1965), or the Bruceton method (R. WILD, E. von COLLANI, 2002). Class divisions are not required for probit analysis or the LANGLIE method. A review of these methods and a comparison of the results obtained with them can be found in M. MALDAGUE (2008).

***Number of shots required.*** When one is determining $v_{50}$ and s directly as described above, the question again arises as to how many shots are required in order to draw reasonably reliable conclusions. Large numbers of tests and simulations have shown that one can obtain acceptable results from 30 shots (with about 15 ± 3 shots perforating and the remainder being stopped) . See Fig. 6.33. Very reliable values are obtained if this test can be repeated several times (see Fig. 6.34, with 5 repetitions).

As such tests are fairly costly, one will have to decide how reliable the results actually need to be. Useful results can also be achieved with 16–20 shots. However, these should be supervised by an experienced examiner.

**Table 6.9.** Parameter values for determining perforation probability

| $P_P$ | $10^{-6}$ | $10^{-5}$ | $10^{-4}$ | $10^{-3}$ | $10^{-2}$ | $2\cdot10^{-2}$ | $5\cdot10^{-2}$ | $10^{-1}$ |
|---|---|---|---|---|---|---|---|---|
| $\alpha_{TP}$ | -4.753 | -4.265 | -3.719 | -3.090 | -2.326 | -2.054 | -1.645 | -1.282 |

***Determining perforation probability.*** Once we have determined $v_{50}$ and the associated standard deviation s, we can calculate the associated threshold velocity $v_{TP}$ for a given perforation probability $P_P$ using normal distribution:

$$(6.5{:}10) \qquad v_{TP} = v_{50} + \alpha_{TP} \cdot s \qquad [\text{m/s}]$$

Values of $\alpha_{wd}$ are taken from standard normal distribution. Table 6.9 lists these values for the most important perforation probabilities.

Conversely, if one knows $v_{50}$ and s for a given velocity $v_{TP}$, one can determine the corresponding perforation probability $P_P$.

$$\text{With} \qquad \alpha_{TP} = \frac{v_{TP} - v_{50}}{s} \qquad [-]$$

we obtain:

$$(6.5{:}11) \qquad P_P = \frac{1}{\sqrt{2 \cdot \pi}} \cdot \int_{-\infty}^{\alpha_{TP}} e^{-\frac{\xi^2}{2}} \cdot d\xi \qquad [-]$$

The values can also be taken from normal distribution tables or calculated using a normal distribution program.

### 6.5.6 Assessing the potential effect behind the protection

Protective materials and equipment may stop projectiles and yet still produce injuries on the protected side. Fragments of glass or hard protective material may cause injury. Soft protective materials tend to bulge very easily on the protected side, resulting in blunt force trauma. Testing ballistic protection therefore includes determining the damage potential of a stopped round behind the protection.

Some regulations do not permit any fragmentation at all on the protected side. Others stipulate a maximum energy density for any fragments, which is tested by placing a sheet of thin foil (e.g. aluminium) behind the test specimen. If any fragments perforate the foil, the protection is deemed inadequate.

Protective materials that tend to bulge on the rear side when struck by a projectile are stretched over a soft, deformable medium (such as plasticine or ballistic soap) for testing. The damage potential on the protected side is assessed on the basis of the dent created, which must not exceed a certain volume.

This volume is a measure of the energy transferred; see 6.2.2. However, thresholds that would allow one to determine the severity of blunt force injuries are still largely absent (for a review, see B. P. KNEUBUEHL Ed. 2022).

Because the perforation resistance of soft protective materials is highly dependent on the deformability of the backing material, that material must be precisely specified in order to ensure that tests are reproducible. This is usually done by specifying the penetration depth of a heavy steel ball dropped on the material from a specified height.

# 7 Ricochets

## 7.1 Introduction

### 7.1.1 Preliminary remarks

A ricochet always involves interaction between a projectile and a solid or liquid medium. The process therefore belongs to the field of terminal ballistics; it has already been mentioned in 6.1.5, in connection with the Backman-Goldsmith diagram. That diagram shows that ricochets can occur over a wide range of impact velocities and angles.

Because the trajectory of a projectile does not end at the ricochet, but begins anew, the events leading up to that point are of great interest, as they set the initial conditions for the remainder of the trajectory. That alone would justify this separate chapter devoted to ricochets. Much of what follows is based on parts of B. P. KNEUBUEHL (1999).

### 7.1.2 Terminology

German has four words for this phenomenon: "Abpraller", "Prellschuss", "Gellerschuss" and "Ricochet". Unfortunately, these are open to different interpretations in everyday language and it is difficult to find precise definitions in reference works. French and English, however, use the single term "ricochet" to cover all four German terms.

According to the standard French dictionary *Le Petit Robert*, the word "ricochet" was already in use in 1611, and its origins are said to be traceable to the 13th Century (in the fable of Ricochet, a round in which the word "coq" is repeated). The definition in the Petit Robert reads as follows: "Rebond d'un objet plat lancé obliquement sur la surface de l'eau ou *d'un projectile renvoyé du sol ou d'un corps dur*". (Rebound of an object projected against the surface of water at an angle, or of a *projectile that rebounds from the ground or from a hard object*).

In German, "Ricochet" (or "Rikoschett") is sometimes equated with a tumbling or unstable bullet. However, since bullets can fly unstable without having ricocheted, and ricochets sometimes fly stable, we must define our terms for this chapter.

We shall use the word *ricochet* to refer to a bullet that undergoes a change of



B. P. Kneubuehl, *Ballistics*,
https://doi.org/10.1007/978-3-662-69237-0_7

direction along its trajectory owing to contact with a solid or liquid medium, which may be anything from a blade of grass to a concrete wall. We shall refer to an *unstable projectile* by that name, or we may say that it is *tumbling,* regardless of whether it has ricocheted or not.

## 7.2 Collision theory

### 7.2.1 Types of collision

When two bodies (masses) collide, the process is known as a *collision* or *impact.* While they are in contact, the two bodies exchange part of their energy and momentum. Some of the kinetic energy involved may be converted into mechanical work (reversible deformation) or heat. As a result, the velocity vectors of both bodies change during the collision.

We use the directions of movement of the two bodies and the *shock normal* to differentiate between different types of collision. The shock normal is the straight line perpendicular to the (common) tangent plane at the point of contact of the bodies (see Fig. 7.1). It constitutes the line of action of the collision forces.

A collision in which the velocity vectors of the two bodies lie on a common straight line is referred to as a *straight collision.* In such a collision, no forces act transversely to the direction of movement.

If the centres of gravity of the two masses are moving along different (skew) lines, they can only touch each other if the shortest distance between the two lines is less than the sum of the distances between the centres of gravity and the body surfaces. In this case, we speak of an *oblique collision.*

If the centres of gravity of the two bodies lie on the common shock normal at the time of contact, the collision is *centric*, otherwise it is *eccentric.*

In kinematics terms, this results in four collision processes, all of which are encountered in ballistics (see Table 7.1).

We make a further distinction with regard to the conservation of energy during the collision process. If all of the kinetic energy is conserved, the collision is termed *elastic*. If both bodies are moving at the same velocity after the collision, it is termed *inelastic*. Collisions in which the total kinetic energy decreases, but the bodies still have different velocities after the collision, are termed *semi-elastic.*

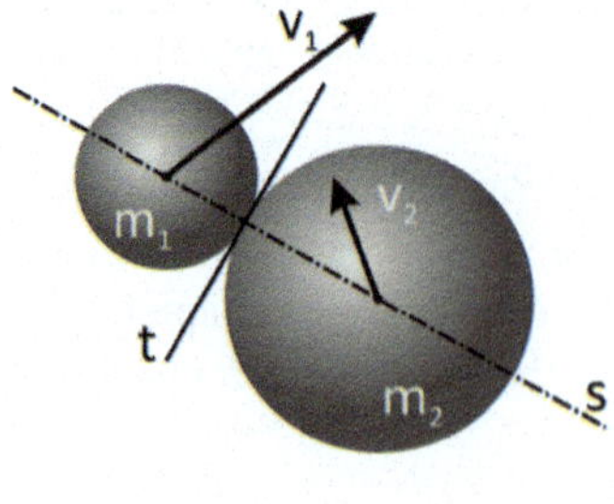


**Fig. 7.1.** Definition of the shock normal.
$m_1$, $v_1$: Mass and velocity of the first body.
$m_2$, $v_2$: Mass and velocity of the second body.
t: Common tangent plane at the point of contact.
s: Shock normal.

**Table 7.1.** Types of collision, with examples from ballistics

| | Straight | Oblique |
|---|---|---|
| Centric | 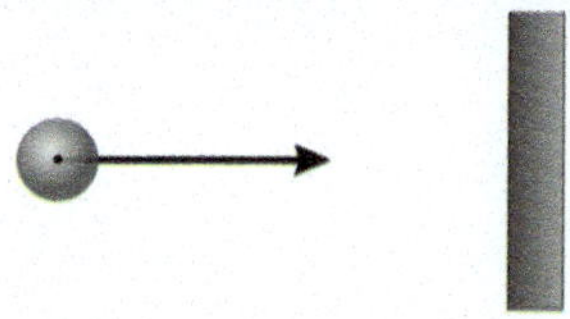 Sphere striking a surface perpendicularly | 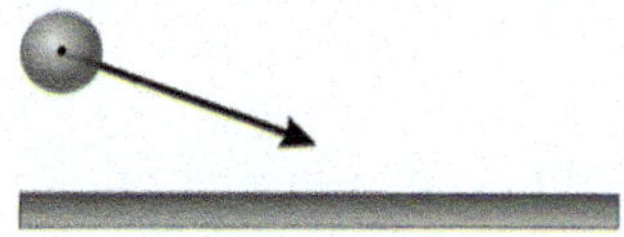 Sphere striking a target surface at an angle |
| Eccentric | 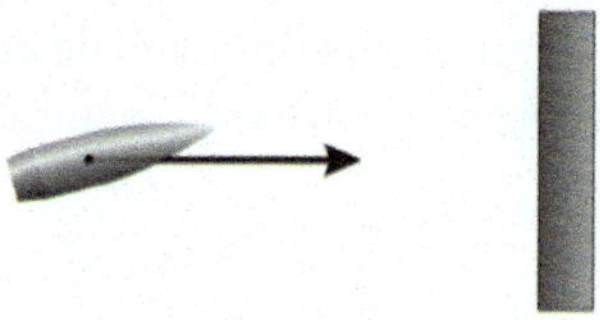 Long projectile striking a surface with a non-zero angle of attack | 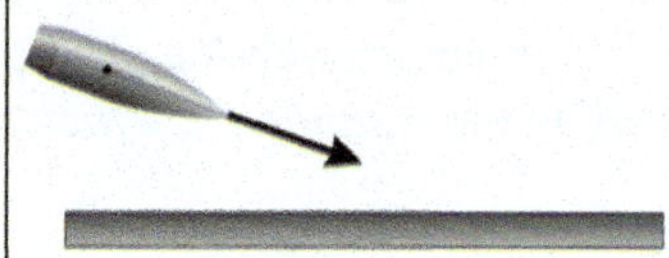 Long projectile striking a surface at an angle |

Most ricochets are oblique, eccentric, semi-elastic collisions.

It is often easier to understand a collision if we assume one of the two bodies to be at rest before the impact. The velocity of the moving body must then be referenced to the body at rest. We distinguish the two bodies using subscript 1 and 2. In the discussion below, Body 2 is at rest before the collision ($v_2 = 0$), v denotes the velocity before and u the velocity after the collision. Momentum and energy after the collision are indicated by a subscript u.

## 7.2.2 Centric, straight collision

The simple case of a centric, straight collision is particularly suitable for gaining an initial overview of the velocity and energy conditions obtaining during a collision. Inelastic collisions are a special case because of the requirement that both bodies have the same velocity after impact. The laws of conservation of energy and momentum give us:

(7.2:1) $$m_1 \cdot v_1 = (m_1 + m_2) \cdot u \qquad [\text{kg·m/s}]$$

(7.2:2) $$\tfrac{1}{2} \cdot m_1 \cdot v_1^2 = \tfrac{1}{2} \cdot (m_1 + m_2) \cdot u^2 + Q \qquad [\text{J}]$$

where Q is that portion of the kinetic energy converted into deformation work and ultimately into heat. For other symbols, please see 2.3.1.

Solving for velocity u and deformation work Q, with $\mu = m_1/m_2$, yields:

(7.2:3) $u = \frac{\mu}{\mu+1} \cdot v_1$ [m/s]

(7.2:4a) $E_u = \frac{\mu}{\mu+1} \cdot E_1$ [J]

(7.2:4b) $Q = \frac{1}{\mu+1} \cdot E_1$ [J]

$E_1$ (the kinetic energy of the impacting body) also equates to total energy, because we are assuming $v_2 = 0$.

If some of the deformation work Q is converted back into kinetic energy, the collision is termed semi-elastic. The proportion of recovered kinetic energy $E_r$ is described by the *restitution coefficient* k:

(7.2:5) $E_r = k^2 \cdot Q$ [J]

This approach is equivalent to the Newtonian collision hypothesis, according to which the relative velocity of the two bodies after the collision has a fixed ratio to their relative velocity before the collision, the absolute value of which corresponds to the restitution coefficient k.

k lies between 0 (for an inelastic collision) and 1 (for an elastic collision). The loss of kinetic energy is therefore:

(7.2:6) $Q_w = Q - E_r = (1 - k^2) \cdot Q = \frac{(1-k^2)}{\mu+1} \cdot E_1$ [J]

$Q_w$ is also the remaining deformation work, which is ultimately converted into heat.

The restitution coefficient k depends on the material, composition and structure of the two bodies, but also on the energy involved.

If a steel sphere hits a steel plate at low velocity, the collision is almost completely elastic (k = 0.95). At high velocities, the steel sphere penetrates the plate (k = 0).

We can derive the following expression for a semi-elastic collision from the laws of conservation of momentum and energy:

(7.2:7) $m_1 \cdot v_1 = m_1 \cdot u_1 + m_2 \cdot u_2$ [kg·m/s]

(7.2:8) $\frac{1}{2} \cdot m_1 \cdot v_1^2 = \frac{1}{2} \cdot m_1 \cdot u_1^2 + \frac{1}{2} \cdot m_2 \cdot u_2^2 + Q_w$ [J]

A brief calculation using Eqn 7.2:6 leads to the following formulae:

(7.2:9a) $u_1 = \frac{\mu - k}{\mu + 1} \cdot v_1$ [m/s]

(7.2:9b) $u_2 = \frac{(1+k) \cdot \mu}{\mu + 1} \cdot v_1$ [m/s]

(7.2:10a) $$E_{1,u} = \left(\frac{\mu - k}{\mu + 1}\right)^2 \cdot E_1$$ [J]

(7.2:10b) $$E_{2,u} = \frac{(1+k)^2 \cdot \mu}{(\mu+1)^2} \cdot E_1$$ [J]

With k = 1, Eqns 7.2:9a, b and 7.2:10a, b constitute the laws of elastic collision and with k = 0, as expected, they become the equations of inelastic collision, Eqns 7.2:3 and 7.2:4a.

### 7.2.3 Centric, oblique collision

At the moment of contact in a centric collision, the force vectors pass through the centres of gravity of the bodies. As a result, no torque acts on them.

The shock normal and the velocity vector of the impacting body (1) form a plane normal to the tangent plane. The coordinate system is chosen such that the x-axis coincides with the line of intersection of the two planes and the y-axis coincides with the shock normal (see Fig. 7.2). Let the impacted body (2) be at rest before the collision and let the velocity vector of the impacting body form an angle φ to the shock normal. With

$$\vec{v}_1 = v_1 \cdot \begin{pmatrix} \sin\phi \\ -\cos\phi \end{pmatrix}$$ [m/s]

we obtain the following equations for the normal and tangential components ($v_{1n}$ and $v_{1t}$ respectively) at the point of contact:

(7.2:11a) $$v_{1n} = -v_1 \cdot \cos\phi$$ [m/s]

(7.2:11b) $$v_{1t} = v_1 \cdot \sin\phi$$ [m/s]

Assuming frictionless collision:

(7.2:12a) $$u_{1t} = v_{1t}$$ [m/s]

(7.2:12b) $$u_{2t} = 0$$ [m/s]

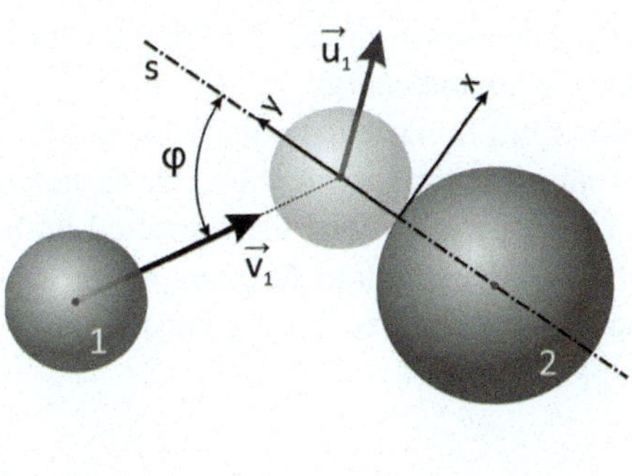


**Fig. 7.2.** Centric oblique collision. s: Shock normal.

Let $u_{1n}$ and $u_{2n}$ be the magnitudes of the velocities after the collision in the direction of the shock normal. Then according to Eqns 7.2:9a, b:

$$(7.2{:}13a)\quad u_{1n} = \frac{\mu - k}{\mu + 1} \cdot v_{1n} \qquad [m/s]$$

$$(7.2{:}13b)\quad u_{2n} = \frac{(1+k)\cdot\mu}{\mu + 1} \cdot v_{1n} \qquad [m/s]$$

From Eqns 7.2:11a and 7.2:12a, b we obtain

$$(7.2{:}14a)\quad \vec{u}_1 = \frac{v_1}{\mu + 1} \cdot \begin{pmatrix} (\mu+1)\cdot\sin\phi \\ -(\mu-k)\cdot\cos\phi \end{pmatrix} \qquad [m/s]$$

$$(7.2{:}14b)\quad \vec{u}_2 = \frac{v_1\cdot\mu\cdot(1+k)}{\mu + 1} \cdot \begin{pmatrix} 0 \\ -\cos\phi \end{pmatrix} \qquad [m/s]$$

The impacted body therefore moves in the direction of the shock normal. Eqn 7.2:14a gives the angle of departure $\beta$ for the impacting body with respect to the tangent plane at the point of impact:

$$(7.2{:}15)\quad \beta = \arctan\left(\frac{\mu - k}{\mu + 1}\cdot\cot\phi\right) \qquad [-]$$

The relative velocity of the two bodies follows from Eqns 7.2:14a, b:

$$(7.2{:}16)\quad \vec{u}_{1-2} = \vec{u}_1 - \vec{u}_2 = v_1 \cdot \begin{pmatrix} \sin\phi \\ k\cdot\cos\phi \end{pmatrix} \qquad [m/s]$$

This equation clearly demonstrates the validity of the Newtonian collision hypothesis (see 7.2.2). It is also noteworthy that relative velocity is independent of the masses of the two bodies. By analogy with Eqn 7.2:6, the loss of kinetic energy is:

$$(7.2{:}17)\quad Q_w = \frac{1-k^2}{\mu + 1}\cdot E_1\cdot\cos^2\phi \qquad [J]$$

### 7.2.4 Eccentric, straight collision

In an eccentric collision, the bodies experience torque, in addition to an exchange of momentum. We therefore need to add the law of conservation of angular momentum to those of the conservation of momentum and energy.

In the case of a straight collision, it is sufficient to write the equations in scalar form. As a further simplification, and to facilitate practical application, it is assumed that the shock normal passes through the centre of gravity of the impacted body (Fig. 7.3). This means that only the impacting body can rotate. Once again, the coordinate system is so chosen that the impacted body is at rest before the collision.

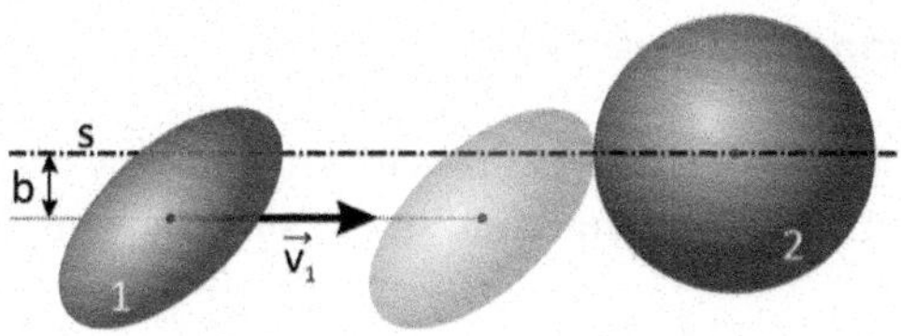


**Fig. 7.3.** Eccentric, straight collision.
s: Shock normal.
b: Eccentricity.

From the laws of conservation and using the Newtonian collision hypothesis, we obtain the following system of equations:

(7.2:18a) $m_1 \cdot v_1 = m_1 \cdot u_1 + m_2 \cdot u_2$ [kg·m/s]

(7.2: 18b) $J_1 \cdot \omega_1 = m_1 \cdot (u_1 - v_1) \cdot b$ [kg·m²/s]

(7.2:18c) $k = -\dfrac{u_1 - u_2 + \omega_1 \cdot b}{v_1}$ [-]

(7.2:18d) $\frac{1}{2} \cdot m_1 \cdot v_1^2 = \frac{1}{2} \cdot m_1 \cdot u_1^2 + \frac{1}{2} \cdot m_2 \cdot u_2^2 + \frac{1}{2} \cdot J_1 \cdot \omega^2 + Q_T$ [J]

where $J_1$ = moment of inertia of the impacting body, $\omega_1$ = its angular velocity after the collision and b = distance between the centre of gravity of the impacting body and the shock normal (the eccentricity of the collision).

Equations 7.2:18a-d form a non-linear system of equations for the velocities $u_1$ and $u_2$ of the two bodies, the angular velocity $\omega_1$ of the impacting body after the collision and the loss of kinetic energy $Q_T$. Solving this system of equations and inserting the dimensionless moment of inertia number

(7.2:19) $\theta = \dfrac{m_1 \cdot b^2}{J_1}$ [-]

yields the following equations:

(7.2:20a) $u_1 = \dfrac{\mu - k + \theta}{\mu + 1 + \theta} \cdot v_1$ [m/s]

(7.2:20b) $u_2 = \dfrac{(1 + k) \cdot \mu}{\mu + 1 + \theta} \cdot v_1$ [m/s]

(7.2:21) $\omega_1 = \dfrac{(1 + k) \cdot m_1 \cdot b}{J_1 \cdot (\mu + 1 + \theta)} \cdot v_1$ [1/s]

(7.2:22) $Q_T = \dfrac{1 - k^2}{\mu + 1 + \theta} \cdot E_1$ [J]

### 7.2.5 Eccentric, oblique collision

The most general and most common collision process is the eccentric, oblique collision. Here, deflection and rotation of the impacting body occur simultaneously (see Fig. 7.4).

We derive the formulae for the velocities and the loss of kinetic energy in a manner analogous to that outlined in 7.2.3 and 7.2.4 (velocity components in the direction of the shock normal, substituted into the system of equations 7.2:18a-d). Using the same symbols as in 7.2.4, this results in the following formulae:

(7.2:23a) $$\vec{u}_1 = \frac{v_1}{\mu+1+\theta}\cdot\begin{pmatrix}(\mu+1+\theta)\cdot\sin\phi\\ -(\mu-k+\theta)\cdot\cos\phi\end{pmatrix},$$ [m/s]

(7.2:23b) $$\vec{u}_2 = v_1\cdot\frac{(1+k)\cdot\mu}{\mu+1+\theta}\cdot\begin{pmatrix}0\\ -\cos\phi\end{pmatrix},$$ [m/s]

(7.2:24) $$\omega_1 = \frac{(1+k)\cdot m_1\cdot b}{J_1\cdot(\mu+1+\theta)}\cdot v_1\cdot\cos\phi\ ,$$ [1/s]

(7.2:25) $$Q_T = \frac{1-k^2}{\mu+1+\theta}\cdot E_1\cdot\cos^2\phi\ ,$$ [J]

(7.2:26) $$\beta = \arctan\left(\frac{\mu-k+\theta}{\mu+1+\theta}\cdot\cot\phi\right),$$ [-]

where $\beta$ is once again the angle of departure with respect to the tangent plane at the point of impact (the angle between $\vec{u}$ and the x-axis, see Fig. 7.4). Here again, the impacted body moves in the direction of the shock normal after the collision.

The angle of deflection $\lambda$ is also an interesting parameter. This is the angle by which the impacting body is deflected from its original direction of movement. To determine this angle, the velocity vector $\vec{u}_1$ is transformed into a coordinate system of which the x-axis coincides with the direction of movement before the collision (see Fig. 7.5). With angular displacement

$$-\left(\tfrac{1}{2}\cdot\pi-\phi\right)$$ [-]

and the associated transformation formulae

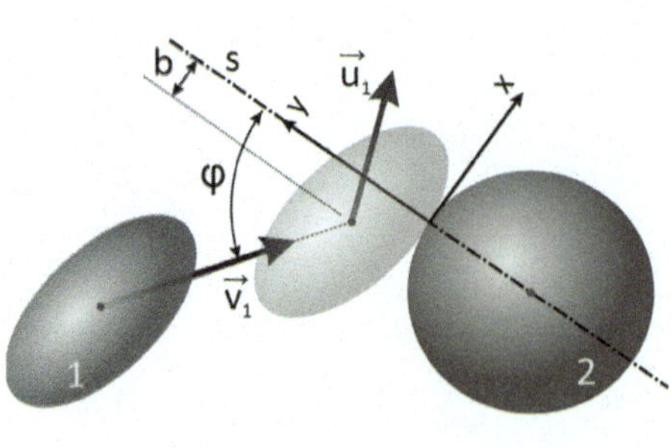


**Fig. 7.4.** Eccentric, oblique collision.
s: Shock normal.
b: Eccentricity.

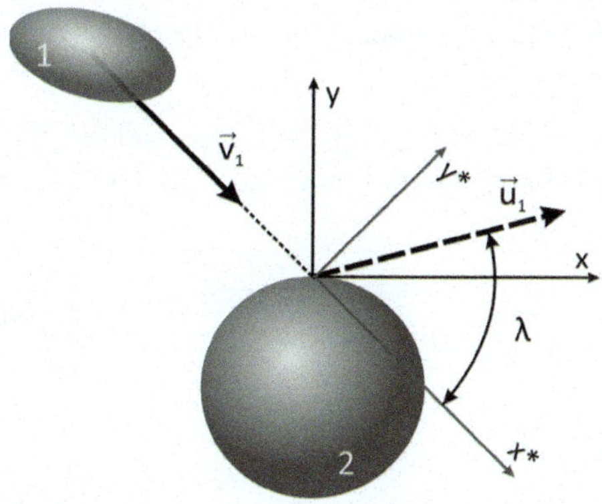


**Fig. 7.5.** Determining the angle of deflection λ.

(7.2:27a) $x^* = x \cdot \sin\phi - y \cdot \cos\phi$ [m]

(7.2:27b) $y^* = x \cdot \cos\phi + y \cdot \sin\phi$ [m]

we obtain the velocity $\vec{u}_1$ in the star system of the impacting body following the collision from Eqn 7.2:23a:

(7.2:28) $$\vec{u}_1^* = \frac{v_1}{\mu + 1 + \theta} \cdot \begin{pmatrix} \mu + \theta + \sin^2\phi - k \cdot \cos^2\phi \\ (1+k) \cdot \sin\phi \cdot \cos\phi \end{pmatrix}$$ [m/s]

from which we obtain the following equation for the angle of deflection λ:

(7.2:29) $$\lambda = \arctan\left( \frac{(1+k) \cdot \sin\phi \cdot \cos\phi}{\mu + \theta + \sin^2\phi - k \cdot \cos^2\phi} \right)$$ [-]

One advantage of the representation chosen in 7.2.2 to 7.2.5 is that all relevant parameters (mass ratio μ, restitution coefficient k and moment of inertia number θ) are dimensionless. This makes it easier to discuss special cases. For example, it can be seen immediately that if the two bodies have the same mass ($\mu = 1$) all laws of impact (with the exception of angular velocity in the case of rotation) are independent of mass.

### 7.2.6 Special cases

In practice, extreme mass ratios often occur, each of which allows a considerable simplification of the laws of impact. We shall therefore examine three special cases, addressing them roughly in order of the frequency with which they occur in relation to ricochets.

A $m_1 << m_2$: The impacting mass is much lighter than the mass that was at rest before the collision. The mass ratio μ is therefore very small ($\mu \rightarrow 0$).

B $m_1 >> m_2$: The impacting mass is much heavier than the impacted mass. The mass ratio μ is therefore very large ($\mu + 1 \rightarrow \mu$).

C $m_1 = m_2$: The impacting and impacted mass are equally heavy. The mass ratio $\mu$ is therefore equal to 1.

*Case A*: If the impacting mass is small compared to the impacted mass and the collision is inelastic, we can ignore $\mu$ and k must be set to 0. Equations 7.2:23 to 7.2:26 and Eqn 7.2:29 then become:

(7.2:30a) $$\vec{u}_1 = \frac{v_1}{1+\theta} \cdot \begin{pmatrix} (1+\theta) \cdot \sin\phi \\ -\theta \cdot \cos\phi \end{pmatrix}$$ [m/s]

(7.2:30b) $$\vec{u}_2 = 0$$ [m/s]

(7.2:31) $$\omega_1 = \frac{m_1 \cdot b}{J_1 \cdot (1+\theta)} \cdot v_1 \cdot \cos\phi$$ [1/s]

(7.2:32) $$Q_T = \frac{1}{1+\theta} \cdot E_1 \cdot \cos^2\phi$$ [J]

(7.2:33) $$\beta = \arctan\left(\frac{\theta}{1+\theta} \cdot \cot\phi\right)$$ [-]

(7.2:34) [-]

*Case B*: If $\mu$ is very large and the collision is elastic, then $\mu + 1 \to \mu$ and $k = 1$. From Eqns 7.2:23 to 7.2:25 and Eqn 7.2:29 we obtain:

(7.2:35a) $$\vec{u}_1 = v_{1x} \cdot \frac{1}{\mu+\theta} \cdot \begin{pmatrix} \mu+\theta \\ -\sin(2\cdot\phi) \end{pmatrix} \qquad \lambda = \arctan\left(\frac{\sin(2\cdot\phi)}{2\cdot(\theta+\sin^2\phi)}\right)$$

[m/s]

(7.2:35b) $$\vec{u}_2 = v_{1x} \cdot \frac{2\cdot\mu}{\mu+\theta} \cdot \begin{pmatrix} 0 \\ \cos\phi \end{pmatrix}$$ [m/s]

(7.2:36) $$\omega_1 = \frac{2\cdot m_1 \cdot b}{J_1 \cdot (\mu+\theta)} \cdot v_1 \cdot \cos\phi$$ [1/s]

(7.2:37) $$Q_T = 0$$ [J]

(7.2:38) $$\lambda = \arctan\left(\frac{\sin(2\cdot\phi)}{\mu+\theta}\right)$$ [-]

As we see from Eqn 7.2:38, the angle of deflection in this special case reaches a maximum at an angle of incidence of 45°. In the case of a centric collision ($\theta = 0$) that maximum depends only on the mass ratio of the two bodies.

*Case C*: In ballistics, impacting and impacted bodies are virtually never of equal mass. Such situations do occur in billiards, snooker and pool, where the intrinsic angular momentum of the impacting ball and its transfer to the impacted ball play an essential role. For a detailed account of the physics of billiards, see A. SOMMERFELD (1977).

With $\mu = 1$, Eqns 7.2:23 to 7.2:26 and Eqn 7.2:29 simplify to:

$$\text{(7.2:39a)} \quad \vec{u}_1 = \frac{v_1}{2+\theta} \cdot \begin{pmatrix} (2+\theta) \cdot \sin\phi \\ -(1-k+\theta) \cdot \cos\phi \end{pmatrix} \quad \text{[m/s]}$$

$$\text{(7.2:39b)} \quad \vec{u}_2 = v_1 \cdot \frac{(1+k)}{2+\theta} \cdot \begin{pmatrix} 0 \\ -\cos\phi \end{pmatrix} \quad \text{[m/s]}$$

$$\text{(7.2:40)} \quad \omega_1 = \frac{(1+k) \cdot m_1 \cdot b}{J_1 \cdot (2+\theta)} \cdot v_1 \cdot \cos\phi \quad \text{[1/s]}$$

$$\text{(7.2:41)} \quad Q_T = \frac{1-k^2}{2+\theta} \cdot E_1 \cdot \cos^2\phi \quad \text{[J]}$$

$$\text{(7.2:42)} \quad \beta = \arctan\left(\frac{1-k+\theta}{2+\theta} \cdot \cot\phi\right) \quad \text{[-]}$$

$$\text{(7.2:43)} \quad \lambda = \arctan\left(\frac{(1+k) \cdot \sin\phi \cdot \cos\phi}{1+\theta+\sin^2\phi - k \cdot \cos^2\phi}\right) \quad \text{[-]}$$

If the collision is also centric ($b = 0$ and $\theta = 0$), the above formulae become those for oblique, centric collision (see 7.2.3).

# 7.3 The phenomenology of ricochets

## 7.3.1 Definitions and symbols

Various terms are required when discussing the question of ricochets. These are summarized in Table 7.2 and partly illustrated in Fig. 7.6.

As for the ballistic coordinate system, a frame of reference with its origin at the point of impact is used. The x-axis is formed by the line of intersection between the plane of ricochet and the plane of impact, the y-axis is perpendicular to the plane of ricochet and the z-axis completes the system to form a right-handed system.

In many cases, we can make geometric simplifications without significantly affecting the generality of the results. For instance, we assume that in the vicinity of the point of impact, the trajectory of the undisturbed incoming projectile and that of the ricocheted projectile each lie in a different plane (the plane of impact and the departure plane). The point of impact and the point of departure are assumed to both lie on the plane of ricochet.

Given these simplifications, we can calculate the vertical angle of deflection λ using simple geometry, from the line of arrival and the departure path. The result (as an angle between skew lines) is:

(7.3:1) $$\lambda = \arccos(\cos\alpha \cdot \cos\beta \cdot \cos\gamma - \sin\alpha \cdot \sin\beta) \quad [-]$$

If $\gamma$ is small (so that we can assume $\cos\gamma \approx 1$), the vertical angle of deflection is the sum of the angle of arrival and the angle of departure.

When calculating the trajectory of a ricochet at greater distances, we must of course take even small horizontal angles of deflection into account.

**Table 7.2.** Terminology and definitions

| Term | Symbol | Description |
|---|---|---|
| Target | | Solid or liquid body with which the projectile comes into contact on its trajectory |
| Point of impact | $P_A$ | Point at which the projectile touches the surface of the target |
| Plane of ricochet | | Tangent plane to the target at the point of impact |
| Plane of impact | | Plane normal to the surface of the target through the undisturbed trajectory of the projectile |
| Line of arrival | | Tangential vector to the undisturbed trajectory of the projectile at the point of impact (parallel to the velocity vector) |
| Angle of incidence | $\alpha$ | Angle between the surface of the target and the line of arrival |
| NATO angle | $\varphi$ | Angle between the line of arrival and the normal to the surface of the target at the point of impact ($\varphi = 90° - \alpha$) |
| Impact velocity | v | Velocity at which the projectile strikes the target |
| Point of departure | $P_B$ | Point at which the projectile leaves the surface of the target |
| Departure plane | | Plane normal to the surface of the target through the undisturbed trajectory of the projectile following ricochet |
| Departure path | | Tangential vector to the trajectory of the ricocheting projectile at the point of departure (parallel to the velocity vector of the ricocheting projectile) |
| Angle of departure | $\beta$ | Angle between the surface of the target and the tangent to the trajectory of the ricocheting projectile at the point of departure |
| Velocity after ricochet | u | Velocity at which the projectile leaves the target |
| Horizontal angle of deflection | $\gamma$ | Angle between plane of impact and departure plane |
| Vertical angle of deflection | $\lambda$ | Angle between line of arrival and departure path |
| Length of ricochet mark | $s_x$ | Distance between point of impact and point of departure |
| Contact time | $t_s$ | Duration of contact between projectile and target |

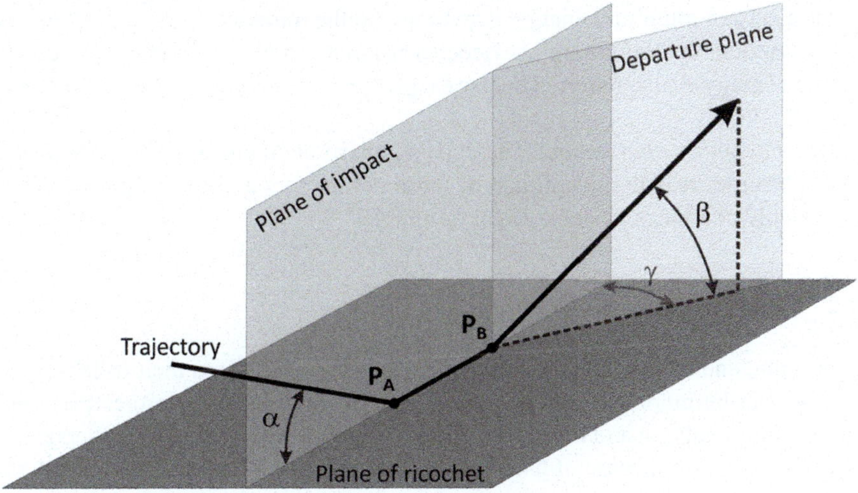


**Fig. 7.6.** Definitions of terms related to ricochet. (See also Table 7.2).

### 7.3.2 Parameters influencing ricochet behaviour

The way in which a projectile ricochets off an object is largely determined by the following parameters:

*Mass of the target.* The ratio of the mass of the impacting body (the projectile) to that of the impacted body (the target) determines the applicable laws of impact (see 7.2.6).

*Material and structure of the target.* Whether or not the projectile can penetrate the target is a decisive factor. The degree to which the projectile can penetrate depends on the target material. If a projectile penetrates, different vertical angles of deflection can occur than with a purely superficial ricochet, the behaviour of which is influenced by the structure of the surface.

*Dimensions of the target.* If the distances from the point of impact to the edges of the target are of the same order of magnitude as the calibre of the projectile, the process will be different than if the projectile strikes the target at a greater distance from the edge. Thin targets can deform and thus produce different ricochet behaviour than thick targets of the same material.

*Projectile material and design.* Projectile material and design have a decisive effect on the angle of departure and the velocity after ricochet.

*Projectile behaviour.* On contact with a target, a projectile can deform, disintegrate or remain unchanged. Which of these cases actually occurs and to what extent the projectile deforms or disintegrates depends on the configuration of the target and the structure of the projectile (see 6.1.5).

*Impact energy.* The energy of the projectile at the moment of impact also influences what happens. For example, the projectile can only penetrate the target if it has a certain minimum energy, depending on the material. If the projectile has less energy than this minimum, its ricochet behaviour will be different. A certain minimum energy is also required to cause a projectile to deform or disintegrate (see also 6.1.5).

*Flight behaviour after impact.* The impact imparts an impulse and a torque shock to the projectile. It is the magnitude of those parameters that determines whether the projectile continues in stable flight or tumbles.

### 7.3.3 Typical ricochet modes

Targets can generally be divided into three categories as regards their influence on ricochets. Furthermore, a distinction is made between deformable projectiles (lead and lead-cored bullets) and dimensionally stable projectiles (e.g. steel bullets).

*Hard targets.* In the case of hard targets, the projectile does not penetrate at all. The track of the projectile runs entirely along the surface of the target and only the projectile can undergo deformation. Typical examples of hard targets include thick steel plates (Fig. 7.7 a) and polished, hard stone or concrete slabs. The angle of departure is practically independent of the angle of incidence, and is only a few degrees.

*Semi-hard targets.* In the case of semi-hard targets, both the target and the projectile are deformed, if the latter is not dimensionally stable. The projectile leaves a track less than 1 calibre deep (Fig. 7.7 b). Examples of semi-hard targets include thin sheet metal, soft concrete and asphalt. The angle of departure can range from a few degrees to the amplitude of the angle of incidence, depending on projectile design and target behaviour. The more the projectile deforms (e.g. in the case of solid lead bullets), the less it damages the target and the smaller its angle of departure.

*Soft targets.* As the angle of incidence increases, the projectile creates an increasingly deep track, with depths in excess of one calibre (see Fig. 7.7 c). When a certain material-specific angle of incidence is exceeded, the projectile penetrates the target completely before exiting again. In such cases, the angle of departure may be significantly greater than the angle of incidence; a 15° angle of incidence can easily result in a 30° angle of departure. Typical soft target materials include wood (including chipboard), sand and water. A projectile striking such materials often undergoes no deformation, regardless of its design.

A bullet striking a thin plate (of steel or aluminium, for instance) at a low angle of incidence may display quite surprising behaviour. Because the contact point is in front of the projectile's centre of gravity, the projectile receives a torque shock and begins to rotate about its centre of gravity. This forces the rear of the projectile through the plate. The projectile is now perpendicular to its direction of movement and is cut in two by the plate, resulting in projectile parts being found on both sides.

An example of such behaviour occurred when a lead bullet (38 Spl.) ricocheted off a skull bone. The front part of the bullet was found outside the skull and the rear part inside the skull, in the brain.

A ricochet conserves a relatively large percentage of its energy, as long as the

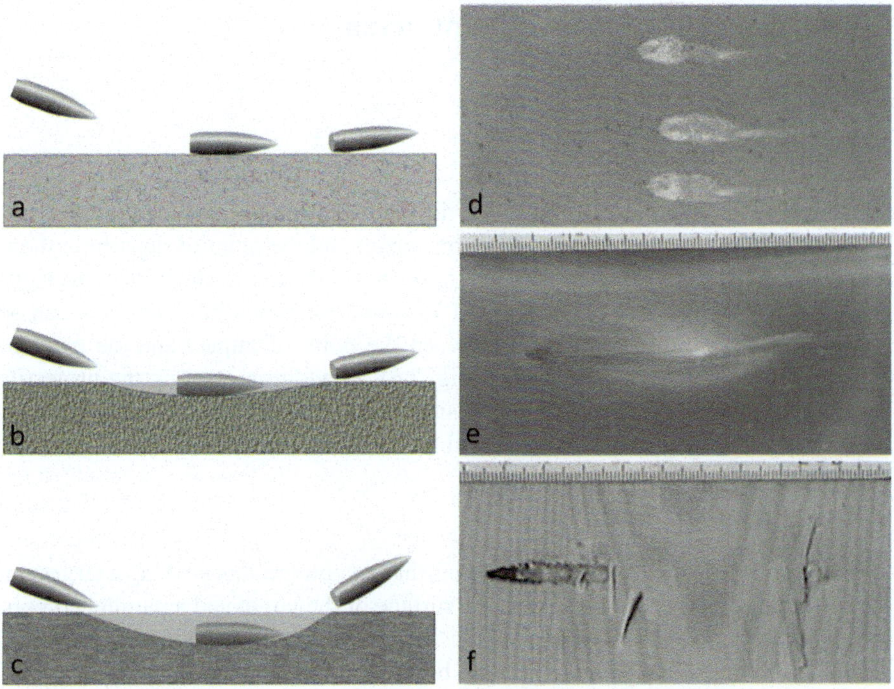


**Fig. 7.7.** Target types and ricochet tracks. *Left:* Trajectory (schematic). *Right:* Real-life examples. **a**, **d**: Hard target, thick steel plate. **b**, **e**: Semi-hard target, thin aluminium sheet (2 mm). **c**, **f**: Soft target, spruce.

projectile retains most of its mass. A projectile will conserve approx. 70–95% of its impact energy after striking a hard target (depending on the angle of incidence). The range may be as wide as 20–95% in the case of a soft target.

### 7.3.4 Projectile striking a hard surface at 90°

A projectile striking a hard surface at 90° without penetrating constitutes a special case. This usually leads to deformation or fragmentation of the projectile and localized destruction of the target. Parts of the projectile or target may experience high levels of acceleration and be projected away from the target.

If a jacketed bullet hits a hard surface at a steep angle and at high velocity, the deformation of the lead core can create so much pressure that the jacket bursts and the jacket and lead fragments fly several dozen metres in practically any direction against the direction of fire. The initial velocity of these fragments is very high; they can seriously endanger the firer and their surroundings at considerable distances (30 m and more). See case study in 3.8.3.2.

Solid bullets or bullets with a hard core may also ricochet towards the firer. Target surfaces with elastic properties (e.g. the plastic layers in bullet-resistant glass) may accentuate this behaviour.

## 7.4 The projectile as a point mass

### 7.4.1 Heavy targets

#### *7.4.1.1 Preliminary remark*

Although different projectile paths with different depths of penetration into the target can be observed between the point of impact and the point of departure, these can be combined in a point mass model to form a uniform ricochet model. Regardless of which of the three cases in Fig. 7.7 occurs, the path of the projectile can be represented as a direct connection between the point of impact and the point of departure (as shown schematically in Fig. 7.6). In addition, treating the projectile as a point mass means that the prerequisites for oblique, centric collision are met, so we can use the equations in 7.2.3, with $\mu \rightarrow 0$.

#### *7.4.1.2 Ricochet theory*

The law of independence of motion means that an oblique shot striking a target can be split into a component normal to the surface of the target and a parallel component, and the two components can be treated separately. Taking $\alpha = \frac{1}{2} \cdot \pi - \varphi$, Eqns 7.2:11a, b for the two components before the collision become:

(7.4:1a) $$v_x = v \cdot \cos \alpha$$ [m/s]

(7.4:1b) $$v_y = - v \cdot \sin \alpha$$ [m/s]

From Eqn 7.2:14a, the normal component after impact is:

(7.4:2) $$u_y = v \cdot k \cdot \sin \alpha$$ [m/s]

The restitution coefficient k is not a constant, but a function of the angle of incidence $\alpha$, as both the projectile and the target tend to deform as velocity components in the normal direction increase, thus reducing the quantity of kinetic energy recovered. However, under certain conditions (e.g. if the projectile is deflected in the target rather than ricocheting) k can also increase at a larger angle of incidence.

In contrast to the frictionless conditions assumed in deriving the laws of impact, a decrease in velocity must also be expected for components parallel to the plane of ricochet when ricochets occur. This deceleration is caused on the one hand by friction during contact time, and on the other hand by deformation of the target or displacement of material along the projectile track.

As we are examining the ricochet process using the point-mass model, it makes sense to combine all effects that influence the parallel velocity component of the projectile into a single "global" friction. From a physical point of view, it also makes sense to assume that this friction $R_x$ is proportional to the contact pressure $F_y$ acting perpendicular to the plane of ricochet, which is assumed to be constant throughout the process

(7.4:3) $$R_x = \sigma \cdot F_y$$ [N]

The proportionality factor σ also depends on the angle of incidence α, as contact area (and hence friction) increases with bullet deformation.

Despite the similar approach, the "friction number" σ is not the same as the conventional coefficient of friction from mechanics. It includes not only the amplitude of the frictional force, but also the duration of the effect and a proportion of the deformation of projectile and target, insofar as these are caused by tangential forces.

Let $t_s$ be the contact time. From Newtons Second law:

(7.4:4) $$R_x \cdot t_s = m \cdot (u_x - v_x)$$ [N·s]

and with (7.4:2):

(7.4:5) $$F_y \cdot t_s = -m \cdot (u_y - v_y) = m \cdot v_y \cdot (1 + k)$$ [N·s]

Substituting Eqns 7.4:4 and 7.4:5 into Eqn 7.4:3 and solving the result for $u_x$, we obtain:

(7.4:6) $$u_x = v_x + \sigma \cdot v_y \cdot (1 + k)$$ [m/s]

The two components of the velocity after impact, $u_x$ and $u_y$, determine the angle of departure β thus:

$$\tan\beta = \frac{u_y}{u_x} = \frac{-k \cdot v_y}{v_x + \sigma \cdot v_y \cdot (1+k)} .$$ [-]

With $\tan\alpha = -v_y/v_x$, $k = k(\alpha)$ and $\sigma = \sigma(\alpha)$, we obtain an expression linking the angle of incidence and the angle of departure:

(7.4:7) $$\tan\beta = \frac{k(\alpha) \cdot \tan\alpha}{1 - \sigma(\alpha) \cdot \tan\alpha \cdot \{1 + k(\alpha)\}}$$ [-]

For a given angle of incidence, the angle of departure is therefore determined by the restitution coefficient k and the friction number σ.

If measurements of the projectile velocity before and after ricochet are available, plus the angles of incidence and departure, we can determine the corresponding values for k and σ thus:

(7.4:8a) $$k(\alpha) = -\frac{u_y}{v_y}$$ [-]

(7.4:8b) $$\sigma(\alpha) = \frac{u_x - v_x}{u_y + v_y}$$ [-]

We can then derive an expression linking k and σ from Eqn 7.4:7:

(7.4:9) $$\sigma = \frac{\cot\alpha - k \cdot \cot\beta}{1 + k}$$ [-]

We can derive bounds for k and σ, though these are approximate in some instances. We can assume σ to be positive, which allows us to derive the following limits for k using Eqn 7.4:9:

$$(7.4:10a) \quad 0 \leq k(\alpha) \leq \frac{\tan\beta}{\tan\alpha}, \qquad \alpha, \beta > 0. \qquad [-]$$

In Eqn 7.4:7, tan β, k and tan α are all positive. It follows that the denominator must also be positive. This leads to a restriction for σ:

$$(7.4:10b) \quad 0 \leq \sigma(\alpha) \leq \frac{\cot\alpha}{1+k(\alpha)}. \qquad \alpha > 0. \qquad [-]$$

We can therefore describe the ricochet behaviour of any projectile-target combination using a pair of these functions. If we know the functions, we can estimate the angle of departure and the velocity after ricochet for any angle of incidence. Figures 7.8 and 7.9 show typical function curves.

If a projectile remains embedded in the (soft) target when the critical angle is reached, then from Eqn 7.4:2 k = 0 (because $u_y = 0$) and consequently from Eqn 7.4:9b σ = cot α. The friction number thus has a limit value that can be useful when plotting the curve.

If it is not possible to determine the velocity of a ricochet, one can measure the length of the contact mark $s_x$ in addition to the angles. This can be taken into account when determining the restitution coefficient k and the friction number σ. From the basic kinematics equations:

$$(7.4:11) \quad s_x = \tfrac{1}{2} \cdot t_s \cdot (u_x + v_x) \qquad [m]$$

$$(7.4:12) \quad u_x = \frac{2 \cdot s_x}{t_s} - v_x \qquad [m/s]$$

By comparing with Eqn 7.4:6, we obtain the following relationship between restitution coefficient k, σ and the length of the ricochet mark $s_x$ ($v_x / v_y = -\cot\alpha$):

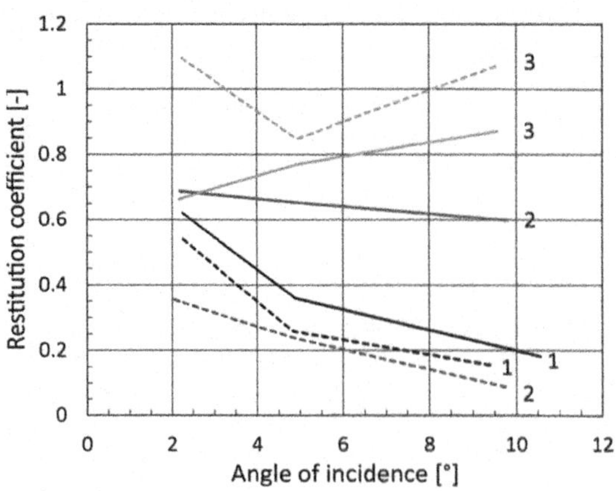


**Fig. 7.8.** Examples of restitution coefficients. Solid line: FMJ bullet in 9 mm Luger. Dotted line: Solid lead bullet in 38 Spl. Targets:
1 Thick steel plate.
2 Concrete slab.
3 Spruce.

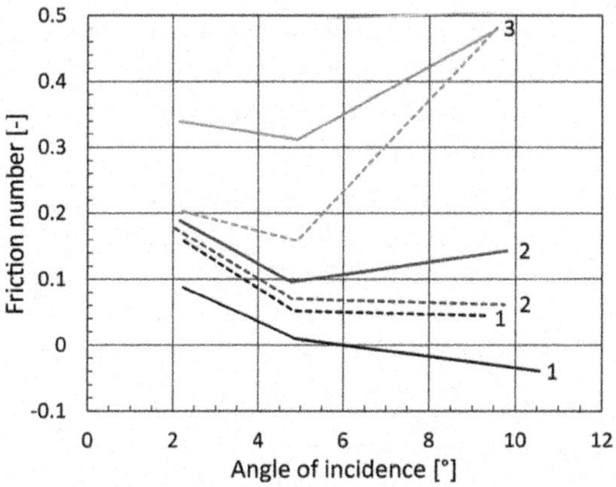


**Fig. 7.9.** Examples of friction numbers. Solid line: FMJ bullet in 9 mm Luger. Dotted line: Solid lead bullet in 38 Spl. Targets:
1 Thick steel plate.
2 Concrete slab.
3 Spruce.
The negative value for the FMJ bullet on the steel plate is interesting. The horizontal velocity component increases at greater angles (deflection).

(7.4:13) $$\sigma \cdot (1+k) = \frac{2 \cdot s_x}{v_y \cdot t_s} + 2 \cdot \cot\alpha$$ [-]

This means that $\sigma \cdot (1+k)$ can be replaced in the denominator of Eqn 7.4:7, and after brief calculation we obtain an equation linking the angle of incidence, angle of departure, length of ricochet mark, restitution coefficient and contact time:

(7.4:14) $$\cot\beta = -\frac{1}{k} \cdot \left( \cot\alpha + \frac{2 \cdot s_x}{v_y \cdot t_s} \right)$$ [-]

If measurements of $\alpha$, $\beta$ and $s_x$ are available, we can solve Eqn 7.4:14 for $t_s$, whereby two additional constraints apply, which are derived from the two extreme cases below:

- The projectile comes to rest at a distance $s_x$, so $u_x = 0$.
- The projectile loses no velocity in the x direction, so $u_x = v_x$.

Equation 7.4:11 gives us the following two inequalities:

(7.4:15) $$\frac{s_x}{v_x} \le t_s \le \frac{2 \cdot s_x}{v_x}$$ [s]

By including the conditions in 7.4:10a,b, we can calculate k and $t_s$ from Eqns 7.4:14 and 7.4:15 by means of optimization. $\sigma$ then follows from Eqn 7.4:9. However, $\sigma$ is very sensitive to fluctuations in k at small angles of departure $\beta$, which makes it somewhat difficult to determine the friction number.

#### *7.4.1.3 Curves of ricochet functions. Approximations.*

In view of physical considerations, and in the light of the results of experiments, curves of restitution coefficient as a function of angle of incidence can take one of two forms:

- In the case of hard targets (concrete or thick steel plates) that the projectile cannot penetrate, the restitution coefficient decreases monotonically with increasing angle of incidence, this decrease becoming more pronounced as projectile deformation increases. It is always less than 1.
- In soft and deformable materials (such as wood and thin sheet metal), the projectile may be deflected during penetration, often without being significantly deformed. This manifests itself in an increase in the restitution coefficient, which may well exceed 1. The restitution coefficient remains limited, however, because of Eqn 7.4:8. If the projectile becomes embedded, k is zero.

As long as the angle of incidence is small ($< 5°$), the friction number increases with increasing angle, regardless of the type of projectile and target. This is because it is primarily the cylindrical part of the projectile that is in contact with the target, but the point of contact moves towards the ogive at greater angles.

If the angle exceeds 5°, the projectile penetrates soft targets, which greatly increases the contact area and hence the friction number. With hard targets, it remains almost constant or increases slightly (if the projectile causes slight spalling in the case of concrete). A jacketed bullet striking a steel plate experiences an increase in its horizontal velocity component; this also results in deflection.

Because of the process described, approximation of the friction coefficient function usually involves using a power function.

By measuring impact velocities, velocities after ricochet and their components, at a number of angles of incidence, we can determine the restitution coefficient and friction number experimentally as a function of the angle of incidence and approximate these two parameters by a function. In principle, it is even possible to use Eqn 7.4:7 to determine the two ricochet functions solely from measured pairs of angle of incidence and angle of departure, plus a best-fit calculation.

## 7.4.2 Lightweight targets

### *7.4.2.1 General*

Another situation frequently encountered in practice is exactly the opposite from a physical point of view: that of a projectile striking a target of significantly smaller mass. This may occur when the projectile touches stalks or thin twigs on its trajectory or collides with drops of liquid (e.g. rain).

Of primary interest here are the angle by which the projectile is deflected from its original trajectory and the loss of velocity that it experiences as a result of the impact. Here, too, we examine the problem using the point-mass model, taking as our basis the equations for a centric oblique collision. We can derive the equation for the angle of deflection from Eqn 7.2:29 by setting $\theta = 0$.

### 7.4.2.2 Maximum angle of deflection

If a bullet perforates a small target exactly in the centre, no deflection is to be expected, for reasons of symmetry. Similarly, there will be no deviation from the original trajectory when the bullet passes the target tangentially. This justifies the assumption that an angle of incidence (or a NATO angle) exists for which the angle of deflection reaches a maximum.

We can re-write Eqn 7.2:29 as follows:

$$(7.4{:}16) \qquad \tan\lambda = f(\alpha) = \frac{(1+k)\cdot\sin\alpha\cdot\cos\alpha}{\mu+\cos^2\alpha - k\cdot\sin^2\alpha} \qquad [-]$$

where $\lambda$ = angle of deflection, $\alpha$ = angle of incidence, ($\alpha = ½\cdot\pi - \varphi$), $\mu$ = mass ratio $m_1/m_2$ and k = restitution coefficient.

This function does indeed have the expected curve (see Fig. 7.10). To determine the maximum angle of deflection, we can differentiate f with respect to $\alpha$ and then set the derivative obtained to zero. This gives us a formula for the angle of incidence $\alpha_m$ that results in a maximum angle of deflection:

$$(7.4{:}17) \qquad \tan\alpha_m = \sqrt{\frac{\mu+1}{\mu-k}}, \qquad \mu > k\,. \qquad [-]$$

If we substitute Eqn 7.4:17 into Eqn 7.4:16, we obtain the following expression for the maximum possible angle of deflection:

$$(7.4{:}18) \qquad \tan\lambda_{max} = \frac{1+k}{2\cdot\sqrt{(\mu-k)\cdot(\mu+1)}}, \qquad \mu > k. \qquad [-]$$

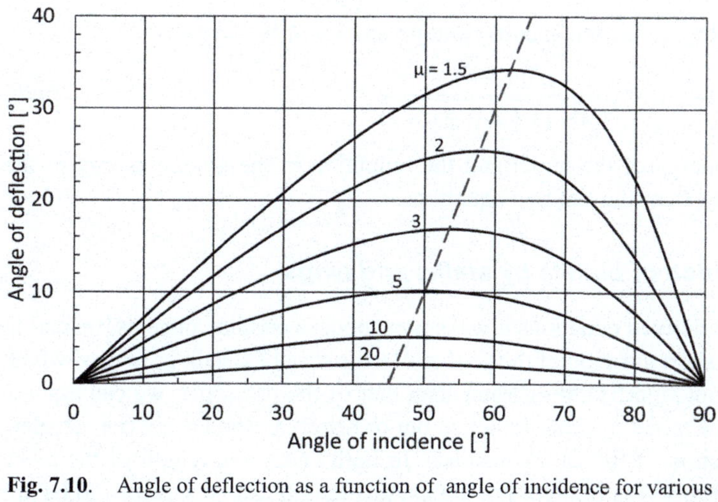


**Fig. 7.10.** Angle of deflection as a function of angle of incidence for various mass ratios $\mu$ and with k = 0.8 (from Eqn 7.4:16). The dashed line passes through the maxima, calculated using Eqn 7.4:19.

If we eliminate k from Eqns 7.4:17 and 7.4:18, we obtain a direct relationship between the maximum angle of deflection and the corresponding angle of incidence

(7.4:19) $$\lambda_{max} = \arctan\left[-\cot(2\cdot\alpha_m)\right] \quad [-]$$

This function takes the form of a straight line with a gradient of 2 and is shown in Fig. 7.10 as a dashed line. The maximum angles of deflection therefore always occur at the same angles of incidence for all values of k and $\mu$.

### 7.4.2.3 Velocity and energy

We can also use the equations for maximum angle of deflection to calculate the expected reduction in velocity and energy.

From Eqn 7.2:14a, we obtain the following expression for the magnitude of projectile velocity after collision (again using the angle of incidence $\alpha$ instead of the NATO angle $\varphi$):

(7.4:20) $$u_1 = v_1 \cdot \sqrt{\cos^2\alpha + \left(\frac{\mu - k}{\mu + 1}\right)^2 \cdot \sin^2\alpha} \quad [\text{m/s}]$$

At the maximum angle of deflection, which occurs at the angle of incidence calculated using Eqn 7.4:17, relative velocity loss and energy loss are given by:

(7.4:21) $$\frac{\Delta v_1}{v_1} = \frac{v_1 - u_1}{v_1} = 1 - \sqrt{\frac{\mu - k}{\mu + 1}} \quad [-]$$

(7.4:22) $$\frac{\Delta E_1}{E_1} = \frac{1 + k}{\mu + 1} \quad [-]$$

From Eqn 7.2:14b, the (instantaneous) initial velocity of the target is:

(7.4:23) $$u_2 = v_1 \cdot \frac{\mu\cdot(1 + k)}{\sqrt{(\mu + 1)\cdot(2\cdot\mu - k + 1)}} \quad [\text{m/s}]$$

We obtain the energy of the target and the reduction in the kinetic energy of the projectile directly from the last two equations.

### 7.4.2.4 Deflection of bullets by stalks and twigs

The use of these derived equations always presupposes certain concepts, related in particular to the short duration of the process. In the case of a stalk or twig of which the diameter is not significantly greater than that of the projectile, we can assume that a piece of the target is either knocked out or briefly pushed away (i.e. accelerated in the direction of the shock normal). In both cases, the length of the piece involved in the process is limited. For stalks, this length can be assumed to be ap-

**Table 7.3.** Maximum angle of deflection (k = 1) and lateral deviation 10 m after ricochet point (dry softwood, $\rho = 550$ kg/m³)

| Bullet | Mass | Stalk, 1 mm | | Twig, 3 mm | | Twig, 5 mm | |
|---|---|---|---|---|---|---|---|
| | [g] | [°] | [cm] | [°] | [cm] | [°] | [cm] |
| 22 L.R. | 2.6 | 0.11 | 1.9 | 0.72 | 12.6 | 1.47 | 25.6 |
| 7.65 Browning | 4.7 | 0.08 | 1.4 | 0.54 | 9.5 | 1.11 | 19.3 |
| 9 mm Luger | 8.0 | 0.06 | 1.0 | 0.38 | 6.6 | 0.77 | 13.4 |
| 38 Spl. | 10.2 | 0.045 | 0.8 | 0.29 | 5.2 | 0.60 | 10.5 |
| 44 Rem. Mag. | 15.6 | 0.035 | 0.6 | 0.24 | 4.2 | 0.49 | 8.5 |
| 45 ACP | 14.9 | 0.04 | 0.7 | 0.26 | 4.5 | 0.52 | 9.1 |
| 5.56 × 45 | 4.0 | 0.07 | 1.2 | 0.46 | 8.1 | 0.95 | 16.5 |
| 7.62 × 51 | 9.5 | 0.04 | 0.7 | 0.27 | 4.7 | 0.55 | 9.5 |
| 8 × 68 S | 12.1 | 0.035 | 0.6 | 0.22 | 3.9 | 0.45 | 7.9 |

proximately twice the projectile diameter; as the diameter of the target increases, it will approach that of the projectile.

The piece participating in the impact process will in all cases be so light that we can assume $\mu >> 1$ and hence also $\mu >> k$. From Eqn 7.4:18, the maximum angle of deflection is therefore:

$$(7.4:24) \qquad \tan \lambda_{max} = \frac{1 + k}{2 \cdot \mu} \qquad [-]$$

The greater the mass ratio $\mu$, the smaller the angle of deflection. Heavy projectiles are therefore deflected less than light ones. The initial velocity of the target follows from Eqn 7.4:23:

$$(7.4:25) \qquad u_2 = v_1 \cdot \frac{1 + k}{\sqrt{2}} \qquad [m/s]$$

It is noteworthy that for $\mu >> 1$ and at the maximum angle of deflection, the velocity of the target does not depend on the mass ratio $\mu$.

An elastic collision will result in the largest possible angle of deflection and the greatest possible loss of kinetic energy by the projectile. Table 7.3 shows maximum angles of deflection and corresponding lateral deviations for a number of projectiles as a function of the thickness of the stalk or twig.

### *7.4.2.5 Raindrops*

Raindrops have a diameter of between 0.5 and 5 mm. Smaller drops are classed as drizzle, while larger raindrops do not occur as they are torn apart by aerodynamic forces. The largest drops weigh around 65 mg and reach a descent velocity of around 12 m/s (see Table 7.4 next page).

When calculating the velocity of descent, it is assumed that the forces acting on the drop (weight and drag) are in equilibrium, i.e. the drop is descending at "terminal velocity".

The collision between a bullet and a raindrop can be regarded as a completely inelastic collision (k = 0). It is most convenient to consider the effect of the raindrop on the trajectory of the bullet using a frame of reference based on the bullet itself.

A raindrop of mass $m_{rt}$ and velocity $v_{rt}$ collides centrically at a velocity $v_r$ and at an angle $\zeta$ relative to the direction of flight with a bullet of mass $m_p$ which, in this frame of reference, is at rest (see Fig. 7.11).

(7.4:26) $$v_r = \sqrt{v^2 + v_{rt}^2}$$ [m/s]

(7.4:27) $$\zeta = \arctan\left(\frac{v_{rt}}{v}\right)$$ [-]

where v = velocity of bullet before impact and $v_{rt}$ = velocity of raindrop.

We can use Eqn 7.2:3 to calculate the velocity of the bullet after the collision in the bullet-fixed system. With $\mu = m_p/m_{rt}$ we can write:

(7.4:28) $$u_r = \frac{1}{\mu + 1} \cdot v_r$$ [m/s]

In the earth-fixed frame of reference, this velocity is superimposed on the original projectile velocity. From Fig. 7.12, we see that the vertical angle of deflection and the velocity of the projectile after collision can be determined using trigonometric functions (the cosine and sine theorems). Because $v_{rt} \ll v$, the exact equations can be simplified for practical calculations. This results in the following equations for projectile velocity after collision u and for the vertical angle of deflection $\lambda$

(7.4:29) $$u = \frac{\mu}{\mu + 1} \cdot v$$ [m/s]

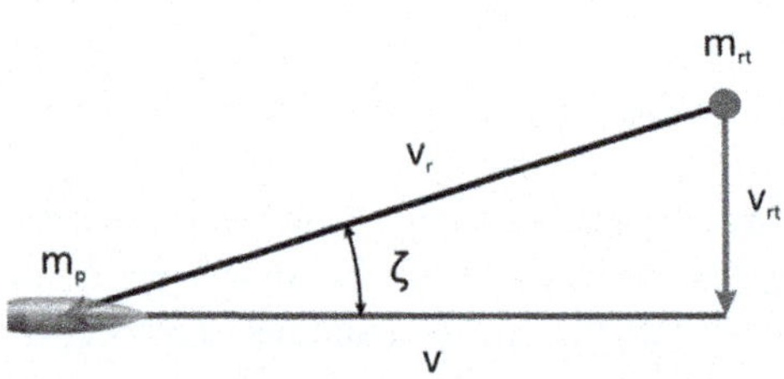


**Fig. 7.11**: Velocity of the raindrop in a body(bullet)-fixed system.

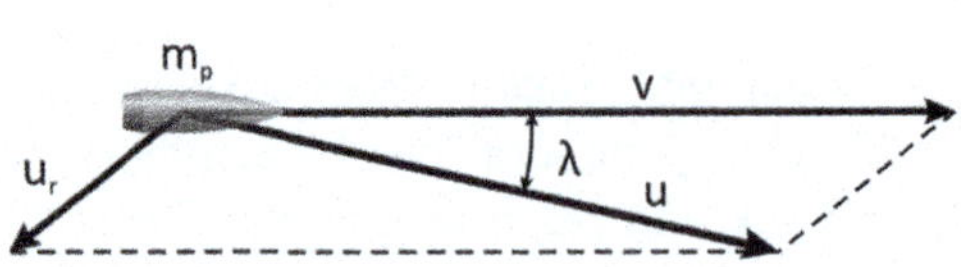


**Fig. 7.12** Determining the velocity and direction of the bullet in the earth-fixed system after the collision.

**Table 7.4.** Diameter, mass and descent velocity of raindrops

| Diameter [mm] | Mass [mg] | $v_{rt}$ [m/s] |
|---|---|---|
| 0.5 | 0.1 | 3.7 |
| 1.0 | 0.5 | 5.2 |
| 1.5 | 1.8 | 6.4 |
| 2.0 | 4.2 | 7.4 |
| 2.5 | 8.2 | 8.2 |
| 3.0 | 14.1 | 9.0 |
| 3.5 | 22.4 | 9.8 |
| 4.0 | 33.5 | 10.4 |
| 4.5 | 47.7 | 11.1 |
| 5.0 | 65.4 | 11.7 |

(7.4:30) $$\sin\lambda = \frac{1}{\mu}\cdot\sqrt{1+\frac{v_{rt}^2}{v^2}}\cdot\sin\zeta \Rightarrow \lambda \approx \frac{1}{\mu}\cdot\zeta$$ [-]

Fast, heavy bullets are less affected by rain than light, slow ones.

If, for example, a 5.56 × 45 (223 Rem.) bullet weighing 4 g collides with a 3 mm raindrop at a velocity of 850 m/s ($m_{rt}$ = 14 mg, $v_{rt}$ = 9 m/s), its velocity decreases by 3 m/s and it is deflected downwards by 0.038 per mil (3.8 mm at 100 m).

As both deflection and loss of velocity will lower the fall of shot, colliding with several drops can cause a noticeable deviation in point of impact.

## 7.5 The projectile as a rigid body

### 7.5.1 General

Assessing the stability of a ricochet requires us to take account of the proper motion of the projectile. For spin-stabilized projectiles in particular, the torque shock generated on contact with the target plays a decisive role. This means that we need to know the geometric dimensions and, in particular, the distance between the shock normal and the centre of gravity.

For spherical projectiles this distance is of course zero, but these are fairly rare today, with shot and certain types of shotgun slug being the most common examples. Almost all current bullets – be they small-, medium- or large-calibre – are 1.5–6 calibres in length. Despite the variety of shapes, the geometry relevant to contact with a target can be divided into three classes: round, ogival and pointed.

### 7.5.2 Position of the shock normal for different bullet shapes

#### *7.5.2.1 Parameters*

To calculate the torque shock that the nose of the bullet experiences on impact, it is necessary to know the lever arm of the corresponding impulse. This can be determined from the geometry of the nose. We shall be presenting the equations for the three nose shapes mentioned above; these can be used for most bullets. The parameters are as follows:

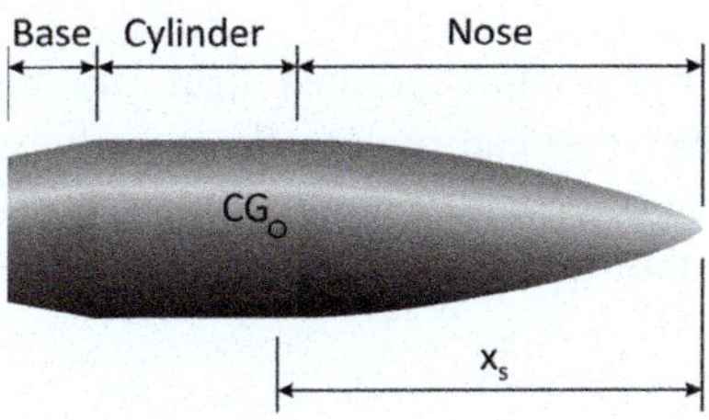


**Fig. 7.13** The parts of a bullet. Several nose shapes are possible. CG: Centre of gravity.

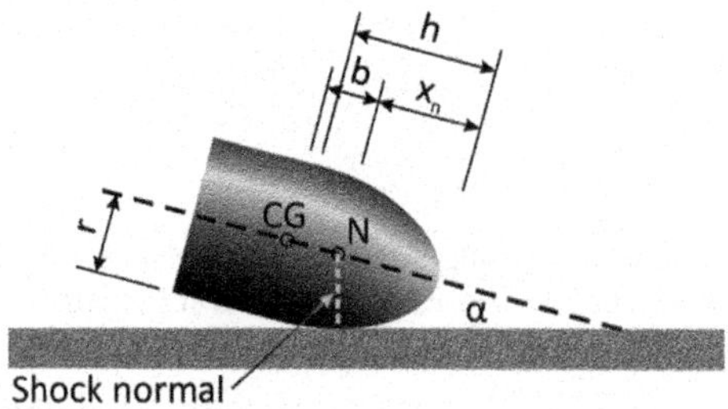


**Fig. 7.14.** Round-nose bullet. The line of impact at the target is the tangent with an angle α to the semi-ellipse with semi-axes h and r.

CG: Centre of gravity.

N: Point of intersection between collision normal and projectile axis (*point of intersection*).

h: Nose length.

d: Calibre, r = ½·d.

$x_S$: Distance between bullet tip and centre of gravity (see Fig. 7.13).

$x_n$: Distance between bullet tip and point of intersection.

b: Distance between point of intersection and centre of gravity $b = x_S - x_n$ (lever arm of the impact force, measured along the projectile axis).

The distance $x_s$ between the centre of gravity and the tip of the bullet is usually determined by measurement. It can also be calculated in the case of simple bullet designs.

### 7.5.2.2 Elliptical round-nose bullets

Many "round-nose" bullets can be approximated by a cylinder with an attached half ellipsoid of rotation. The total length of a handgun bullet is less than 2 calibres, about half of which is cylindrical. Rifle bullets are 3 to 5 calibres long and the nose length h accounts for approximately one third of the total length of the bullet.

We can calculate the distance between point of intersection and bullet tip using the geometry of the ellipse (see Fig. 7.14):

$$(7.5{:}1) \qquad x_n = h - \frac{(h^2 - r^2)\cdot \tan\alpha}{\sqrt{h^2 \cdot \tan^2\alpha + r^2}} \qquad [m]$$

### 7.5.2.3 Parabolic round-nose bullets

The noses of some round-nose bullets are much more pointed than can be approximated with a semi-ellipse. In such cases one can replace the nose by a paraboloid of rotation. We calculate the distance between the point of intersection and the bullet tip using the geometry of the parabola, in a manner analogous to the method for the elliptical nose.

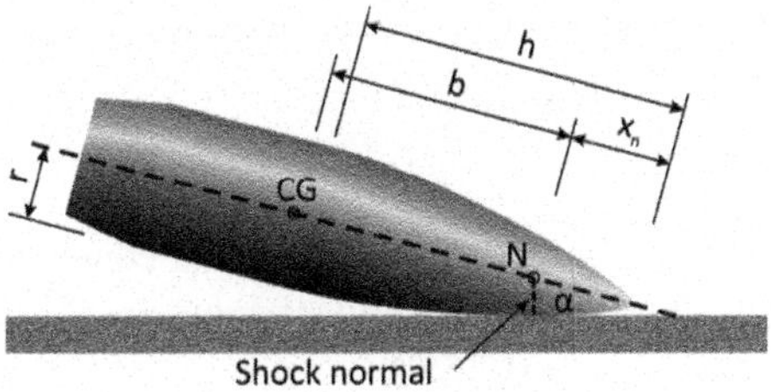


**Fig. 7.15.** Pointed bullet. The line of impact at the target is a tangent with the angle α to the arc of which the radius can be determined from h and r.

(7.5:2) $$x_n = \frac{r^2}{2 \cdot h} \cdot \left(1 + \frac{1}{2 \cdot \tan^2 \alpha}\right) \qquad [m]$$

As the transition from the nose to the cylindrical part of such a bullet is not smooth, a fillet radius is generally used. If the contact point lies in this area, we can use the equation for the circular arc ogive if necessary (see next section).

#### 7.5.2.4 Pointed bullets

The noses of most pointed projectiles take the form of circular arc ogives. The transition to the cylindrical part is usually smooth in the case of bullets and at an angle of a few degrees for artillery shells. We can therefore calculate the ogive radius (at least approximately) from the nose length and the projectile diameter.

The distance between the point of intersection and the tip of the bullet follows from the geometry of the circle, whereby (using the symbols from Fig. 7.15):

(7.5:3) $$x_n = h - \frac{(h^2 - r^2) \cdot \tan \alpha}{2 \cdot r} \qquad [m]$$

Other nose shapes can be approximated by circular arc ogives (e.g. the Haack nose) or must be calculated separately in the case of special meridian functions (e.g. the ¾-power shape).

#### 7.5.2.5 Cylindrical, coned and flat-nose bullets

In the case of a coned or flat-nose bullet, if the half point angle is less than the angle of incidence (as is the case in most ricochet scenarios), it can be treated in the same way as a cylindrical bullet (such as a wadcutter or semi-wadcutter).

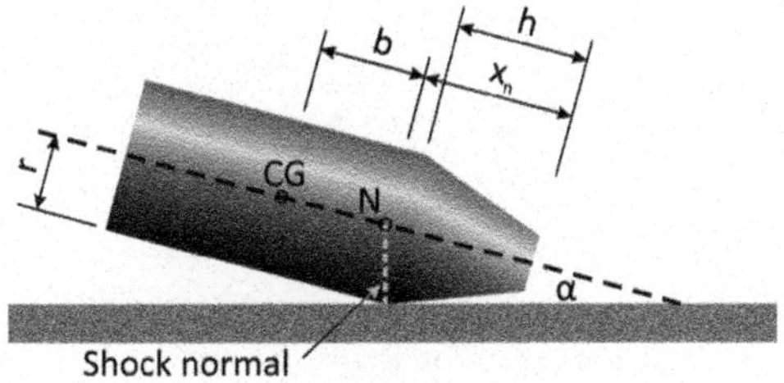


**Fig. 7.16.** Projectile with tapered nose.

The distance between the point of intersection and the tip of the bullet (or, in the case of a cylindrical bullet, its leading face) is then given by:

(7.5:4) $$x_n = h + r \cdot \tan \alpha$$ [m]

where of course $h = 0$ in the case of a cylindrical bullet. In the case of a coned or flat-nose bullet, if half the point angle is less than the angle of incidence, the expression to be used is the following (rather than Eqn 7.5:4):

(7.5:4a) $$x_n = r_{sp} \cdot \tan \alpha$$ [m]

where $r_{sp}$ is the radius of the nose flat.

### 7.5.3 The torque shock acting on the projectile

If, for a short time $\Delta t$, a force acts at a distance from the centre of rotation on the figure axis of an otherwise force-free gyro, a torque shock $\vec{T}$ occurs and as a result the angular momentum changes:

(7.5:5) $$\Delta\vec{L} = \vec{T} \cdot \Delta t = (\vec{F} \times \vec{r}) \cdot \Delta t$$ [kg·m²/s]

where $\vec{r}$ = vector from the centre of rotation to the contact point of the force and $|\vec{r}| = b$.

The change in angular momentum has the same direction as the torque. It can be split into a component parallel to the existing angular momentum and a component perpendicular to it. The parallel component changes the magnitude but not the direction of the angular momentum. It therefore results in an increase or decrease in angular velocity, depending on the sign. The vertical component produces an abrupt change in the direction of angular momentum. The figure axis moves in the direction of the force $\vec{F}$ immediately after the collision, but then performs a nutation movement in which it circles the new direction of the angular momentum $\vec{L}_n$ on the surface of a cone (see Fig. 7.17). It has therefore been deflected on average by the angle

(7.5:6) $$\tan \varepsilon = \frac{|\Delta\vec{L}|}{|\vec{L}|}$$ [-]

perpendicular to the force in the direction of the torque.

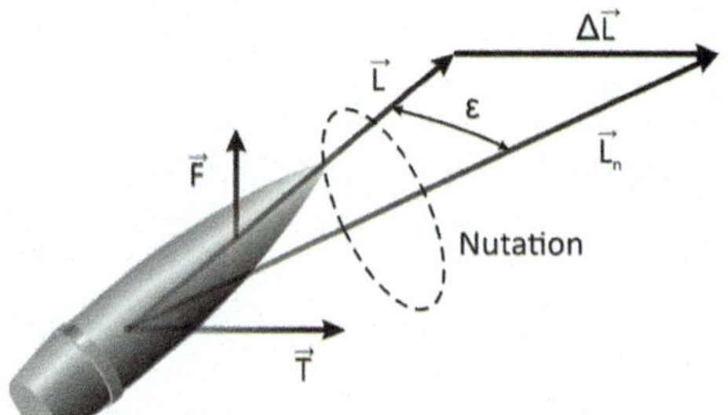


**Fig. 7.17.** The effect of a torque shock. See text for details.

In the case of a projectile, the vector of the torque is always perpendicular to the direction of force and to the figure axis. The torque shock hence changes the direction of angular momentum but not its magnitude.

When a projectile strikes a target, an impulse is applied to the projectile axis in the direction of the shock normal, the magnitude of which is as follows (from Eqn 7.4:5):

$$F_y \cdot t_s = -m \cdot (u_y - v_y) = m \cdot v \cdot (1+k) \cdot \sin\alpha \qquad [\text{N·s}] \tag{7.5:7}$$

This impulse can be decomposed into a component in the direction of the projectile axis and a component perpendicular to it. The first reduces the velocity, the second imparts a torque shock with respect to a transverse axis through the centre of gravity.

We can then calculate the change in angular momentum – the torque shock – from Eqn 7.5:7 by multiplying by the distance between the shock normal and the centre of gravity. Using the symbols from 7.5.2, we can write:

$$\Delta L = F_y \cdot t_s \cdot b \cdot \cos\alpha$$

$$= m \cdot v_y \cdot (1+k) \cdot b \cdot \cos\alpha$$

$$\Delta L = \tfrac{1}{2} \cdot m \cdot v \cdot (1+k) \cdot b \cdot \sin(2 \cdot \alpha) \qquad [\text{kg·m}^2/\text{s}] \tag{7.5:8}$$

Where m = mass of projectile, v = velocity, k = restitution coefficient, b = distance between shock normal and centre of gravity and $\alpha$ = angle of incidence.

From Eqn 7.5:6, the angle between the projectile axis and the axis of angular momentum is given by the quotient of the change in angular momentum and the angular momentum. Using Eqns 7.5:8 and 2.2:25, we obtain:

$$\tan\varepsilon = \frac{m \cdot v \cdot (1+k) \cdot (x_s - x_n) \cdot \sin(2 \cdot \alpha)}{2 \cdot J_a \cdot \omega} \qquad [-] \tag{7.5:9}$$

where $J_a$ = axial moment of inertia of the projectile and $\omega$ = its angular velocity.

After the collision, the projectile is still exposed to aerodynamic forces; the axis of angular momentum will therefore perform a precession around the trajectory tangent in addition to the nutation (see Fig. 7.18). In addition, depending on nutation and angle of deflection, the initial angle between the tangent to the trajectory and the angular moment vector will be considerable. Whether the nutation decays after a short time depends largely on this angle.

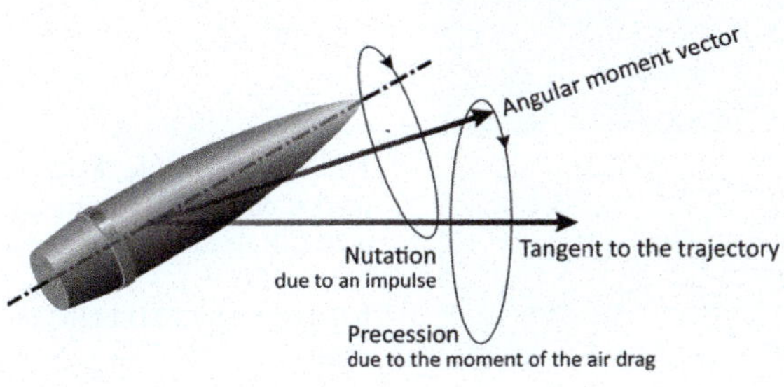


**Fig. 7.18.** Motion following collision. Pronounced nutation caused by the torque shock is superimposed on the precession due to the aerodynamic moment. (Not drawn to scale).

If we consider the nutation of the projectile in isolation and disregard aerodynamic force, we can estimate the speed of rotation (and hence the angular velocity) of the nutation. For a free gyroscope:

$$(7.5{:}10) \qquad \nu_n = \frac{|J_a - J_q|}{J_q} \cdot \frac{\omega}{2 \cdot \pi} \qquad [1/s]$$

This result explains a somewhat surprising observation. When a single projectile ricochets off a hard target, it may leave two marks (SELLIER, 1987). In the case of a crime scene, this can lead to mistaken conclusions regarding the number of shots fired (see Fig. 7.19).

The tail of the projectile also describes a circular motion during nutation, because the projectile rotates about its centre of gravity as a free body. If the frequency and nutation angle of this motion are sufficiently large, the tail of the bullet may strike the target near the primary ricochet track, giving the impression of a second shot.

For example, the nutation of a 22 L.R. bullet has an (estimated) frequency of rotation of around 500 $s^{-1}$. At $v = 300$ m/s, the projectile therefore takes approx. 0.6 m to rotate through 360°, so it is quite possible for it to strike the surface of the target a second time after about 15 cm (Fig. 7.19 a).

### 7.5.4 When does a ricochet fly stable?

A physical system is stable if it always seeks to return to a state of equilibrium despite the effects of a disturbance. In ballistics, this means that a stable projectile always aligns its axis with the tangent to the trajectory, despite the constant influence of the aerodynamic moment. Ballistics stability theories (gyroscopic stability and the Molitz stability triangle) only address the conditions under which the design of a projectile ensures stable flight in air (see 5.6).

The question of the maximum torque shock on the projectile axis (or rather the axis of angular momentum) after which the projectile will resume stable flight following impact can only be answered with considerable effort (solving Euler's gyroscopic equations with the appropriate initial conditions).

Many years of experience indicate, however, that normal bullets become unstable if the angle of attack exceeds approximately 15° after impact.

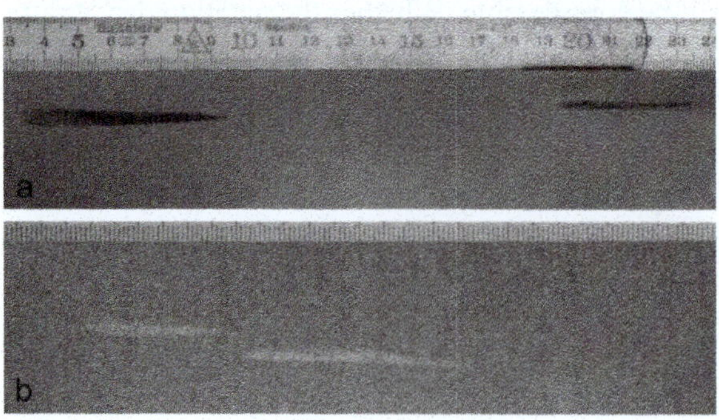


**Fig. 7.19.** Double ricochet marks: one bullet, two tracks.
**a** 22 L.R. lead bullet on sheet brass.
**b** 9 mm Luger FMJ bullet on concrete.

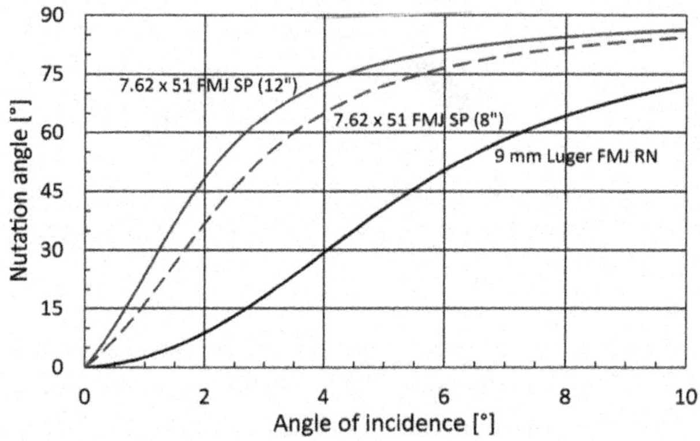


**Fig. 7.20.** Relationship between angle of incidence and nutation angle (as an estimate of the angle of attack after impact). See text for details.

The exact angle naturally depends on the design of the bullet, and the speed of rotation will be an important factor. The figure of 15° assumes a normal bullet design with $J_q/J_a < 10$ and a gyroscopic stability number of between 1.5 and 2.

In a first approximation, we can equate the initial angle of attack after ricochet to the nutation angle according to Eqn 7.5:6, as the angle of attack caused by precession in a projectile that was stable before collision is usually less than about 3°. Under these assumptions, we can establish a relationship between the angle of incidence and the angle of attack immediately after the impact, taking into account both bullet geometry and angular velocity. Fig. 7.20 shows two examples of such a relationship (one for a 9 mm Luger bullet, the other for 7.62 × 51 rifle bullet (both FMJ).

From the graph, we can see that the 9 mm Luger bullet can resume stable flight after impact if the angle of incidence does not exceed approx. 2.5°. The rifle bullet will fail to recover stability even at an angle of incidence of less than 1°. Both values correspond surprisingly well to practical experience. It is also worth noting the influence of an increase in angular velocity, which noticeably delays the onset of destabilization e.g. a twist length of 8" (203 mm) instead of 12" (305 mm).

If we have the necessary geometric and physical data, we can plot these curves for any projectile.

## 7.5.5 Proper motion and flight attitude of a ricochet

### *7.5.5.1 Proper motion*

The previous section shows that a ricochet can result in a large angle of attack even if the angle of incidence is small. The question now arises of the movements the projectile will undergo if its axis and direction of movement form a large angle.

Fig. 7.21 shows the start of such a movement. It shows four holes (in thin cardboard) made by a single 7.5 mm bullet after grazing a cylindrical wooden rod shortly after leaving the muzzle. The orientation of the holes and their different lengths indicate that the bullet axis was undergoing conical rotation and that the bullet was pitching.

**Fig. 7.21.** Holes made in sheets of cardboard by a single projectile, 2, 4, 6 and 8 m after grazing a cylindrical wooden rod. Bullet: 7.5 x 55 Swiss. See text for details. See also the holes punched through cardboard sheets by an unstable round in Fig. 5.44.

If we use a frame of reference with its origin at the centre of gravity of the projectile and initially disregard drag, the projectile performs the typical nutation movement of a prolate gyroscope (a symmetrical gyroscope with the moments of inertia $J_1 = J_2 > J_3$). Based on geometric gyrostatics (W. GREINER 1989, K. MAGNUS 1971), this can be represented by two circumferential surfaces of a right cone (from POINSOT, see Fig. 7.22). The tips of both are located at the centre of gravity of the projectile. One, the polhode cone, is thought of as being anchored to the projectile, while the axis of the other, the herpolhode cone, coincides with the angular momentum vector. The nutation movement results from the surface of the polhode cone rolling across the surface of the herpolhode cone. The common convex surface line forms the axis of rotation, which is therefore constantly changing. The projectile axis, axis of rotation and axis of angular momentum are therefore always in the same plane. As it moves around the axis of angular momentum, the projectile axis also sweeps over a conical surface, the nutation cone surface.

In reality, this nutation movement is superimposed on the precession of the projectile, which is caused by the constantly acting momentum of the aerodynamic force. As the angle of attack increases, drag increases and so does the precession angle. Angles of attack of 90° and more (up to 180°) are therefore quite possible. If this occurs, it should be noted that if the axis of the projectile rotates through an

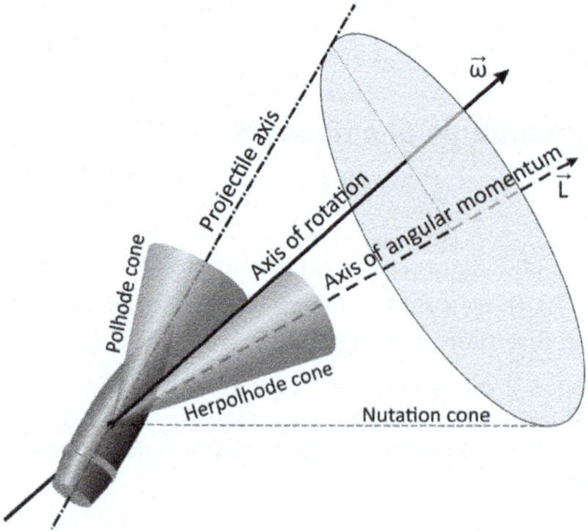


**Fig. 7.22.** Nutation of a ricochet. The surface of the polhode cone, which is anchored to the projectile, rolls across the surface of the herpolhode cone. The common convex surface line forms the axis of rotation of the system.

angle of more than 90° to the tangent to the trajectory, the direction of rotation of the projectile in the body-fixed system is reversed due to the conservation of angular momentum (see 5.6.4.6 and the example described there).

### *7.5.5.2 Flight attitude and stability*

The question now arises as to whether the impact-induced nutation decays as the bullet continues its flight and whether the projectile axis aligns itself with the trajectory tangent. It is not possible to answer this question by means of gyrostatics, as the real system – with its flow and frictional forces – is extremely complex.

If no more forces acted on the projectile after impact, the new angular momentum vector would be maintained, and hence also the nutation.

However, it is possible to measure the trajectory and to draw qualitative conclusions about the flight attitude and stability of the projectile from the results. We know from 5.3.2.1 and Eqn 5.3:20 that drag is represented as follows in the point-mass model:

(7.5:11) $$F_D = C_D \cdot \frac{1}{2} \cdot \rho \cdot v_0^2 \cdot A_0 \qquad [N]$$

where $C_D$ = drag coefficient and $A_0$ is the reference area. If angle of attack δ increases as a result of nutation, the area A(δ) opposing the flow increases, which means that drag also increases. If we measure the velocity of the projectile after ricochet using Doppler radar and determine the deceleration and drag coefficient from that, the drag coefficient will reflect the changes in the incident flow surface if we retain $A_0$ as the reference area:

(7.5:12) $$F_D = C_D \cdot \frac{1}{2} \cdot \rho \cdot v_0^2 \cdot \frac{A(\delta)}{A_0} \cdot A_0 \qquad [N]$$

The Doppler radar measurement will thus reflect the following $C_D$ function:

(7.5:13) $$C_D^* = C_D \cdot \frac{A(\delta)}{A_0} \qquad [-]$$

If this $C_D$* function approximates the original function of the projectile, this can only mean that A(δ) is now once more close to the cross-section of the stable projectile (i.e. the projectile axis is close to the trajectory tangent). However, this can occur with either the nose or the tail of the projectile leading.

If the bullet continues its path tail-first following ricochet, which is perfectly possible, the $C_D$ function for the bullet when it resumes undisturbed flight will of course be different.

Fig. 7.23 shows $C_D$ measurements for two ricocheted 7.62 × 51 bullets, both of which apparently flew approximately stable again after a certain distance. Similarly, Fig. 7.24 shows measured $C_D$ curves for two ricochets involving 5.56 × 45 bullets, one of which (1) possibly continued its flight tail-first.

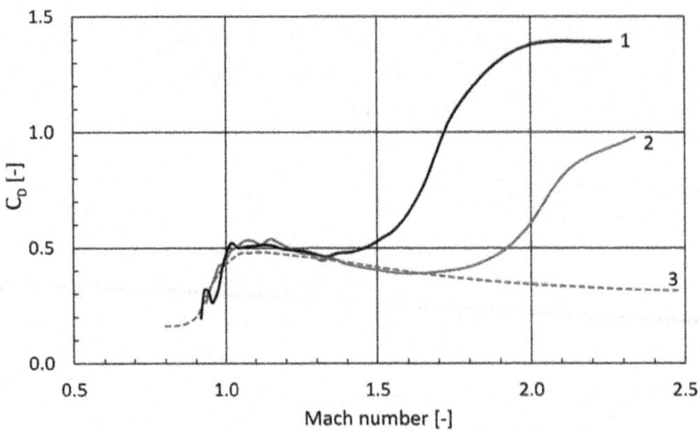


**Fig. 7.23.** $C_D$ curves of two 7.62 × 51 bullets that have ricocheted off a wooden rod. Both curves (1 and 2) initially coincide with that of the $C_D$ curve for the stable projectile (3), with flight becoming less stable around the speed of sound.

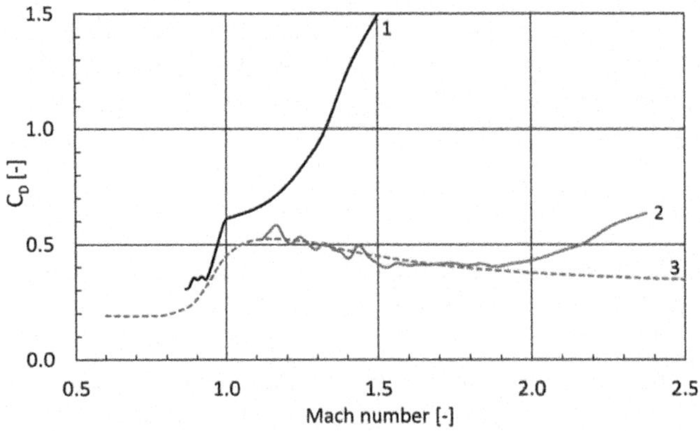


**Fig. 7.24.** $C_D$ curves of two 5.56 × 45 bullets that have ricocheted off a wooden rod. After re-stabilization, curve (1) remains slightly above the $C_D$ curve for the stable projectile (3), while curve (2) partially coincides with it, with somewhat unstable flight.

We know that bullets may re-stabilize tail-first when they have been fired steeply into the air, as they reach high angles of attack due to a lack of tractability in the apex and re-stabilize in the descent branch. This is regularly confirmed by gunshot wounds from falling bullets (see Fig. 5.26).

## 7.6 The trajectory of a ricochet

### 7.6.1 Sectional density and drag coefficient

As already explained in 7.5.5, a ricochet performs a very complicated proper motion about its centre of gravity, which consists mainly of the nutation induced by the impact. Its angle of attack therefore fluctuates periodically between 0° and a maximum angle that can easily reach 180°. This maximum angle depends on the angle of incidence and the torque shock.

A very large angle of attack is generally to be expected at angles of incidence of 3° and more in the case of handgun bullets and 1.5° and more in the case of rifle bullets (see 7.5.4).

When calculating the trajectory of a ricochet, we must decide what sectional density and drag coefficient function to use. We can obtain an approximate solution to this problem using the periodicity of the motion, the oscillation length of which is measured in metres.

For long ranges, one can calculate the sectional density using the mean shadow area of the projectile as it rotates about a transverse axis. For the drag coefficient, we can use the mean of the coefficients for cylinders subjected to longitudinal and to transverse flow. For short bullets (from handguns), the $C_D$ value for a sphere is also an acceptable choice.

Calculations are generally not required at short range (up to a few metres). If calculations should prove necessary, one can take an average sectional density and a constant $C_D$ of the order of 2.

Calculations of this sort yield only approximate results, but these are perfectly adequate for practical purposes.

### 7.6.2 Maximum range

Calculating the maximum range of a ricochet is particularly important in connection with safety zones at shooting ranges. The greatest risk is posed by bullets that do not shatter when they ricochet.

When a projectile fragments, the energy remaining is distributed over several projectiles. In addition, the sectional density of the individual parts of the projectile is much lower, which results in greatly reduced range. Nevertheless, fragments from soft core bullets can travel up to 300 m.

To determine the range of a ricochet, we must first know its angle of departure and velocity after ricochet. Section 7.4 described ways of determining these parameters. B. P. KNEUBUEHL (1999) presents extensive experimental results for various handgun bullets. We can approximately calculate the first phase of the trajectory, during which the projectile undergoes pronounced nutation, using the information in 7.6.1.

As we saw in 7.5.5.2, the projectile can re-stabilize after a certain distance, i.e. its axis is re-aligned with the tangent to its trajectory. It may then fly with either nose or tail forward.

We can calculate this second phase, during which the projectile will be flying at subsonic speeds, with the usual sectional density and a $C_D$ value corresponding to the attitude of the bullet.

Pointed bullets flying tail-first may well have a lower $C_D$ value than when in normal nose-first flight, owing to the low base drag; base suction effects will be largely absent.

We can use the point-mass model from 5.4.3 for such calculations. As this stable flight phase significantly increases the range of a ricochet, we must always include it when determining safety zones.

## 7.7 Further ricochet theories

### 7.7.1 SELLIER's theory

K. SELLIER (1971) (also reprinted in K. SELLIER, 1977) briefly discussed ricochets in a paper on the subject of projectile deflection and deformation. In that paper, he decomposed the motion of the projectile into a component perpendicular to the surface of the target and a component tangential to it. The perpendicular component is handled according to the laws governing a centric, straight, semi-elastic collision, while the tangential component is retained.

The latter assumption allows the restitution coefficient k (denoted by $\varepsilon$ in SELLIER) to be determined solely from the angle of incidence and angle of departure:

$$(7.7{:}1) \qquad k = \frac{\tan\beta}{\tan\alpha} \qquad [-]$$

Since k cannot be greater than 1 owing to the conservation of energy and the assumption that the tangential component is constant, this theory can only explain ricochet phenomena if the angle of departure is smaller than the angle of incidence. It does not cover cases where angle of departure is greater than angle of incidence, as is quite common in practice.

### 7.7.2 JAUHARI's theory

M. JAUHARI (1971) developed a detailed theory on the phenomenology of the ricochet. JAUHARI decomposed both impact velocity and velocity after ricochet into a tangential and a normal component and introduced independent reduction factors $\mu$ and $\nu$ (labelled $\alpha$ and $\beta$ in JAUHARI) in both directions:

$$(7.7{:}2a) \qquad u_x = \mu \cdot v_x = \mu \cdot v \cdot \cos\alpha \qquad [m/s]$$

$$(7.7{:}2b) \qquad u_y = \nu \cdot v_y = \nu \cdot v \cdot \sin\alpha \qquad [m/s]$$

The parameter $\nu$ has the same value as the restitution coefficient k (see 7.4.1.2).

The angle of departure can be calculated from the angle of incidence and the two parameters:

$$(7.7{:}3) \qquad \tan\beta = \frac{\nu}{\mu} \cdot \tan\alpha \qquad [-]$$

Since $\mu$ and $\nu$ are independent of each other, this model – in contrast to SELLIER's theory – also allows for angles of departure greater than the angle of incidence.

JAUHARI used the length and depth of the bullet track on the target as additional information for determining the two parameters. Assuming a constant deceleration (from $v_x$ to $u_x$ in the tangential direction and from $v_y$ to 0 in the perpendicular direction), he derived the following two equations for μ and ν, using the law of conservation of momentum:

$$(7.7{:}4a)\qquad \mu = -\frac{\tan\alpha}{2\cdot\tan\beta}-\frac{1}{2}\pm\frac{\tan\alpha}{2\cdot\tan\beta}\cdot\sqrt{\left(1-\frac{\tan\beta}{\tan\alpha}\right)^2+\frac{4\cdot s_x\cdot\tan\beta}{s_y}}\qquad [-]$$

$$(7.7{:}4b)\qquad \nu = -\frac{\tan\beta}{2\cdot\tan\alpha}-\frac{1}{2}\pm\frac{1}{2}\cdot\sqrt{\left(1-\frac{\tan\beta}{\tan\alpha}\right)^2+\frac{4\cdot s_x\cdot\tan\beta}{s_y}}\qquad [-]$$

where $s_y$ = depth of bullet track. All other symbols are as per Table 7.2.

In principle, this makes it possible to determine the two parameters μ and ν for a given projectile-target combination from empirical geometric data. For physical reasons, however, it is clear that the parameters are not constants, but depend on the angle of incidence.

Two comments regarding JAUHARI's theory:

- It is not clear why JAUHARI restricts the parameter ν to the interval (0,1). In certain materials (e.g. wood) it is quite possible that the perpendicular component of the velocity after ricochet is greater in magnitude than the corresponding component of the impact velocity (a combination of ricochet and deflection takes place). This can also be demonstrated experimentally.
- Equations 7.7:4a and 7.7:4b presuppose that the depth of the track $s_y$ is greater than zero. However, there will be no track depth to measure at small angles of incidence on targets made of hard materials (such as thick steel plates or smooth concrete). In such cases, it will not be possible to determine μ and ν.

### 7.7.3 BIRKHOFF's theory for water surfaces

In 1944, G. BIRKHOFF et al. published a work on the ricochet of spheres on water. The authors made the following assumptions:

- The surface element of which the normal vector is at an angle ζ to the direction of movement (this corresponds to the NATO angle) experiences a pressure of:

$$(7.7{:}5)\qquad p = \frac{1}{2}\cdot\rho\cdot v^2\cdot\cos^2\zeta\qquad [\text{Pa}]$$

- This pressure acts only on the spherical surface that lies below the undisturbed water surface.

No physical justification is given for these assumptions. BIRKHOFF only mentioned that this would provide a plausible estimate of the forces involved. However, his approach is strongly reminiscent of NEWTON's flow hypothesis, according to which resistance is caused exclusively by the transfer of momentum to the body surface subjected to flow (see also 5.7.3.3). The back of the body remains force-free.

However, NEWTON's pressure formula does not include the factor ½, for which BIRKHOFF provides no explanation. W. JOHNSON and S. R. REID (1975) therefore varied this coefficient and used it to adjust to measurement results.

Based on these two assumptions, one can calculate the pressure distribution over the wetted area of the surface of the sphere, which gives us the equation for motion perpendicular to the trajectory tangent as a function of the angle of incidence. The critical angle for re-emerging from the water is reached when the sphere is completely submerged, its trajectory is horizontal and the lateral force disappears. The critical angle of incidence is therefore:

(7.7:6) $$\alpha_{krit} = 17.5 \cdot \sqrt{\frac{\rho_W}{\rho_K}} \qquad [°]$$

where $\rho_W$ = density of water and $\rho_K$ = density of the material of which the sphere is made. Surprisingly, the critical angle depends neither on impact velocity nor on calibre.

JOHNSON and REID discussed the derivation of this equation in detail in their article. They also set up a system of differential equations for the trajectory of the sphere and calculated the trajectories for an aluminium sphere. Furthermore, they showed that Eqn 7.7:6 can be applied to other liquid-like materials (e.g. sand) if the following condition is met:

(7.7:7) $$\frac{\rho_m \cdot v_m^2}{Y_m} >> 1 \qquad [-]$$

where $\rho_m$ = density, $Y_m$ = yield point of the material and $v_m$ corresponds approximately to the velocity of the particles of matter during the impact process.

Various experiments have been conducted and described regarding the ricochet of spheres in water (and also in sand), with Eqn 7.7:6 proving quite accurate. Please see B. P. KNEUBUEHL (1999) for a detailed compilation.

# 8 Special applications of ballistics

## 8.1 Introduction

In addition to the classic applications of ballistics in defence technology, there are many practical applications of ballistics methods in the civilian sector. Typical examples include sports in which an object is hit, thrown or pushed and then flies a certain distance through the air. This includes not only golf, tennis and basketball, but also football and those athletics events that involve throwing. The angle of departure for maximum flight distance, initial velocity and elevation when aiming a basketball at a basket or putting spin on a football to create a double-curved trajectory are ballistics questions that the player almost always solves purely intuitively, on the basis of experience.

Ballistics knowledge is frequently used in forensics, to which all the sub-disciplines in this book are of relevance. A sound knowledge of propellant compositions, the firing process and the way the projectile exits the muzzle play an important role when it comes to determining firing distance at close range or identifying a shooter using gunshot residue analysis. Several textbooks and articles on the associated techniques are available, such as K. Sellier (1989, 1991), A. Gallusser et al. (2014), C. Frings and F. Rabe (2016).

The chapter on exterior ballistics will be useful when establishing the source of a stray bullet or of a bullet that has caused damage or injury after missing its target. The literature contains very little discussion of this application of ballistics to forensics; we shall therefore examine the topic in this chapter, using a number of real-life examples.

A similar problem arises in bloodstain analysis, the subdiscipline of forensics that attempts to reconstruct the details of a crime from traces of blood found at the scene. We shall also devote a section to this very specific use of exterior ballistics.

Safety assessments of shooting ranges and the elucidation of unusual accidents and incidents require knowledge and experience of both exterior and terminal ballistics. Exterior ballistics is of value primarily when identifying the source of a stray bullet, while terminal ballistics can help one identify the causes, sometimes revealing quite unexpected mechanisms of cause and effect.



B. P. Kneubuehl, *Ballistics*,
https://doi.org/10.1007/978-3-662-69237-0_8

## 8.2 Forensic applications

Co-author: Dr. sc. forens. Fabiano RIVA

### 8.2.1 Trajectory reconstruction

#### 8.2.1.1 Short and long trajectories

When reconstructing the trajectory of a projectile, one must first decide whether it was "short" or "long", because the approach differs significantly between the two cases.

A "short" trajectory can be approximated by a straight line. The only reason that a trajectory deviates from a straight line is acceleration due to gravity. The sole criterion for deciding whether a trajectory is "short" is therefore the time of flight, which determines how long the projectile is exposed to gravity. For example, taking a drop of 1 cm as the threshold for classing a trajectory as "short" equates to a time of flight of 0.0452 s. The deceleration of the projectile will play only a subordinate role over this short time of flight. The expression for the maximum distance of a "short" trajectory is hence:

$$(8.2{:}1) \qquad D = 0.0452 \cdot v_0 \qquad [\mathrm{m}]$$

Table 8.1 gives maximum ranges at which one can assume a straight trajectory, for a number of typical calibres.

M. G. HAAG and L. C. HAAG (2011) and R. M. GARDNER (2005) give 18–30 m as the maximum range at which one can label a trajectory "short". The above equation closely matches this.

It will be necessary to investigate longer – sometimes very long – trajectories primarily in connection with hunting accidents, the (unnecessary and dangerous) firing of shots into the air during celebrations and errors at shooting ranges.

A long trajectory is far more difficult to reconstruct than a short one; it will be necessary to assemble all the parameters required for a trajectory calculation. Some are easy to determine, while acquiring others is more labour-intensive, and even if one can obtain them there will still be a degree of uncertainty. For some data, one can only make as realistic an estimate as possible.

#### 8.2.1.2 Data from the scene of the incident

Reconstructing a long trajectory from the data for one of its points – usually the point of impact – is closely related to the initial value problem (see 5.4.3.3). However, while the initial value problem is usually solved using standard data, reconstruction is based on the real-life data for the shot.

***Projectile data.*** If the projectile has been found, then in most cases the first of the important parameters are available, namely projectile diameter (and thus in many cases calibre), mass, shape and composition. Any deformation of the project-

**Table 8.1**. Maximum range at which a straight trajectory can be assumed

| Calibre | D [m] | $v_0$ [m/s] | Calibre | D [m] | $v_0$ [m/s] |
|---|---|---|---|---|---|
| 4.5 mm Diabolo | 8 | 175 | 12/70 Brenneke | 19 | 420 |
| 45 Auto | 11 | 245 | 7.62 × 39 | 32 | 710 |
| 38 Spl. SWC | 13 | 290 | 7.5 mm GP 11 | 34 | 750 |
| 22 L.R. | 15 | 330 | 7.62 mm NATO | 38 | 830 |
| 9 mm Luger | 16 | 350 | 223 Rem. Sierra MK | 40 | 885 |

tile must be examined to determine whether it occurred exclusively at the point of impact or whether the projectile came into contact with other objects on its trajectory.

The absence of deformation does not necessarily mean that the projectile had no contact with a solid object before coming to rest. For example, a bullet can perforate soft wood without undergoing any change (see 6.3.5), although this will affect its stability.

***Angle of trajectory***. To calculate a trajectory, one must know the orientation of a tangent to it (i.e. an angle of trajectory) at some point. In the most favourable of cases, it will be possible to establish this direction from two points (e.g. a hole in a window and the impact point on a wall). It may also be possible to calculate direction from a channel through a solid material (such as plaster or wood) or from an elliptical hole in a plate. In the last two cases, one first determines the angle relative to the wall or plate and then its orientation. How accurately to measure this angle depends on the specific situation and must be decided on a case-by-case basis.

For the case of a channel in plasterboard, M. G. HAAG and L. C. HAAG (2011) give a mean error of ± 5° for the measured line of arrival.

It becomes much more difficult to determine the angle of trajectory when the bullet has penetrated a human body. It will be possible to identify a wound channel and hence a line of arrival in relation to the body surface during the postmortem examination, but little will be known about the person's posture at the time of the shot (F. RIVA et al., 2017, W. KERKHOFF et al., 2018). The same applies to bullets that have struck objects in motion (such as vehicles).

If the projectile and the angle of descent are known, one already has a fairly good indication of the distance, as long as one has a graph relating the projectile-specific angle of descent to range (see Fig. 8.1 next page). Such a graph can easily be prepared for the most common calibres and for the altitude at which the incident occurred. Independently of the incident-specific data (muzzle velocity and weather conditions), one can get quite close to the launch point as long as one can determine the direction reasonably reliably.

***Impact velocity***. Determining the impact velocity of the projectile is somewhat more difficult. The angle of descent and the approximate range derived from that parameter also provide an estimate of impact velocity. At the point where the bullet is found, the only information about the energy it was carrying on impact comes from the bullet itself and its effects (penetration depth, extent of damage and projectile deformation). One needs experience of terminal ballistics (and of wound ballistics in the case of injury) to extract a usable value from this.

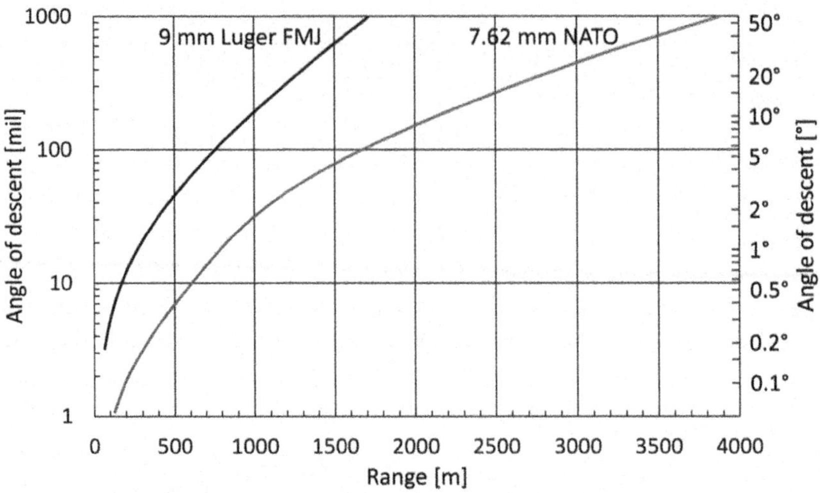


**Fig. 8.1.** Plot of angle of descent against range, for producing a quick estimate of the distance from which a round was fired.

***Obstructions***. The calculated angle of descent and the direction of fire result in a presumed trajectory, which must be inspected for possible obstructions (trees, masts, wires, etc.) in the immediate vicinity of the scene. This can give an indication as to whether the bullet was deflected before impact.

### 8.2.1.3 Obtaining further data

In addition to the data from the scene, one must obtain other data to reconstruct the trajectory.

***Drag coefficient $C_D$***. Even if the projectile is damaged, one can usually recognize its original shape. If it is not immediately possible to assign the bullet to a known calibre, it may be possible to do so by comparing it with bullets from an ammunition collection.

If it is a common type, its $C_D$ curve will be available. If it is not, one can use the $C_D$ curve of a projectile as similar in shape as possible. The calibre and mass are not important, but the *slenderness ratio* (ratio of nose length to overall length) should be similar. If no suitable data are available and no measurements can be carried out (e.g. using radar), one can use MAYEVSKI's method of a standard projectile plus ballistic coefficients (BC) (see 5.4.2.4). However, this presupposes that the bullet manufacturer's documentation includes the BC of the projectile for the velocity range in question.

The BC is a constant that directly transposes the fixed $C_D$ curve of the standard projectile to the true $C_D$ values of the projectile under investigation. However, because the $C_D$ curve of the projectile under investigation can differ significantly from that of the standard projectile, it is only possible to make this adjustment step by step, as a function of velocity. MAYEVSKI's representation of drag therefore assigns up to five BC values to a projectile.

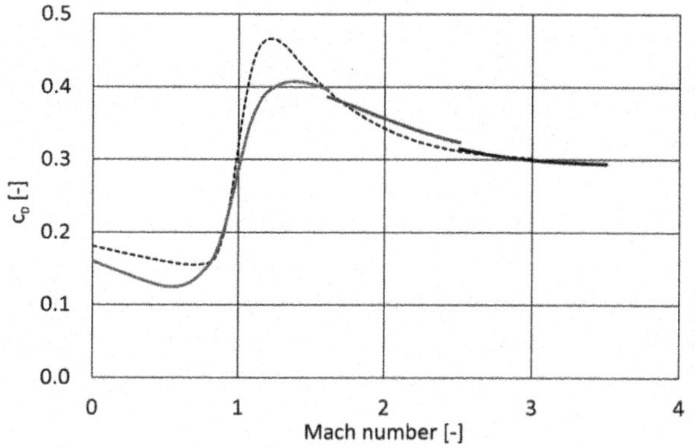


**Fig. 8.2.** Comparison between the $C_D$ curve generated using BC and a standard projectile (solid line) and the actual $C_D$ curve of a projectile (dotted line).

Bullet manufacturers that work with this system generally base their documentation on the standard G1 projectile, although they occasionally use the G7 for slender projectiles. This makes it possible to reduce the number of BC values.

Fig. 8.2 shows a $C_D$ curve based on a standard projectile with three BC values and compares it with the $C_D$ curve for the pointed bullet in question (dotted line). The BC method is quite usable at short and medium ranges, as long as one knows the BC.

***Muzzle velocity***. As a rule, the weapon from which the round originated is not initially available and one does not know the initial velocity of the trajectory. In such cases, one has to make do with the nominal muzzle velocity of the calibre in question. When using that value, one should bear in mind that weapons of different barrel lengths are common in many calibres and that certain calibres can be used in both handguns and long weapons. The condition of the barrel (wear, etc.) and that of the ammunition (temperature, propellant humidity, etc.) can also have a significant influence on initial velocity.

If one has access to the weapon that is believed to have been used, one can measure the muzzle velocity of 6–10 shots fired from it. Whenever possible, one should use the ammunition involved in the incident, to establish its velocity dispersion. If this is no longer available, one should procure ammunition of the same or similar age and batch.

***Meteorological data***. Reconstruction will be noticeably more accurate if the drag used for calculations reflects the conditions obtaining at the time. This means that one must establish the approximate air density at the time of the incident. It is usually possible to obtain the necessary data (air pressure and temperature) from a meteorological office. For a trajectory with a high apex (which equates to a long time of flight, recognizable by a large drop angle), it is also important to know wind velocity and direction at all the altitudes through which the bullet would have passed.

For flat trajectories (up to about 1000 m), the time of flight is short compared to the time it takes for wind velocity to change. The trajectory therefore passes through a wind profile of different but fixed wind velocities. It is virtually impossible to establish this profile, so for trajectory calculations one replaces it by a constant wind velocity that results in the same deviations.

***Topography***. Topography is very important, and often receives too little attention when a trajectory is being reconstructed. The topographical data for the area can usually be obtained from the authorities responsible for mapping, in the form of topographical maps or three-dimensional datasets.

The importance of the landscape contours lies in the fact that the ballistics data collected at the point of impact (projectile and angle of incidence) and the resulting initial values (muzzle velocity and elevation) can correspond to multiple, very different firing locations, especially if the horizontal angle of deflection is not known precisely. Fig. 8.3 shows a real example in which the relief changed significantly within the measuring uncertainty of the horizontal angle of deflection at the point of impact, almost doubling the possible range.

***Calculating the trajectory***. The reconstructed trajectory is calculated using a computer program based on the mass point model (see 5.4.3). This determines the main parameters of the trajectory (position, velocity, time of flight, energy and angle of trajectory) from the data mentioned above.

Various programs are available on the market, some more elaborately programmed and more accurate than others. What matters is that the software integrates one of the systems of differential equation described in 5.4.3 by a numerical method, rather than (for instance) using polynomial approximations of the trajectory parameters.

#### 8.2.1.4 Visualization

It is not the intention of this section to give rules for the visual representation of a reconstruction; the various cases are too different. One will have to choose the visual representation most suited to the incident. The following considerations are merely general suggestions and examples.

Short trajectories can be shown as straight lines on a plan view of the scene. The spatial aspect can be taken into account by showing additional plan and side views,

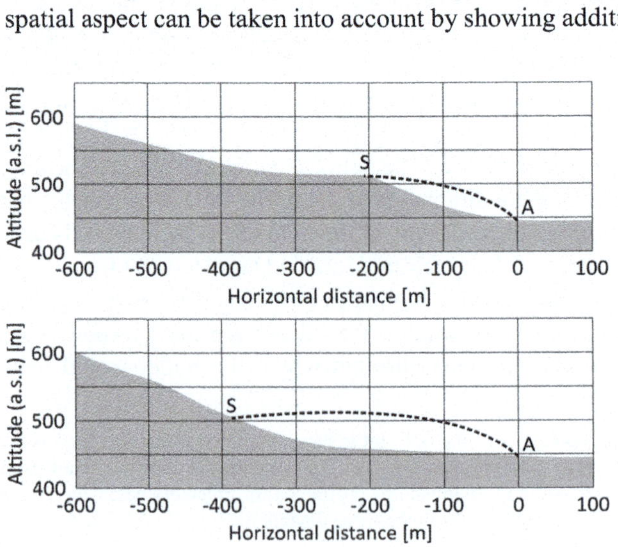


**Fig. 8.3**. The influence of topography. The range of error in measuring the horizontal angle of deflection at the scene, A, may allow of different firing locations S. The diagrams show a real-life case.

to which one can usefully add photographs of the scene. With rather more effort, it is possible to create three-dimensional graphics for short trajectories that allow measuring uncertainties regarding the angles to be indicated from the point of impact back along the trajectory (e.g. in the form of a cone).

Reconstructions of long trajectories often require much more complex visualizations, because they have to show the profile of the terrain. One needs to examine trajectory height and possible lateral deviation due to crosswind, to verify that the bullet could not have struck the ground during its flight. A three-dimensional pictorial graphic can often prove helpful in such cases.

This can also be represented by linked two-dimensional drawings using a plan and side view.

It will often be necessary to plot the results of reconstruction calculations on a topographical or cadastral map, from the point of impact in the direction of the presumed firing location. The uncertainties in horizontal angle of deflection and range result in a zone that contains the firing location to a high degree of probability (see Fig. 8.4 for an example).

## 8.2.2 Reconstructing the trajectory of an unstable projectile

### *8.2.2.1 Causes of instability*

In forensic ballistics, one regularly encounters unconventional cases in which it is not possible to reconstruct the trajectory by the usual means. Such situations may arise when the angle of attack is greater than approximately 15° (see 7.5.4). One can categorize such a projectile as "unstable".

With sufficient spin, such an "unstable" projectile may realign its longitudinal axis with the tangent to the trajectory (i.e. stabilize itself again) later in its flight.

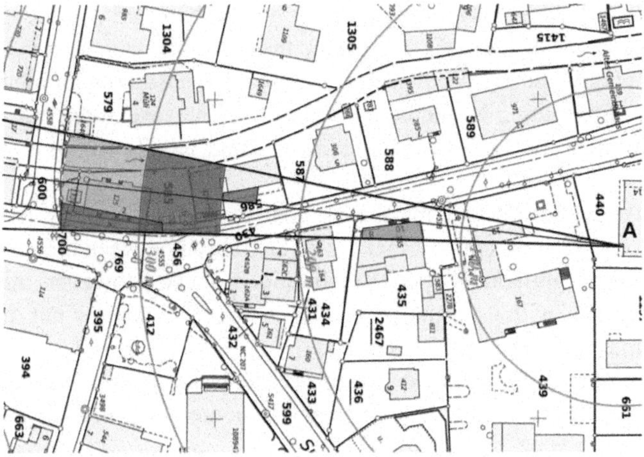


**Fig. 8.4.** Real-life case, shown on a cadastral map. One can mark possible and impossible starting points for the trajectory, together with obstructions. The letter A indicates the point of impact.

Two physical factors can cause a projectile to assume a large angle of attack. One is insufficient gyroscopic stability (see 5.6.3 and 5.6.4.6), the other is a brief torque shock acting on the projectile (described in 7.5.3).

***Insufficient stability*** is in most cases a consequence of insufficient angular velocity, or occasionally of a mass shift (e.g. in tracer rounds) that changes the moment of inertia of the projectile. If a bullet is spinning too slowly, the cause is always the barrel from which it was fired.

Because of its design (in which the moment of inertia plays a decisive role), each projectile requires a certain minimum speed of rotation for its gyroscopic stability number to attain the required value. It must therefore be fired from a barrel with a twist length that satisfies the equations of 3.4.3. If the twist length is too great, angular velocity will be insufficient and the projectile will execute a gyroscopic movement with a large precession angle that may well increase over its flight (see 5.6.4.6).

One typical example is the 5.56 mm NATO cartridge with the SS109/M855 projectile, which requires a twist length of 178 mm. Because the cartridge dimensions are compatible, this projectile can also be fired from a 223 Rem. rifle with a twist length of 305 mm. However, it will then have a pronounced precession and large angle of attack (see Fig. 5.45).

Cartridges are sometimes used in weapons of a calibre larger than that for which they were designed. In such cases, the bullet does not engage the rifling at all, and tumbles on leaving the barrel (see Fig. 5.44 for an example). Its muzzle velocity will be lower than normal, and it will decelerate rapidly. If the difference in diameter is large enough, the bullet may become skewed within the barrel and suffer damage or even become stuck.

It is very difficult to even get a cartridge to function in a weapon that is too large for it. There have been numerous cases in which 7.65 Parabellum cartridges have been used in a 9 mm Luger weapon, or 9 mm Luger cartridges in a weapon designed for 40 S&W. In one suicide attempt, 7.65 mm cartridges were loaded in an old 8 mm Pieper revolver; the bullets became stuck in the barrel.

The gyroscopic stability number normally increases with range, because the ratio $\omega/v$ becomes greater. The precession-related angle of attack therefore decreases. However, a bullet may become unstable when it enters the transonic velocity range. As we saw in 5.6.4.6, not only the drag coefficient but also the overturning moment coefficient can increase. Depending on the bullet design, this increase in overturning moment may be sufficient for the bullet to become unstable, albeit perhaps only temporarily.

A projectile may also experience a considerable loss of angular velocity if it passes through a sufficient thickness of soft material. The friction between the projectile and the material can significantly reduce its rotational speed. However, this effect always occurs together with a torque shock and is therefore usually not directly recognizable as the cause of destabilization.

***Torque shocks***. If a projectile is subjected to a torque shock (i.e. an impulse with a line of action outside the centre of gravity), the angular momentum vector is deflected from the projectile axis (see 7.5.3). The resulting nutation (shown in Fig. 7.22) is superimposed on normal precession, resulting in a fairly large angle of attack that varies over time.

The precession induced by drag alone results in only a small angle of attack for a normally stabilized projectile, which is already taken into account in the drag coefficient (and in the BC) in the point-mass model.

A projectile can be subjected to a torque shock in various ways. The disturbance to which it is exposed immediately after leaving the barrel is unavoidable. The gases emerging from the muzzle overflow the projectile from behind (see 4.2.1) and subject it to a torque shock owing to the aerodynamic force that is briefly applied. The projectile reacts with a nutation movement. With a normally stable projectile, this additional movement decays relatively quickly. However, if the projectile is spinning too slowly, the nutation has an additional destabilizing effect.

The most common case of destabilization owing to a torque shock is when the projectile comes into contact with a solid or liquid medium. In such instances, it makes no difference whether the projectile perforates the medium or ricochets off its surface. The equations in 7.5.3 show that the angle of attack will be greater at higher values of torque shock (caused by higher velocity and/or a steeper angle of incidence).

When a projectile perforates a solid material, the degree to which it is disturbed depends on the density of the material, the length of the shot channel and whether it tumbles within the material. Low-density substances (including biological tissue) will result in smaller angles of attack, unless the projectile turns perpendicular to its direction of travel prior to exit. The projectile will quickly realign its axis with the trajectory tangent after passing through such materials, as it is able to maintain gyroscopic stability.

#### 8.2.2.2 Sectional density and drag coefficient

We need to know the sectional density and drag coefficient of a projectile to calculate its trajectory using the point-mass model. It is difficult to determine either of these for destabilized projectiles, although quite similar nutation and precession occur following ricochet under the same conditions. When reconstructing the trajectory of a ricochet, we do not usually know the magnitude of the impact or the subsequent angle of attack. We can obtain an approximate solution to this problem using the periodicity of the motion, the oscillation length of which is measured in metres. Here too, there is a distinction to be made between short and long ranges.

At short ranges (see 8.2.1.1), the trajectory can be considered straight, owing to the short time of flight. However, it is often necessary to plot velocity as a function of distance to assess the dangerosity (or effectiveness) of the projectile. As we know neither the sectional density nor the drag coefficient for the nutating projectile, we generally take the default value for the former and assign the increase in drag entirely to the coefficient.

L. C. Haag (2007) indicates that the coefficient will increase by a factor of 10–20, based on radar measurements of handgun bullets. Experiments with rifle bullets revealed increases by a factor of 3 to 4 (see Fig. 7.23).

If a bullet flies for a long distance after ricocheting, it is important to know whether it stabilized again or maintained a large angle of attack. In the first case, the decreasing drag at constant sectional density is reflected in the drag coefficient, which falls from a high value while the projectile is yawing to a lower value when it resumes approximately normal flight (see example in Figs. 7.23 and 7.24).

If the projectile nutates or tumbles during its entire flight, it can be treated as a fragment (see 5.4.8). The increase in sectional density is taken into account and a drag coefficient for a classic fragment (cube or cylinder) is used. For short bullets, the $C_D$ value for a sphere is also an acceptable choice.

Calculations of this sort yield only approximate results, but these are perfectly adequate for practical purposes.

#### 8.2.2.3 Reconstruction

***Short ranges***. At short ranges, we can generally treat the trajectory of the projectile as a straight line (see 8.2.1.1). In doing so, we must also take account of ricochet behaviour. If, for example, we know the angle of departure (from the ricochet mark and the point of impact), we can determine the angle of incidence and thus the line of fire fairly accurately.

***Medium and long ranges***. It is very difficult to reconstruct the trajectory of a ricochet that continues for an extended distance. Errors in estimates of sectional density and drag coefficient have a more significant effect with longer times of flight. In particular, whether the projectile stabilizes again or continues to fly unstable until impact has a major influence.

However, one can limit the unknown trajectory to a certain range by calculating one trajectory with deceleration that is definitely too low and a second with deceleration that is too high, bearing in mind the considerations in the previous section with regard to drag. This makes it possible to at least state the minimum and maximum distances from which the shot must have been fired, for instance.

The meteorological conditions prevailing at the time of the event have a significant influence on long trajectories. One must take account of air pressure and temperature (to determine air density), plus the vectorial wind velocity in the area of the expected trajectory.

### 8.2.3 Case studies

Cases that require the reconstruction of undeflected shots include errors by soldiers or other users of shooting ranges and shots fired into the air during celebrations. Although it is repeatedly pointed out that almost all falling projectiles can cause life-threatening injuries, many gun users still appear to believe they are harmless (see also 5.4.6.5).

***Case 1***. On the occasion of a national holiday, a father and his daughter were standing on the shore of a lake watching the fireworks when the child was hit by a 7.62 × 39 Kalashnikov bullet that penetrated her shoulder at a steep angle and lodged next to her aorta after penetrating approximately 18 cm. From the very large angle of descent and the depth of penetration, the round had clearly been fired steeply into the air; as it was not possible to establish the horizontal angle of deflection at the point of impact, it was not possible to determine the direction from which the round had come.

Following an appeal by the police, a number of people came forward and admitted to firing into the air that evening. They included the firer of the round in question, so the firing location and muzzle velocity were known. He had fired his weapon at an elevation of 1231 mil (69.2°). The projectile rose to a height of 1810 m and reached the ground again after a flight time of 38.5 s. The distance was 1406 m, the final velocity 113 m/s (energy 51 J) and the angle of descent 1529 mil (86°). All the trajectory data corresponded to the relatively minor injury that the girl had suffered. Fortunately, she survived with no serious consequences.

***Case 2***. On returning from shopping one Saturday, a woman found the walls of the bedroom on the second floor of her detached house covered in a layer of black/grey. Police who were called to the scene found an almost horizontal bullet hole in the front of the timber-built house. They also found a bullet (in a box) that had previously perforated a torch powered by zinc-carbon batteries. This explained the grey deposit, but not the origin of the projectile.

A military training area around 4 km away was initially suspected to be the source. However, it quickly became clear that a horizontal bullet hole could only be created near the apex of the trajectory and that a distance of 4 km was therefore out of the question. Nor could a long-range shot have caused so much damage to the façade of the house, the torch and the box. The direction of fire could be easily determined from the long channel in the façade, and reconstructing the trajectory back from the apex led to a narrow area on the road that led past the local shooting range. Since shooting practice had been held that Saturday, the only possible culprit was someone who was on their way home along that road and unintentionally or deliberately fired off one last round.

***Case 3***. A family was eating when something struck the façade of their house below the kitchen window, making a loud noise. Upon inspection, a suspected bullet hole was discovered in a panel of the cladding. The police found a rifle bullet behind the panel that had penetrated about 5 cm into the wood. As there was a dispute with a neighbour living 50 m away, he was initially suspected of having fired the shot. However, a shot fired from only 50 m would have penetrated more deeply, so the bullet must have come from further away. Here too, a military train-

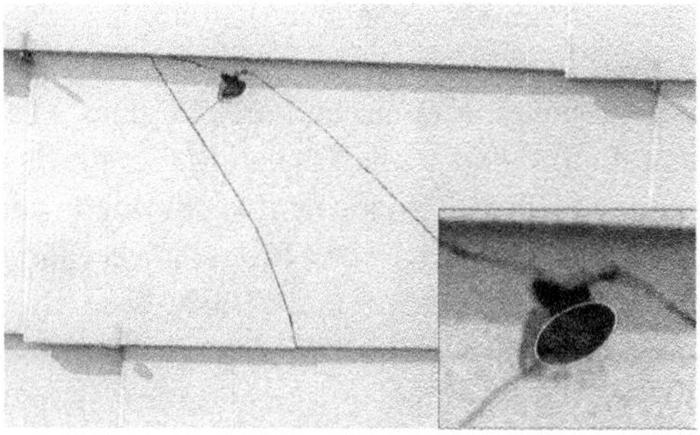

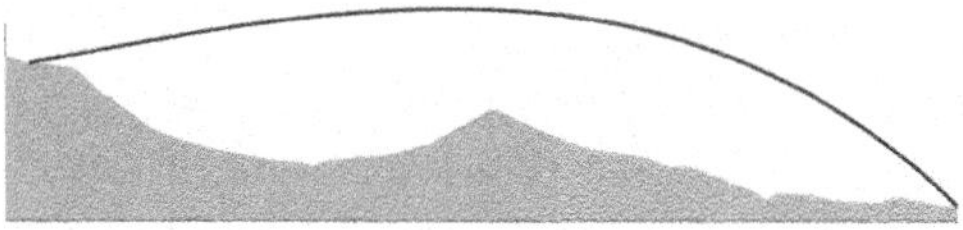

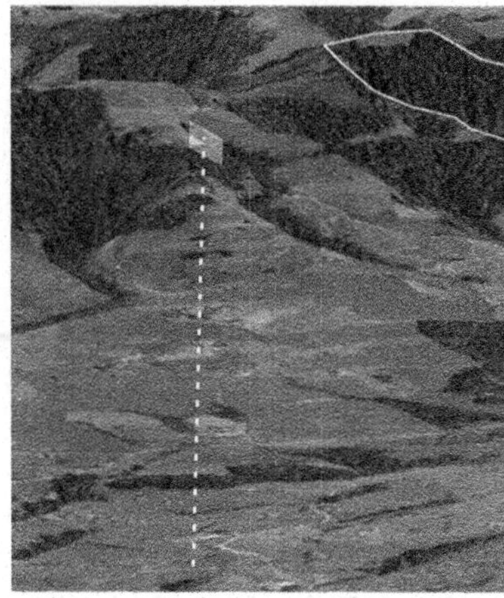

**Fig. 8.5**. (*Top*) Bullet hole in the façade of a house. It was possible to determine the angle of incidence approximately from the elliptical form of the hole.

**Fig. 8.6**. (*Bottom*) Possible trajectory, shown in relation to the terrain cross-section.

**Fig. 8.7**. Direction of fire and the area from which the round is suspected to have been fired. The military training area is outlined in white.

ing area was within range. The origin of the shot was identified three weeks after the event.

The damaged façade panel had already been replaced, but was still in the waste container. After the pieces were assembled, an elliptical, sharp-edged bullet hole was found (Fig. 8.5), which allowed the angle of incidence to be determined with sufficient accuracy. From the direction of the shot relative to the wall of the house and the orientation of the house, it was possible to determine a mean direction of fire, with a certain lateral and fall angle error. Using trajectory reconstructions (see Fig. 8.6), it was hence possible to establish a distance and a lateral area from which the round must have come, with a high degree of certainty. That area was located well away from the military training area (see Fig. 8.7), and the details were passed on to the local police for further action.

***Case 4***. A scaffolder was working about 7 m above ground when he felt a pain. He had been hit by a 22 L.R. bullet that perforated his liver from the back right and stopped under the skin of his abdominal wall on the front left. The bullet appeared to be undamaged, so this was clearly a direct shot. The victim's injury, which had been documented by 3D computer tomography, was used to establish the approximate angle of incidence and impact velocity.

The penetration of a bullet into ballistics gelatine as a function of velocity (see B. P. KNEUBUEHL et al., 2022) is very similar to that of the same bullet in biological tissue.

Measurements of the penetration depth in gelatine for the projectile in question resulted in the graph of Fig. 8.8. This indicated that the bullet had been travelling at

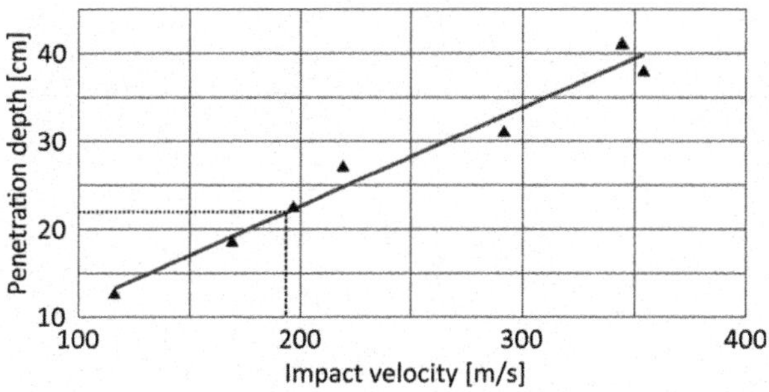


**Fig. 8.8.** Results of test shots into gelatine conducted with a view to establishing impact velocity from the depth to which the bullet had penetrated the victim.

approximately 200 m/s when it struck the victim. Determining the line of arrival was more difficult. The deviation experienced by the projectile inside the body (which can also be seen in the gelatine) must be taken into account. Furthermore, the victim was only able to describe his location and posture at the time of the event approximately, which meant that one could not directly establish the exact line of arrival from the wound channel. (F. RIVA et al., 2017, W. KERKHOFF et al., 2018).

Trajectory calculations incorporating meteorological data from the time of the incident yielded possible firing distances. To calculate the actual distance, it was necessary to know the muzzle velocity of the round when it was fired from the as-yet unknown weapon. By using a topographical map and taking account of the terrain profile (see Fig. 8.3), it was possible to identify the area from which the round had probably originated (see Fig. 8.9).

***Case 5***. Shooting ranges are a common source of deflected shots. This narrows down the likely firing location to a limited area, and experience has shown that the critical points of such an area are usually known.

However, this does not obviate the need to verify that the bullet could not have come from elsewhere (see Case 3 in the previous section).

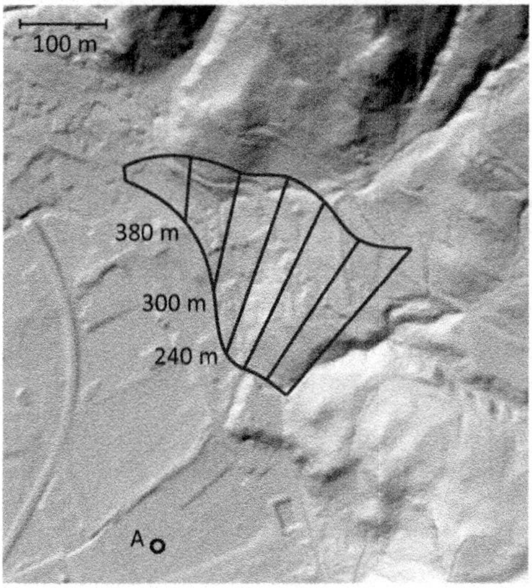


A person was standing on the edge of a village football field when they felt a blow to the side of their chest. A 9 mm Luger bullet lay at their feet, and there was a superficial graze on their chest. A pistol range was located at a distance of about 400 m, 23 m below. It was therefore necessary to establish whether the bullet could have come from that range and what path it could have taken.

**Fig. 8.9.** Contour map showing the area from which the bullet must have been fired. The exact location would depend on muzzle velocity.

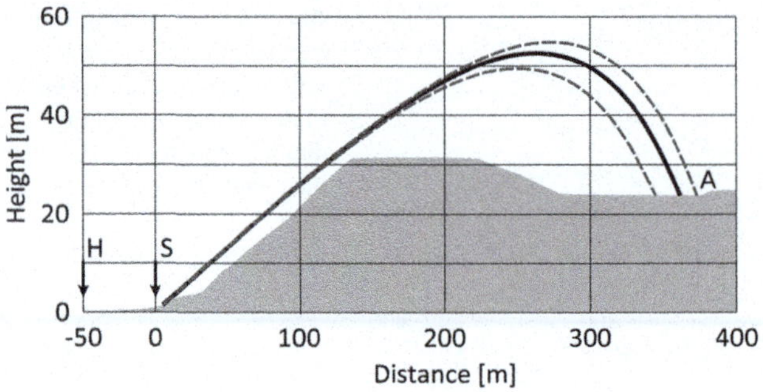


**Fig. 8.10.** Terrain profile for Case 2. A: Point of impact. H: Covered firing point. S: Targets and bullet catcher. Dashed lines indicate trajectory variants for different velocities. The solid line indicates the most probable trajectory.

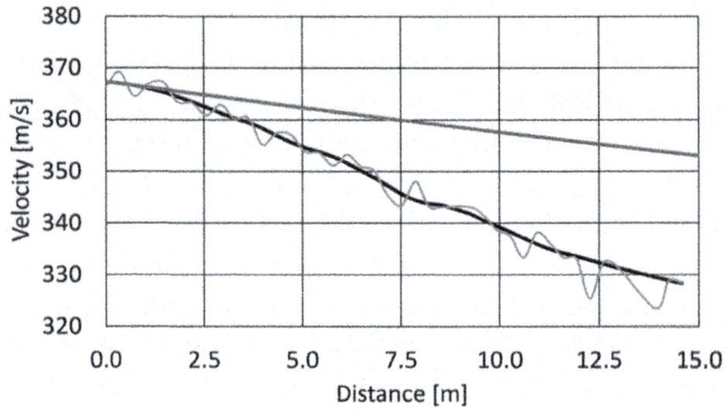


**Fig. 8.11.** Deceleration of a pistol bullet following ricochet (black line) compared to deceleration of the same bullet over a normal trajectory (grey line). The wavy line represents the original radar measurement.

Fig. 8.10 shows a terrain section, including the covered firing point H and the targetline S.

A direct shot from the firing point was out of the question, as it would have had to be fired at an angle of at least 10° in order to pass over the hill in front. The distance would have been at least 1300 m. This left only the possibility of a ricochet from the area of the targets and bullet catcher, or a round of unknown origin.

The bullet showed virtually no marks from the rifling of the weapon; it was as if they had been polished away. However, its surface was covered in fine scratches. The bullet must have passed through soft soil or fine sand. This appeared to clarify the starting point of its trajectory. An inspection of the shooting range revealed obvious deficiencies in the maintenance of the bullet catcher.

Trajectory calculations were performed to confirm the assumptions made, using the drag coefficient (or BC) based on radar measurements of a pistol bullet following ricochet (see Fig. 8.11). The fit calculations yielded plausible values for velocity and elevation, with a credible value for the deceleration of the bullet in the ground. Fig. 8.10 shows trajectories for three possible initial velocities. According to calculations, the impact velocity was around 40 m/s (6.4 J); this also corresponded very closely to the injury suffered.

### 8.2.4 Ballistics in bloodstain analysis

#### *8.2.4.1 The problem*

One of the most important problems in analysing bloodstains is determining where the drops originated and were projected from. The line of arrival can be determined from the elliptical form of the mark left by a blood drop, and that line of arrival must form a tangent to the trajectory of the blood drop at the point of impact. The normal projection of this tangent in the horizontal plane always passes through the normal projection of the starting point of the drop. Multiple normal projections of this type therefore intersect in an area that represents the normal projection of the area from which the drop came. However, there is no equally reliable information regarding the location of this initial area, unless the drops have travelled such short distances that their trajectories can be approximated by straight lines.

In this case, it will be necessary to treat drops as following a straight-line trajectory even when they fall from much greater heights than specified in Eqn 8.2:1 (see 8.2.1).

In many bloodstain patterns, however, the lines of arrival for some stains lie far above the presumed source (see Fig. 8.12). It may be important to know whether these tracks originated from that source or could possibly have a different starting point. To determine this, one must carry out a ballistics calculation of the trajectory of the drop. Given the spherical form of blood drops, the point-mass model is perfectly suitable.

A trajectory can be calculated using the point-mass model if we know the direction of flight and the velocity of the projectile at any point along its trajectory, in addition to its mass and shape. The drag coefficient of a sphere at low velocities and Reynolds numbers can be taken from Fig. 8.13. With the exception of velocity, we know all these parameters sufficiently accurately for a drop of blood at the point of impact.

**Fig. 8.12.** Blood spatters on a wall following an attack. The lines of arrival are unusual in several cases.

Errors in determining droplet diameter – and hence mass – (see C. WILLIS et al., 2001, W. F. ROWE, 2006) play a subordinate role, as it is not the mass or the calibre that mainly determines the results of trajectory calculations, but the sectional density, and this is related to diameter linearly rather than cubically. Furthermore, these calculations only serve to confirm or refute assumptions regarding the source of the blood drops. Direct measurements of diameter are sufficiently accurate.

However, knowing the maximum possible velocity of a blood drop would make it possible to decide by means of a ballistics calculation whether it could have come from a particular source.

### 8.2.4.2 The maximum velocity of a drop of blood

Bloodstains are usually categorized according to the velocity of the object that caused them (T. BEVEL, R. M. GARDNER, 1997, S. H. JAMES, W. G. ECKERT, 1999). A distinction is made between "low-velocity stains" caused by blood drops travelling at up to 1.5 m/s, "medium-velocity impact spatter" from drops travelling at 1.5 to 7.6 m/s and "high-velocity impact spatter" resulting from drops moving at over 30 m/s. Droplet size is related to velocity. Droplets causing low-velocity stains can be 4 mm or more in diameter, those resulting in medium-velocity impact spatter measure 1–4 mm and high-velocity droplets have diameters of 1 mm or less.

Drops of liquid flying through the air are spherical, at least at low velocities. This is a consequence of the surface tension specific to every liquid, which creates a characteristic internal pressure inside the drop. As velocity increases, so does the stagnation pressure caused by drag. If stagnation pressure exceeds internal pressure, the droplet separates into several smaller droplets.

This means that there is a clear relationship between the size (or diameter) of a drop of liquid and its maximum possible speed. If we know the diameter of a drop

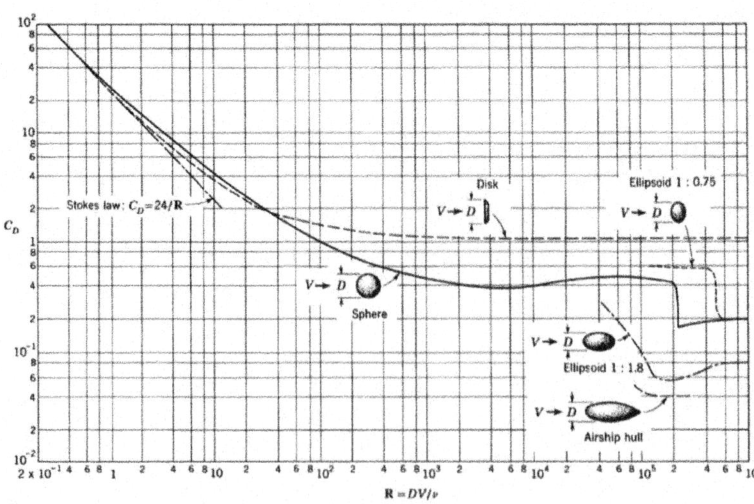


**Fig. 8.13.** Drag coefficients for various body shapes at low Reynolds numbers (see list of figures for source).

of blood, it is possible to deduce its maximum possible initial velocity at the moment it formed. The actual initial velocity could be lower than this, but not higher.

A drop of liquid begins to break up owing to external (aerodynamic) forces when the ratio between the aerodynamic force and the force induced by surface tension attains a certain value. This ratio is described by the dimensionless Weber number (L. P. HSIANG, G. M. FAETH, 1995, T. G. THEOFANTOUS, 2011, W. WÜRZ, U. GAISBAUER, 2014), which is defined as follows:

$$(8.2{:}2) \qquad We = \frac{d_b \cdot \rho \cdot v^2}{\sigma_s} \qquad [-]$$

where $\sigma_s$ = surface tension, $d_b$ = diameter, $\rho$ = air density and v = droplet velocity.

Experiments in free fall have shown that droplets begin to oscillate and break up when the Weber number rises beyond about 13 (L. P. HSIANG, G. M. FAETH, 1995). SCIPP gives a figure of $\sigma_s = 0.058$ for the surface tension of blood. We can therefore write:

$$(8.2{:}3) \qquad \frac{d_b \cdot \rho \cdot v^2}{0.058} = 13 \qquad [-]$$

$$\text{or} \qquad v^2 = 0.754 \cdot \frac{1}{\rho \cdot d_b} \qquad [m/s]$$

$$(8.2{:}4) \qquad v = 0.868 \cdot \sqrt{\frac{1}{\rho \cdot d_b}} \qquad [m/s]$$

Trajectory calculations generally assume normal air density at sea level, which is 1.225 kg/m³. Substituting this value into Eqn 8.2:4, we obtain:

$$(8.2{:}5) \qquad v = 0.785 \cdot \sqrt{\frac{1}{d_b}} \qquad [m/s]$$

We only need to use Eqn 8.2:3, with the true air density measured at the scene, if we expect velocity to be close to the maximum value. Table 8.2 lists maximum possible velocities for a number of droplet diameters.

**Table 8.2.** Maximum blood droplet velocities associated with various droplet diameters

| d [mm] | $v_{max}$ [m/s] |
|---|---|
| 1 | 24.8 |
| 2 | 17.6 |
| 3 | 14.3 |
| 4 | 12.4 |
| 5 | 11.1 |
| 6 | 10.1 |

#### *8.2.4.3 Application*

It often happens that the projection of the line of arrival coincides with the projection of the presumed source, but the height of the line of arrival does not correspond to any possible starting point. In such cases, one can use the droplet diameter estimated from the elliptical shape of the stain to calculate the mass of the droplet. From this, one can then perform a best-fit ballistics trajectory calculation with the drag coefficient of a sphere at small Reynolds numbers (see Fig. 8.13). Starting from the point of impact, trajectories are calculated (with negative increments) using the height of the assumed point of departure and the velocity. If one finds a geometrically compatible trajectory that begins at the assumed starting point and arrives at the stain at the angle measured, one then verifies that the associated initial velocity is lover than the maximum possible velocity for this droplet diameter. If that is not the case, the stain cannot have originated from the presumed point of departure.

This procedure shows whether it is ballistically and physically plausible that a stain originated at a particular point, primarily in the case of sharply curved trajectories.

Knowing the maximum possible velocity of drops of blood as a function of their diameter significantly enhances the analysis of bloodstain patterns. We can use trajectory calculations to decide whether we can assign a bloodstain to one of the known sources or should consider another source.

#### 8.2.4.4 Case studies

***Case 1***. A woman was killed by blows to the head using a heavy hammer and the assumption was that she had been asleep at the time of the attack. The usual method indicated four sources of blood spatter, all located in the area of the pillow. These four sources also indicated likely movements after she was struck for the first time and awoke. 20 stains could not be linked to any of those sources. They had to be investigated using ballistic techniques, as the perpetrator claimed that the woman had been sitting on the edge of the bed (see Fig. 8.14). However, all 20 stains could

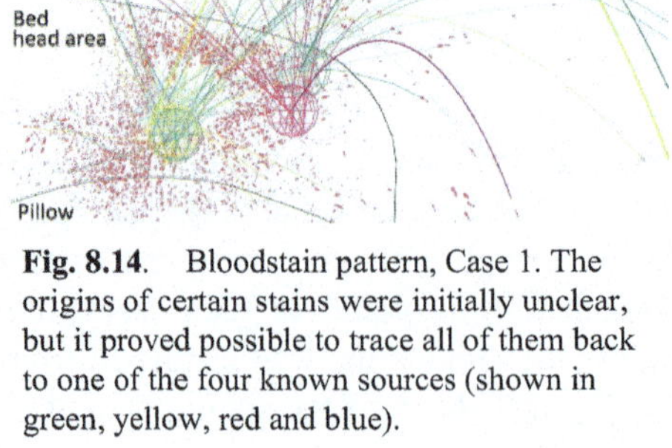


**Fig. 8.14**. Bloodstain pattern, Case 1. The origins of certain stains were initially unclear, but it proved possible to trace all of them back to one of the four known sources (shown in green, yellow, red and blue).

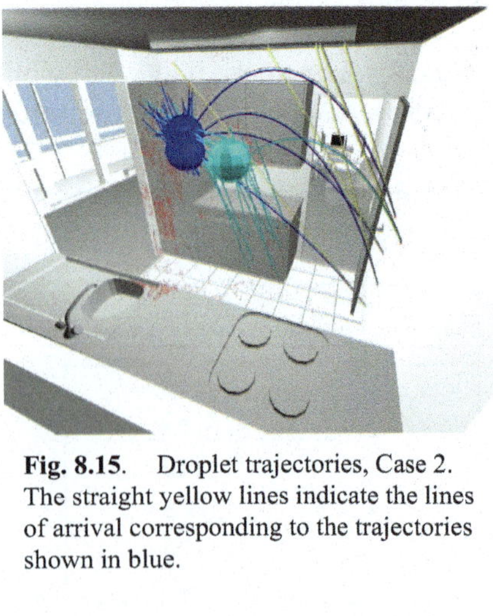

**Fig. 8.15**. Droplet trajectories, Case 2. The straight yellow lines indicate the lines of arrival corresponding to the trajectories shown in blue.

be traced back to the four known sources by computing ballistics trajectories for velocities lower than the maximum possible velocities shown in Table 8.2. It was also possible to show that if the victim had been sitting as the perpetrator claimed, the corresponding stains could only have been created by drops travelling at velocities greater than the maximum possible values. It was hence possible to refute the perpetrator's claims by means of ballistics.

***Case 2***. Following another homicide, a large number of bloodstains were found in a kitchen, most of them emanating from three sources. However, some stains on the more distant wall could not initially be traced back to any of these sources (Fig. 8.15). It was possible to fit trajectories with initial velocities of between 3.5 and 6.5 m/s to each of these stains. Those velocities were well below the maximum possible values, so these stains could also be linked to one of the three known sources.

## 8.3 Safety on shooting ranges

### 8.3.1 Safety and dangerosity

The safety of shooting ranges is a key concern, and involves a number of ballistics questions that can usually be investigated independently. The first question is how far projectiles or pieces of projectile can travel when they ricochet or fragment after striking the target area. One must also take account of trajectories resulting from firer error. This can be achieved using the exterior ballistics methods described in Chapters 5 and 7. If an outdoor range is not large enough to cover all these situations, protective measures must be taken. These must be designed on the basis of terminal ballistics experience (see Chapter 6) and appropriate experiments. However, the essential question is how to define "safety".

One cannot avoid also introducing the concept of dangerosity (described in detail in B. P. KNEUBUEHL et al, 2022, and B. P. KNEUBUEHL, 2013). Projectiles are no longer dangerous if they do not cause injuries – even minor ones – and their impact generally goes unnoticed. In the case of small fragments and pellets (including lead shot up to about 4 mm in diameter), there is often a significant difference between maximum range and safety distance.

Safety distance is approximately 75% of maximum range for 4 mm lead shot (0.38 g) and approximately 65% for 3 mm shot (0.16 g). Steel spheres and fragments cease to be dangerous at even shorter ranges, owing to their lower sectional density. The safety distance for a 4 mm steel sphere equates to 65% of its maximum range, for instance.

If one can be certain that only small projectiles will enter a target area, it is perfectly possible to base the danger area on the safety distance rather than the maximum range if space is limited.

The maximum range must be taken as the safety distance for all larger projectiles and fragments, as these can cause life-threatening injuries in the entire area. This is particularly the case for shots fired steeply into the air (see 5.4.6.5 and Fig. 5.26).

### 8.3.2 Maximum range

It is possible to calculate maximum ranges quite well using ballistic techniques. In so doing, one must take the most disadvantageous values for all parameters that affect range. This applies in particular to air density, wind in the direction of fire and muzzle velocity. One must also distinguish between shooting ranges used for weapons with flat trajectories and artillery target areas, where only scatter and the maximum range of fragments need be considered.

Maximum ranges of direct-fire weapons are usually so great that very large range areas or very good bullet catchers are required, especially if any ricochets in the target area are also to remain within the zone from which other persons are excluded. Such sites are often not available, which has prompted the development of short-range training ammunition. This type of ammunition simulates the real-life trajectory closely up to the usual operational range, but has a shorter total range than a standard cartridge.

Short-range projectiles are usually based on either ballistics or aerodynamics. The ballistics approach consists of firing a very light projectile at a very high muzzle velocity. The trajectory is initially very flat (flatter than that of the original projectile). Owing to its low sectional density, the projectile experiences high deceleration and therefore has a short range (see Fig. 5.23, Trajectory 1).

A plastic training (PT) bullet in 7.62 mm NATO weighs 0.7 g and has a muzzle velocity of 1480 m/s. At up to 50 m, the point of impact is the same as with the original projectile. At 100 m the point of impact is 19 cm low. Maximum range is only approximately 300 m. Increasing the mass to 1.5 g would ensure that the trajectory matched the original at ranges of up to approximately 100 m, but the maximum range would still only be approximately 575 m instead of the approximately 3.8 km range of the normal FMJ bullet.

The aerodynamic approach involves using a tubular projectile. If the projectile is designed correctly, flow around it at high velocities causes an annular compression shock at the leading edge that does not touch the interior of the projectile even at the rear (Fig. 8.16 a). Air flows through the tube. Sectional density is high, as only the annular face opposes the direction of movement. The drag coefficient is low and corresponds approximately to that of a plate oriented end-on to the flow.

As velocity decreases, the Mach angle increases. When velocity falls to a certain level, the shock wave inside the tube is reflected from the walls. This prevents flow through the tube (the tube is "aerodynamically closed") and the projectile behaves like a very light cylinder (Fig. 8.16 b). Sectional density becomes low and the drag coefficient high. Drag increases by a factor of 7 to 9, depending on the shape of the projectile.

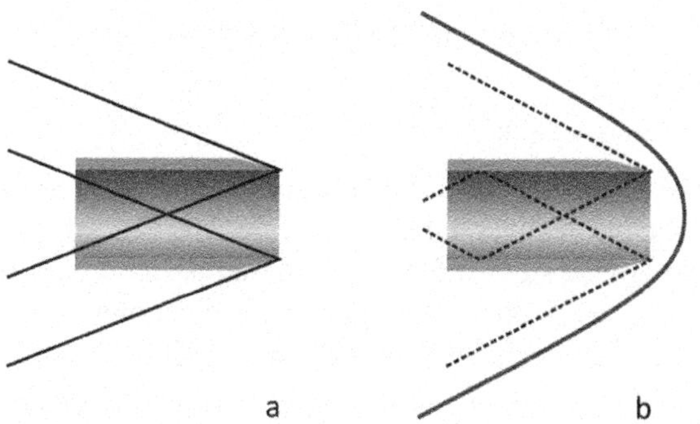


**Fig. 8.16.** Tubular projectile under two different flow conditions. **a.** High velocity. Air flows through the tube, the sectional density is high and $C_D$ is low. **b.** After reaching the boundary layer separation point, the tube is closed by the shock wave reflections. Sectional density is low and $C_D$ high.

The ratio between the sectional densities before and after the transition point is given by:

$$(8.3:1) \qquad \frac{q_n}{q_v} = 1 - \frac{d_i^2}{d^2} \qquad [-]$$

where $q_v$ = effective sectional density before transition, $q_n$ = effective sectional density after transition, $d_i$ = inside diameter and $d$ = calibre.

The Mach number at which transition takes place depends on the length of the projectile $\ell$ and its inside diameter

$$(8.3:2) \qquad M = \sqrt{1 + \frac{\ell^2}{d_i^2}} \qquad [-]$$

Tubular projectiles require a sabot that seals the barrel during firing. The sabot is discarded axially and hence presents no particular problems.

Long-rod training rounds with a perforated cone guidance system are also based on this principle. At high velocities, the air flows through the perforated cone; after the transition point, it acts as a compact high-resistance cone guide mechanism.

### 8.3.3 Rounds that fall short of the target

It regularly happens that a projectile hits the ground well before reaching its target. Before conducting live-firing exercises that will involve firing small-arms rounds over the heads of persons in front, it is important to check that none of the weapons to be used has a worn barrel. A worn barrel can cause bullets to tumble, greatly reducing their range.

A rifle may produce a perfect group when cold or warm, but fire only unstable rounds once it has heated up from firing bursts. Once it cools down, it functions normally again.

Tracer rounds may also fall short of the target, at least in the development phase. This may be due to the tracer charge at the tail causing a shift in the centre of gravity and/or to the lead core starting to melt owing to the heat that the tracer generates as it burns. There have been cases of tracer charges becoming skewed and of molten lead dripping out of the projectile. Both will destabilize the projectile, causing it to strike the ground short of its target.

Please see 5.6.4.6 for an example of a tracer round falling short, plus Fig. 5.47.

Mortar bombs are usually fired from smooth-bore tubes and are fin-stabilized. They always fly with a certain pitching motion about a transverse axis. The fins are also slightly angled, causing the shell to rotate slowly about its longitudinal axis. This is intended to compensate for small eccentricities. If the pitch frequency comes into resonance with the rotation frequency, the angle of attack can increase to such an extent that the projectile crashes. A small change in the angle of attack of the fins will immediately solve this problem.

In a real-life case of this kind, range was reduced by two thirds.

Complete destabilization of a projectile can also occur if, for some reason, it contains a mass (e.g. a liquid) that is free to move around inside, and that mass only gradually begins to spin as the projectile accelerates in the barrel. Such a system of two unequal rotating masses is unlikely to be stable, and the projectile will tumble.

During an artillery test in which inert warmed projectiles with base bleed were fired into a lake, a material that was liquid at the prescribed test temperature was mistakenly used as a substitute for the explosive. As a result, the shells fell into the lake at a distance of 10 km instead of the expected 25 km.

When a projectile falls short of its target, this is often because it has lost stability owing to technical problems or errors. These occur rarely and are usually unpredictable. They are not decisive when assessing the safety of a range, but should be kept in mind.

### 8.3.4 Rebounds

There is a significant risk that a bullet will be deflected in a bullet catcher filled with sand, gravel or similar material owing to a pressure difference between opposite sides of the projectile. In the case of an inclined bullet catcher or one consisting of material that has been poured into place, the projectile will be deflected upwards (see 8.2.3.2 for an example). In a closed bullet catcher, projectiles can hit the ceiling and rebound towards the firer (see Fig. 8.17 for a sketch of the principle).

This happens not only with bullets, but also with inert artillery shells used for interior ballistics studies. To avoid such rebounds, cylindrical projectiles with tail and guide sections identical to the original are generally used.

Bullets and parts of bullets often rebound dangerously on smaller ranges, especially indoors. Plastic bullet catchers harden with increasing numbers of hits and with age, so that projectiles no longer penetrate but rebound (Fig. 8.18). Projectiles that disintegrate on impact with a steel screen produce fragments that can fly a long way back towards the firer and carry enough energy to cause injury. High-tensile steels

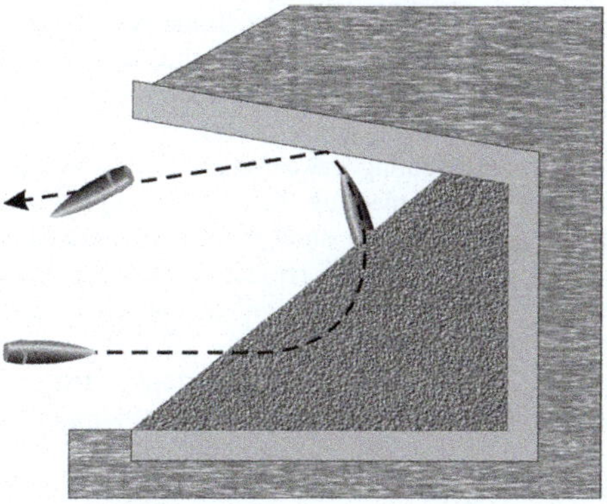

**Fig. 8.17.** How an artillery shell can rebound from a gravel trap. Bullets can behave similarly in sand.

that are constantly exposed to impacting projectiles become brittle and can splinter (Fig. 8.19). This makes it even more likely that projectiles will disintegrate.

Floor, wall and ceiling coverings in shooting ranges must consist of materials that eliminate the risk of ricochets and rebounds as far as possible, and the inside of a range must be designed with this hazard in mind. The VPAM has issued a guideline for testing range materials and designs (see VPAM, 2012).

Rebounding projectiles can cause serious injuries at short distances on outdoor ranges if shots are fired into hard materials such as rocks or steel-plate targets. Military bullets will burst when subjected to such impacts, with the resulting fragments flying sideways and back towards the firer at high speed (see 3.8.3.2 for a case study).

A flat-trajectory projectile that contains explosives is carrying an energy source that enables it to propel its components in any direction. Because the initial velocity of the fragments relative to the projectile is added vectorially to the projectile velocity, they have a velocity component in the original direction of flight. However, the base of certain types of projectile may be propelled back towards the firer, because the fragmentation velocity is greater than the velocity of the projectile. Because it is spinning, this fragment can assume a stable flight attitude and fly several hundred metres back towards the firer.

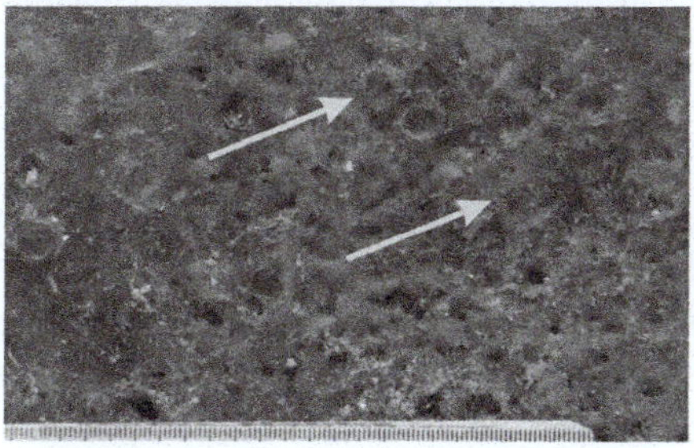

**Fig. 8.18.** Plastic bullet trap. The arrows indicate hardened areas from which bullets have rebounded.

**Fig. 8.19.** Edge of a bullet catcher made of steel slats. Chipping at the edges increases the risk of rebounding bullets.

***Case study***: A surprising rebound incident occurred with an 81 mm mortar. Initially, the cause appeared unrelated to ballistics. This case also shows how difficult it is to understand and correctly assess all the possible connections and interrelationships when dealing with ammunition and ballistics.

To reduce damage in the range target area, 81 mm mortar training was carried out using practice bombs, which ignited a small black powder charge on impact, marking the point of impact with smoke and at the same time breaking the projectile up into a small number of large fragments. Probably for cost reasons, the original grey cast-iron shell casings were replaced by casings made of malleable cast iron, which is somewhat tougher. The casings disintegrated noticeably less, sometimes remaining almost intact, which facilitated clearing-up operations.

During an open day for families, a mortar unit was firing their mortars into a target area with clay soil at a range of 320 m (elevation 1260 mil). During the demonstration, a person standing some 40 m *behind* the mortar positions was killed by a mortar bomb striking them in the head. Reconstruction of the event with extensive firing tests showed that the mortar bomb exploded in the target area as normal, but that instead of causing the projectile to disintegrate, the black powder charge fired it out of the channel it had bored in the clay, as if from a gun barrel. After a short distance, the fin-stabilized shell assumed its normal flight attitude and flew (without a fuze) along an almost standard trajectory for about 360 m against the direction of fire (with an apex of approx. 270 m), landing 40 m behind the firing point, where it hit a person who just happened to be standing there.

Nobody could have imagined that the decision to change the casing material would cause later on to a fatal accident.

This last case study clearly shows how, despite the physical models and the many facts we have learned about ballistics through experiments, unexpected and unpredictable events can still occur. One cannot be careful enough when making changes to weapons and munitions.

## 8.3.5 Ricochets

### *8.3.5.1 Ricochets in the field*

Military live-firing training areas are often located in open terrain with no buildings. Because the larger types of small- and medium-calibre bullet have a considerable range, bullet catchers may be located at some distance from the firers. Structures to protect users against ricochets are generally not desirable for training reasons. Rounds fired too low can therefore strike the ground in the area between the weapon and the target. Because the angle of incidence decreases as distance increases, there is a risk of ricochets.

The small angle of incidence means that the projectile often travels just under the surface for anything from a few tens of centimetres to over a metre and leaves the (soft) ground at an angle of departure greater than its angle of incidence (see 7.3.3). Because the projectile continues to spin almost as it did before impact, such ricochets can stabilize again (as already described in 7.5.5.2) and continue for a surprising distance, often ending up outside the restricted area.

Ranges located in enclosed valleys and surrounded by high mountains pose no problems. If this is not possible and the restricted area is limited, short-range projectiles (see 8.3.2) or fragmentation projectiles can be used, the latter being designed to break up into several pieces on contact with the ground. Such fragments have a significantly lower sectional density and do not fly stable, so their range is considerably shorter.

Hunters operate not on enclosed ranges but in the open countryside (except perhaps when wild boar or foxes invade residential areas). As a result, they often shoot in areas with woodland or forest cover. Furthermore, the ranges involved are usually great enough that ricochets are to be expected in the event of misses.

If one assumes that ricochets can occur on natural ground at angles of impact under 10°, a standing firer on level ground is already at risk of a ricochet at a range of approx. 9 m; if they are sitting on a 3 m tower, the minimum range increases to approximately 26 m.

The German Federal Institute for Agriculture and Food commissioned a detailed study on the ricochet behaviour of hunting bullets during the first decade of the 21st Century. Six different hunting bullets (three with lead cores and three lead-free) in three calibres each were fired at tree trunks, soft and hard ground and stone slabs, at various angles. The tests (involving over 1500 shots) took place at the Deutsche Versuchsanstalt für Jagd- und Sportwaffen (German testing institute for hunting and target weapons, DEVA). Please see B. P. KNEUBUEHL (2011) for results and a full evaluation.

### 8.3.5.2 Ricochets from range equipment and protective structures

Civilian shooting ranges that are not located in enclosed spaces require protective structures to ensure that no projectiles can leave the range, either directly or indirectly. First of all, a good bullet catcher is required to safely catch undeflected shots. A common approach is to use an earth bank or concrete wall, with a bullet trap box or a pile of squared or round timber between each target and the bank or wall. The bullet trap boxes generally consist of steel boxes with a plastic plate on the side nearest the firer. They are either filled with a material (rubber chips) that prevents the projectiles from fragmenting or else contain a steel structure in which the projectile energy is dissipated.

If banks of earth are used as bullet catchers on their own (without wood piles) and are not regularly maintained, they become hollowed out where the largest numbers of bullets strike, and have a

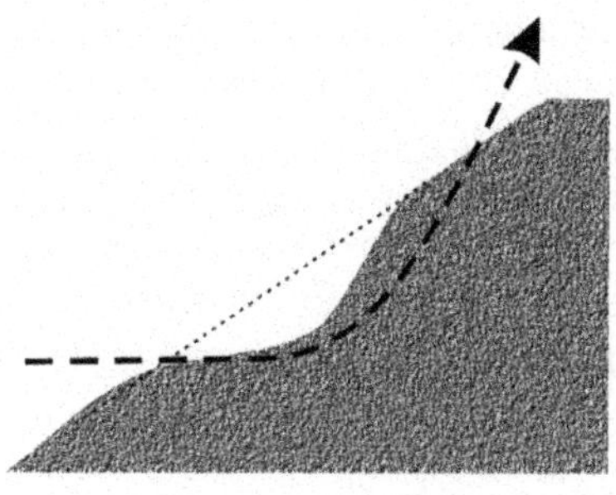

**Fig. 8.20.** Ricochets from an embankment at points of heavy shot load.

significantly shallower gradient at those points. Ricochets may occur at these points, and will usually fly very steeply upwards (see Figs 8.20 and 8.17).

High-level screens are usually required in the area between the firing point and the targets to prevent a shooter from overshooting the protective wall behind the targets with a direct shot. The closer to the firing point such screens are installed, the smaller they will be. Suitably-designed screens can even be installed in the immediate vicinity of the firer.

In many (soft) soils, a ricochet will have an angle of departure significantly greater than the angle of incidence. Low-level screens are often installed to prevent a ricochet from the ground between the firing point and the target from passing over the protective wall.

One should bear in mind that fixtures installed between the firing point and the targets not only serve to absorb ricochets (and undeflected shots), but can in fact generate ricochets. When installing screens, one must therefore ensure that all possible ricochets from them will also remain within the shooting range.

***Case study 1***. A 300 m outdoor range was well protected by a high embankment behind the targets and by two high-level screens, the first at a distance of approximately 15 m to protect against excessively high shots, the second at 60 m to prevent shots from the covered firing point from passing directly over the embankment (see Fig. 8.21). One day, however, a bullet passed through the door of a garage about 700 m behind the embankment and left a dent in a vehicle standing inside. It was not easy to trace the path of this bullet, but eventually a fresh ricochet mark was discovered on one of the posts supporting Screen 1, so low down that the ricochet could pass under the second screen and over the embankment (see Fig. 8.21, Path 3).

A low-level screen was installed at Screen 1 to prevent further ricochets occurring in this fashion. As that created new ricochet points, the height of the protective embankment was increased. A few months later, however, a bullet from this range struck a building about 100 metres closer than the first. Inspection revealed that the low-level screen had mistakenly been covered with a wooden board (see Fig. 8.22 a). A round that penetrates the edge of such a board will exit from the top after about 10 cm, at a steep angle. Such a channel was indeed found; the bullet had passed over the second screen. The low-level screen was replaced by one consisting of a steel plate (clad with wood) instead of a wooden board (see Fig. 8.22 b). At the time of writing, over 30 years later, there have been no further reports of stray bullets.

If an FMJ bullet hits the steel plate with its tip below the edge, it fragments; if the tip is above the edge, it is deflected at a slight angle. According to calculations, the angle is less than 7.5°. In practice, angles of less than 6° were measured (B. P. KNEUBUEHL, E. MAISSEN, 1984). The wooden or plastic cladding prevents the fragments from rebounding.

In addition to screens, other structures may be placed in the area between the firing point and the targets, e.g. for noise reduction or in connection with shooting activi-

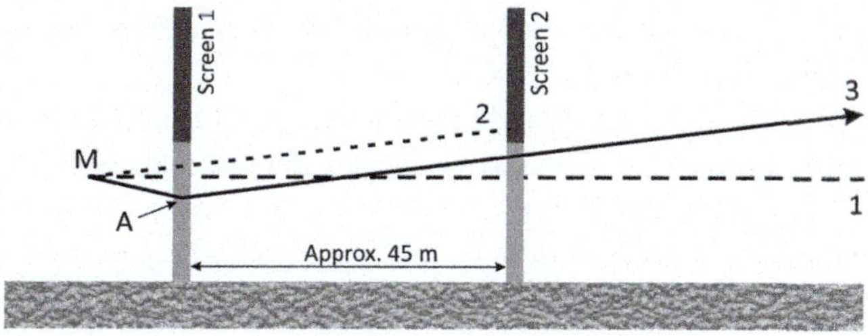


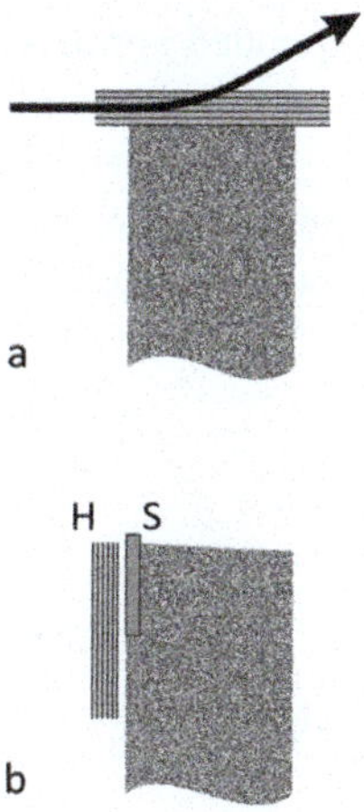


**Fig. 8.21**. (*Above*) Ricochet from a post supporting a high-level screen.
1: Normal line of fire. 2: Line of fire resulting from weapon pointing too high. 3: Line of fire resulting from weapon pointing too low.
(M: Muzzle. A: Ricochet point on side of post).

**Fig. 8.22**. **a**. Bullet deflected by a wooden plank fitted to the top of the low-level screen. **b**. Steel plate (S) covered with wood or plastic (H).

ties. For example, mechanically operated targets running on rails require equipment in the target area to position and hold them there. If a projectile hits the front edge of one of the rails forming part of such an installation, it may ricochet at a very steep angle of departure.

***Case study 2***. A bullet was found 30 m behind the high concrete protective wall of a range on which the targets were moved between target area and firing point on rails. During inspection of the range, a fresh ricochet mark was found on one of the rails (see Fig. 8.23 a and b). The bullet found (see Fig. 8.24) and the manner in which it had been deformed corresponded very closely to the marks at the point of impact. The angle of departure must have been very large, as the bullet flew over the 4.5 m concrete wall located just 2 m behind the point of impact. Trajectory calculations narrowed this angle down to between 73° and 84°, which also corre-

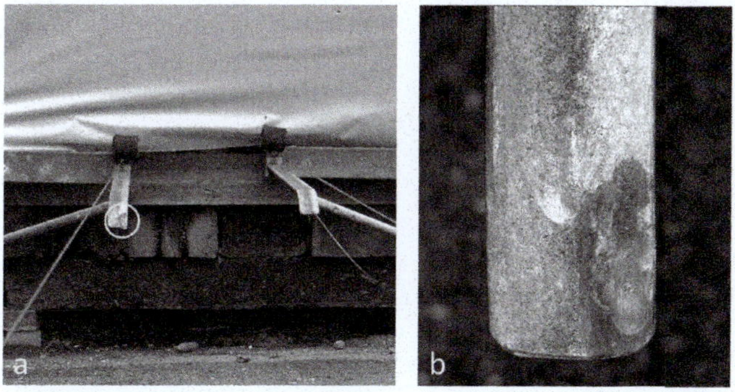


**Fig. 8.23**. **a**. Rails for a target installation. The circle indicates the point of origin of the ricochet. **b**. Detail of the rail, showing ricochet marks (the new mark is on the right).

**Fig. 8.24**. Bullet found outside the range.

sponded approximately to the vertical angle of deflection. Such ricochets can be prevented by installing a roof over the target line.

Large vertical angles of deflection usually indicate that the bullet has lost a high percentage of its energy to the impact. In the above example, the residual energy of the ricocheted projectile was only a few joules.

# Bibliography

## 2 Physical basics


DAUGERTHY RL, FRANZINI JB, FINNEMORE EJ (1989) Fluid Mechanics with Engineering Applications, MacGraw-Hill Book Company, Singapore

GERTHSEN C (1966) Physik, Springer-Verlag, Berlin

GREINER W (1989) Theoretische Physik Band 2, Mechanik Teil 2, Verlag Harri Deutsch, Thun

KUCHLING H (1986) Taschenbuch der Physik, Verlag Harri Deutsch, Thun, 15. Aufl

KALIDE W (1980) Technische Strömungslehre, Carl Hanser Verlag, München

MAGNUS K (1971) Kreisel, Theorie und Anwendungen, Springer-Verlag, Berlin, Heidelberg, New York

OSWATITSCH K (1976) Grundlagen der Gasdynamik, Springer-Verlag, Wien

SCHREIER W (ed.) (1991) Geschichte der Physik, Deutscher Verlag der Wissenschaften, Berlin

SZABÒ I (1972) Höhere Technische Mechanik, Springer-Verlag, Berlin

WIEGHARDT K (1974) Theoretische Strömungslehre, BG Teubner, Stuttgart

ZIEREP J (1979) Grundzüge der Strömungslehre, G Braun, Karlsruhe

ZIEREP J (1991) Theoretische Gasdynamik, G Braun, Karlsruhe


## 3 Interior ballistics


AKHAVAN J (2008) The Chemistry of Explosives, The Royal Society of Chemistry, Cambridge

BOLLÉ E, SEITZ G (1941) Einführung in die innere Ballistik, Friedrich Vieweg & Sohn, Braunschweig

BRÖMEL H (2010), QuickLOAD für Windows® 3.6, Software für Innenballistische Berechnung, Hartmut G Brömel, Babenhausen

CARLUCCI DE, JACOBSON SS (2008) Ballistics, Theory and Design of Guns and Ammunition, CRC Press, Boca Raton

CHARBONNIER PJ (1908) Balistique intérieure, O Doin, Paris

CIP (2007) Texts and tables of the “Commission Internationale permanente pour l'Épreuve des armes portatives”, Brussels

CORNER J (1950) Theory of the Interior Ballistics of Guns, John Wiley & Sons, Inc, New York

CRANZ C (1926) Lehrbuch der Ballistik Bd. II, Innere Ballistik – Die Bewegung des Geschosses durch das Rohr und ihre Begleiterscheinungen, Springer-Verlag, Berlin

CRAWFORD BL, HUGGETT C, FARRINGTON D, WILFONG RE (1947) Direct Determination of Burning Rates of Propellant Powders, Analytical Chemistry, 19, 9, 630–633

DE HALLER P (1948) Une solution graphique du problème de Lagrange en balistique intérieure, Bull tech Suisse romande 74, 1

DEVA (ed.) (2017) Wiederlade-Handbuch der Deutschen Versuchsanstalt für Jagd- und Sportwaffen e V, Altenbeken

ENGEL S, KOCH A (2003) Gurney Velocity from Expansion Tests, armasuisse, Wissenschaft & Technologie, Thun

GERMERSHAUSEN R (ed.) (1980) Waffentechnisches Taschenbuch, Rheinmetall GmbH, Düsseldorf






B. P. Kneubuehl, *Ballistics*,
https://doi.org/10.1007/978-3-662-69237-0

GERMERSHAUSEN R, WITT W, MELCHIOR E (1971) Thermodynamisches Modell der Innenballistik, feste Treibstoffe mit konstanter Oberfläche abbrennend, Wehrtechnik 7, 281–288
GOSSOT F, LIOUVILLE R (1914) Mém poudres et salpêtre, 17, 1
HEYDENREICH W (1908) Lehre vom Schuss, Mittler u. Sohn, Berlin
HUGONIOT H (1889) Sur la propagation du mouvement dans les corps et spécialement dans les gaz parfaits IV, Mouvement d'un fluide dans un tuyeau cylindrique, J école polytech 58, 1–144
KAPUR JN (1958) Lagrange's Ballistic Problem for Unorthodox (H/L, RCL) Guns and Solid-Fuel Rockets, Proc Nat Inst Sci India 24A, 31–39
KNEUBUEHL BP (1995) Schwarzpulverballistik, Int Waffen Magazin Spezialheft 3
KNEUBUEHL BP (1998) Physikalisch-ballistische Grundlagen zur Wirksamkeit von Gas- und Schreckschusswaffen, Bericht Nr 1436, armasuisse Wissenschaft & Technologie, Thun
KNEUBUEHL BP (1999) Splitterballistik und Splitterwundballistik, Bericht Nr 1486, armasuisse Wissenschaft & Technologie, Thun
KNEUBUEHL BP (ed.), COUPLAND RM, ROTHSCHILD MA, THALI MJ (2022) Wound Ballistics. Basics and Applications (SW RAWCLIFFE, Trans.), Springer-Verlag, Berlin, Heidelberg, 2nd ed. (Original work 4th German edition, published in 2022)
KNEUBUEHL BP (2013) Geschosse Gesamtausgabe, Ballistik, Messtechnik, Treffsicherheit, Wirkungsweise, Verlag Stocker-Schmid, Dietikon, 3. Aufl
KOCH A, ARNOLD N, ESTERMANN M (2002) A Simple Relation between the Detonation Velocity of an Explosive and its Gurney Energy, Propellants, Explos, Pyrotech 27, 1–4
KÖHLER J, MEYER R, HOMBURG A (2008) Explosivstoffe, Wiley-VCH Verlag GmbH, Weinheim, 10. Aufl
KUTTERER RE (1959) Ballistik, Friedr Vieweg & Sohn, Braunschweig
KUHN C (1866) Über den elektroballistischen Chronographen von LE BOULENGÉ und über die gebräuchlichen elektroballistischen Apparate, Polytech. J 179 Nr XI, 30–50
LE BOULENGÉ PE (1864) Mémoire sur un chronographe électro-balistique, Paris
LOVE AEH, PIDDUCK FB (1922) Lagrange's Ballistic Problem, Phil Trans Roy Soc 222, 167–226
MURAOUR H, AUNIS G (1949) Étude de la vivacité de combustion de poudres colloidales, Mém de l'Art fr T 25, 4e fasc 859–866
OERLIKON Taschenbuch (1981), Werkzeugmaschinenfabrik Oerlikon-Bührle AG (ed.), Zürich
OSWATITSCH K (1976) Grundlagen der Gasdynamik, Springer-Verlag, Wien
PIDDUCK FB (1937) The Limiting Problem of Internal Ballistics, J Appl Phys 8, 144
POISSON SD (1832) Formules relatives au mouvement du boulet dans l'intérieur du canon, extraites des manuscrits de LAGRANGE, J de l'école polytech. 13, cahier 21:187–204
RIEMANN B (1860) Über die Fortpflanzung ebener Luftwellen endlicher Schwingungsweite, Abhandlungen der Königlichen Gesellschaft der Wissenschaften zu Göttingen, 8, 43–65
SCHMITZ O (1913) Neue Versuche der Friedr Krupp AG über innere Ballistik und Artillerie, Artill Monatsh Nr 84:482ff und Nr 86: 85ff
SELLIER KG (1982) Schusswaffen und Schusswirkungen, 2. Aufl, Schmidt-Römhild-Verlag, Lübeck
SUGOT G (1928) Balistique intérieure théorique et tables numériques, Gauthier-Villars et Cie, éditeurs, Paris
VALLIER E (1899) Comptes Rendues 128, S 1305, 129, S 258, Paris
VIEILLE P (1893) Étude sur le mode de combustion des matières explosives, Mém des poudres 6:256–391
VON BÖCKH P (2004) Fluidmechanik, Springer-Verlag, Berlin, Heidelberg, New York
WITT W, MELCHIOR E (1974) Thermodynamisches Modell der Innenballistik, feste Treibstoffe progressiv oder degressiv abbrennend, Teil 1, Wehrtechnik 6, 222–225
WITT W, MELCHIOR E (1974) Thermodynamisches Modell der Innenballistik, feste Treibstoffe progressiv oder degressiv abbrennend, Teil 2, Wehrtechnik 8, 288–292
WOLFF W (2006), Raketen und Raketenballistik, Reprint 5. Aufl, Elbe-Dnjepr-Verlag, Klitzschen
ZUKAS JA, WALTERS WP (1998) Explosive Effects and Applications, Springer-Verlag, New York

## 4 Intermediate ballistics

CARLUCCI DE, JACOBSON SS (2008) Ballistics, Theory and Design of Guns and Ammunition, CRC Press, Boca Raton

FARRAR CL, LEEMING DW (1983) Military Ballistics, A Basic Manual, Brassey's Defence Publishers, Oxford

GERMERSHAUSEN R (ed.) (1980) Waffentechnisches Taschenbuch, Rheinmetall GmbH, Düsseldorf

KNEUBUEHL BP (1998) Physikalisch-ballistische Grundlagen zur Wirksamkeit von Gas- und Schreckschusswaffen, Bericht Nr 1436, armasuisse Wissenschaft & Technologie, Thun

OERLIKON Taschenbuch (1981), Werkzeugmaschinenfabrik Oerlikon-Bührle AG (ed.), Zürich

## 5 Exterior ballistics

ATHEN H (1958) Ballistik, Quelle & Meyer, Heidelberg

AUBIN D (2010) 'I'm just a mathematician' Why and How Mathematicians Collaborated with Military Ballisticians at Gâvre, Université Pierre et Marie Curie (UPMC) HAL 00639895

BENOIST A (1922) Méthode spéciale de détermination des éléments topographiques du tir, Revue d'Artillerie, tome 89, Berger-Levrault, Paris, 366–384

BILHARZ H (1944) Bemerkung zu einem Satz von HURWITZ, Z angew Math Mech 24, 2, 77–82

BRUHN G (1961) Querkräfte auf langsam pendelnde Rotationskörper im Überschallflug, Z Flugwiss 9, 285–299

CARLUCCI DE, JACOBSON SS (2008) Ballistics, Theory and Design of Guns and Ammunition, CRC Press, Boca Raton

CARTOUX E (1974) Balistique extérieure, deuxième partie, Sciences et technique de l'armement, 48, 4e fasc, 774-811

CRANZ C (1896) Compendium der theoretischen äußeren Ballistik, BG Teubner, Leipzig

CRANZ C (1925) Lehrbuch der Ballistik Bd. I, Äußere Ballistik oder Theorie der Bewegung des Geschosses von der Mündung der Waffe ab bis zum Eindringen in das Ziel, Springer-Verlag, Berlin

DIDION I (1852) Cours élémentaire de balistique, Librairie militaire de J Dumaine, Paris

EGGERS AJ, RESNIKOFF MM, DENNIS DH (1957) Bodies of Revolution Having Minimum Drag at High Supersonic Airspeeds, http://naca.central.cranfield.ac.uk/reports/1957/naca-report-1306.pdf

ENGELN-MÜLLGES G, NIEDERDRENK K, WODICKA R (2011) Numerik-Algorithmen, Springer-Verlag, Berlin, Heidelberg, New York

EULER L (1755) Recherche sur la véritable courbe que décrivent les corps jettés dans l'air ou dans un autre fluide quelconque, Mém de l'Académie Royale des Sciences de Berlin IX, 321–352

FLECK V, MOLITZ H (1977) Contribution to the Study of the Influence of Nonlinear Effects on the Trajectory of Spinning Shells, Proceedings of the 3rd International Symposium on Ballistics, Karlsruhe

FLECK V (2009) Einführung in die Außenballistik mit Übungen, Lehrgang „Ballistik von Geschossen" der Carl-Cranz-Gesellschaft (CCG), Saint-Louis

FARRAR CL, LEEMING DW (1983) Military Ballistics – A Basic Manual, Brassey's Publishers Ltd, Oxford

GERMERSHAUSEN R (ed.) (1980) Waffentechnisches Taschenbuch, Rheinmetall, Düsseldorf

GLUCHOFF A (2011) Artillerymen and mathematicians: Forest Ray Moulton and changes in American exterior ballistics, 1885–1934, Historia Mathematica 38, 506–547

GREINER W (1989) Theoretische Physik Band 2, Mechanik Teil 2, Verlag Harri Deutsch, Thun

HAACK W (1941) Geschossformen kleinsten Wellenwiderstandes, Bericht 139 der Lilienthalgesellschaft S 14–28, https://web.archive.org/web/20070927020313/http://www.lima-wiederladetechnik.de/Ballistik/Lilienthal-139-1-Haack.pdf

HAACK W (1959) The Calculation of Stability and Damping of Spin-Stabilized Projectiles, in NELSON WC (ed.) Selected Topics on Ballistics, Pergamon Press, London

HANTSCHE G, RINEHART J (1952), Air Drag on Cubes at Mach Number 0.5 to 3.5, J Aeronaut Sci Vol 19, Feb

HATCHER JS (1962) Hatcher's Notebook, The Military Publishing Company, Third Edition

HAUCK G (1972) Äußere Ballistik oder die Theorie der Bewegung des Geschosses von der Mündung der Waffe ab, Militärverlag der DDR, Berlin

HAUCK G (1990) Der Flug ungelenkter Geschosse und Raketen, Militärverlag der DDR, Berlin, 3. Aufl

HUND F (1996) Geschichte der physikalischen Begriffe, Spektrum Akademischer Verlag Heidelberg, Berlin, Oxford

HURWITZ A (1895) Bedingungen, unter welchen eine Gleichung nur Wurzeln mit negativen reellen Teilen besitzt, Math Ann 46:273–285

INGALLS JM (1918) Ingalls' Ballistic Tables, revised under the Direction of the Ordnance Board 1917, Government Printing Office, Washington

KLOSE B (2016) Meteorologie, Springer-Verlag, Berlin, Heidelberg, 3. Aufl

KNEUBUEHL BP (1976) Über die Herkunft und Gültigkeit der Didion'schen Formel zur Berechnung des Windeinflusses bei Geschossen und Raketen, interner Bericht, armasuisse, Wissenschaft & Technologie, Thun

KNEUBUEHL BP (1979) Seitenwindsimulation, Wehrtechnik 6, 36–37

KNEUBUEHL BP (1982) Physical and Mathematical Backgrounds of the Stability Theory of a Spinning Shell, J of Ballistics Vol 6, No 2:1385–1392

KNEUBUEHL BP ($1983_1$) Calculation of the Stability and the Tractability Factors of a Spinning Shell, Proceedings of the 7th Int Symposium on Ballistics, The Hague

KNEUBUEHL BP ($1983_2$) $c_w$-Bestimmung in Freifluganlagen, Kursmanuskript, CCG-Kurs „Außenballistik", Weil am Rhein

KNEUBUEHL BP (1986) What is the maximum length of a spin stabilized projectile? Proceedings of the 9th Int Symposium on Ballistics, Shrivenham

KNEUBUEHL BP (1999) Splitterballistik und Splitterwundballistik, Bericht Nr 1486, armasuisse Wissenschaft & Technologie, Thun

KNEUBUEHL BP (2002) Splinefunktionen und ihre Anwendung, Bericht Nr 1680, armasuisse Wissenschaft & Technologie, Thun

KNEUBUEHL BP (ed.), COUPLAND RM, ROTHSCHILD MA, THALI MJ (2022) Wound Ballistics. Basics and Applications (SW RAWCLIFFE, Trans.), Springer-Verlag, Berlin, Heidelberg, 2nd ed. (Original work 4th German edition, published in 2022)

KNEUBUEHL BP (2013) Geschosse Gesamtausgabe, Ballistik, Messtechnik, Treffsicherheit, Wirkungsweise, Verlag Stocker-Schmid, Dietikon, 3. Aufl

KUTTERER RE (1959) Ballistik, Friedr Vieweg & Sohn, Braunschweig

KUYPERS F, UCKE CH (1994) Steh' auf Kreisel! Physik in unserer Zeit, 5/94, 214–215

LAUBE G, KNEUBUEHL BP (1981) A Fast Running Solution of the Exterior Boundary Value Problem, Proceedings of the 6th Int Symposium on Ballistics, Orlando FL

LEDOUX T (2011) Quaternionen, Ausarbeitung im Rahmen des Seminars „Geometrie SS 2011", http://www.mathematik.uni-dortmund.de/~lschwach/SS11/Seminar_II/Quaternionen.pdf

LESCHKE K, STEINKRAUß T (2011) Quaternionen: von Hamilton, Basketbällen und anderen Katastrophen, http://didaktik.mathematik.hu-berlin.de/files/2011_quaternionen.pdf

LIESKE RF, MCCOY RL (1964) Equations of Motion of a Rigid Projectile, Ballistic Research Laboratories, Report No 1244, Aberdeen Proving Ground, Maryland, http://www.dtic.mil/dtic/tr/fulltext/u2/441598.pdf

LIESKE RF, REITER ML (1966) Equations of Motion for a Modified Point Mass Trajectory, Ballistic Research Laboratories, Report No 1314, Aberdeen Proving Ground, Maryland http://www.dtic.mil/dtic/tr/fulltext/u2/485869.pdf

LUST M (2001) Quaternionen – mathematischer Hintergrund und ihre Interpretation als Rotation, https://userpages.uni-koblenz.de/~cg/veranst/ws0001/sem/Lust_quaternion.pdf

MAY A, WITT WR (1953) Free-Flight Determinations of the Drag Coefficients of Spheres, J Aeronaut. Sci Vol 20 Sept, 635–638
MAYEVSKI N (1872) Traité de balistique extérieure, Gauthier-Villars, Paris
MCCOY RL (1999) Modern Exterior Ballistics, Schiffer Military History, Atglen
MCSHANE EJ, KELLEY JL, RENO FV (1953) Exterior Ballistics, The University of Denver Press
MOLITZ H (1957) Das sekundäre Hauptproblem der äußeren Ballistik, Wehrtech Mh 54, 8, 333–342
MOLITZ H, STROBEL R (1963) Äußere Ballistik, Springer-Verlag, Berlin, Heidelberg, New York
MUNK MM (1924) The Aerodynamic Forces on Airship Hulls, NACA Report 184, https://ntrs.nasa.gov/archive/nasa/casi.ntrs.nasa.gov/19930091249.pdf
OERLIKON Taschenbuch (1981) Werkzeugmaschinenfabrik Oerlikon-Bührle AG (ed.), Zürich
PITON-BRESSANT AH (1852) Formules des portées, [publisher unknown]
PONTRAJAGIN LS (1965) Gewöhnliche Differenzialgleichungen, Deutscher Verlag der Wissenschaften, Berlin
RALSTON A, WILF HS (1967) Mathematische Methoden für Digitalrechner, R Oldenbourg Verlag, München/Wien S 203 ff
REINSCH C (1967) Smoothing by Spline Functions, Numer Math 10, 177-183
ROBINS B (1745) Neue Grundsätze der Artillerie enthaltend die Bestimmung der Gewalt des Pulvers nebst einer Untersuchung über den Unterschied des Widerstands der Luft in schnellen und langsamen Bewegungen, aus dem Englischen übersetzt und mit den nötigen Erläuterungen und vielen Anmerkungen versehen von Leonhard EULER, A Haude, Berlin
ROTH-DESMEULES E (1955) Zur Berechnung der Geschossabweichung unter dem Einfluss eines Seitenwindes, ZAMP 6, 6:494–497
RUTISHAUSER H (1976) Vorlesungen über numerische Mathematik, Band I, Birkhäuser-Verlag, Basel
SCHLICHTING H (1965) Grenzschicht-Theorie, 5. Auflage, Verlag G Braun, Karlsruhe
SCHMIDT WR, HERPFER E (1973) Zusammenstellung experimenteller Unterlagen zum Heckwiderstand von Flugkörpern im Transschall- und im Überschallbereich und Vergleich mit theoretischen Berechnungsmethoden, BMVg-FBWT 73-27
SONNTAG D (1990) Important new Values of the Physical Constants of 1986, Vapour Pressure Formulations based on ITS-90 and Psychrometer Formulae, *Z Meteorol* 40, 5:340–344
STOER J, BULIRSCH R (1973) Einführung in die numerische Mathematik, Springer-Verlag, Berlin, Heidelberg, New York
STÜMKE H (1969) Grundzüge der Flugmechanik und Ballistik, Friedr Vieweg & Sohn, Braunschweig
SUTTERLIN R (1965) Cours de munitions, Livre III, Les projectiles, chap IV
SZABÒ I (1972) Höhere Technische Mechanik, Springer-Verlag, Berlin
SZABÒ I (1987) Geschichte der mechanischen Prinzipien, Birkhäuser Verlag, Basel, Boston, Berlin, 3. Aufl
VALLERIANI M (2013) Metallurgy, Ballistics and Epistemic Instruments, The *Nova Scientia* of Nicolò TARTAGLIA, A new Edition, Edition Open Access http://www.edition-open-sources.org/sources/6/index.html
WHITEFORD CW, REGAN JM (1961) The Determination of the Striking Velocity of Steel Fragments by Their Mass and Penetration Into Witness Material, BRL Memorandum Report No 1333, April 1961
WOLFF W (2006), Raketen und Raketenballistik, Reprint 5. Aufl, Elbe-Dnjepr-Verlag, Klitzschen
YOUSSEF EM, RADWAN MA, MOSTAFA HE, HADHOUD MK (2015) Low Signature Advanced Base Bleed Grains, IJSER Vol 6, 12
ZIEREP J (1991) Theoretische Gasdynamik, G. Braun, Karlsruhe

## 6 An outline of terminal ballistics


ALEKSEEVSKII VP (1966) Penetration of a Rod into a Target at High Velocity, Combustion, Explosion and Shock Waves 2, 63-66

BACKMAN ME, GOLDSMITH W (1978) The mechanics of penetration of projectiles into targets, Int J Engng Sci, Vol 16, 1–99

BHATNAGAR A (ed.) (2016) Lightweight Ballistics Composites, Woodhead Publishing (Elsevier) Oxford

BILLON HH (2007) A New Method for Calculating the Critical Penetration Velocity $V_0$, DSTO-TN-0791, Defence Science and Technology Organisation, Fishermans Bend, Victoria 3207, Australia

BUCKINGHAM E (1914) On physically similar systems; illustrations of the use of dimensional equations, Physical Review 4, S 345–376

EN (1998) EN 1523, Windows, Doors, Shutters and Blinds – Bullet Resistance – Test Method, CEN European Committee for Standardization, Brussels

EN (1999) EN 1063, Glass in Building – Security Glazing – Testing and Classification of Resistance against Bullet Attack, CEN European Committee for Standardization, Brussels

FINNEY DJ, TATTERSFIELD F (1952) Probit Analysis: A Statistical Treatment of the Sigmoid Response Curve, JASA, Vol 47, No 260, 687–691

GABEAUD ML (1935) Essai sur la théorie de la perforation des blindages, Mémorial de l'artillerie française, XIV, 85–92

GABEAUD ML (1947) Sur la perforation des blindages, Mémorial de l'artillerie française, XXI, 97–112

KNEUBUEHL BP (2003) Ballistic Protection, http://www.vpam.eu/fileadmin/Wissenswertes/Ballistic-protection.pdf

KNEUBUEHL BP (ed.), COUPLAND RM, ROTHSCHILD MA, THALI MJ (2022) Wound Ballistics. Basics and Applications (SW RAWCLIFFE, Trans.), Springer-Verlag, Berlin, Heidelberg, 2nd ed. (Original work 4th German edition, published in 2022)

LAIBLE RC (Ed.) (1980) Ballistic Materials and Penetration Mechanics, Elsevier Scientific Publishing Company, Amsterdam, Oxford, New York

LANGLIE HJ (1965) A Reliability Test Method for "One Shot" Items, Ford Aerospace Communications Corporation, Aeronautic Division

MAUCHANT D, RILEY MA, RICE KD, FORSTER AL, LEBER D, SAMAROV D (2011) Analysis of Three Different Regression Models to Estimate the Ballistic Performance of New and Environmentally Conditioned Body Armor, NIST 7760 National Institute of Standards and Technology

MALDAGUE M (2008) Evaluation of some methods in order to determine $v_{50}$, Proceedings of Personal Armour System Symposium, Brussels

NIJ (2008) Ballistic Resistance of Body Armor, NIJ Standard 0101.06, National Institute of Justice, Washington DC, https://www.ncjrs.gov/pdffiles1/nij/223054.pdf

ROSENBERG Z, DEKEL E (2012) Terminal Ballistics, Springer-Verlag, Berlin, Heidelberg

TATE A (1967) A Theory for the Deceleration of Long Rods after Impact, J Mech Phys Solids, Vol 15, 6, 387-399

VPAM[1] (2021) General Basics of Ballistic Material, Design and Product Tests – Requirements, Test Levels and Test Procedures, https://www.vpam.eu/pruefrichtlinien/aktuell/apr-2006/

VPAM[1] (2006) Personal Protective Equipment. Ballistic Protective Vests – Requirements, Classifications and Testing Methods, https://www.vpam.eu/pruefrichtlinien/aktuell/bsw-2006/

WILD R, VON COLLANI E (2002) Modelling of Explosives Sensitivity, Part 1, The Bruceton Method, Economic Quality Control Vol 17, No 1, 113–122

ZUKAS JA, NICHOLAS T, SWIFT HF, GRESZCZUK LB, CURRAN DR (1992) Impact Dynamics, Krieger Publishing Company, Malabar FL


[1] Association of Test Laboratories for Attack Resistant Materials and Constructions.

## 7 Ricochets

BIRKHOFF G, BIRKHOFF GD, BLEICK WE, HANDLER EH, MURNAGHAN FD, SMITH TL (1944) Ricochet off Water, A.M.P. Memo. 42.2M (According to JOHNSON, REID (1975), this publication is not available, so they explained and discussed this theory in detail in their paper)

GREINER W (1989) Theoretische Physik Band 2, Mechanik Teil 2, Verlag Harri Deutsch, 5. Aufl

JAUHARI M (1971) Mathematical Model for Bullet Ricochet, J Crim Law Criminol Pol Sci 61, 469-473

JOHNSON W, REID SR (1975) Ricochet of Spheres off Water, J Mech Eng Sci, 17, 2, 71–81

KNEUBUEHL BP (1999) Das Abprallen von Geschossen aus forensischer Sicht, Dissertation, Université de Lausanne, Faculté de Droit, École des Sciences Criminelles

KUYPERS F, UCKE CH (1994) Steh‘ auf Kreisel! Physik in unserer Zeit, 5/94, 214–215

MAGNUS K (1971) Kreisel, Theorie und Anwendungen, Springer-Verlag, Berlin, Heidelberg, New York

REY-DEBOVE J, REY A (1994) Le nouveau Petit Robert, Dictionnaires le Robert, Paris

SELLIER K (1971) Über Geschoßablenkung und Geschoßdeformation, Z Rechtsmed, 69, 217–251

SELLIER K (1977) Schusswaffen und Schusswirkungen II, Verlag Schmidt-Römhild, Lübeck

SELLIER K (1987) Doppelte Rikochettier-Marken, Z Rechtsmed, 90, 241–244

SOMMERFELD A (1977) Vorlesungen über Theoretische Physik, Band I, Mechanik, Nachdruck der 8. Aufl, Verlag Harri Deutsch, Thun, Frankfurt/Main

## 8 Special applications of ballistics

BEVEL T, GARDNER RM (1997) Bloodstain Pattern Analysis, CRC Press, Boca Raton, p 75

BUCK U, KNEUBUEHL BP, NÄTHER S, ALBERTINI N, SCHMIDT L, THALI M (2011) 3D bloodstain pattern analysis: ballistic reconstruction of the trajectories of blood drops and determination of the centres of origin of the bloodstains, Forensic Sci Int 2011 Mar 20;206(1-3):22–28

FRINGS C, RABE F (2016) Grundlagen der Kriminaltechnik Bd I und II, Verlag Deutsche Polizeiliteratur, GmbH, Hilden

GALLUSSER A, BONFANTI M, SCHÜTZ F (2014) Expertise des armes à feu, Presses polytechniques et universitaires romandes, Lausanne, 2ème ed

GARDNER RM (2005) Practical Crime Scene Processing and Investigation, CRC Press, Boca Raton, Florida

HAAG LC (2002) Balloting Bullets: Exterior and Wound Ballistics of Balloted Bullets, Physical Characteristics of Balloted Bullets and Associated Cartridge Cases, AFTE Journal, 34(4), 375–378

HAAG LC (2007) Wound Production by Ricocheted and Destabilized Bullets. Am J Forensic Med Pathol, 28, 4–12

HAAG MG, HAAG LC (2011) Shooting Incident Reconstruction, 2nd edition, Academic Press, San Diego

HSIANG LP, FAETH GM (1995) Drop deformation and breakup due to shock-wave and steady disturbances, International Journal of Multiphase Flow 21(4), 545–560

JAMES SH, ECKERT WG (1999) Interpretation of Bloodstain Evidence at Crime Scenes, 2nd Ed p 15–17

KERKHOFF W, BOLCK A, ALBERINK I, MATTIJSSEN EJAT, HERMSEN R, RIVA F (2018) Pistol bullet deflection through soft tissue simulants, Forens Sci Int 289, 270–276

KNEUBUEHL BP (ed.), COUPLAND RM, ROTHSCHILD MA, THALI MJ (2022) Wound Ballistics. Basics and Applications (SW RAWCLIFFE, Trans.), Springer-Verlag, Berlin, Heidelberg, 2nd ed. (Original work 4th German edition, published in 2022)

KNEUBUEHL BP (2013) Geschosse Gesamtausgabe – Ballistik, Messtechnik, Wirksamkeit, Treffsicherheit, Verlag Stocker-Schmid, Dietikon-Zürich

LITZ B (2011) Applied Ballistics for Long-Range Shooting, 2nd edition, Applied Ballistics, LLC
OCSW-WSAD (1968) Office of the Chief of Staff (Army) Washington DC, Weapons Systems Analysis Directorate, Report of the M16 Rifle Review Panel, Ammunition Development Program, Vol. 5, Appendix 4, PN, http://www.dtic.mil/dtic/tr/fulltext/u2/a953114.pdf
RIVA F, KERKHOFF W, BOLCK A, MATTIJSSEN EJAT (2017) Possible influences on bullet trajectory deflection in ballistic gelatine, Forens Sci Int 271 107–112
ROWE WF (2006) Errors in the determination of the point of origin of bloodstains, Forens Sci Int 161, 47–51
SCIPP, Santa Cruz Institute of Particle Physics, Surface Tension (Manuscript) p 3, http://scipp.ucsc.edu/~haber/ph5B/bubble.pdf
SELLIER K (1988) Schussentfernungsbestimmung, Verlag Schmid-Römhild, Lübeck
SELLIER K (1991) Shot Range Determination, Forensic Science Progress, Springer Verlag, Berlin, Heidelberg, New York
THEOFANOUS TG (2011) Aerobreakup of Newtonian and Viscoelastic Liquids, Annual Review of Fluid Mechanics, Vol 43, Palo Alto, Annual Reviews, 43: 661–690
VPAM (2012) Materialien und Konstruktionen zur Vermeidung von ab- und rückprallenden Geschossen, VPAM, Vereinigung der Prüfstellen für angriffshemmende Materialien und Konstruktionen, http://www.vpam.eu/index.php?id=112
WILLIS C, PIRANIAN AK, DONAGGIO JR, BARNETT RJ, ROWE WF (2001) Errors in the estimation of the distance of fall and angles of impact blood drops, Forens Sci Int 123, 1–4
WÜRZ W, GAISBAUER U (2014) Bedeutung von Ähnlichkeitszahlen (Kennzahlen) Manuskript Institut für Aerodynamik und Gasdynamik, Universität Stuttgart

# Illustration credits

| Chap. | Figures | Source |
|---|---|---|
| **2** | 1–10, 12–19 | Beat P. Kneubuehl, CH-3603 Thun |
| | 11 | armasuisse, Wissenschaft & Technologie, CH-3602 Thun |
| **3** | 1–7, 16–18, 20–21, 26–31, 33–35a, 36–42 | Beat P. Kneubuehl, CH-3603 Thun |
| | 8–15 | Beat P. Kneubuehl, CH-3603 Thun, based on documents supplied by Nitrochemie, CH-3752 Wimmis, with their kind permission |
| | 19 | From A. E. H. LOVE and F. B. PIDDUCK (1922) http://rsta.royalsocietypublishing.org/content/roypta/222/594-604/167.full.pdf, CCC License Id 4386440963596 |
| | 22–25, 32 | From B. P. KNEUBUEHL (1998), Bericht Nr. 1436, armasuisse, Wissenschaft & Technologie, CH-3602 Thun |
| | 35b | Verein Schweizer Armeemuseum, CH-3600 Thun |
| **4** | 1, 2b, 3b, 4–6, 14, 19–20 | armasuisse, Wissenschaft & Technologie, CH-3602 Thun |
| | 2a, 3a, 7–8, 12–13, 15–17 | Beat P. Kneubuehl, CH-3603 Thun |
| | 9 | Institut für Rechtsmedizin der Universität Bern, Abt. Forensische Physik / Ballistik |
| | 10 | http://www.defensie.nl/onderwerpen/materieel/inhoud/communicatie-en-informatiesystemen/doelgegevens-evaluatie-cel-dec (modified by B. P. Kneubuehl) |
| | 11 | http://news.soc.mil/Photo%20Archive/2009/2009/album/slides/090502-A-4552H-026.html (modified by B. P. Kneubuehl) |
| | 18 | A. Mägert, CH-3800 Spiez, by kind permission |
| **5** | 1, 3–25, 27, 31–36, 39–44, 46–51, 54–57, 61–63, 66–68, 70–83 | Beat P. Kneubuehl, CH-3603 Thun |
| | 2 | http://www.e-rara.ch/zut/wihibe/content/titleinfo/6331451 |
| | 26 | R. Coupland, International Committee of the Red Cross, CH-1200 Geneva, by kind permission |
| | 28–29, 52–53, 58–59 64–65, 69 | armasuisse, Wissenschaft & Technologie, CH-3602 Thun |
| | 30 | Verein Schweizer Armeemuseum, CH-3600 Thun |
| | 37, 38 | Redrawn by B. P. Kneubuehl from V. FLECK, H. MOLITZ (1977), by kind permission of the authors |
| | 45 | Institut für Rechtsmedizin, ZFPB, CH-3012 Bern |
| | 60 | From W. R. SCHMIDT, E. HERPFER (1973), re-drawn by B. P. Kneubuehl, by kind permission of the Fachinformationszentrums der Bundeswehr, Bonn |


B. P. Kneubuehl, *Ballistics*,
https://doi.org/10.1007/978-3-662-69237-0

| Chap. | Figures | Source |
|---|---|---|
| **6** | 1–2, 4–6, 16–35, 37–39 | Beat P. Kneubuehl, CH-3603 Thun |
| | 3 | From M. E. BACKMAN, W. GOLDSMITH (1978), CCC License Id: 4386431241800, re-sized by B. P. Kneubuehl and with minor changes to terminology |
| | 7–15 | W. Odermatt, CH-3612 Steffisburg |
| | 36 | Deutsches Bundesarchiv, Bild 102-12610 / CC-BY-SA 3.0 |
| **7** | 1–22 | B. P. Kneubuehl, CH-3603 Thun |
| | 23, 24 | armasuisse, Wissenschaft & Technologie, CH-3602 Thun |
| **8** | 1–2, 5–7, 16–18, 20–22 | B. P. Kneubuehl, CH-3603 Thun |
| | 3–4, 8–9, 10–11 | F. Riva, École des Sciences Criminelles, Université de Lausanne, CH-1015 Lausanne-Dorigny |
| | 12, 14–15 | Institut für Rechtsmedizin, CH-3012 Bern |
| | 13 | From L. PRANTL (ed.) Ergebnisse der Aerodynamischen Versuchsanstalt zu Göttingen, II. Lieferung, 1923, Nachdruck 2009, https://doi.org/10.17875/gup2009-101, Universitätsverlag Göttingen, by kind permission of the publisher |
| | 19 | Kantonspolizei Aargau, KTD, CH-5000 Aarau, by kind permission |
| | 23–24 | P. Drapela, armasuisse Wissenschaft & Technologie CH-3602 Thun |

# Index




B. P. Kneubuehl, *Ballistics*,
https://doi.org/10.1007/978-3-662-69237-0

## V



## W



## Y



## Z

The manufacturer's authorised representative in the EU is Springer Nature Customer Service Centre GmbH, Europaplatz 3, 69115 Heidelberg, Germany. If you have any concerns regarding our products, please contact ProductSafety@springernature.com

Printed and bound by CPI Group (UK) Ltd, Croydon, CR0 4YY
30/09/2026
02225454-0012